# APPLICATIONS OF SUPRAMOLECULAR CHEMISTRY

# APPLICATIONS OF SUPRAMOLECULAR CHEMISTRY

Edited by
## Hans-Jörg Schneider

**CRC Press**
Taylor & Francis Group
Boca Raton   London   New York

CRC Press is an imprint of the
Taylor & Francis Group, an **informa** business

CRC Press
Taylor & Francis Group
6000 Broken Sound Parkway NW, Suite 300
Boca Raton, FL 33487-2742

First issued in paperback 2016

© 2012 by Taylor & Francis Group, LLC
CRC Press is an imprint of Taylor & Francis Group, an Informa business

No claim to original U.S. Government works

ISBN 13: 978-1-138-19926-2 (pbk)
ISBN 13: 978-1-4398-4014-6 (hbk)

**Visit the Taylor & Francis Web site at**
**http://www.taylorandfrancis.com**

**and the CRC Press Web site at**
**http://www.crcpress.com**

# Contents

# Foreword

Supramolecular chemistry has experienced extraordinary development over the last 40 years or so, being at the triple meeting point of chemistry with biology and physics. It has grown into a major field of investigation in chemical sciences and has fueled numerous developments at its interfaces with biology and physics, leading to the emergence and progressive establishment of supramolecular science and technology.

Supramolecular chemistry has progressed over the years through three overlapping phases. The first phase covered the investigation of molecular recognition processes and their corollaries, supramolecular reactivity, catalysis, and transport events; it relied on design and preorganization. An important extension, particularly in the present context, concerned the design of functional supramolecular devices built on photoactive, electroactive, or ionoactive components operating, respectively, with photons, electrons, or ions. Such functional entities represent entries into molecular photonics, electronics, and ionics that deal with the storage, processing, and transfer of materials, signals, and information at the molecular and supramolecular levels.

In the second phase, supramolecular chemistry actively explored systems undergoing self-organization—that is, systems capable of spontaneously generating well-defined, functional supramolecular architectures by self-assembly from their components, under the control of molecular recognition processes.

The design of molecular information–controlled, "programmed," and functional self-organizing systems also provides an original approach to nanoscience and nanotechnology. The generation of well-defined, functional molecular and supramolecular architectures of nanometric size through self-organization may offer a very powerful alternative or complement to nanofabrication and nanomanipulation.

In the third phase, supramolecular chemistry was recognized as an intrinsically dynamic chemistry in view of the lability of the interactions connecting the molecular components of a supramolecular entity and the resulting ability of supramolecular species to exchange their constituents. Extension of such dynamic features to molecular chemistry was achieved through the introduction into a molecular entity of covalent bonds that may form and break reversibility to allow a continuous change in constitution by reorganization and exchange of building blocks. Taken together, these developments led to the definition of a constitutional dynamic chemistry as a unifying concept covering both the molecular and supramolecular levels. It takes advantage of dynamic diversity to allow variation and selection and operates on dynamic constitutional diversity in response to either internal or external factors to allow for adaptation and evolution. Vast perspectives are thus opening, pointing toward the emergence of adaptive and evolutive chemistry.

Over the years, the intense research activity along these broad themes has generated numerous review articles, special issues of journals and books, and frequent meetings.

As a field, it is strongly rooted in its concepts and is becoming established. The implementation of the basic knowledge acquired provides opportunities toward more and more applications. This is indeed the case for supramolecular chemistry. The time is thus opportune for the present volume, which gathers thorough presentations of the numerous actually realized or potentially accessible applications of supramolecular chemistry by a number of the leading figures in the field. The variety of topics covered is witness to the diversity of the approaches and the areas of implementation. We must be thankful to the authors for providing such a broad and timely panorama of the field and to the editor for assembling such an eminent roster of contributors.

**Jean-Marie Lehn**
*Strasbourg*

# Contributors

**Antonella Accardo**
Department of Biological Sciences
CIRPeB University of Naples
  "Federico II"
and
IBB CNR
Naples, Italy

**M. Teresa Albelda**
Departament de Química Inorgànica
ICMol, Facultat de Química
Universitat de València
Burjassot, Spain

**Shamindri Arachchige**
Department of Chemistry
Virginia Tech
Blacksburg, Virginia

**Laura Baldini**
Dipartimento di Chimica Organica e
  Industriale
Università di Parma
Parma, Italy

**Dirk Beckmann**
Max Planck Institute for Polymer
  Research
Mainz, Germany

**Chandra K. C. Bikram**
Department of Chemistry
University of North Texas
Denton, Texas

**Karen J. Brewer**
Department of Chemistry
Virginia Tech
Blacksburg, Virginia

**Hans-Jürgen Buschmann**
Deutsches Textilforschungszentrum
  Nord-West e.V.
Krefeld, Germany

**Alessandro Casnati**
Dipartimento di Chimica Organica e
  Industriale
Università di Parma
Parma, Italy

**Brent E. Dial**
Department of Chemistry and
  Biochemistry
University of South Carolina
Columbia, South Carolina

**Francis D'Souza**
Department of Chemistry
University of North Texas
Denton, Texas

**Juan C. Frías**
Departament de Química Inorgànica
ICMol
Facultat de Química
Universitat de València
Burjassot, Spain

**Enrique García-España**
Departament de Química Inorgànica
ICMol
Facultat de Química
Universitat de València
Burjassot, Spain

**Róbert E. Gyurcsányi**
Department of Inorganic and Analytical
    Chemistry
Budapest University of Technology and
    Economics
and
Hungarian Academy of Sciences
Budapest, Hungary

**Rainer Haag**
Institut für Chemie und Biochemie
Freie Universität Berlin
Berlin, Germany

**Mir Wais Hosseini**
Laboratoire de Chimie de Coordination
    Organique
Institut Le Bel
University of Strasbourg
Strasbourg, France

**Wlodzimierz Kutner**
Department of Physical Chemistry of
    Supramolecular Complexes
Institute of Physical Chemistry
Polish Academy of Sciences
and
Faculty of Mathematics and Natural
    Sciences
School of Science
Cardinal Stefan Wyszynski University
Warsaw, Poland

**Jean-Marie Lehn**
Laboratoire de Chimie
    Supramoléculaire/ISIS
Strasbourg Cedex, France

**Hsin-Chieh Lin**
Department of Chemistry
Brandeis University
Waltham, Massachusetts

**Ernö Lindner**
Department of Biomedical Engineering
The University of Memphis
Memphis, Tennessee

**Andreas Mohr**
Institut für Chemie und Biochemie
Freie Universität Berlin
Berlin, Germany

**Giancarlo Morelli**
Department of Biological Sciences
CIRPeB University of Naples
    "Federico II"
and
IBB CNR
Naples, Italy

**Klaus Müllen**
Max Planck Institute for Polymer
    Research
Mainz, Germany

**Agnieszka Pietrzyk-Le**
Department of Physical Chemistry of
    Supramolecular Complexes
Institute of Physical Chemistry
Polish Academy of Sciences
Warsaw, Poland

**Ernö Pretsch**
Institute of Biogeochemistry and
    Pollutant Dynamics
ETH Zürich
Zürich, Switzerland

**Francesco Sansone**
Dipartimento di Chimica Organica e
    Industriale
Università di Parma
Parma, Italy

**Hans-Jörg Schneider**
FR Organische Chemie der Universität
    des Saarlandes
Saarbrücken, Germany

**Volker Schurig**
Institute of Organic Chemistry
University of Tübingen
Tübingen, Germany

**Ken D. Shimizu**
Department of Chemistry and
    Biochemistry
University of South Carolina
Columbia, South Carolina

**Diego Tesauro**
Department of Biological Sciences
CIRPeB University of Naples
    "Federico II"
and
IBB CNR
Naples, Italy

**Rocco Ungaro**
Dipartimento di Chimica Organica e
    Industriale
Università di Parma
Parma, Italy

**Jing Wang**
Department of Chemistry
Virginia Tech
Blacksburg, Virginia

**Bing Xu**
Department of Chemistry
Brandeis University
Waltham, Massachusetts

# 1 Introduction and Overview

*Hans-Jörg Schneider*

## CONTENTS

## INTRODUCTION

### OVERVIEW ON SOME SPECIAL APPLICATIONS

Noncovalent interactions are the basis of the most impressive functions of living systems. After some early explorations started by Cramer in the 1950s[1] chemists have begun, in recent decades, to systematically explore the unlimited possibilities and promise of such interactions in the framework of supramolecular chemistry. Several books and many reviews describe the principles and achievements of this very fast developing field.[2] As outlined by J.-M. Lehn in the foreword, there are already several phases in this still relatively new chapter of science, ranging from understanding of the relevant interaction mechanisms[3] to manifold possible applications.[4] It has been pointed out that three periods characterize the introduction of a new technology: "(I) In the beginning, there is a period of exaggerated expectations, during which exciting—but sometimes irreproducible—results and unrealistic claims are made. (II) When these high expectations go unmet, a period of disappointment sets in. (III) There is then a return to the fundamental aspects of the technology; science is linked with applications; new tools are developed; and real commercial investment begins."[5] It seems safe to say that supramolecular chemistry has now reached phase III of such development, which warrants publication of a new book outlining applications. It is hoped that the present monograph provides help for researchers and engineers who want to make more practical use of supramolecular complexations and, in particular, for students who are taking corresponding courses. Until now, there are, besides an older volume,[4] only some reviews and countless original papers that discuss practical applications of noncovalent interactions. In the present

1

monograph, an attempt is made to summarize promising applications of these new technologies with contributions of leading experts in different fields. The different chapters highlight how far the development has gone until now without trying to achieve a comprehensive description.

Some fields are only touched upon, particularly if they are closer to other established technologies for which corresponding monographs are already available. One might miss a chapter on noncovalent interactions in biological systems, which, however, are a central issue of biological and medicinal chemistry and dealt with in many books[6] and papers, in particular, on drug design.[7] In Chapter 15, nevertheless, scientists from Parma discuss supramolecular actions on proteins and nucleic acids. Noncovalent interactions dominate intriguing new materials based on proteins and nucleic acids and their hybrids, frequently in nanoparticles.[8] In the following introductory sections, we can only highlight some possible applications of noncovalent interactions that are not discussed in separate chapters of the present book, which are closer to practical applications. Many of the new applications discussed below also bear more on engineering aspects; nevertheless, an attempt is made to provide the most recent references for the interested reader.

*Supramolecular catalysis* is only mentioned here as most practical examples belong to traditional metal-coordination chemistry, which began decades ago with the Ziegler–Natta complexes for olefin polymerization. The intriguing catalytic cases based on purely organic complexations, for instance, with cyclodextrins, are aptly discussed in recent monographs[9] and in several reviews.[10–13] Also, many models for supramolecular catalysis involve quite special substrates, which are made only for the purpose of such studies. We will discuss here only two cases that are closer to potential applications, both involving hydrolysis in excess water and, thus, free from the problem of inhibition of the catalyst by the product unless the product has an unusually high affinity toward the catalyst.[11,13] Figure 1.1 illustrates how, besides ion pairing, stacking interactions between aryl groups of catalyst and substrate contribute to an efficient cleavage of dinitrophenylphosphate BNPP, which is a frequently used model for chemical weapons and some insecticides. With EuIII as

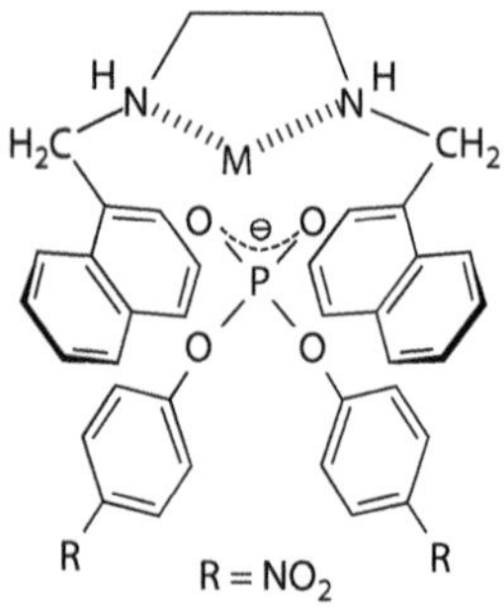

**FIGURE 1.1** Catalysis of phosphate ester cleavage; rate enhancement $k_{cat}/k_{uncat} = 10^7$ (M = EuIII). (From Negi, S. et al., *Tetrahedron Letters*, 43, 411–414, 2002. With permission.)

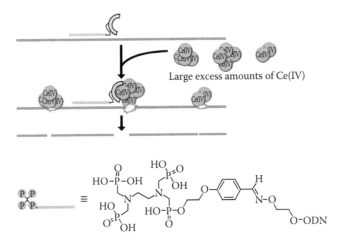

**FIGURE 1.2** Site-selective DNA hydrolysis with CeIV–oligonucleotide complexes, based on nucleobase recognition. (Lönnberg, T. et al.: *Chem. Europ. J.* 2010. 16. 855–859. Copyright Wiley-VCH Verlag GmbH & Co. KGaA. Reproduced with permission.)

a metal ion of the [BNPP] hydrolysis, half-life time is shortened from about 100.000 hours to a few minutes, which holds promise for corresponding detoxification processes.[14] Another case that relates to artificial restriction enzymes is the site-selective hydrolysis of DNA, bearing promise for applications of enzymatic DNA manipulation in molecular biology and new therapeutic methods.[15] The complex form ethylenediamine-N,N,N′,N′-tetrakis (methylenephosphonic acid) (EDTP) and CeIV ions are used here for the highly site-selective hydrolysis of, for example, an 85-mer DNA, based on oxidizing a CeIII/EDTP complex that is bound to a 20-mer- oligodeoxyribonucleotide (ODN). Using two EDTP–ODN conjugates, a five-base gap for site-selective scission occurs in the middle of a single-stranded DNA (Figure 1.2).

## Molecular Machines

Another exciting development of supramolecular chemistry concerns molecular motors; these are discussed already in some books[16] and reviews,[17] emphasizing systems activated electrochemically,[18] chemically,[17,19] or by light.[20,21] Figure 1.3 illustrates how, for instance, a chemical or electrochemical signal can trigger a mechanical action on a microcantilever based on shuttling between electron-rich and -poor stations.[17]

Photooxidation of the ruthenium complex in the rotaxane shown in Figure 1.4a initiates electron transfer, which changes the charges of the methylviologen MV stations with subsequent moving of the crown ether shuttle toward the more electron-deficient station. In Figure 1.4b, the macrocyclic shuttle and the stations HB1 and HB2 interact by hydrogen bonding powered by UV light via an external electron relay.[21] Even light-driven molecular rotary motors were designed that can reverse their motion from clockwise to counterclockwise.[22]

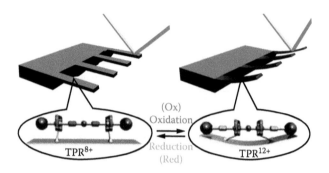

**FIGURE 1.3** Conversion of chemical or electrochemical energy into mechanical energy by bending a microcantilever. Tethered palindromic [3]rotaxane (TPR8+) contains two tetrathiafulvalene (TTF) stations near the termini of the dumbbell and two 1,5-dioxynaphthalene (DNP) stations at the center. Two electron-deficient cyclobis (paraquat-p-phenylene) (CBPQT4+) rings bear disulfide tethers such that TPR8+ can be assembled onto the gold surface. Oxidation of TTF by either chemical or electrochemical activation causes the shuttling of the CBPQT4+ rings to the DNP sites, shortening the inter-ring distance from 4.2 to 1.4 nm in a doubly bistable [3]rotaxane with contraction of the microcantilever. (From Boyle, M.M. et al., *Chem. Science*, 2, 204–210, 2011. Reproduced by permission of The Royal Society of Chemistry.)

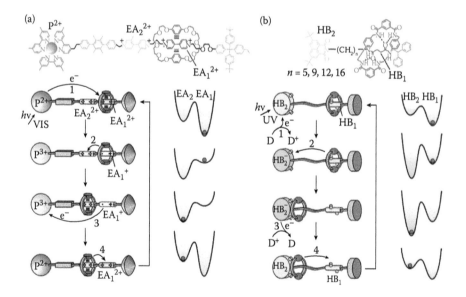

**FIGURE 1.4** Light-induced electron transfer in two different rotaxanes (a) and (b) movements of a macrocyclic shuttle. Curves on the right show an idealized representation of the potential energy profile for each molecular structure illustrated on the left. (Credi, A. et al.: *Chem. Phys. Chem.* 2010. 11. 3398–3403. Copyright Wiley-VCH Verlag GmbH & Co. KGaA. Reproduced with permission.)

*Self-assembly* between organic and/or inorganic molecules relies on noncovalent interactions and is the basis of many new technologies, which can be only partially touched upon in the present monograph. The ways in which molecules self-assemble can be programmed at the molecular level by their interaction mechanisms. In Chapter 10, Hosseini describes how self-assembly in the solid state leads, by rationally designed crystal engineering, to molecular networks or lattices in two or three dimensions.

## Porous Materials

*Metal-organic frameworks (MOFs) and porous materials* can selectively bind many small molecules by noncovalent forces and have gained enormous attention, for instance, for separation, storage, and release of gases. The field, which relates also to zeolite chemistry, is highlighted in many reviews[23-27] and will be characterized here only with some typical examples.

Zeolites are crystalline tectosilicates with a very sharp pore-size distribution with pore sizes limited to approximately 15 $Å^3$. Crystalline-ordered MOF networks are formed from an organic electron donor, such as α,ω-dicarboxylate and metal cations (Figure 1.5). The pore size can be tuned by organic linkers such as dicarboxylates, pyrazine, or 4,4-bipyridine to afford pores that, inside the solid materials, exhibit extremely large surface areas. Two-dimensional grid sheets form with $Cu^{2+}$, $Zn^{2+}$, and $Co^{2+}$ as a metal center; reaction with the linker yield pillared 3D cubic MOFs.[23] With 1,4-benzenedicarboxylate, for instance, one obtains one-dimensional micropores of about 4.0-Å diameter, which discriminate linear and branched hydrocarbons. Thin films made of MOFs can stand alone as membranes; they lend themselves to applications such as luminescence, quartz crystal microbalance-based sensors, and optoelectronics, in addition to separations.[28] MOFs can also serve as potential drug carriers; depending on their pore size, they can absorb and release large amounts of drugs such as ibuprofen or procainamide.[29]

*Mesoporous and microporous, noncrystalline materials* with well-defined pores are obtained by "soft" sol–gel processes in solution with structure-directing agents or surfactants and the subsequent addition of silica precursors.[30] Polycondensation of the precursor results in a mesostructured composite; calcination or solvent extraction of the template leads to a mesoporous silica phase, exhibiting pore diameters from 2 to 5 nm and large inner surface areas of typically 1000 $m^2$/g. Aerosol or spray processing

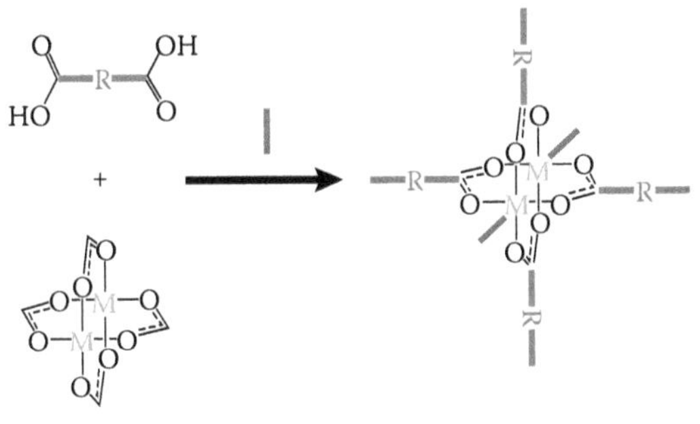

**FIGURE 1.5**  3D MOF building block assembled from paddle-wheel clusters M2(CO2)4 and dicarboxylates. (Reprinted with permission from Chen, B.L. et al., *Accts. Chem. Res.,* 43, 1115–1124. Copyright 2010 American Chemical Society.)

with sol–gel chemistry, self-assembly, and multiple templating offers an appealing and relatively easy way to create nanostructured hybrid porous materials (Figure 1.6).[31] The presence of the pore-confined surfactant provides a hydrophobic medium to which the organic compounds have a high affinity, resulting in significant sorption capacities toward organic adsorbates.[32] Mercaptopropyl-, diethylenetriamine-, or cyclam-functionalized monolayers on ordered mesoporous silica can extract toxic metal ions such as mercury or copper. Mesoporous silicas containing propylammonium moieties may be used for binding several anions such as arsenate, chromate, molybdate, selenate, nitrate, and phosphate, as well as pesticides. Carbamoylphosphonates and other phosphonate ligands bind radionuclides effectively.[33] Periodic mesoporous organosilicas (PMOs) made with chiral organosilane precursors allow chiral separations and catalysis.[34] Self-assembly of alkyl chains at the solid–liquid interface is controlled essentially by van der Waals forces and can lead to multicomponent 2D nanomaterials.[35]

In many hybrid nanomaterials,[36] supramolecular interactions between organic and inorganic components are the basis for the design of ordered structures, also in the form of gels,[37] micelles,[38] etc. Such materials offer new directions for sensor design.[39] The association between organic amphiphiles such as surfactants can be tuned to yield not only covalent binding, for example, by photoinduced reactions, but also often better by reversible noncovalent forces to manifold promising supramolecular materials.[40] Amphiphiles bearing bis-ureas or guanosine residues, for example, assemble even in water not only by hydrophobic interactions but also by

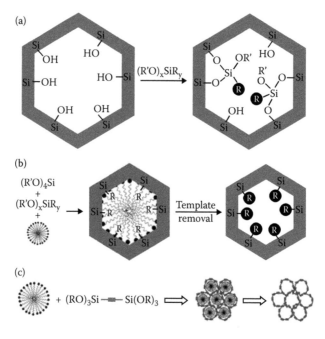

**FIGURE 1.6** Main strategies for organically functionalized mesoporous silica: (a) post-synthesis grafting; (b) co-condensation route; (c) periodic mesoporous organosilicas from bridged silsesquioxanes (RO)3Si—Si(OR)3. (From Walcarius, A. et al., *J. Mater. Chem.*, 20, 4478–4511, 2010. Reproduced by permission of The Royal Society of Chemistry.)

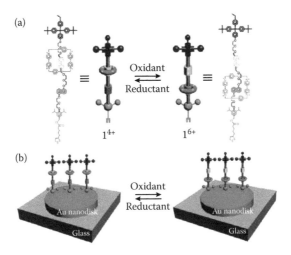

**FIGURE 1.7** (a) Structures and redox-induced switching of a dithiolane-terminated bistable [2]rotaxane. (b) Schematic illustration of molecular mechanical switching at the surface of a gold nanodisk. (From Klajn, R. et al., *Chem. Soc. Rev.* 39, 2203–2237, 2010. Reproduced by permission of The Royal Society of Chemistry.)

hydrogen bonding; those with permanent charges assemble by ion pairing contributions.[41] Noncovalent interactions in amphiphilic peptides allow one to design a large variety of nanosized materials, frequently in the form of gels, with promising applications in catalysis.[42] *Dendrimers* can bind guest molecules not only in their interior but also at their branch termini periphery. In addition, they can form a large variety of nanosized materials, including gels, liquid crystals, nanotubes, films, membranes, etc. Again, the reader is referred to recent reviews.[43]

*Molecular switches* can be part of self-assembled monolayers and still function in a similar manner as in solution, and control optical, fluorescent, electrical, and magnetic properties or the release of small molecules.[44,45] Figure 1.7 illustrates how such systems can function by redox-switched movements[44] (for a related example, see Chapter 8).

## Self-Assembled Monolayers (SAMs)

The assembly of preformed nanoparticles allows one to generate new multidimensional nanostructures, again based on noncovalent interactions.[46] Nanofabrication encompasses the patterning of nanostructures with molding, embossing, printing, scanning probe lithography, and edge lithography. Organic nanostructures were used as templates to mask the deposition of metals and for the growth of metal nanoparticles, and can be used also for dip-pen nanolithography.[47] The use of nanoparticles in microelectronics is also discussed in Chapter 9. Other than top-down techniques such as photolithography, bottom-up approaches use interactions between molecules or colloidal particles to assemble discrete nanoscale structures in two and three dimensions and, thus, can reach much smaller dimensions. The reader is referred to the literature, which also describes the engineering aspects involved in fabrication, characterization, and application of these techniques, including microelectronics.[31,47]

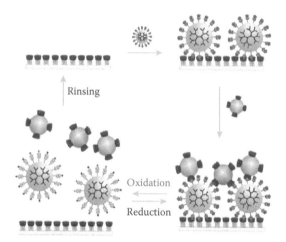

**FIGURE 1.8**   SAM with redox-active ferrocene 6-thio-ß-CD-complexes reversibly attached with a dendrimer "glue" to a monolayer. (Reprinted with permission from Ling, X. Y. et al., *Chem. Materials,* 20, 3574–3578. Copyright 2008 American Chemical Society.)

Surfaces can be functionalized with a large number of receptor molecules;[48] this and the high epitope density on surfaces significantly increase by multivalency the effective molarity. For instance, an apparent association constant of up to $K = 10^{10}$ [$M^{-1}$] was measured with cyclodextrin SAMs.[49] Thus, thermodynamically and kinetically stable assemblies are possible at interfaces, which allow localized surface assemblies, or patterning, as well as materials by a layer-by-layer buildup metalation that can lead to new printboards for microelectronics.[50] The coupling on surfaces can rest not only on covalent binding, for instance, via reversible –S-S-bridge formation, but also on metal coordination,[51] electrostatic interactions,[52] hydrogen bonding,[53] or the well-known strong biotin–streptavidin complexes.[54] Noncovalent binding has the advantage of controllable molecular recognition abilities with the possibility for error correction. Figure 1.8 illustrates how ferrocene (Fc) complexed at a per-6-thio-ß-CD monolayer can, by Fc oxidation, electrochemically be reversibly desorbed from ß-CD SAMs, which are bound in a dendrimer "glue" onto the surface; reduction then leads back to the bound nanoparticle.[55]

## ACKNOWLEDGMENTS

The editor is most grateful to the authors of all the book chapters, who tirelessly secured a high standard of presentation. Professors Werner Nau and Robert Strongin kindly reviewed the author's own chapters and made valuable suggestions.

## REFERENCES

1. Cramer, F. *Angew. Chem.* 1952, *64*, 136–136.
2. Lehn, J.-M. *Supramolecular Chemistry: Concepts and Perspectives,* Wiley-VCH, Weinheim, 1995; Beer, P. D., Gale, P. A., Smith, D. K. *Supramolecular Chemistry,* Oxford University Press, 1999; Schneider, H.-J., Yatsimirski, A. *Principles and*

*Methods in Supramolecular Chemistry*, Wiley, Chichester, 2000; Steed, J. W., Atwood, J. L. *Supramolecular. Chemistry*, Wiley, New York, 2000; Cragg, P. *A Practical Guide to Supramolecular Chemistry*, Wiley, New York, 2005; Ariga, K., Kunitake, T. *Supramolecular Chemistry: Fundamentals and Applications*, Springer, Berlin, 2005; Steed, J. W., Turner, D. R., Wallace, K. J. *Core Concepts in Supramolecular Chemistry and Nanochemistry*, Chichester, Wiley, 2007.

3. Schneider, H.-J. *Angew. Chem. Int. Ed. Engl.*, 2009, *48*, 3924–3977.
4. Reinhoudt, D. N., ed., *Supramolecular Materials and Technologies*, Wiley Interscience, 1999; Jones, W., Rao, C. N. R., eds., *Supramolecular Organization and Materials Design*, Cambridge University Press, 2002.
5. Whitesides, G. M., Lipomi, D. J. *Faraday Discuss*, 2009, *143*, 373–384.
6. Ferhst, A. *Structure and Mechanism in Protein Science*, W. H. Freeman, San Francisco, 1999.
7. Zürcher, M., Diederich, F. *J. Org. Chem.*, 2008, *73*, 4345–4361; Smith, D. K. *J. Chem. Education*, 2005, *82*, 393–400.
8. Teller, C., Willner, I. *Curr. Opinion. Biotech.*, 2010, *21*, 376–391; Teller, C., Willner, I. *Trends Biotech*, 2010, *28*, 619–628; Willner, I., Shlyahovsky, B., Zayats, M., Willner, B. *Chem. Soc. Rev.*, 2008, *37*, 1153–1165.
9. van Leeuwen, P. W. N. M., ed., *Supramolecular Catalysis*, Wiley VCH Weinheim, 2008; Breslow, R., ed., *Artificial Enzymes*, Wiley VCH Weinheim, 2005.
10. D'Souza, F., Ito, O. *Chem. Comm.*, 2009, 4913–4928.
11. Zhang, J., Meng, X. G., Zeng, X. C., Yu, X. Q. *Coord. Chem. Rev.*, 2009, *253*, 2166–2177.
12. Sandee, A. J., Reek, J. N. H. *Dalton Trans.*, 2006, 3385–3391.
13. Kovbasyuk, L., Kramer, R. *Chem. Rev.*, 2004, *104*, 3161–3187.
14. Negi, S., Schneider, H. M. *Tetrahedron Letters*, 2002, *43*, 411–414.
15. Lönnberg, T., Aiba, Y., Hamano, Y., Miyajima, Y., Sumaoka, J., Komiyama, M. *Chem. Europ. J.*, 2010, *16*, 855–859.
16. Balzani, V., Venturi, M., Credi, A. *Molecular Machines*, Wiley-VCH, 2003; Sauvage, J. P., ed., *Molecular Machines and Motors*, Springer, 2001; Kelly, T. R. ed., *Molecular Machines*, Springer, 2005.
17. Boyle, M. M., Smaldone, R. A., Whalley, A. C., Ambrogio, M. W., Botros, Y. Y., Stoddart, J. F. *Chem. Science*, 2011, *2*, 204–210.
18. Durot, S., Reviriego, F., Sauvage, J. P. *Dalton Trans.*, 2010, *39*, 10,557–10,570.
19. Saha, S., Stoddart, J. F. *Chem. Soc. Rev.*, 2007, 77; Leigh, D. A., Zerbetto, F., Kay, E. R. *Angew. Chem. Int. Ed. Engl.*, 2007, *46*, 72–191.
20. Balzani, V., Credi, A., Venturi, M. *Chem. Soc. Rev.*, 2009, *38*, 1542–1550.
21. Credi, A., Venturi, M., Balzani, V. *Chem. Phys. Chem.*, 2010, *11*, 3398–3403.
22. Ruangsupapichat, N., Pollard, M. M., Harutyunyan, S. R., Feringa, B. L. *Nature Chemistry*, 2011, *3*, 53–60.
23. Chen, B. L., Xiang, S. C., Qian, G. D. *Accts. Chem. Res.*, 2010, *43*, 1115–1124.
24. Zhao, D., Timmons, D. J., Yuan, D. Q., Zhou, H. C. *Accts. Chem. Res.* 2011, *44*, 123–133.
25. Meek, S. T., Greathouse, J. A., Allendorf, M. D. *Advanced Materials*, 2011, *23*, 249–267.
26. Zacher, D., Schmid, R., Woll, C., Fischer, R. A. *Angew. Chem. Intern. Ed. Engl.*, 2010, *50*, 176–199.
27. Okamoto, K., Chithra, P., Richards, G. J., Hill, J. P., Ariga, K. *Intern. J. Molec. Sci.*, 2009, *10*, 1950–1966.
28. Shekhah, O., Liu, J., Fischer, R. A., Woll, C. *Chem. Soc. Rev.*, 2011, *40*, 1081–1106.
29. Keskin, S., Kizilel, S. *Ind. Engin. Chem. Research*, 2009, *50*, 1799–1812; Huxford, R. C., Della Rocca, J., Lin, W. B. *Curr. Opin. Chem. Biology*, 2011, *14*, 262–268.
30. Ruiz Hitzky, E., Ariga, K., Lvov, Y. M. *Bio-inorganic Hybrid Nanomaterials: Strategies, Syntheses, Characterization and Applications*, Wiley-VCH, 2007; Vallet-Regi, M., Arcos, D. *Biomimetic Nanoceramics in Clinical Use: From Materials to Applications*, RSC Cambridge, 2008; Ethirajan, A., Landfester, K. *Chem. Europ. J.*, 2010, *16*, 9398–9412.

31. Boissiere, C., Grosso, D., Chaumonnot, A., Nicole, L., Sanchez, C. *Advanced Materials,* 2011, *23,* 599–623.
32. Thomas, A. *Angew. Chem. Intern. Ed. Engl.,* 2010, *49,* 8328–8344; Hoffmann, F., Fröba, M. *Chem. Soc. Rev.,* 2011, *40,* 608–620.
33. Walcarius, A., Mercier, L. *J. Mater. Chem.,* 2010, *20,* 4478–4511.
34. Kim, K., Banerjee, M., Yoon, M., Das, S. *Funct. Metal-Organic Framework,* 2010, *293,* 115–153.
35. Ciesielski, A., Palma, C. A., Bonini, M., Samori, P. *Advanced Materials,* 2010, *22,* 3506–3520; Kudernac, T., Lei, S. B., Elemans, J., De Feyter, S. *Chem. Soc. Rev.,* 2009, *38,* 3505–3505.
36. Rurack, K., Martínez-Máñez, R., eds., *The Supramolecular Chemistry of Organic–Inorganic Hybrid Materials,* Wiley, Hoboken, 2010.
37. Hirst, A. R., Escuder, B., Miravet, J. F., Smith, D. K. *Angew. Chem. Int. Ed.,* 2008, *47,* 8002–8018.
38. Ge, Z., Liu, S. *Macromol. Rapid Commun.,* 2009, *30,* 1523–1532.
39. Martinez-Manez, R., Sancenon, F., Hecht, M., Biyikal, M., Rurack, K. *Anal. Bioanal. Chem.,* 2011, *399,* 55–74.
40. Wang, Y., Xu, H., Zhang, X. *Advanced Materials,* 2009, *21,* 2849–2864; Zhang, X., Wang, C. *Chem. Soc. Rev.,* 2011, *40,* 94–101; Yagai, S., Kitamura, A. *ibid.* 2008, *37,* 1520–1529; Zhou, Y., Yan, D. *Chem. Comm.,* 2009, 1172–1188.
41. Zayed, J. M., Nouvel, N., Rauwald, U., Scherman, O. A. *Chem. Soc. Rev.,* 2010, *39,* 2806–2816.
42. Cavalli, S., Albericio, F., Kros, A. *Chem. Soc. Rev.,* 2010, *39,* 241–263.
43. Tomalia, D. A. *Soft Matter,* 2010, *6,* 456–474; Astruc, D., Boisselier, E., Ornelas, C. *Chem. Rev.,* 2010, *110,* 1857–1959; Shen, M. W., Shi, X. Y. *Nanoscale,* 2010, *2,* 1596–1610; Caminade, A. M., Majoral, J. P. *Chem. Soc. Rev.,* 2010, *39,* 2034–2047.
44. Klajn, R., Stoddart, J. F., Grzybowski, B. A. *Chem. Soc. Rev.,* 2010, *39,* 2203–2237.
45. Nguyen, T. D., Tseng, H. R., Celestre, P. C., Flood, A. H., Liu, Y., Stoddart, J. F., Zink, J. I. *Proc. Nat. Acad. Sci. USA,* 2005, *102,* 10,029–10,034.
46. Kinge, S., Crego-Calama, M., Reinhoudt, D. N. *Chem. Phys. Chem.,* 2008, *9,* 20–42.
47. Gates, B. D., Xu, Q. B., Stewart, M., Ryan, D., Willson, C. G., Whitesides, G. M. *Chem. Rev.,* 2005, *105,* 1171–1196; Xu, Q. B., Rioux, R. M., Dickey, M. D., Whitesides, G. M. *Accts. Chem. Res.,* 2008, *41,* 1566–1577; Siegel, A. C., Tang, S. K. Y., Nijhuis, C. A., Hashimoto, M., Phillips, S. T., Dickey, M. D., Whitesides, G. M. *Accts. Chem. Res.,* 2010, *43,* 518–528; Wang, Y. H., Mirkin, C. A., Park, S. J. *Acs Nano,* 2009, *3,* 1049–1056; Pan, T. R., Wang, W. *Ann. Biomed. Engineering,* 2011, *39,* 600–620; Salaita, K., Wang, Y. H., Mirkin, C. A. *Nature Nanotech,* 2007, *2,* 145–155.
48. Gomar-Nadal, E., Puigmarti-Luis, J., Amabilino, D. B. *Chem. Soc. Rev.,* 2008, *37,* 490–504.
49. Ludden, M. J. W., Reinhoudt, D. N., Huskens, J. *Chem. Soc. Rev.,* 2006, *35,* 1122–1134.
50. Perl, A., Reinhoudt, D. N., Huskens, J. *Advanced Materials,* 2009, *21,* 2257–2268; Willner, I. *Small,* 2009, *5,* 28–44.
51. Rubinstein, I., Vaskevich, A. *Israel J. Chem.,* 2010, *50,* 333–346.
52. Lefort, M., Popa, G., Seyrek, E., Szamocki, R., Felix, O., Hemmerle, J., Vidal, L., Voegel, J. C., Boulmedais, F., Decher, G., Schaaf, P. *Angew. Chem. Intern. Ed. Engl.,* 2010, *49,* 10,110–10,113.
53. Binder, W. H., Zirbs, R., Kienberger, F., Hinterdorfer, P. *Polymers Advanced Techn.,* 2006, *17,* 754–757.
54. Rasal, R. M., Hirt, D. E. *Macromol. Bioscience,* 2009, *9,* 989–996; Wenz, G., Liepold, P. *Cellulose,* 2007, *14,* 89–98.
55. Ling, X. Y., Reinhoudt, D. N., Huskens, J. *Chem. Materials,* 2008, *20,* 3574–3578.

# 2 Inorganic Analytes and Sensors

*Enrique García-España, M. Teresa Albelda, and Juan C. Frías*

## CONTENTS

## INTRODUCTION

Molecular sensor design is an active field of supramolecular chemistry.[1] Actually, many of the basic concepts in this discipline have been derived from early selective-binding studies of alkali metal cations by natural and synthetic macrocyclic ligands (crown ethers and cryptands).[2] Sensing applications of supramolecular chemistry rely on exploiting the forces involved in the formation of noncovalent host–guest complexes; the more complementary the binding sites of the host to those of the guest, the higher the binding energy. Molecular recognition is still a challenging topic nowadays, and many efforts have been devoted to the development of supra-molecular sensors able to selectively detect analytes of chemical or biological sig-nificance. This fascinating field exploits the recognition abilities of supramolecular receptors to yield analytical tools characterized by high sensitivity, specificity, and selectivity.[3] In fact, the term "supramolecular analytical chemistry" has been coined recently to describe the application of molecular sensors to analytical chemistry.[4]

Sensors based on molecules designed along supramolecular principles are wide-spread in the scientific literature, and in most cases, those devices contain a molecular recognition element (able to interact specifically with a target) coupled to a transducer (able to convert the recognition event into a measurable signal).[5–8] The mechanism of signal transduction involves a chemical interaction to produce a detectable signal. One of the most convenient and simple means of chemical detection is the genera-tion of an optical signal—for example, changes in absorption or emission bands of the chemosensor in the presence of the target analyte. Colorimetric sensors have the advantage of allowing simple on-site real-time detection without instruments. Although this type of sensor can detect analytes through simple color changes by visual observation, fluorimetric detection provides higher sensitivities. For this rea-son, the largest group of chemical sensors described in this chapter comprises those devices based on the use of fluorescent probes or labels. Fluorescence-based sensors offer significant advantages such as high sensitivity (capable of detecting very low concentration levels of the analyte $< 10^{-7}$ M), rapid response, and simple instrumen-tation.[5–7] In other cases, signaling of the analyte binding process takes place with an electrochemical response. Electrochemical sensors provide low detection limits, a wide linear response range, and good stability and reproducibility.[8] Depending on the sensing mechanism, electrochemical sensors can be classified as potentiometric, conductimetric, and voltammetric. Although many promising scientific results have emerged, the future of sensors calls for autonomous and sensitive mobile devices. Nanotechnologies will enable solutions for sensors that are inexpensive, ecologically sustainable, robust in harsh environmental conditions, and stable over a long period of time. Improvements in reproducibility, calibration, and manufacturability are still some of the challenges for developing chemical sensory devices.[9]

The field of supramolecular analytical chemistry opens up new molecular designs and approaches for the use of synthetic receptors in analytical sciences. The aim of this chapter is to give an overview of this fascinating field with a particular insight into some examples of receptors designed for the recognition of inorganic analytes. The term "inorganic analytes" includes metallic and nonmetallic species, different oxidation states, and the analysis of speciated metals. This chapter also contains some references for gas-sensing supramolecular materials.

## CATION SENSING

The design of highly selective, sensitive, and low detection-limit chemosensors for metal ions has been actively investigated because detection of cations is, in fact, of great interest for several applications. Metal ions play important roles in many bio-logical processes. For example, the fluid balance in the body, crucial for all vital func-tions, is maintained largely by sodium, potassium, and chloride ions.[10] Tight control of electrolytes (sodium, potassium, calcium, and magnesium) is critical for normal muscle contraction, nerve-impulse transmission, heart function, and blood pressure. Calcium and magnesium are known for their structural roles as they are essential for the development and maintenance of bones and teeth.[11,12] Iron plays a major role in oxygen transport and storage.[13] Cellular energy production requires iron, copper, and zinc, which act as enzyme cofactors in the synthesis of many proteins, hormones, neurotransmitters, and genetic material. Zinc is also essential for many other body functions, such as growth, development of sexual organs, and reproduction.[14]

Beyond the essentiality of metal ions in life, cations find an unconventional application, encompassing diagnostics and therapeutics, in healthcare.[15] Gold compounds have been used for the treatment of rheumatoid arthritis and other autoimmune diseases for more than 75 years.[16] Platinum(II) complexes[17] have been used in anticancer drugs since the 1960s, and other metal complexes (iron, gallium, germanium, tin, bismuth, titanium, ruthenium, rhodium, gold, etc.) have been shown to be effective against tumors.[18] Lithium drugs are used for the treatment of neurological disorders and silver compounds have been used for years as antimicrobial agents.[19] Lanthanides have physicochemical properties that are well suited for diagnostic imaging procedures as contrast-enhancing agents for magnetic resonance imaging. For example, gadolinium chelates are used as contrast media for conventional angiography and venography.[20] (See Chapter 16 of this book)

However, not all metal ions are involved in essential biological reactions; some heavy-metal ions (mercury, lead, tin, arsenic, and aluminum) are toxic and dangerous to health or to the environment.[21] Consequently, early detection of these species is desirable.

## ALKALI AND ALKALINE EARTH CATION TARGETS

Complexation chemistry of alkali cations developed rapidly with the discovery that valinomycin and other natural antibiotics are able to transport potassium across biological membranes efficiently and selectively.[22] These results inspired the work of Pedersen, Lehn, and Cram,[2] who developed the synthesis of crown ethers, cryptands, and spherands.[2] Several reviews[23] and monographs[24] have appeared on this subject, showing the cation binding properties of these molecules. Oxygen atoms uniformly distributed in the macrocycle skeleton establish electrostatic interactions with a guest metal ion located in the crown ether cavity. Furthermore, the ring size of the macrocycle plays an important role in regulating the selectivity for metal ions. Design of novel crown ether–based chemosensors is generally directed to the modification of the signaling unit appended to the macrocycle (Scheme 2.1).

Compound **1** was the first and simplest photoinduced electron transfer (PET) cation sensor[25] and was designed so that the anthracene fluorescence, which is efficiently quenched by electron transfer from the lone pair of nitrogen atoms, is recovered after incorporation of alkali metal cations into the azacrown ether unit.[26] Its fluorescence quantum yield increases from 0.003 to 0.14 upon binding of $K^+$ in methanol. The results are consistent with the basis of hole size correspondence in which the 18-membered crown ether cavity fits the bound cation.[27] Reducing the macrocyclic cavity, Gunnlaugsson et al. synthesized a chiral diaza-9-crown-3 derivative

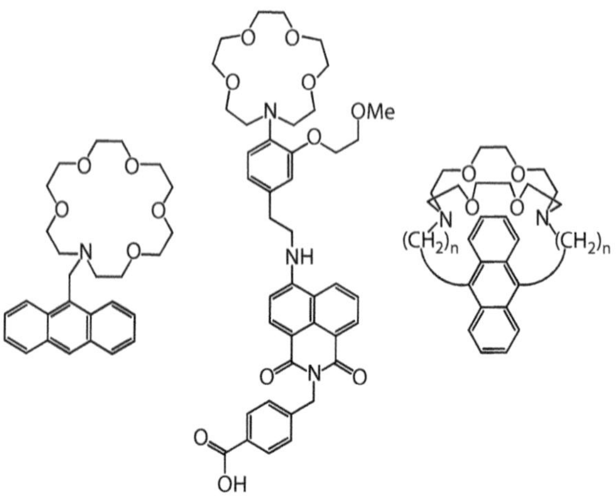

**SCHEME 2.1**   Crown ether–based chemosensors for alkali cations.

with excellent selectivity for Li$^+$ over other alkaline cations in acetonitrile because of size discrimination.[28] The 15-membered macrocyclic compound **2** shows a real-life application of these compounds when immobilized in a hydrophilic polymer layer.[29] It was designed as an optical sensor suitable for practical measurement of Na$^+$ in serum and blood samples. The slope of the sensor, defined as the percentage signal change per millimolar change in sodium, is close to 0.5%/mM in the typical clinically significant range of 120 to 160 mM. Compounds **1** and **2** are examples of fluorescent PET cation sensors and illustrate how a metal ion regulates the on/off fluorescence emission intensity. However, we have not mentioned the role of the spacer in those systems. PET rates are dependent on the electron-transfer distance. As an example of this, Sankaran et al. reported a series of compounds that consist of a pyrenyl group as acceptor, an amidobenzocrown ether group as donor, and alkyl spacers. Those systems behave either as simple PET or as ratiometric (binding of the analyte to the sensor induces a shift in excitation or emission maxima, and internal calibration can be achieved by measuring the ratio of free sensor to complexed form)[30] sensors toward alkali metal ions based on the length of the spacer that connects the donor and acceptor groups.[31]

In an attempt to obtain better selectivity, cryptand-based PET sensors were designed. In contrast to crown ethers, which essentially form two-dimensional complexes, cryptands form enveloping complexes in which the metal ion is completely desolvated. Anthracene cryptands (**3**, Scheme 2.1) present drastic changes in fluorescence emission upon complexation by alkali-metal cations and protons; the fluorescence spectrum displays a hypsochromic shift, and the quantum yield is enhanced.[32] Another example of a chemosensor based on a [2.2.2] cryptand structure with a 4-methylcoumarin appended unit was developed by Golchini et al. for monitoring levels of K$^+$ in blood samples and across biological membranes.[33] Doludda and coworkers used a smaller cavity size (cryptand [2.2.1]), which makes the receptor well suited for measuring Na$^+$.[34]

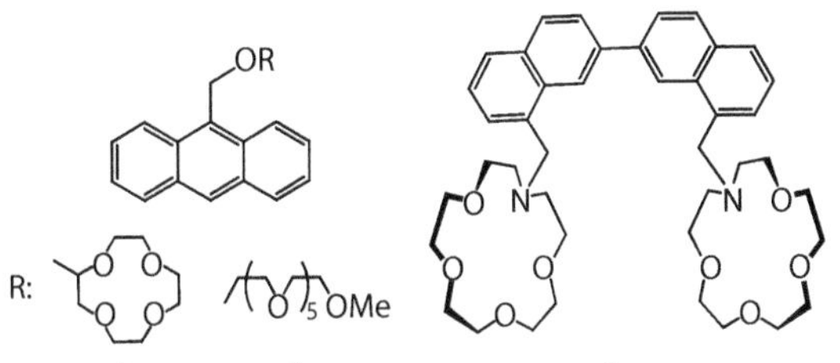

**SCHEME 2.2**  Crown ethers specially designed for the recognition of alkaline earth metal ions.

Alkaline earth metal ions are also strongly recognized by crown ethers. Compounds **4** and **5** (Scheme 2.2) interact with alkaline earth salts more effectively than with alkali salts, bringing about a substantial fluorescence enhancement.[35] The ether-linked host **4** recognized calcium cations preferentially in the presence of barium cations in contrast to host **5**, which preferred $Ba^{2+}$ to $Ca^{2+}$; these preferences were not affected by the kind of counteranion. These results, therefore, may be attributed to the size of the cation. $Ca^{2+}$ fits better in the macrocyclic cavity of 12-crown-4 ether, and the flexible pseudocrown ether of the polyether moiety prefers the barium cation. Kondo et al. studied the sandwich complex formation by bis-azacrown ethers and a 2,2′-binaphthalene moiety (**6**, Scheme 2.2).[36] Compound **6** showed a unique response to $Ba^{2+}$ upon the addition of alkali and alkaline earth metal cations. Selectivity was obtained as a result of the large size of the metal ion, which induced a conformational change through the formation of the intramolecular sandwich complex.

Besides crown ether derivatives, different approaches were described to design and synthesize receptors for alkaline and alkaline earth sensing. For example, carbonyldipyrrinone **7** (Scheme 2.3) possesses two carbonyl groups optimally placed for cation coordination.[37] The response was highly selective to $Ca^{2+}$ with respect to the other cations tested because of the optimal size of the calcium ion. Calix[4]arene diester receptor **8** binds sodium strongly in acetone ($K = 3500$ $M^{-1}$), in lithium less strongly ($K = 470$ $M^{-1}$), and displays no affinity for the larger alkali-metal ions.[38]

Metal nanoparticles are emerging as important colorimetric reporters because of their high extinction coefficients, which are several orders of magnitude larger than those of organic dyes. Obare et al. designed 4-nm gold nanoparticles coated with a 1,10-phenanthroline ligand that binds selectively to $Li^+$.[39] Formation of a 2:1 ligand–metal complex causes gold nanoparticles to aggregate, resulting in visible color changes. This system can detect concentrations of lithium ion of 10–100 mM in an aqueous medium.

## COPPER AND ZINC SENSING

The design of highly selective, sensitive, and low detection-limit chemosensors for copper ion has always been of particular importance as copper is an essential trace element in biological systems, and it is a widely used industrial metal. Sensors

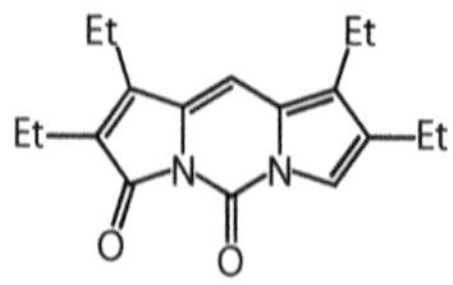

**SCHEME 2.3**  Receptors for alkaline and alkaline earth sensing.

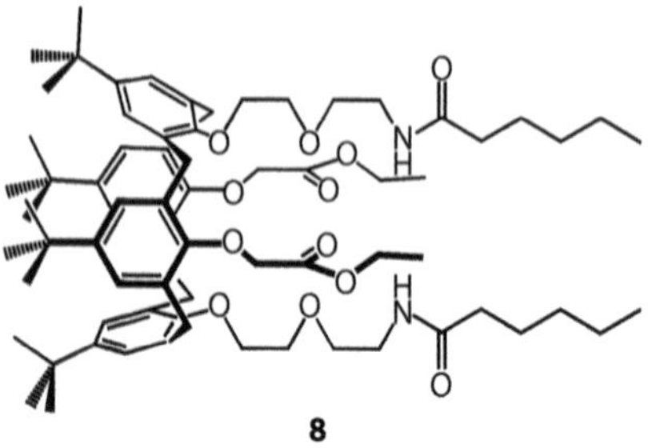

based on the ion-induced changes in fluorescence appear to be particularly attractive because of the simplicity and high detection limit of the fluorescence technique. In addition, they are specifically useful in this case as copper(II) has a particularly high thermodynamic affinity for typical N,O-chelate ligands and fast metal-to-ligand binding kinetics.[40] Qi et al. reported a fluorescent chemodosimeter for $Cu^{2+}$ based on a 4,4-difluoro-4-bora-3a,4a-diaza-s-indacene (BODIPY) derivative (**9**, Scheme 2.4).[41] This useful fluorophore possesses high excitation coefficients, high fluorescent quantum yields, and high stability against light and chemical reactions. Ligand **9** showed an extreme selectivity for $Cu^{2+}$ over other metal ions examined. The efficiency in the recognition event takes place via a selective hydrolysis of the acetyl groups. Receptor **10**, developed by Goswami et al., achieved an excellent colorimetric response (from orange to purple) and a dramatic enhancement of fluorescence intensity (almost a 25-fold increase) upon $Cu^{2+}$ complexation via an internal charge transfer mechanism.[42] The association constant as determined by fluorescence titration was found to be $2.54 \times 10^4$ $M^{-1}$. The series of compounds **11a** and **b**, having naphthalimide and dansyl units, exhibited unusual selectivity for $Cu^{2+}$ as compared to other metal ions and signaled the binding event through inhibition of energy transfer–mediated emission intensity.[43] The uniqueness of these dyads is that they form stable 2:1 stoichiometric complexes involving sulfonamide functionality. Formation of $Cu^{2+}$ complexes alters the interaction between the naphthalimide and dansyl chromophores

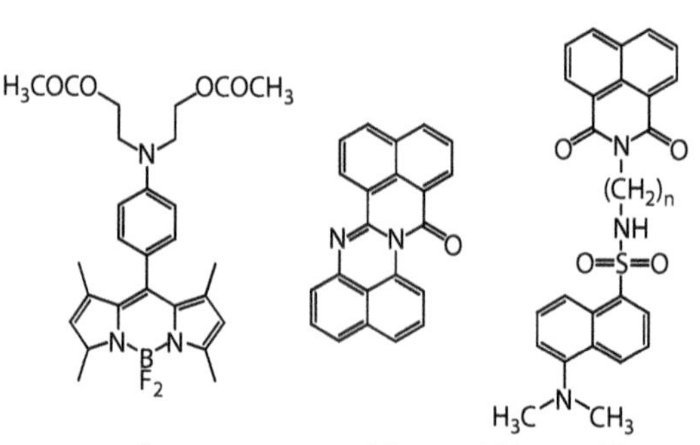

**SCHEME 2.4**  Chemosensors for $Cu^{2+}$ including fluorophore units.

leading to the disruption of fluorescence-resonance energy transfer, thereby enabling the visual fluorescence ratiometric detection of copper ions.

Other fluorophores were also used in order to achieve sensors with colorimetric and fluorescent features for copper ions. For instance, Wang et al. prepared a chemosensor with a merocyanine moiety as a signaling unit able to monitor the variation of $Cu^{2+}$ in living cells.[44] Jung et al. reported a coumarin-based fluorogenic probe bearing a 2-picolyl unit that was employed for the fluorescence detection of the changes in intracellular $Cu^{2+}$ in cultured cells.[45] Other coumarin-based colorimetric chemosensors developed by Sheng et al. were demonstrated to be a valuable step ahead in terms of practical applications.[46] Those compounds exhibited good sensitivity and selectivity for $Cu^{2+}$ over other cations both in aqueous solution and on paper-made test kits. The change in color is very easily observed by the naked eye (from yellow to red) in the presence of $Cu^{2+}$. Rhodamine seems an excellent model for the design and development of fluorescent chemosensors. A recent review describes new types of rhodamine derivatives for their use in biological applications as well as silica mesoporous hybrid materials for the selective detection of specific ions.[47] Zhao and coworkers synthesized a rhodamine-based chemosensor that displayed a highly sensitive turn-on fluorescent response toward $Cu^{2+}$ in aqueous solution with an 80-fold fluorescence intensity enhancement.[48] With the experimental conditions optimized, the probe exhibited a dynamic response range for $Cu^{2+}$ from $8.0 \times 10^{-7}$ to $1.0 \times 10^{-5}$ M with a detection limit of $3.0 \times 10^{-7}$ M. The color and fluorescence changes of the chemosensor are remarkably specific for copper in the presence of other heavy- and transition-metal ions. It was used also for the imaging of $Cu^{2+}$ in living cells with satisfying results.

As an example of Cu(I) selective probes, Taki et al. developed highly sensitive fluorescein derivatives.[49] The tetradentate ligand tris[(2-pyridyl)methyl]amine (TPA) is connected to a reduced form of a fluorescein platform through a benzyl ether linkage. Detection of intracellular $Cu^{+}$ is based on the cleavage of the C–O bond of benzyl ether by copper(I) ions under physiological reducing conditions. The reaction product, O-methylfluorescein, emits intense green fluorescence allowing the visualization of $Cu^{+}$ present in living cells.

Chemical immobilization of small fluorescent molecules onto a surface is a convenient approach to preparing assemblies with specific sensing properties. Yin et al. reported the synthesis of thermoresponsive polymer microgels, incorporating picolinamine units and fluorescent reporters.[50] At 20°C, these microgels can selectively bind $Cu^{2+}$ over other metal ions, leading to prominent quenching of fluorescence-emission intensity. At a microgel concentration of $3.0 \times 10^{-6}$ g/mL, $Cu^{2+}$ detection limits considerably improve from 46 to 8 nM upon elevating detection temperatures from 20°C to 45°C. Choi et al. reported the immobilization of ethylenediamine-functionalized gold particles on a glass slide for the selective and sensitive detection of $Cu^{2+}$ by plasmonic resonance energy transfer nanospectroscopy.[51] When a copper ion binds the matching ligand, $d$ orbitals are split, which can generate a new absorption band in the metal–ligand complex in the visible range. This method can detect concentrations of $Cu^{2+}$ down to 1 nM.

As mentioned previously, zinc is an essential trace element, acting as a structural component of proteins or in the catalytic site of enzymes. It is involved in many neurobiological processes in the body, playing a role in numerous diseases. Therefore, there is much interest in the development of zinc chemosensors, especially those used for its detection *in vivo*. Fluorescent sensor molecules again offer useful information about chelatable $Zn^{2+}$ in cellular systems, and fluorescent imaging has been proven to be the most suitable technique for *in vivo* zinc monitoring. There have been many excellent advances in this field recently collected in a review published by Jiang and Guo.[52] Here, we will briefly discuss some representative examples of rational design of $Zn^{2+}$ fluorescent chemosensors.

Among the most widely used Zn(II) fluorescent sensors are quinoline derivatives. Zhang et al. synthesized a water-soluble and highly selective carboxamidoquinoline with an alkoxyethylamino chain as receptor (Compound **12**, Scheme 2.5). The signal response of **12** was approximately an eightfold increase in fluorescent quantum yield and a 75-nm red shift of fluorescence emission in buffer aqueous solutions upon binding of $Zn^{2+}$. Preliminary studies of $Zn^{2+}$ sensing behavior in living yeast cells revealed chemosensor **12** as a potential imaging agent of zinc in living tissue or in cells. Compound **13** constitutes an example of a simple, naked-eye, and ratiometric fluorescent sensor. It displays high selectivity for $Zn^{2+}$ (with a significant visual color change from colorless to yellow) and can be used under visible light excitation.[53] An approximately 14-fold $Zn^{2+}$ selective chelation-enhanced fluorescence response is attributed to its strong coordination ability (log $K = 8.45$). However, poor water solubility reduces the possibility of it being used as a zinc sensor in the living system. The incorporation of cyclodextrins in compound **14** avoids the problem of sparing solubility of zinc probes in a neutral aqueous solution.[54] This sensor showed a strong binding affinity (log $K = 12.4$) and exhibited a good fluorescence response to $Zn^{2+}$ over a wide pH range. Significantly, its sensing ability was not affected by other cations. In addition, the capacity to include organic and biological substrates within the hydrophobic cavity of cyclodextrins may enable this compound to adhere to the surface of tissues or cells by including accessible surface molecules in the cavity. Royzen et al. reported the ratiometric detection of $Zn^{2+}$ by the tripodal compound **15**.[55] This ligand forms a highly stable 1:1 complex with zinc showing excellent chelation-enhanced steady-state fluorescence properties. Results indicated subpicomolar sensitivity, and the detection limit of **15** was from 10 fM to 1 pM (log $K = 13.29$). The selectivity of **15** for Zn(II) and the time-resolved fluorescence data not only provide for the accurate detection of this ion but also can discriminate between other fluorescent forms of the probe and its metal complex.

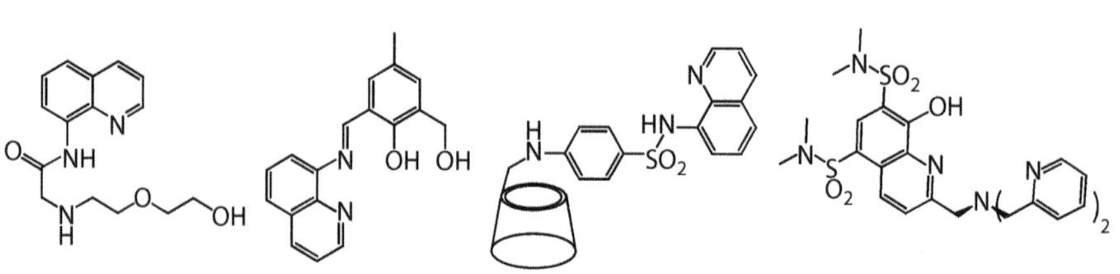

**SCHEME 2.5**   Fluorescent sensor molecules for $Zn^{2+}$ detection.

One of the problems of some available fluorescent zinc sensors is they have difficulty distinguishing zinc and cadmium because $Cd^{2+}$ is in the same group of the periodic table and has similar properties to $Zn^{2+}$. Xu et al. developed a $Zn^{2+}$-selective fluorescent sensor with the ability to exclude the interference of cadmium.[56] The receptor, with a naphthalimide fluorophore, displayed an enhanced fluorescence (22-fold) as well as a red shift in emission from 483 to 514 nm upon zinc complexation. Meanwhile, binding of cadmium resulted in an enhanced (21-fold) blue shift in emission from 483 to 446 nm. Furthermore, this sensor is cell permeable and can be applied to trace zinc ions during the development of a living organism. Wang et al. used a different strategy to avoid the interference of cadmium observed for zinc sensors.[57] They synthesized a water-soluble sensor based on a quinoline platform with femtomolar sensitivity for zinc ions. The chemosensor displayed a 14-fold enhanced quantum yield upon zinc complexation with high selectivity over cadmium and other divalent metal ions in the presence of ethylene diamine tetraacetic acid (EDTA). The strong chelator EDTA is able to scavenge unfavorable transition metals without interfering with the ability of this compound to sense zinc ions.

## Sensing Other Metal Ions

A recent review published by Quang and Kim addresses literature covering chromogenic and fluorogenic materials as systems for sensing heavy-metal ions in both solution and biological samples over the last five years.[58] Here, we will analyze some examples of fluorescent molecular sensors for the detection of noble, transition, and heavy-metal cations.

The noble element silver has been widely used in the fields of catalysis, optoelectronics, medicine, bioactive materials, photography, superconductivity, and water purification. Some examples of receptors able to selectively detect silver are illustrated in Scheme 2.6. Compound **16** was prepared by Li et al. using a thiacalix[4] arene scaffold as macrocyclic host for selective $Ag^+$ complexation. The introduction of proton-ionizable groups (three aminopyridyl and one carboxylate group in

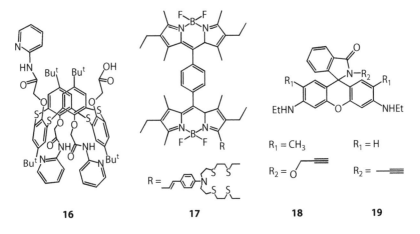

|     16     |     17     |     18     |     19     |

**SCHEME 2.6**   Some examples of receptors able to selectively detect silver and gold.

1,3-alternate conformation) onto the lower rim of the macrocycle enhanced selectivity and affinity for Ag⁺.[59] Ligand **17** incorporates two boradiazaindacene fluorophore units—one of them derivatized with a chromophore moiety.[60] It constitutes an outstanding ratiometric chemosensor because the binding of $Ag^+$ resulted in a blue shift (41 nm) of the emission spectra, and the emission ratio of intensities at 630 and 671 nm changed from 0.25 to 1.42 for the free and the complexed chemosensor, respectively. The binding constant of silver in tetrahydrofuran was determined to be $1.7 \times 10^5$ $M^{-1}$. Chatterjee et al. developed a rhodamine derivative as a fluorogenic and chromogenic probe for $Ag^+$ ions that is also applicable for the detection of silver nanoparticles (AgNPs).[61] The probe showed remarkably high selectivity over other metal ions, and the detection limit was estimated to be 14 ppb. Furthermore, the sensing protocol was applied for the quantification of AgNPs in consumer products (e.g., hand sanitizer gel and fabric softener).

Recently, gold and gold salts have been widely applied in the fields of nanomaterials and catalysis. Furthermore, functionalized gold nanoparticles have been extensively used as drug and gene delivery systems, biosensors, and bioimaging agents. Determination and quantification of gold in different kinds of samples is often a challenging task for analysts. Instrumental techniques include atomic absorption spectrometry; UV/vis spectrometry and derivatization reactions followed by spectrophotometric measurements; inductively coupled plasma-atomic emission spectrometry and mass spectrometry; nuclear activation; and sensitive electrometric methods.[62] However, fluorescent sensing systems for gold ions are hitherto unexplored. Chemosensors based on rhodamine-derived alkyne probes that show turn-on fluorescence change only toward gold species among other metal ions examined.[63,64] Because gold has high alkynophilicity, coordination to the triple bond induces a spirolactam ring opening and heterocycle formation upon gold coordination, resulting in both color and fluorescence changes. Compound **18** can sense $Au^{3+}$ selectively at the 50-nM level in aqueous media and could be used for fluorescence imaging of $Au^{3+}$ in living cells as was also successfully demonstrated.[64] Compound **19** responds to both gold species (Au(I)/Au(III)) in aqueous media, but fluorescence enhancement in the case of $Au^+$ (140 times) is larger than in the case of $Au^{3+}$ (100 times).[63] The detection limit of probe **19** for $Au^+$ was estimated to be 0.4 ppm.

Regarding sensing iron species, very few sensors have been reported despite its importance in many biochemical processes and environmental applications. $Fe^{3+}$ is a well-known fluorescence quencher because of its paramagnetic nature, which makes it difficult to develop a turn-on fluorescent sensor. Another point to consider is that detection of $Fe^{3+}$ is always accompanied by interference from $Cu^{2+}$ and $Cr^{3+}$, which affects their selectivity. Scheme 2.7 includes some representative examples of iron chemosensors. Ligand **20**, synthesized from rhodamine B and diethylenetriamine, exhibited high selectivity for $Fe^{3+}$ over other commonly coexistent metal ions in both ethanol and water solutions.[65] Upon the addition of 1 equiv of Fe(III), the spirocyclic ring was opened, generating a change of color (colorless to purple) and enhancement of fluorescence (154-fold). Sulphonephthalein derivative **21** in the presence of chlorophenol red and $H_2O_2$ was demonstrated to be a highly selective detection system for $Fe^{2+}$ in aqueous media (detection limit = $1.61 \times 10^{-6}$ mol $L^{-1}$).[66] The spectral response gave rise to a solution color change from yellow to colorless that is visible

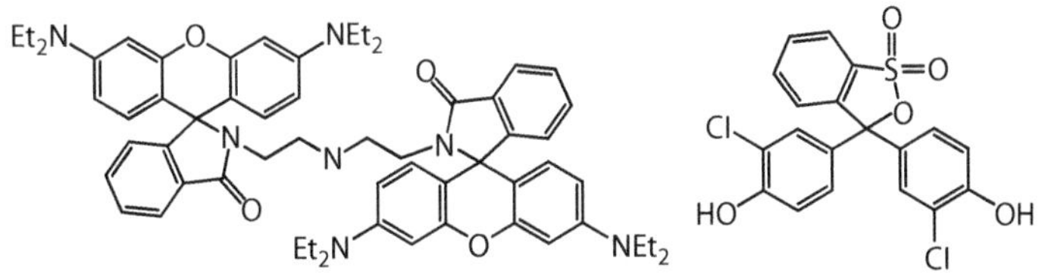

**SCHEME 2.7** Sensors reported for the sensing of iron species.

to the naked eye. This system employed the Fenton reaction for the generation of a hydroxyl radical (reaction of $Fe^{2+}$ with $H_2O_2$ by means of a single electron transfer). Once formed, the hydroxyl radical oxidizes the chromophore leading to complete decolorization.

Cadmium and lead are toxic heavy-metal pollutants in the environment. Long-term exposure to cadmium can cause adverse health effects because it strongly displaces zinc from its proper sites in the body. Lead's toxicity is largely a result of its capacity to mimic calcium and substitute it in many of the fundamental cellular processes that depend on calcium. These factors make their detection and quantification a significant area of research. Compound **22** in Scheme 2.8 possesses a prominent fluorescence enhancement only for $Cd^{2+}$ at pH 7.0 in an aqueous methanol solution (binding constant: $\log \beta = 8.8 \pm 0.1$).[67] The results clearly suggest that the specific semirigid structure could selectively accommodate cadmium according to the ionic radius, which causes a chelation enhanced fluorescence (CHEF) effect. Chemosensor **23**, based on a quinoline fluorophore, uses a ratiometric displacement approach for sensing cadmium.[68] The fluorescent emission maximum of the zinc complex ($Zn^{2+}$–**23**) is blue shifted with the addition of $Cd^{2+}$. The detection limit was calculated to be $2.38 \times 10^{-6}$ M, indicating that $Zn^{2+}$–**23** could be used for sensing cadmium in the micromolar range. A practical application of cadmium sensors can be found in a paper reported by Cheng et al.[69] The authors used the BODIPY fluorophore to develop a fluorescent sensor able to monitor cadmium in biological systems. The fluorescence intensity was significantly enhanced approximately 195-fold, and the quantum yield increased almost 100-fold. Moreover, its fluorescence intensity enhanced in a linear fashion with a concentration of $Cd^{2+}$ and thus can be potentially

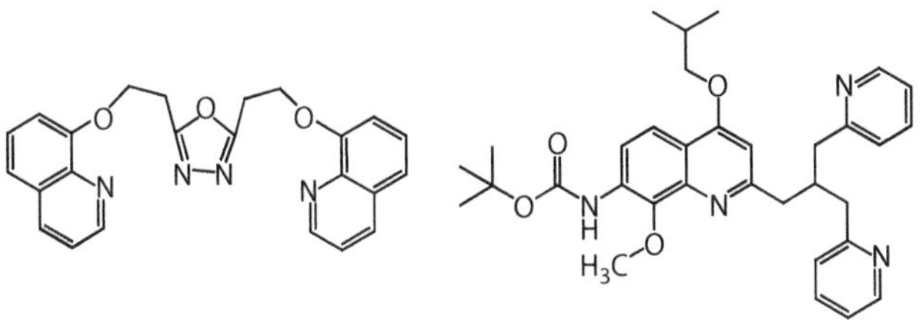

**SCHEME 2.8** Fluorescent sensors for cadmium detection.

used for cadmium quantification. Living cell image experiments demonstrate cell membrane uptake, suggesting its potential use for imaging $Cd^{2+}$ in living cells and *in vivo*.

Representative examples of lead probes are illustrated in Scheme 2.9. Chemosensor **24** is composed of a *tert*-butylthiacalix[4]arene bearing two facing amide groups and a crown-6 ring as cation recognition sites.[70] The remarkable fluorescence quenching induced by $Pb^{2+}$ is ascribed not only to reverse PET from pyrene units to carbonyl oxygen atoms but also to a heavy-metal effect. Compound **24** has a high selectivity for $Pb^{2+}$ (calculated association constant: $2.57 \pm 10^5$ $M^{-1}$ in acetonitrile/chloroform) over other metal ions, which is probably achieved by the selective encapsulation of $Pb^{2+}$ by the crown ring. Ligand **25** was developed by He et al. as a turn-on fluorescent sensor for selective detection of $Pb^{2+}$ in water and in living cells.[71] It combines a fluorescein-type scaffold with attractive optical properties and biological compatibility with a dicarboxylate pseudocrown receptor designed to satisfy the size and charge requirements of the $Pb^{2+}$ cation. The addition of 15 ppb of $Pb^{2+}$, the maximum EPA limit for allowable level of lead in drinking water, to a 5 µM solution of **25** yields a $15 \pm 2\%$ increase in fluorescence intensity. Subsequent experiments probed the ability of **25** to track $Pb^{2+}$ levels in living cells.

New hybrid nanomaterials have been designed with the ability to sense lead with high efficiency. For example, Alizadeh et al. developed a colorimetric method for the fast detection of lead at ambient temperatures using monoazacrown functionalized gold nanoparticles (AuNPs) in aqueous media.[72] A solution of modified AuNPs showed a color change from brown to purple in response to the addition of $Pb^{2+}$. The recognition mechanism is attributed to the $Pb^{2+}$-induced aggregation via sandwich complexation of azacrown ether-capped gold nanoparticles. Son and coworkers employed a BODIPY-derivative moiety as fluorophore/chromophore and nanocrystals of $Fe_3O_4@SiO_2$ core/shell nanoparticles as solid support.[73] This system exhibited high affinity and high selectivity for $Pb^{2+}$ over other competing metal ions tested (association constant: $6.3 \times 10^4$ $M^{-1}$; detection limit: 1.5 ppb). Moreover, the BODIPY-functionalized nanoparticles successfully detected lead in living cells through a reversible mechanism.

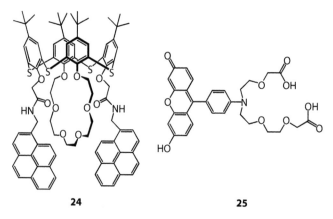

**24**                                  **25**

**SCHEME 2.9**   Representative examples of lead probes.

In biological systems, manganese can replace $Mg^{2+}$ as the activating ion for a number of $Mg^{2+}$-dependent enzymes, although some enzyme activity is usually lost. For example, $Mn^{2+}$ effectively binds ATP and allows hydrolysis of the energy molecule by most ATPases. Liang and Canary followed a soft atom poisoning strategy for the design of manganese probes.[74] The substitution of carboxylate groups of the commercially available calcium chelate BAPTA with pyridines resulted in much stronger $Mn^{2+}$ selectivity over $Ca^{2+}$. Solution and *in vitro* experiments demonstrated that these sensing compounds have good selectivity toward $Mn^{2+}$ with "on" fluorescence.

To conclude this part, we will highlight some supramolecular systems especially developed for the detection of mercury ions. $Hg^{2+}$ is, environmentally, one of the most important metal ions whose toxicity, even at very low concentrations, has long been recognized. Therefore, design and implementation of efficient mercury detection tools still represents an important challenge for researchers. In 2008, Nolan and Lippard published a review that provides a comprehensive account of the remarkable progress in the synthesis and application of optical sensors for mercury.[75] Recently, another review has summarized developments in analytical techniques and methods for the determination and speciation of mercury in natural waters.[76] Some water-soluble mercury chemosensors are exemplified in Scheme 2.10.

Chemosensor **26**, based on a tripodal cyclotriveratrylene chromophore, was developed for the colorimetric determination and visual detection (a color change from yellow to red-orange) of $Hg^{2+}$ ions with a detection limit of 0.5 µM.[77] The possibility of adsorbing compound **26** onto silica yielded a prototype of a strip-test assay. Once the strip was immersed in a sample solution, the response time was of the order of a few seconds, and the color change remained unaffected for at least 24 hours. The detection limit (5 µM) is 10 times larger compared to the solution UV/vis method as a result of the effect of the immobilization of **26** onto the solid phase. Ros-Lis et al. developed compound **27** with two independent subchromophores for a highly selective and sensitive dual chromo-fluorogenic determination of mercury.[78] Formation of a squaraine dye triggered by a Hg(II)-induced loss of the thioether moiety resulted in a dramatic change of color (colorless to blue) only in the presence of mercury ions. This chromogenic indication reaction allows $Hg^{2+}$ ions to be detected

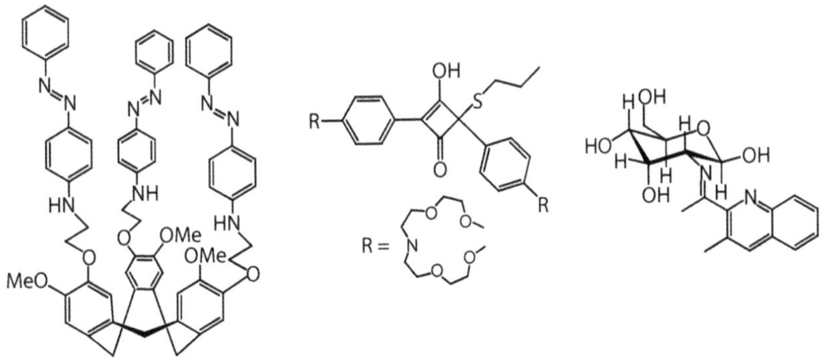

|     26     |     27     |     28     |

**SCHEME 2.10**  Some examples of water-soluble mercury chemosensors.

spectrophotometrically down to 20 ppb in aqueous solutions. Molecule **28** incorporates the well-known fluorophore quinoline and a water-soluble D-glucosamine moiety.[79] This soluble chemosensor for $Hg^{2+}$ exhibited high sensitivity, and the detection limit was calculated to be 0.5 μM in natural water. The conformation of the sugar constrains the $N_2O_2$ chelating donor in a tetrahedral geometry from which high selectivity over transition metals was achieved.

Great effort has been made in the design and fabrication of mercury detectors based on solid materials. Recently, Knecht and Sethi have published a review indicating the benefits of using gold nanoparticles for the colorimetric detection of $Hg^{2+}$.[80] For instance, a colorimetric method based on DNA-functionalized gold nanoparticles (DNA-AuNPs) detects $Hg^{2+}$ ions at concentrations as low as 100 nM (20 ppb) in aqueous samples.[81] The basis of the method, which was developed by Lee et al., relies on thymidine–$Hg^{2+}$–thymidine coordination chemistry and complementary DNA-AuNPs with deliberately designed T-T mismatches. In the absence of $Hg^{2+}$ ions, gold nanoparticles functionalized with one of two complementary DNA sequences aggregate and form duplexes leading to a red color solution. The presence of $Hg^{2+}$ ions permits the binding of the two types of DNA sequences at the point of the T-T mismatch and causes the otherwise red nanoparticle solution to appear purple.

Besides gold nanoparticles, different types of solid supports are used as sensing systems for the detection of mercury. Functional structured inorganic–organic hybrid materials consisting of boehmite-silica core-shell nanoparticles and anthracene-containing amines covalently attached to the nanoparticle surface were developed as selective fluorescent chemosensors for mercury ions.[82] The system functionalized when the monoamine chain reaches a detection limit of 0.2 ppb in water.

The last example relies on the determination of methylmercury in real samples. Climent et al. developed an organically capped mesoporous inorganic material for selective $CH_3Hg^+$ sensing through signal amplification inspired by gated ion channels and pumps.[83] The pores of the hybrid material are loaded with a large amount of dye molecules, which are only liberated upon $CH_3Hg^+$-induced opening of the pores. Sensitivity of this method reaches the ppb range.

## RARE EARTH-METAL ION SENSING

Lanthanides are extremely versatile elements as demonstrated by their use in lasers, medical diagnostics (screening for Down syndrome), anticounterfeiting efforts (European Union bank notes), nuclear reactors, fertilizer, pain medications (relief of pain from bone cancer), chemical catalysis, and many other applications. Because of their intrinsic properties (long-lived excited states, good quantum yield), lanthanide ions are often used as signaling moieties in supramolecular sensors with significant advantages over analogous fluorescent systems. Lanthanide-based luminescent switches and sensors have been the scope of many recent reviews. To name just a few, the work of dos Santos et al. highlights new developments in the field of supramolecular lanthanide luminescent sensors and self-assemblies;[84] the work of Bünzli and Piguet relies on synthetic strategies used to insert these ions into monometallic and polymetallic molecular edifices;[85] Shinoda and Tsukube summarized examples of luminescent lanthanide complexes as analytical tools in anion sensing, pH indication, and protein recognition;[86] and Bünzli and coworkers reviewed the applications of lanthanide luminescence for biomedical analyses and imaging.[87] Despite such industrial, biological, and therapeutic importance, selective sensing of lanthanide ions using chromogenic sensor molecules has not grown as a research area as one would expect. Recent examples of chemosensors for lanthanide metal ions include systems **29** through **32** (Scheme 2.11).

Compound **29**, bearing a diazostilbene chromophore and a benzo-15-crown-5 ether moiety, emits strong fluorescence only in the presence of specific ions ($Pr^{3+}$, $Nd^{3+}$, and $Eu^{3+}$) while the emission wavelength is associated with a particular ion providing high sensitivity and resolution.[88] Chemosensor **30** was synthesized to evaluate a binding interaction with rare earth ions by means of absorption and emission spectrophotometry.[89] Significant emission enhancements were detected in the presence of $Sc^{3+}$, $La^{3+}$, $Pr^{3+}$, $Sm^{3+}$, $Gd^{3+}$, $Tb^{3+}$, $Yb^{3+}$, and $Lu^{3+}$. Quenching of fluorescence was observed in the presence of $Ce^{3+}$ and $Eu^{3+}$. Compound **31**, based on a dansyl derivative, showed high sensitivity and selectivity toward $Tb^{3+}$ ions over a wide range of metal ions in acetonitrile solution.[90] The detection limit for $Tb^{3+}$ was found to be $1.4 \times 10^{-7}$ M. A rare example of a reversible recognition of $Nd^{3+}$ over other lanthanide ions by receptor **32**, having a diazo group as signaling unit, was reported by Das et al.[91] Enhancement in luminescence intensity was observed upon addition of 1 equiv of $Nd^{3+}$ because of the interrupted PET process. Eventually, in the presence of excess $Nd^{3+}$, a complex with 1:2 (ligand–metal) stoichiometry was formed with a red shifted emission maximum and a lower emission quantum yield. This was attributed to an intramolecular charge transfer transition. Furthermore, a sharp change in color could be visually detected allowing a colorimetric detection of $Nd^{3+}$ ion.

As an example of solid supports used in lanthanide ion sensing, Lisowski and Hutchison developed a selective and sensitive molecular sensor based on malonamide-functionalized gold nanoparticles.[92] These AuNPs selectively bind lanthanide ions in

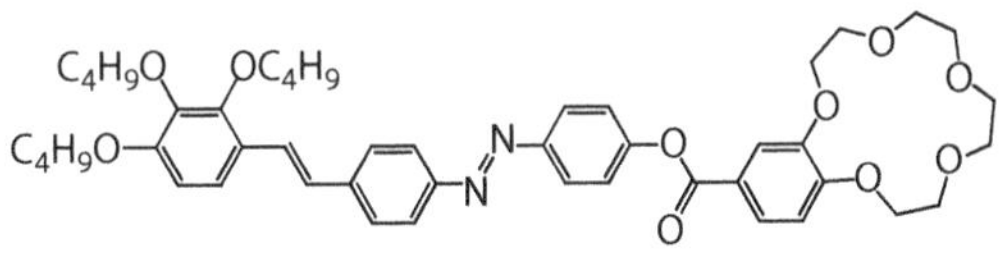

**29**

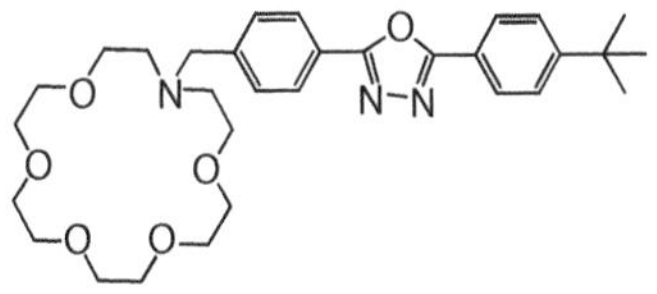

**30**

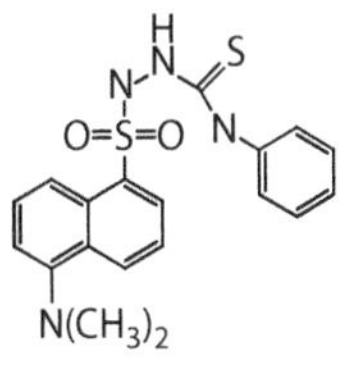

**31**

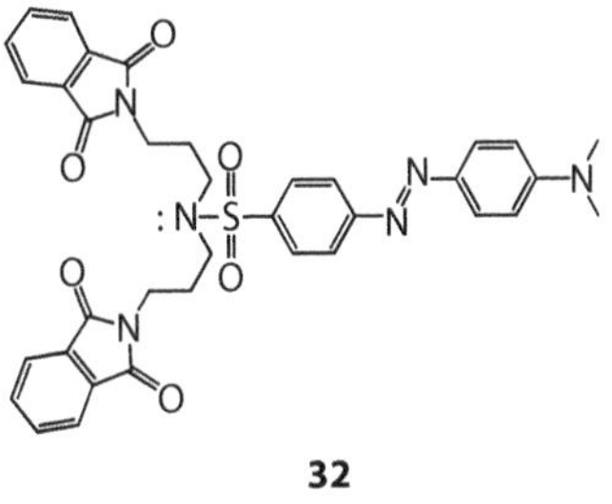

**32**

**SCHEME 2.11**  Fluorescent systems used for lanthanide ion sensing.

water over a variety of potentially interfering metal ions, causing an instantaneous color change in the nanoparticle from red to blue, allowing a colorimetric detection of lanthanides. This sensing system can detect the presence of $Eu^{3+}$ at concentrations as low as 50 nM, which is the lowest reported detection for lanthanide ion colorimetric sensing.

## ANION SENSING

The development of organic chemosensing molecules for anionic species only came up several years after the topic of metal ion sensing was already clearly established. A very early contribution to this topic was performed by Hosseini et al. in 1988 that reported on a polyazadioxamacrocycle containing appended acridine rings that experienced either an increase in fluorescence following the interaction with 5′-adenosine triphosphate or slight decreases when interaction with tripolyphosphate or pyrophosphate occurred.[93] One of the first reports claiming for the use of molecular receptors for sensing anions is a paper by Huston et al. published in 1989.[94] Anthryl polyamine receptors were prepared for the signaling of molecular recognition events involving carboxylate, sulfate, and phosphate groups. The authors observed CHEF effects following the interaction of a variety of added anions such as acetate, citrate, phosphate, and sulfate. Furthermore, the magnitude of the observed CHEF was dependent on the anion added. In the case of hydrogen phosphate, the observed CHEF effect was attributed to an intracomplex proton transfer from the anion to the nonprotonated secondary amine group of triprotonated receptor **33** (Figure 2.1). Such a proton transfer would prevent the PET from the amine lone pair to the excited fluorophore to occur. Nevertheless, this first approach to phosphate sensing seemed to be far from being usable because, as the authors point out, being $K_d = 150$ mM at pH = 6,

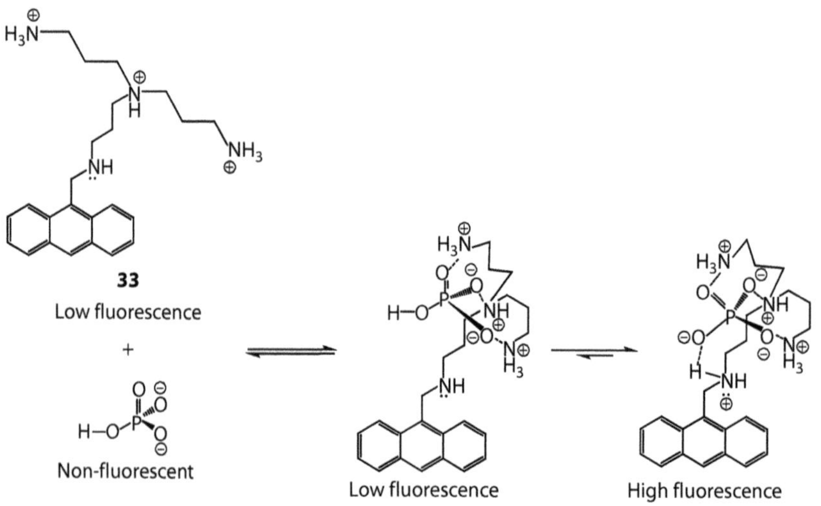

**FIGURE 2.1** Anthrylpolyamine receptor designed for the signaling of phosphate. (Adapted from Huston, M. E. et al., 1989, *J. Am. Chem. Soc.* 111, 8735–8737.)

the midpoint of the decade-wide transition occurred for a phosphate concentration of 150 mM, which is too high for many applications.

In an effort to lower the concentration range of operation, micromolar $K_d$, and to discriminate between phosphate and pyrophosphate anions, the same group synthesized the convergent receptor **34** (Figure 2.2). This receptor binds pyrophosphate over 2000 times more strongly than phosphate, permitting the real-time monitoring of pyrophosphate hydrolysis.

These studies were preceded by researchers studying the changes in the electrochemical and/or photochemical behavior of metallocyanide anions as a result of their interaction with polyammonium receptors. Although formation of ion pairs between hexacyanoferrate(III) anions and quaternary ammonium salts was evidenced as early as 1965 by the shifts produced in the $^1$H NMR signals of the ammonium salts following the interaction with the anions,[95] it was Peter et al.[96,97] who contributed definitely to the development of this topic, studying the changes produced in the cyclovoltamperograms of $[Fe(CN)_6]^{4-}$ aqueous solutions upon addition of increasing amounts of the hexaaza and octaazamacrocycles **35** and **36** (Scheme 2.12).

Addition of the macrocycles brings about anodic shifts whose magnitude depends on the stoichiometries of the complexes formed and on their stabilities. Within this topic, the interaction of the macrocycles [3$k$]aneN$_k$ ($k$ = 7–12) (**37–42**) and of their open-chain counterparts (**43–46**) in their protonated forms with $[Fe(CN)_6]^{4-}$ and $[Co(CN)_6]^{3-}$ was explored by the Supramolecular Chemistry Groups of Florence and Valencia along the years 1985–1995.[98–105]

The analysis of the association constants of these systems indicated that the main binding force is charge–charge, although, in some instances, hydrogen bonding seems also to be significantly contributing to adduct stabilization. The quasireversible response of the $[Fe(CN)_6]^{3-}/[Fe(CN)_6]^{4-}$ couple was also employed to perform displacement assays of the other guest. To solutions containing [21]

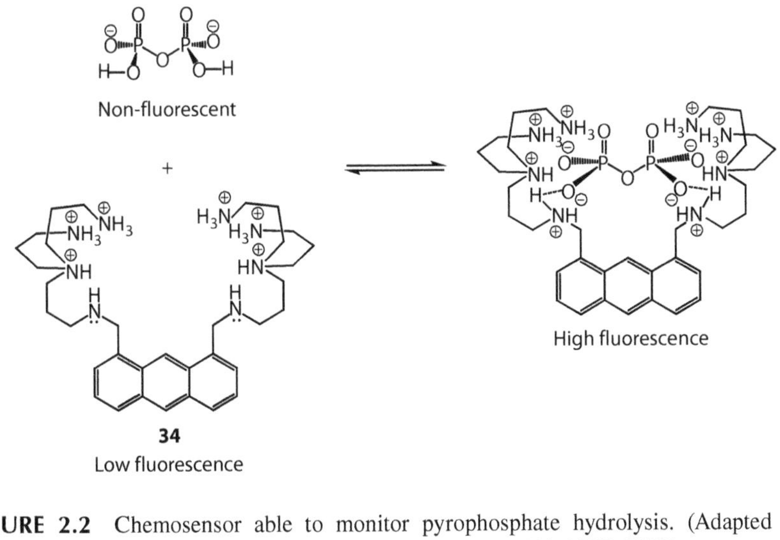

**FIGURE 2.2** Chemosensor able to monitor pyrophosphate hydrolysis. (Adapted from Vance, D. H. and Czarnik, A. W., 1994, *J. Am. Chem. Soc.* 116, 9397–9398.)

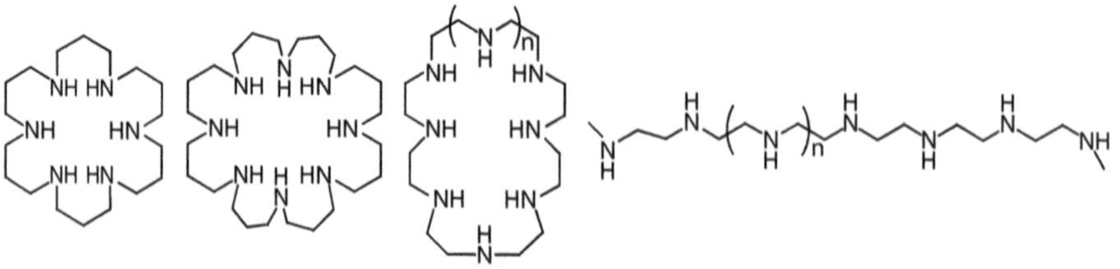

| 35 | 36 | 37 n = 1 [21]ane N$_7$ | 43 n = 1 Me$_2$hexaen |
|---|---|---|---|
| | | 38 n = 2 [24]ane N$_8$ | 44 n = 2 Me$_2$heptaen |
| | | 39 n = 3 [27]ane N$_9$ | 45 n = 3 Me$_2$octaen |
| | | 40 n = 4 [30]ane N$_{10}$ | 46 n = 4 Me$_2$nonaen |
| | | 41 n = 5 [33]ane N$_{11}$ | |
| | | 42 n = 6 [36]ane N$_{12}$ | |

**SCHEME 2.12**    Several examples of azamacrocycles and their open-chain counterparts.

aneN$_7$ and a sufficient [Fe(CN)$_6$]$^{4-}$ amount to warrant the formation of the 1:1 [21] aneN$_7$:[Fe(CN)$_6$]$^{4-}$ adduct, different carboxylate anions were added.[106] The carboxylate substrate interacting most would be the one displacing the most electrochemical parameters of the [Fe(CN)$_6$]$^{3-}$/[Fe(CN)$_6$]$^{4-}$ couple. These studies also showed that, when the macrocycle has a large enough size to lodge the anionic guest within its cavity, significant reinforcement in electrostatic attraction and changes in properties arise. One early example of this situation was evidenced by Lehn and Balzani in the analysis of the yield of the photoaquation reaction of hexacyanocobaltate(III) in the presence of macrocycles **35** and **36**. These authors observed that the quantum yield of the reaction was reduced to one-third of the initial value in the presence of H$_8$(**36**)$^{8+}$, suggesting the inclusion of [Co(CN)$_6$]$^{3-}$ through its equatorial plane within the macrocyclic framework.[107–109] Within this topic, it is interesting to remark that Beer and Keefe reported in 1989 the first redox-responsive class of anion receptors based on the redox-active cobaltocenium moiety (**47**, Scheme 2.13).[110]

These initial studies showing how anion recognition may lead to changes in optical or electrochemical properties of either the anion or guest species were followed, particularly in the last two decades, by a large number of studies with molecules that accompany the binding of anions with changes in optical properties so that they can be used as anion chemosensors.[111–124]

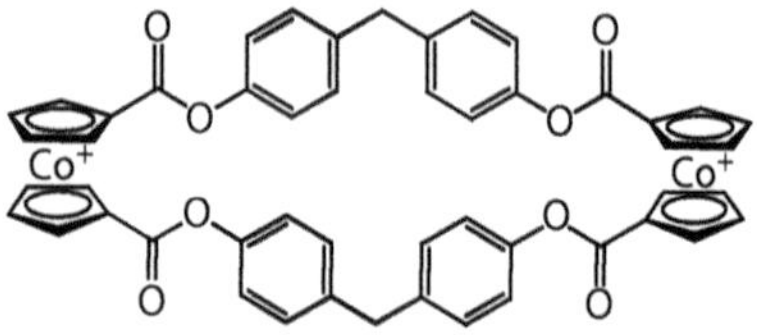

47

**SCHEME 2.13**    Redox-responsive example of anion receptor for [Co(CN)$_6$]$^{3-}$.

## ANION SENSING DEVICES

In general, anion chemosensing devices can be categorized into those using reversible anion binding and the so-called chemodosimeters, which use an irreversible chemical reaction for anion detection. Within the first category, there is a first subgroup of molecules in which the binding site and the reporting or signaling unit are covalently bound. In a second subgroup, a displacement strategy is used in which a noncovalently linked reporter molecule is displaced when the target anion binds the receptor; the displacement brings about noticeable changes in signal magnitude. Electrochemical and optical signals are the two responses most often used in anion-detection devices.

Electrochemical anion detection strategies include either extraction of a charged guest into a membrane by a nonelectroactive host and detection of the membrane potential, namely, anion-selective electrodes (ISEs) and chemically modified field effect transistors (CHEMFETs), or receptors containing electroactive groups that change their electrochemical behavior following the anion interaction.[125,126]

## CHROMOGENIC AND FLUOROGENIC CHEMOSENSORS

### Halide and Pseudohalide Detection

Halide anion recognition was one of the first aspects investigated within the field of anion-coordination chemistry. The first approaches were based on substituting the hydration sphere of the halide or pseudohalide anions by charged ammonium or guanidinium groups strategically disposed within the structure of a bicyclic receptor.[127-131]

Since then, many receptors with functionalities able to interact with spherical anions, very often in nonaqueous solvents or in solvent mixtures containing certain amounts of water, have been developed.[117] These receptors have been often linked reversibly or irreversibly to chromogenic or fluorogenic reporter units to join any of the sensing strategies depicted in previous sections. Here, we are going to describe several representative and recent examples describing chemosensors for halide- and pseudohalide-type anions.

The first example we want to recall is the covalent appendage of fluorescent probes to calyx[4]pyrrole structures described by Miyaji et al. for the recognition of halide anions.[132,133] Fluorescent receptors having calixpyrrole structures conjugated with anthracene fluorophores were shown to be able to detect the presence of $F^-$, $Cl^-$, $Br^-$, and $H_2PO_4^-$ anions in $CH_3Cl_3$ or $CH_3CN$ solution by a quenching mechanism.[132] A new generation of compounds (**48–50**, Scheme 2.14) in which a calyx[4]pyrrole unit was functionalized with electron-deficient anthraquinone moieties through conjugated C–C triple bond was developed to achieve naked-eye detection of anions in dichloromethane. The observed color changes followed the sequence predicted by the association constants $F^- > Cl^- > H_2PO_4^- \gg Br^- \approx I^- \sim HSO_4^-$ and were particularly noticeable in the case of $F^-$. The color changes were attributed to charge transfer from the electron-rich complexed anion to the electron-deficient sink constituted by the anthraquinone units.

A somehow related oxyindolophyrine receptor (**51**, Scheme 2.14) was prepared by Furuta et al.[134] Among the interesting chemical properties of this cycle was its

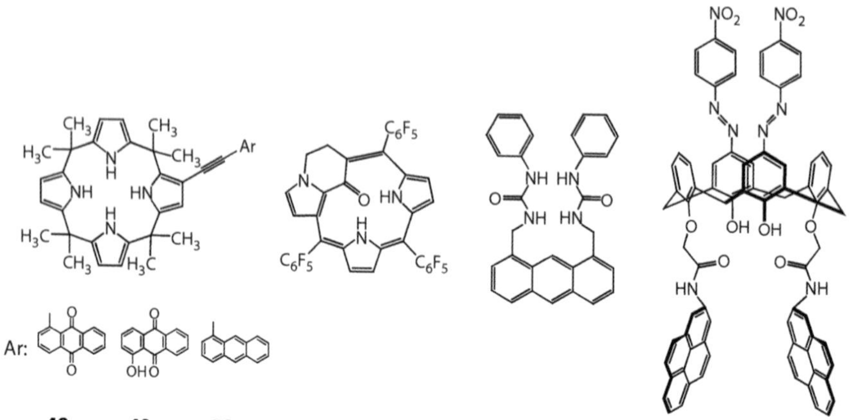

**SCHEME 2.14**  Chemosensor structures for halide and pseudo-halide detection.

capacity to bind fluoride anions using hydrogen-bonding interaction with the inner pyrrolic NH groups. Upon addition of Bu$_4$NF in CHCl$_3$, the Soret band of **51** shifted from 425 to 431 nm, and the emission band at 631 nm was suppressed. The association constant of $1.4 \times 10^4$ M$^{-1}$ obtained from the emission changes at 25°C was comparable to that reported for calyx[4]pyrrole ($1.7 \times 10^4$ M$^{-1}$ in CH$_2$Cl$_2$).[135] Interestingly enough, the authors reported that the chemosensing characteristics of **51** toward F$^-$ were not interfered with by the other halides Cl$^-$, Br$^-$, and I$^-$.

Organic chromophores linked to recognition units containing either urea or thiourea have become very popular as anion chemosensors.[117] One example applied to fluoride sensing was provided by Kim and Yoon with receptor **52** (Scheme 2.14).[136] These authors reported that **52**, bearing two phenyl urea groups at the 1,8-position of anthracene, shows selective quenching with fluoride via a PET mechanism in experiments conducted in acetonitrile-dimethylsulfoxide (9:1 v/v) solvent mixture. The authors indicated that although **52** also exhibits a small quenching effect for Cl$^-$ and Br$^-$, the selectivity for F$^-$ over Cl$^-$ was approximately 120-fold. An interesting system for fluoride sensing through dual fluorescence and visible light absorption changes was proposed by Kim and coworkers (**53**).[137] Receptor **53** was made appending through amide bonds on one side of a calix[4]arene platform, two pyrene moieties, and, on the other two, 4-nitrophenylazo groups. The NHs of the amide groups donate hydrogen bonds to the fluoride anion, changing the characteristic excimer emission of pyrene at 480 nm to a new emission peak at 460 nm from a static excimer.

Receptors including alternative binding sites and chromophoric units are included in Scheme 2.15. Receptors **54** through **59**, containing maleimide dyes with two symmetrical NH binding sites, have been shown to exhibit a distinct color change and fluorescence quenching effect for fluoride, cyanide, and dihydrogen phosphate anions.[138]

The intense fluorophore borodiazaindacene (BODIPY) has been introduced in anion receptors in a number of cases. For instance, Bozdemir et al. showed BODIPY derivatives **58** and **59** with silyl-protected phenolic functionalities signal fluoride

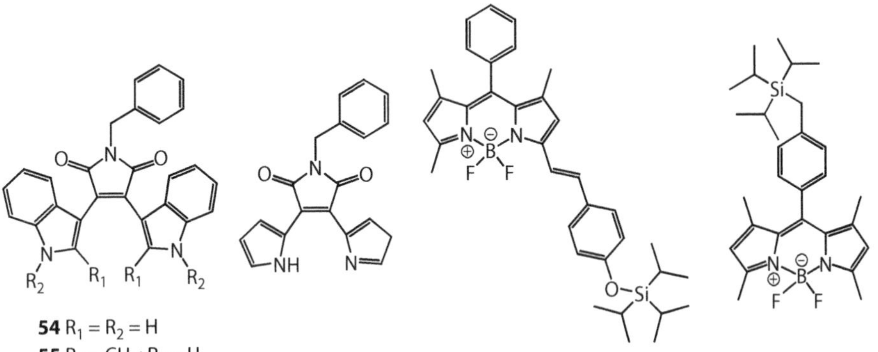

**54** $R_1 = R_2 = H$
**55** $R_1 = CH_3; R_2 = H$
**56** $R_1 = R_2 = CH_3$                    **57**                **58**                **59**

**SCHEME 2.15** Receptors containing maleimide dyes for the signaling of fluoride, cyanide, and dihydrogen phosphate anions.

anions both in solution and in a poly(methylmethacrylate) matrix.[139] Both compounds behave as chemodosimeters, and deprotection of the triisopropylsilyl groups by the action of fluoride leads, in the case of **60**, to a decrease in the emission intensity resulting from the formation of a phenoxy substituent at the meso 8 position of BODIPY, which is a strong PET donor. In the case of **59**, the deprotection reaction facilitated by the fluoride anions generates an intramolecular charge transfer (ICT) donor in full conjugation with BODIPY dye. The great selectivity of these reactions for fluoride deserves to be emphasized.

A last approach to fluoride sensing that we would like to include is the use of preformed metal receptors. One of these is presented in Scheme 2.16. Macrocycle salen–$UO_2$ complex **60** is able to bind fluoride anions with an affinity constant of the order of six logarithmic units in DMSO as calculated by UV/Vis spectroscopy.[140] The fifth binding position of the metal, where anion binding occurs, is directed toward the macrocyclic cavity. Therefore, the access is controlled by the size of the guest, and this makes **60** an extremely highly selective fluoride receptor. Although acetate and phosphate have great affinities for the $UO_2$ sites, the access to the binding site is prevented by their large size.

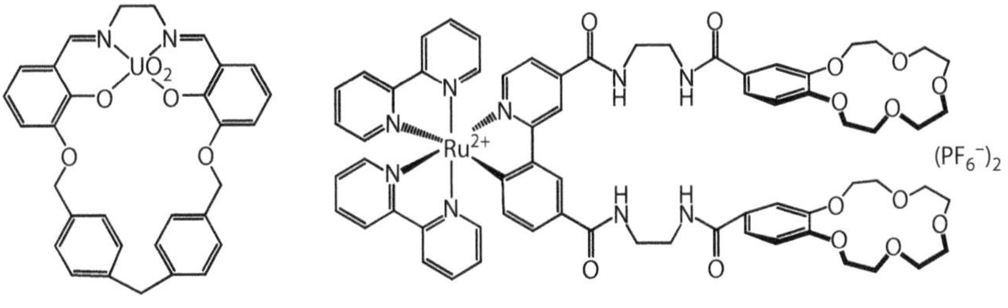

**60**                                              **61**

**SCHEME 2.16** Macrocycle salen–$UO_2$ complex able to bind fluoride and a sandwich receptor for chloride anions.

As a summary, fluoride sensing has been widely studied although there is a general lack of data regarding the analysis of interferences and detection limits. A comprehensive review concerning fluoride recognition and sensing can be found in reference 141.

Chloride selective detection by chromogenic or fluorogenic means is, however, not as popular as fluoride sensing. One very elegant example of this chemistry was provided with receptor **61** (Scheme 2.16) by Beer and Dent.[142] The authors proved that the recognition and sensing properties of either **61** or the analogous Re(I) complex could be tuned via the binding of K$^+$ ions. In the absence of K$^+$, the receptor is selective for H$_2$PO$_4^-$ over Cl$^-$, whereas coordination of K$^+$ leads to formation of a sandwich complex with the metal ion placed between the crown ether moieties, which defines a cavity that reverts the selectivity toward the chloride anions.

Cadmium selenide quantum dots modified with 4-substituted pyridine type ligands containing thiourea groups have been described as fluorescent probes for iodide by Li and coworkers (**62** in Scheme 2.17).[143] Quenching of the luminescence emitted by these nanoparticles allows for detection of I$^-$ amounts as low as $1.5 \times 10^{-9}$ M, thus affording a very sensitive detection system for this chemical species. A disadvantage of these quantum dots is their insolubility in water, which limits their application for environmental purposes.

Another interesting system for iodide recognition was based on AuNPs modified with fluorescein-5-isothyocyanate (FITC) groups (**64**, Scheme 2.17).[144] Such modified particles were demonstrated to be capable of sensing I$^-$ based on a fluorescence resonance energy transfer mechanism occurring from the FITC substituents to the AuNPs. Moreover, IO$_3^-$ detection could be also achieved if a previous reduction to iodide with ascorbic acid was undertaken. Detection limits were as low as 10.0 and 50.0 nM for I$^-$ and IO$_3^-$, respectively. Systems based on organic fragments containing secondary amines and pyrene fragments as fluorophores covalently linked to γ-aluminum oxyhydroxide (boehmite) nanoparticles have been recently prepared (**64** and **65**).[145] The fluorescence of colloidal aqueous solutions of such particles bears selective quenching in the presence of iodide. The system can operate perfectly in

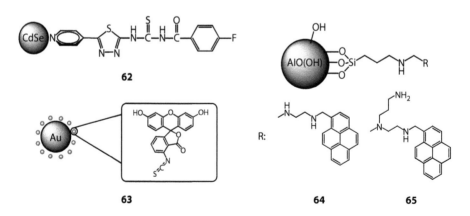

**SCHEME 2.17** Nanoparticles for I$^-$ and IO$_3^-$ sensing.

water for the selective sensing of iodide down to the nanomolar level and does not experience any alteration with time.

## Cyanide and Azide Sensing

Cyanide recognition and sensing has a great interest because of the many environmental and biochemical connotations this anion has. Therefore, different sensing systems have been analyzed for their capacity to detect this linear-shaped monovalent anion.

For instance, Tomasulo et al. reported several chromogenic oxazines able to behave as chemodosimeters for cyanide anion detection (**66**, Scheme 2.18).[146] Cyanide detection lies in the opening of a [1,3]oxazine ring and the concomitant formation of a 4-nitrophenylazophenolate chromophore. The result is the quantitative formation of a cyanamine and the appearance of a band in the visible region corresponding to a 4-nitrophenylazophenolate chromophore. The authors have shown that this particular behavior was not followed in the presence of halide anions, which are common interferents of cyanide in conventional sensing schemes.

Borodiazaindacene (BODIPY) derivative **67** was developed by Ekmecki and coworkers as a colorimetric and fluorescent probe for the detection of cyanide anions.[147] On contact with cyanide ions, **67** displays a large decrease in the emission intensity and a reversible color change from red to blue. Moreover, highly fluorescent polymeric films can be prepared by doping poly(methylmethacrylate) with **67**. Specific recognition of cyanide from an aqueous solution by imidazole containing chromophores **68** and **69** through reversible H-bonding adduct formation was proved by Saha and coworkers.[148] Adduct formation produces an instant visually detectable color and fluorescence change. The receptors can detect cyanide anions much lower than 0.2 ppm, the permissible concentration limit for drinking water according to the EPA. Moreover, **68** has been used for detecting cyanide in *Pseudomonas putida*, a microorganism that utilizes cyanide as its sole source of carbon and nitrogen.

Scheme 2.19 collects two systems that take advantage of a preformed metal complex for the recognition of cyanide anions. While addition of a $Cu^{2+}$ aqueous solution to **70** produces a quenching of the fluorescence, a further addition of

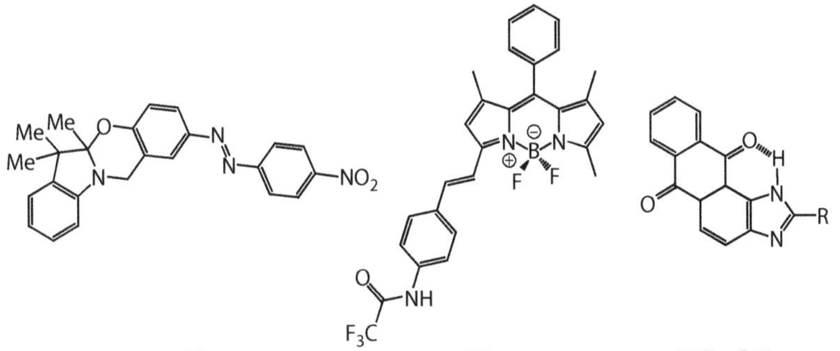

**66**        **67**        **68** R: $-C_6H_5$
                            **69** R: Ferrocene

**SCHEME 2.18**  Chemosensors for the specific recognition of cyanide.

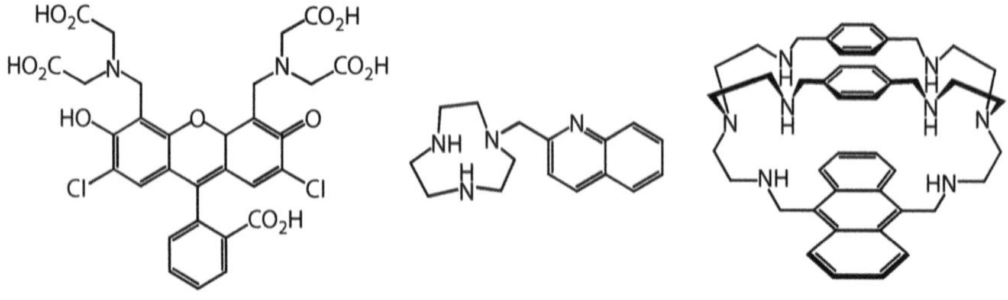

**SCHEME 2.19**   Fluorescent sensors for cyanide and azide anions.

cyanide produces green fluorescence resulting from metal decomplexation to give $Cu(CN)_2$.[149] This sensing system was further applied to the microfluid platform in which chemosensor $Cu^{2+}$–**70** displayed green fluorescence on the addition of cyanide anions. *In vivo* assays have been performed with the bacteriovorous nematode *C. elegans*, suggesting that this system is highly suitable for *in vivo* imaging of cyanide. Compound **71** also operates through the selective color changes experimented by its $Cu^{2+}$ complex in the presence of $CN^-$ in acetonitrile and water.[150]

The last example is a classical system reported by Fabbrizzi and coworkers.[151] Cryptand **72** is able to recognize azide ions by changes in its fluorescence when the anion is added to aqueous solutions of its binuclear $Zn^{2+}$ complex.

## Oxoanion Detection

In 2007, Reyheller and Kubik reported a bis(cyclopeptide) receptor (**73**, Scheme 2.20)[152] built up by two peptide units containing alternating *L*-proline and 6-aminopicolinic acids connected by 4,4′-bis(dimethylamino)biphenyl linkers whose fluorescence was selectively quenched in the presence of sulfate anions. This process occurred in 1:1 water–methanol mixtures even in the presence of a 100-fold excess of chloride. Calculations showed that coordination of the anion to the receptor does not significantly alter the linker conformation from its most stable arrangement.

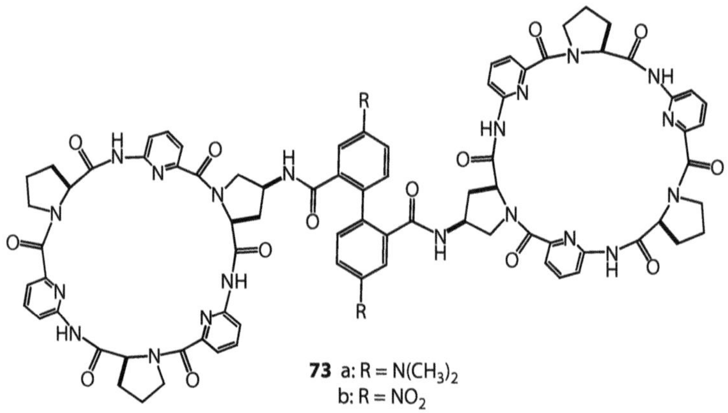

**73** a: R = $N(CH_3)_2$
b: R = $NO_2$

**SCHEME 2.20**   Bis(cyclopeptide) receptor reported for the selective detection of sulfate.

By appropriately modulating the solution pH, aluminum γ-oxohydroxide nano-particles grafted with organic fragments containing a squaramide function and a quaternized terminal ammonium group (**74**, Scheme 2.21) have shown capability for the selective detection of sulfate anions in water.[153] Displacement assays with bromocresol green revealed that sulfate anions were the only ones able to displace the indicator at pH around 4. While halide and pseudohalide monoanions failed to displace the indicator, phosphate just produced a very limited color change. It has to be noted that, at this pH value, the predominant form of phosphate in solution is the diprotonated one, behaving therefore as a monocharged anion.

Although phosphate anions of the nucleotide family will be presented in the section devoted to organic analytes, we would like to introduce here a few examples of systems able to recognize simple phosphate or polyphosphate anions. In 1999, Xie and coworkers reported the flexible tripodal receptor **75** (Scheme 2.21) that could exert discrimination of $HPO_4^{2-}$ anions from other anions such as $SO_4^{2-}$, $Br^-$, or $I^-$.[154] The protonation of the tertiary amino group, charge–charge interactions, and hydrogen bonding with the urea groups play a major role in the formation of the host–guest compound. Azenbacher et al. reported in 2000 a second generation of calixpyrrole anion sensors that were designed to have a rigid aromatic spacer to fix the distance between the quencher (anion) and the reporting unit (dye).[155] Their signaling units were selected to be excited with visible light and to be compatible with certain amounts of water in the solvent mixture; the authors reported stability data in $CH_3CN$ with 0.01% or 4 v/v water contents. Although, as expected, the largest quenching response was achieved for fluoride, these sensing systems were demonstrated to be selective for phosphate and pyrophosphate relative to chloride, something that had not been evidenced previously for any other calixpyrrole receptor. Selective chromogenic sensing of competitive sulfate and phosphate anions by nonselective hosts was reported by Piña et al.[156] To achieve such a goal, the authors used two ensembles based on the squaramide receptor **76** (Scheme 2.21) and the acid-base indicators cresol red and bromocresol green. UV/Vis microplate reader determinations show that these sensing devices are responsive to sulfate and phosphate in water, changing the coloration of the solution. However, other anions present in natural waters do not result in color changes. The detection limits were 3 ppm for $SO_4^{2-}$ and 5 ppm for $HPO_4^{2-}$.

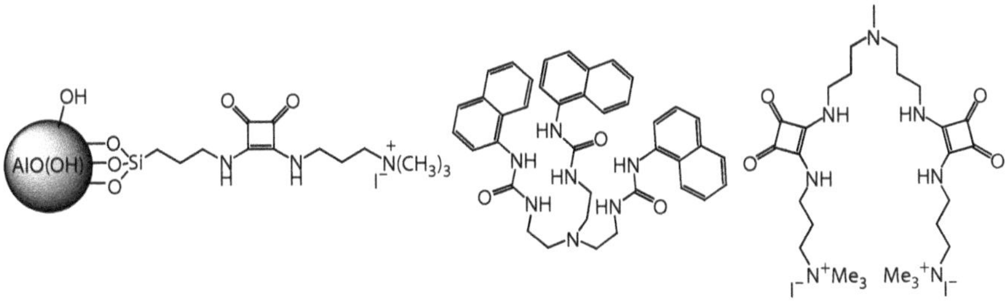

**74**                                    **75**                                    **76**

**SCHEME 2.21**    Squaramide-based nanoparticle for the selective detection of sulfate anions in water. Other systems able to recognize simple phosphate or polyphosphate anions.

Mesoporous MCM41 derivative UVM-7-polyamine hybrid materials with nanoscopic anion-binding pockets for the colorimetric sensing of phosphate have been reported by Comes et al.[157] The porous material was made to react with 3-aminopropyltriethoxysilane to get the hybrid material. Indicator displacement assays carried out in HEPES buffer using methyl red, carboxyfluorescein, and methylthymol blue show a selective response to phosphate that was attributed to a larger binding strength of this anion with the anchored polyammonium receptors (Figure 2.3).

An interesting case of pyrophosphate and dicarboxylic acid sensing was provided by Gunnlaugsson et al. a few years ago.[158] The authors designed a receptor in which an anthracene chromophore was intercalated between two thiourea binding sites (**78**, Scheme 2.22). Interaction with anions, and in particular with pyrophosphate, activated the quenching of the fluorophoric group.

An elegant system for the recognition of phosphate anions from other inorganic anions is provided by a glycosylated amino acetate hydrogel having nanosized fibers with well-developed hydrophobic domains and microsized cavities filled with water.[159] These hydrophobic domains are fundamental for achieving discrimination between hydrophilic and hydrophobic anions. The dinuclear $Zn^{2+}$ complex receptor

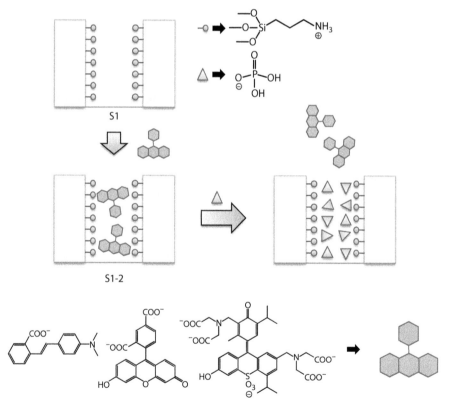

**FIGURE 2.3** Mesoporous hybrid material for the colorimetric sensing of phosphate. (Adapted from Comes, M. et al., 2008, *Chem. Commun.*, 3639–3641.)

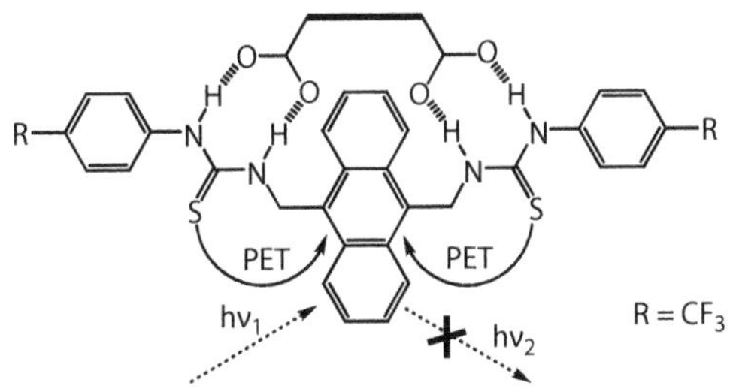

**SCHEME 2.22**  Receptor designed for pyrophosphate and dicarboxylic acid sensing.

entrapped in the supramolecular gel matrix is sufficiently mobile to travel toward the hydrophobic fibers when capturing more hydrophobic anionic guests such as phenyl phosphate or toward the aqueous cavities when interacting with more hydrophilic anions such as phosphate or ATP.

## GAS SENSING

Gas sensors are widely used in many areas such as environmental monitoring, medical and health applications, industrial safety and control, security, chemical analysis, and the automotive industry. A major goal in gas sensing is the development of smart single-chip gas sensors, wherein many sensors, each with unique chemical properties, are integrated together and their output signals processed simultaneously to act as an "electronic nose." Innovations in the sensor design or fabrication are the key to achieving the required specifications without compromising stability or long-term durability. An overview of the general principles toward the design and synthesis of receptor molecules for gases was published by Rudkevich in 2004.[160] The same author in different reviews highlights the principles and techniques for molecular encapsulation as applied to environmentally, biologically, and commercially important gases.[161,162] During the past several years, a shift in sensor technology toward more sensitive recognition layers, increasingly complex architectures, and reduced size have appeared because of the emergence of nanotechnology.[163,164]

### HYDROGEN

Because hydrogen is the lightest of all gases and a small molecule, $H_2$ can diffuse rapidly from the source of a leak. Hydrogen also has relatively wide flammability limits (4–75 vol% in air) and, at certain concentrations, an extremely low ignition energy. $H_2$ is used as fuel, and its explosion after leakage into air is a major concern for gas sensors. In fact, one tenth of the lower explosion limit for gas is taken as an alarming level for gas sensors.

Several types of sensors for hydrogen are described in the literature, and some of the initial efforts are typified by the already commercialized hot wire–type[165] and electrochemical-type[166] sensors. However, those sensors, in their early stage of development, suffered from drawbacks such as high power consumption, poor hydrogen selectivity, and high operating temperatures. Metal–oxide–semiconductor (MOS) frameworks have emerged as sensors able to measure and monitor trace amounts of gases. MOS frameworks are relatively inexpensive compared to other sensing technologies, robust, lightweight, and long-lasting, and they benefit from high material sensitivity and quick response times. For instance, ZnO is one of the most promising and useful materials, especially used for $H_2$ sensing. A single ZnO nanorod developed by Lupan et al. permitted to obtain sensitivity close to 4% at 200 ppm $H_2$ in the air at room temperature. The nanosensor has a high selectivity for $H_2$ because its sensitivity for $O_2$, $CH_4$, CO, ethanol, or LPG is less than 0.25%.[167] Another nanorod sensor based on tin oxide ($SnO_2$) is able to detect 100 ppm $H_2$ at room temperature and shows high gas response, good repeatability, and reversibility at an elevated operating temperature of 200°C.[168] Self-assembled $SnO_2$ thin films prepared by Renard et al. using the sol-gel method from organically bridged ditin hexaalkynides can detect hydrogen gas from 50°C to 200°C at the 200–10,000 ppm level.[169] Low-dimensional Pd nanostructures such as Pd thin films, Pd nanowires, and Pd nanoparticles have emerged to meet the requirement of fast, sensitive, and reliable detection of hydrogen gas.[170] Yang et al. developed Pd nanowires capable of reaching a detection limit down to 2 ppm with excellent reproducibility and baseline stability at room temperature.[171] PdO thin-film sensors demonstrated an ultrahigh sensitivity ($\sim$4.5 × $10^3$%) and a fast response time at room temperature.[172] Recently, Pd functionalized carbon nanotubes (CNTs) have also shown high sensitivity to $H_2$.[173] The last example relies on a microporous manganese formate with permanent porosity, high thermal stability, and highly selective gas sorption properties that was developed by Dybtsev and coworkers.[174] It selectively adsorbs $H_2$ and $CO_2$ but not $N_2$ and other gases with larger kinetic diameters, which appears to be a result of the small aperture of the channels.

## CARBON MONOXIDE AND CARBON DIOXIDE

CO is especially dangerous because, with no odor or color, it is undetectable by humans. It is a gas produced by incomplete hydrocarbon burning and accompanies almost all combustion processes. CO can displace $O_2$ from hemoglobin, interfering with oxygen transport. In the case of carbon dioxide, the level of atmospheric $CO_2$ has risen to become one of the most pressing environmental concerns of our age. Urgent strategies to reduce global atmospheric concentrations of greenhouse gases have prompted action from national and international governments and industries, and a number of high-profile collaborative programs have been established.

Very recently, a sensor for CO based on rhodium complexes has been described (**78**, Scheme 2.23).[175] This binuclear rhodium complex is capable of showing a clear color change to the naked eye at 50 ppm, a concentration at which CO starts to become toxic. However, CO sensors are mainly based on metal oxide semiconductor materials and solid electrolyte sensors. Functionalized CNTs have shown good CO sensitivity.[173,176]

Different materials have been employed for carbon dioxide sensing. Cyclodextrin-based carbonate nanosponges synthetized by Trotta and coworkers showed the ability to encapsulate $CO_2$ even at atmospheric pressure and room temperature.[177] $CO_2$ has been fixed by copper(II) complexes of a terpyridinophane aza receptor

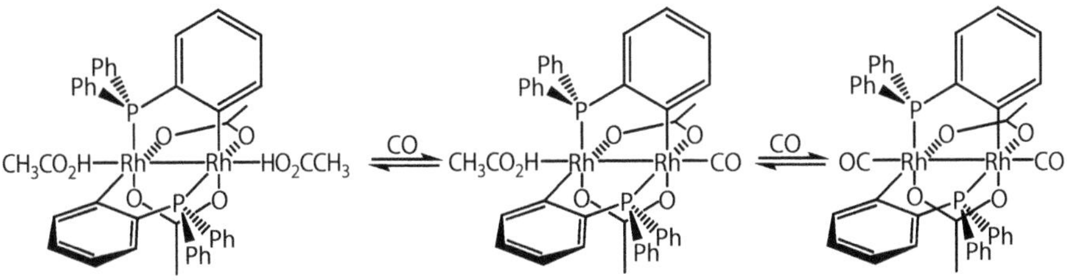

**78**

**SCHEME 2.23** Sensor for CO based on rhodium complexes.

developed by García-España et al. mimicking the fixation of $CO_2$ by the Rubisco enzyme (**79**, Scheme 2.24).[178,179] Ratiometric sensing of $CO_2$ was achieved by the ionic liquid 1-ethyl-3-methylimidazolium tetrafluoroborate (EMIMBF$_4$) containing an ethyl cellulose matrix.[180] EMIMBF$_4$-doped sensor films exhibited an enhanced linear working range between 0% and 100% $pCO_2$. The signal changes were fully reversible, and the shelf life of the sensor was extended from 15 to 95 days. A new class of materials called zeolitic imidazolate frameworks (ZIFs) exhibits exceptional uptake capacities for $CO_2$ and can selectively separate $CO_2$ from industrially relevant gas mixtures.[181,182] These porous crystals with extended three-dimensional structures are constructed from tetrahedral metal ions (e.g., Zn, Co) bridged by imidazolate derivatives (**80–82** in Scheme 2.24 are three imidazolate structures used in the construction of different ZIFs). For example, 1 L of ZIF-69 can store 82.6 L of $CO_2$ at 273 K. A zeolite-like material based on manganese formate selectively adsorbs $CO_2$ and $H_2$.[174] A cobalt adeninate metal–organic framework (MOF) exhibiting pyrimidine- and amino-decorated pores showed high heat adsorption for $CO_2$, high $CO_2$ capacity, and impressive selectivity for $CO_2$ over $N_2$. The authors attributed these favorable $CO_2$ adsorption properties to the presence of the Lewis basic amino and pyrimidine groups of adenine and the narrow pore dimensions.[183] Furthermore, carbon nanotubes have aroused great interest for their application in gas sensing since

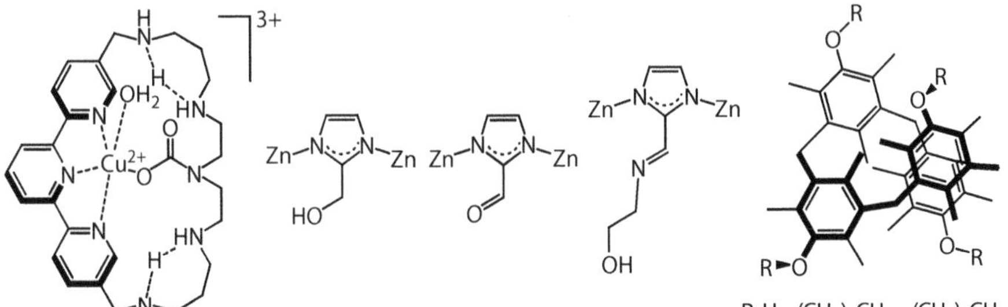

**79**  **80**  **81**  **82**  **83**

R: H, –(CH$_2$)$_5$CH$_3$, –(CH$_2$)$_2$CH$_3$

**SCHEME 2.24** Different materials employed for carbon dioxide sensing.

their discovery in 1991. Modified CNTs with polymer matrices also demonstrated excellent sensitivity to $CO_2$.[173,184]

## AMMONIA, NITRIC OXIDE, AND NITROGEN DIOXIDE

Ammonia is a strongly reducing gas that is known to adsorb substantially on carbon blacks[185] and on phthalocyanines,[186] with strong influences on the conductivity of thin films of the latter. Mesoporous thin films of $TiO_2$–$P_2O_5$ nanocomposite have shown a rapid, reversible, and somewhat selective response to $NH_3$ from 100 ppb to 10 ppm.[187] Carbon nanotubes are again the best sensors for $NH_3$. Modification with polymer composites has yielded lower sensitivity limits.[188]

$NO_X$ gases are nitrogen oxides—NO, $NO_2$, $N_2O_3$ (NO·$NO_2$), $N_2O_4$ ($NO_2$·$NO_2$), and $N_2O_5$. These are toxic atmospheric pollutants, originating in large quantities from fuel combustion and large-scale industrial processes. $NO_X$ gases are involved in the formation of ground-level ozone. They form toxic chemicals and acid rains in the atmosphere and also participate in global warming. $NO_X$ are active in various nitrosation processes with biomolecules and tissues, causing cancers and other diseases. Kang et al. developed supramolecular receptors based on calix[4]arenes derivatives for sensing and fixation of $NO_X$ gases and specifically targeting $NO_2$/$N_2O_4$.[189]

Nitric oxide (NO) plays a key role as the endothelium-derived relaxing factor that is involved in processes of vasoconstriction, signal transmission, apoptosis, immunity, and gastrointestinal motility, and in a large number of pathologies. A sensor of Nafion, CNTs, chitosan, and gold nanoparticles has been recently used. This sensor revealed a high sensitivity and selectivity to NO gas.[190] Single-walled carbon nanotube field effect transistors were electrochemically decorated with metal nanoparticles such as Pt, Pd, Au, or Ag showing unique electronic characteristics upon exposure to 10 ppm of NO gas.[191] A large number of organic or metal-based fluorescent probes are described by Tonzetich et al. as NO sensors.[192]

$NO_2$ is one of the six most common air pollutants. It is primarily produced from internal combustion–engine emissions. It is also one of the most dangerous gases from environmental and health perspectives. ZnO nanotubes have exhibited high sensitivity to $NO_2$ at a low temperature of 30°C.[193] Flexible supramolecular building blocks such as calixarenes are capable of sensing parts per million levels of $NO_2$ (**83**, Scheme 2.24).[194] Nanowires of $SiO_2$ are capable of detecting levels of $NO_2$ down to parts per billion.[195] Mesoporous materials are appealing candidates because of their large surface area and porosity. A tin-modified MCM-41 film has reported enhancing properties for $NO_2$ concentrations of 1 ppm.[196] The high versatility of CNTs makes them suitable sensors for $NO_2$. Bare, single-walled carbon nanotubes (SWCNTs) have shown ppb detection[197] while modified SWCNTs can increase sensitivity and selectivity up to 100 ppt.[198]

## OXYGEN AND OZONE

Oxygen is an oxidizing gas indispensable for living creatures. Ruthenium derivatives such as [Ru(II)-tris(4,7-diphenyl-1,10-phenanthroline)] and [Ru(dpp)$_3$]$^{2+}$ forming part of materials prepared by a sol-gel process are used as sensor materials.[199]

Silicon nanowires have shown decreased conductance when $O_2$ is exposed to the wires.[200] Nanoparticles of $SnO_2$ modified with Pd, $CeO_2$, or $ZrO_2$ mixed with C have produced the same sensing results as the former. Chopra et al. have detected $O_2$ with a thin film of CNTs.[173]

Ozone is a strong oxidizing toxic gas that is involved in the "summer smog" in the lower troposphere. Layers of $In_2O_3$ and $WO_3$ and also a promising material known as indium acetyl-acetonate (InAcAc) are sensitive to oxidizing gases.[201] ZnO nano-rods have been recently described as sensing materials for ozone.[202] Several kinds of dyes including azo dye, anthraquinone dye, and triphenylmethane dye have also been reported in studies with ozone.[203]

## SULFUR DIOXIDE AND HYDROGEN SULFIDE

$SO_2$ is one of the major gases responsible for air pollution. Nanostructured metal oxide–based semiconductor sensors based on vanadium-doped $SnO_2$ have shown that they can detect $SO_2$ down to 5 ppm.[204] Organoplatinum compounds have been developed as efficient, selective, and robust sensor devices for $SO_2$ (**84**, Scheme 2.25).[205] $SO_2$ can also be detected with CNT sensors although the reports for this gas are scarce.

$H_2S$ is a dangerous gas because of its flammability and toxicity. Palladium-doped SWCNTs have been able to detect 50 ppm $H_2S$ in air.[206]

## FLUORINE AND CHLORINE

Halogen gases such as fluorine and chlorine have been sensed through a solid-state electrochemical cell based on PEO-based electrolytes at 35°C.[207]

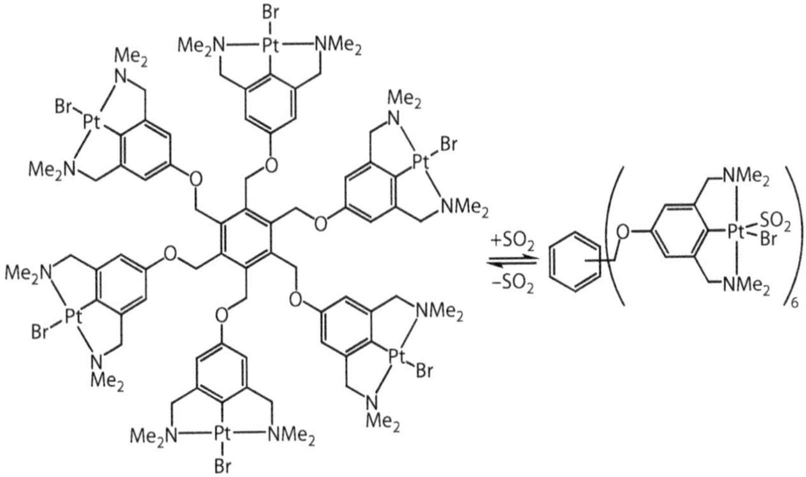

**84**

**SCHEME 2.25**   Organoplatinum compound developed as selective sensor for $SO_2$.

## ACKNOWLEDGMENTS

We thank Consolider Ingenio 2010, CSD-2010-00065 and Prometeo 2011/008 for financial support. One of us, MTA, wants to thank *Ayuntamiento de Valencia* for her *Carmen y Severo Ochoa* fellowship.

## REFERENCES

1. Czarnik, A. W. *Chem. Biol.* 1995, *2*, 423–428.
2. Pedersen, C. J. *J. Am. Chem. Soc.* 1967, *89*, 2495–2496; Simmons, H. E.; Park, C. H. *J. Am. Chem. Soc.* 1968, *90*, 2428–2429; Dietrich, B.; Lehn, J.-M.; Sauvage, J. P. *Tetrahedron Lett.* 1969, 2885–2892; Cram, D. J.; Cram, J. M. *Science*, 1974, *183*, 803–809.
3. Bissell, R. A.; de Silva, A. P.; Gunaratne, H. Q. N.; Lynch, P. L. M.; Maguire, G. E. M.; McCoy, C. P.; Sandanayake, K. R. A. S. *Topics Curr. Chem.* 1993, *168*, 223–264.
4. Anslyn, E. V. *J. Org. Chem.* 2007, *72*, 687–699.
5. Valeur, B. *Molecular Fluorescence: Principles and Applications*, 2002, Weinheim-VCH.
6. Desvergne, J.-P.; Czarnik, A. W. *Chemosensors of Ion and Molecular Recognition*, 1997, Dordrecht: Kluwer Academic Publishers.
7. de Silva, A. P.; Gunaratne, H. Q. N.; Gunnlaugsson, T.; Huxley, A. J. M.; McCoy, C. P.; Rademacher, J. T.; Rice, T. E. *Chem. Rev.* 1997, *97*, 1515–1566.
8. Bakker, E. *Anal. Chem.* 2004, *76*, 3285–3298.
9. Walt, D. R. *ACS Nano.* 2009, *3*, 2876–2880.
10. Wardlaw, G. M. *Perspectives in Nutrition*, 1999, Boston: WCB McGraw-Hill; Whitney, E. N.; Rolfes, S. R. *Understanding Nutrition*, 1996, New York: West Publishing.
11. Forsen, S.; Kordel, J. *Calcium in Biological Systems*, 1997, New York: John Wiley & Sons.
12. Ebel, H.; Gunther, T. *J. Clin. Chem. Clin. Biochem.* 1980, *18*, 257–270; Maguire, M. E.; Cowan, J. A. *Biometals.* 2002, *15*, 203–210.
13. Sigel, A.; Sigel, H. *Metal Ions in Biological Systems.* Vol. 35: *Iron Transport and Storage in Microorganisms, Plants and Animals*, 1998, New York: Marcel Dekker, Inc.
14. Li, Y. V.; Hough, C. J.; Sarvey, J. M. *Do we need zinc to think?* Sci. STKE 2003, pe19; Frassinetti, S.; Bronzetti, G.; Caltavuturo, L.; Cini, M.; Croce, C. D. *J. Environ. Pathol. Toxicol. Oncol.* 2006, *25*, 597–610.
15. Thompson, K. H.; Orvig, C. *Dalton Trans.* 2006, *6*, 761–764.
16. De Wall, S. L.; Painter, C.; Stone, J. D.; Bandaranayake, R.; Wiley, D. C.; Mitchison, T. J.; Stern, L. J.; DeDecker, B. S. *Nat. Chem. Biol.* 2006, *2*, 197–201.
17. Jamieson, E. R.; Lippard, S. J. *Chem. Rev.* 1999, *99*, 2467–2498; Wang, X. *Anticancer Agents Med. Chem.* 2010, *10*, 396–411; Montaña, A. M.; Batalla, C. *Curr. Med. Chem.* 2009, *16*, 2235–2260; Hall, M. D.; Mellor, H. R.; Callaghan, R.; Hambley, T. W. *J. Med. Chem.* 2007, *50*, 3403–3411.
18. Lange, T. S.; McCourt, C.; Singh, R. K.; Kim, K. K.; Singh, A. P.; Luisi, B. S.; Alptürk, O.; Strongin, R. M.; Brard, L. *Drug Des. Devel. Ther.* 2009; *3*, 17–26; Ott, I.; Gust, R. *Arch. Pharm.* 2007, *340*, 117–126; Chen, D.; Milacic, V.; Frezza, M.; Dou, Q. P. *Curr. Pharm. Des.* 2009, *15*, 777–791; van Rijt, S. H.; Sadler, P. J. *Drug Discov. Today* 2009, *14*, 1089–1097.
19. Camins, A.; Crespo-Biel, N.; Junyent, F.; Verdaguer, E.; Canudas, A. M.; Pallàs, M. *Curr. Drug Metab.* 2009, *10*, 433–447; Ray, S.; Mohan, R.; Singh, J. K.; Samantaray, M. K.; Shaikh, M. M.; Panda, D.; Ghosh, P. *J. Am. Chem. Soc.* 2007, *129*, 15,042–15,053.
20. Merbach, A. E.; Toth, E. *The Chemistry of Contrast Agents in Medical Magnetic Resonance Imaging*, 2001, Chichester: John Wiley & Sons.

21. Orvig, C.; Abrams, M. J. *Chem. Rev.* 1999, *99*, 2201–2204.
22. Pinkerton, M.; Steinrauf, L. K.; Dawkins, P. *Biochem. Biophys. Res. Commun.* 1969, *35*, 512–518; Tosteson, D. C.; Cook, P.; Andreoli, T.; Tieffenberg, M. *J. Gen. Physiol.* 1967, *50*, 2513–2525.
23. Gokel, G. W.; Leevy, W. M.; Weber, M. E. *Chem. Rev.* 2004, *104*, 2723–2750; Izatt, R. M.; Bradshaw, J. S.; Pawlak, K.; Bruening, R. L.; Tarbet, B. J. *Chem. Rev.* 1992, *92*, 1261–1354; Takagi, M.; Ueno, K. *Top. Curr. Chem.* 1984, *121*, 39–65.
24. Ferdani, R.; Gokel, G. W. *Ionophores. Encyclopedia of Supramolecular Chemistry.* 2004, 760–766; Yoshio, M.; Noguchi, H. *Analytical Letters* 1982, *15*, 1197–1276; Minkin, V. I.; Dubonosov, A. D.; Bren, V. A.; Tsukanov, A. V. *Arkivoc.* 2008, Part (iv), 90–102.
25. de Silva, A. P.; Gunaratne, H. Q. N.; Gunnlaugsson, T.; Huxley, A. J. M.; McCoy, C. P.; Rademacher, J. T.; Rice, T. E. *Chem. Rev.* 1997, *97*, 1515–1566.
26. de Silva, A. P.; de Silva, S. A. *J. Chem. Soc., Chem. Commun.* 1986, 1709–1710.
27. Wang, Z.; Chang, S. H.; Kang, T. J. *Spectrochim. Acta Part A.* 2008, *70*, 313–317.
28. Gunnlaugsson, T.; Bichell, B.; Nolan, C. *Tetrahedron Lett.* 2002, *43*, 4989–4992.
29. He, H. R.; Mortellaro, M. A.; Leiner, M. J. P.; Fraatz, R. J.; Tusa, J. K. *Anal. Chem.* 2003, *75*, 549–555.
30. Demchenko, A. P. *J. Fluoresc.* 2010, *20*, 1099–1128.
31. Sankaran, N. B.; Nishizawa, S.; Watanabe, M.; Uchida, T.; Teramae, N. *J. Mater. Chem.* 2005, *15*, 2755–2761.
32. Fages, F.; Desvergne, J.-P.; Bouas-Laurent, H.; Marsau, P.; Lehn, J.-M.; Kotzyba-Hibert, F.; Albrecht-Gary, A.-M.; Al-Joubbeh, M. *J. Am. Chem. Soc.* 1989, *111*, 8672–8680.
33. Golchini, K.; Mackovic-Basic, M.; Gharib, S. A.; Masilamani, D.; Lucas, M. E.; Kurtz, I. *Am. J. Physiol. Renal Physiol.* 1990, *258*, F438–F443.
34. Doludda, M.; Kastenholz, F.; Lewitzki, E.; Grell, E. *J. Fluoresc.* 1996, *6*, 159–163.
35. Iwata, S.; Matsuoka, H.; Tanaka, K. *J. Chem. Soc., Perkin Trans. 1.* 1997, 1357–1360.
36. Kondo, S.; Kinjo, T.; Yano, Y. *Tetrahedron Lett.* 2005, *46*, 3183–3186.
37. Coskun, A.; Deniz, E.; Akkaya, E. U. *J. Mater. Chem.* 2005, *15*, 2908–2912.
38. Lankshear, M. D.; Dudley, I. M.; Chan, K.-M.; Beer, P. *New J. Chem.* 2007, *31*, 684–690.
39. Obare, S. S.; Hollowell, R. E.; Murphy, C. J. *Langmuir.* 2002, *18*, 10,407–10,410.
40. Krämer, R. *Angew. Chem. Int. Ed.* 1998, *37*, 772–773.
41. Qi, X.; Jun, E. J.; Xu, L.; Kim, S.-J.; Hong, J. S. J.; Yoon, Y. J.; Yoon, J. *J. Org. Chem.* 2006, *71*, 2881–2884.
42. Goswami, S.; Sen, D.; Das, N. K.; Hazra, G. *Tetrahedron Lett.* 2010, *51*, 5563–5566.
43. Jisha, V. S.; Thomas, A. J.; Ramaiah, D. *J. Org. Chem.* 2009, *74*, 6667–6673.
44. Wang, H.-H.; Xue, L.; Fang, Z.-J.; Li, G.-P.; Jiang, H. *New J. Chem.* 2010, *34*, 1239–1242.
45. Jung, H. S.; Kwon, P. S.; Lee, J. W.; Kim, J. I.; Hong, C. S.; Kim, J. W.; Yan, S.; Lee, J. Y.; Lee, J. H.; Joo, T.; Kim, J. S. *J. Am. Chem. Soc.* 2009, *131*, 2008–2012.
46. Sheng, R.; Wang, P.; Gao, Y.; Wu, Y.; Liu, W.; Ma, J.; Li, H.; Wu, S. *Org. Lett.* 2008, *21*, 5015–5018.
47. Kim, H. N.; Lee, M. H.; Kim, H. J.; Kim, J. S.; Yoon, J. *Chem. Soc. Rev.* 2008, *37*, 1465–1472.
48. Zhao, Y.; Zhang, X.-B.; Han, Z.-X.; Qiao, L.; Li, C.-Y.; Jian, L.-X.; Shen, G.-L.; Yu, R.-Q. *Anal. Chem.* 2009, *81*, 7022–7030.
49. Taki, M.; Iyoshi, S.; Ojida, A.; Hamachi, I.; Yamamoto, Y. *J. Am. Chem. Soc.* 2010, *132*, 5938–5939.
50. Yin, J.; Guan, X.; Wang, D.; Liu, S. *Langmuir.* 2009, *25*, 11,367–11,374.
51. Choi, Y.; Park, Y.; Kang, T.; Lee, L. P. *Nat. Nanotechnol.* 2009, *4*, 742–746.
52. Jiang, P.; Guo, Z. *Coord. Chem. Rev.* 2004, *248*, 205–229.
53. Zhou, X.; Yu, B.; Guo, Y.; Tang, X.; Zhang, H.; Liu, W. *Inorg. Chem.* 2010, *49*, 4002–4007.

54. Liu, Y.; Zhang, N.; Chen, Y.; Wang, L.-H. *Org. Lett.* 2007, *9*, 315–318.
55. Royzen, M.; Durandin, A.; Young Jr., V. G.; Geacintov, N. E.; Canary, J. W. *J. Am. Chem. Soc.* 2006, *128*, 3854–3855.
56. Xu, Z.; Baek, K.-H.; Kim, H. N.; Cui, J.; Qian, X.; Spring, D. R.; Shin, I.; Yoon, J. *J. Am. Chem. Soc.* 2010, *132*, 601–610.
57. Wang, H.-H.; Gan, Q.; Wang, X.-J.; Xue, L.; Liu, S.-H.; Hua, J. *Org. Lett.* 2007, *9*, 4995–4998.
58. Quang, D. T.; Kim, J. S. *Chem. Rev.* 2010, *110*, 6280–6301.
59. Li, X.; Gong, S.-L.; Yang, W.-P.; Li, Y.; Chen, Y.-Y.; Meng, X.-G. *J. Incl. Phenom. Macrocycl. Chem.* 2010, *66*, 179–184.
60. Coskun, A.; Akkaya, E. U. *J. Am. Chem. Soc.* 2005, *127*, 10,464–10,465.
61. Chatterjee, A.; Santra, M.; Won, N.; Kim, S.; Kim, J. K.; Kim, S. B.; Ahn, K. H. *J. Am. Chem. Soc.* 2009, *131*, 2040–2041.
62. Barefoot, R. R.; Van Loon, J. C. *Anal. Chim. Acta.* 1996, *334*, 5–14.
63. Egorova, O. A.; Seo, H.; Chatterjee, A.; Ahn, K. H. *Org. Lett.* 2010, *12*, 401–403.
64. Yang, Y.-K.; Lee, S.; Tae, J. *Org. Lett.* 2009, *11*, 5610–5613.
65. Xiang, Y.; Tong, A. *Org. Lett.* 2006, *8*, 1549–1552.
66. Kuan, Z.; ChengYue, L.; LiPing, H.; Fang, L. *Sci. China Chem.* 2010, *53*, 1398–1405.
67. Tang, X.-L.; Peng, X.-H.; Dou, W.; Mao, J.; Zheng, J.-R.; Qin, W.-W.; Liu, W.-S.; Chang, J.; Yao, X.-J. *Org. Lett.* 2008, *10*, 3653–3656.
68. Xue, L.; Liu, Q.; Jiang, H. *Org. Lett.* 2009, *11*, 3454–3457.
69. Cheng, T.; Xu, Y.; Zhang, S.; Zhu, W.; Qian, X.; Duan, L. *J. Am. Chem. Soc.* 2008, *130*, 16,160–16,161.
70. Kim, S. K.; Lee, J. K.; Lim, J. M.; Kim, J. W.; Kim, J. S. *Bull. Korean Chem. Soc.* 2004, *25*, 1247–1250.
71. He, Q.; Miller, E. W.; Wong, A. P.; Chang, C. J. *J. Am. Chem. Soc.* 2006, *128*, 9316–9317.
72. Alizadeh, A.; Khodaei, M. M.; Karami, C.; Workentin, M. S.; Shamsipur, M.; Sadeghi, M. *Nanotechnology.* 2010, *21*, 315503 (8pp).
73. Son, H.; Lee H. Y.; Lim, J. M.; Kang, D.; Han, W. S.; Lee, S. S.; Jung, J. H. *Chem. Eur. J.* 2010, *16*, 11,549–11,553.
74. Liang, J.; Canary, J. W. *Angew. Chem. Int. Ed.* 2010, *122*, 7876–7879.
75. Nolan, E. M.; Lippard, S. J. *Chem. Rev.* 2008, *108*, 3443–3480.
76. Leopold, K.; Foulkes, M.; Worsfol, P. *Anal. Chim. Acta.* 2010, *663*, 127–138.
77. Nuriman; Kuswandi, B.; Verboom, W. *Anal. Chim. Acta.* 2009, *655*, 75–79.
78. Ros-Lis, J. V.; Marcos, M. D.; Martínez-Máñez, R.; Rurack, K.; Soto, J. *Angew. Chem. Int. Ed.* 2005, *44*, 4405–4407.
79. Ou, S.; Lin, Z.; Duan, C.; Zhang, H.; Bai, Z. *Chem. Commun.* 2006, 4392–4394.
80. Knecht, M. R.; Sethi, M. *Anal. Bioanal. Chem.* 2009, *394*, 33–46.
81. Lee, J.-S.; Han, M. S.; Mirkin, C. A. *Angew. Chem. Int. Ed.* 2007, *46*, 4093–4096.
82. Delgado-Pinar, E.; Montoya, N.; Galiana, M.; Albelda, M. T.; Frías, J. C.; Jiménez, H. R.; García-España, E.; Alarcón, J. *New J. Chem.* 2010, *34*, 567–570.
83. Climent, E.; Marcos, M. D.; Martínez-Máñez, R.; Sancenón, F.; Soto, J.; Rurack, K.; Amorós, P. *Angew. Chem. Int. Ed.* 2009, *121*, 8671–8674.
84. dos Santos, C. M. G.; Harte, A. J.; Quinn, S. J.; Gunnlaugsson, T. *Coord. Chem. Rev.* 2008, *252*, 2512–2527.
85. Bünzli, J.-C. G.; Piguet, C. *Chem. Soc. Rev.* 2005, *34*, 1048–1077.
86. Shinoda, S.; Tsukube, H. *Analyst.* 2011, *136*, 431–435.
87. Eliseeva, S. V.; Bünzli, J.-C. G. *Chem Soc Rev.* 2010, *39*, 189–227; Bünzli, J.-C. G. *Chem. Rev.* 2010, *110*, 2729–2755.
88. Bekiari, V.; Judeinstein, P.; Lianos, P. *J. Fluoresc.* 2003, *104*, 13–15.
89. Yu, T.; Meng, J.; Zhao, Y.; Zhang, H.; Han, X.; Fan, D. *Spectrochim. Acta A.* 2011, *78*, 396–400.

90. Ganjali, M. R.; Veismohammadi, B.; Hosseini, M.; Norouzi, P. *Spectrochim. Acta A.* 2009, *74,* 575–578.

91. Das, P.; Ghosh, A.; Das, A. *Inorg. Chem.* 2010, *49,* 6909–6916.

92. Lisowski, C. E.; Hutchison, J. E. *Anal. Chem.* 2009, *81,* 10,246–10,253.

93. Hosseini, M. W.; Blacker, A. J.; Lehn, J.-M. *Chem. Commun.* 1988, 596–598.

94. (a) Huston, M. E.; Akkaya, E. U.; Czarnik, A. W. *J. Am. Chem. Soc.* 1989, *111,* 8735–8737; (b) Vance, D. H.; Czarnik, A. W. *J. Am. Chem. Soc.* 1994, *116,* 9397–9398.

95. Larsen, D. W.; Wahl, A. C. *Inorg. Chem.* 1965, *4,* 1281–1286.

96. Peter, F.; Gross, M.; Hosseini, M. W.; Lehn, J.-M.; Sessions, R. B. *J. Chem. Soc., Chem. Commun.* 1981, 1067–1069.

97. Peter, F.; Gross, M.; Hosseini, M. W.; Lehn, J.-M. *J. Electroanal. Chem.* 1983, *144,* 279–292.

98. García-España, E.; Micheloni, M.; Paoletti, P.; Bianchi, A. *Inorg. Chim. Acta Lett.* 1985, *102,* L9–L11.

99. Bianchi, A.; García-España, E.; Mangani, S.; Micheloni, M.; Orioli, P.; Paoletti, P. *J. Chem. Soc., Chem. Commun.* 1987, 729–730.

100. Bencini, A.; Bianchi, A.; García-España, E.; Giusti, M.; Mangani, S.; Micheloni, M.; Orioli, P.; Paoletti, P. *Inorg. Chem.* 1987, *26,* 3902–3907.

101. Bencini, A.; Bianchi, A.; Dapporto, P.; García-España, E.; Micheloni, M.; Paoletti, P.; Paoli, P. *J. Chem. Soc., Chem. Commun.* 1990, 753–755.

102. Bencini, A.; Bianchi, A.; Micheloni, M.; Paoletti, P.; Dapporto, P.; Paoli, P.; García-España, E. *J. Inclusion Phenom. Mol. Recognit. Chem.* 1992, *12,* 291–304.

103. Bencini, A.; Bianchi, A.; Dapporto, P.; García-España, E.; Micheloni, M.; Ramírez, J. A.; Paoletti, P.; Paoli, P. *Inorg. Chem.* 1992, *31,* 1902–1908.

104. Aragó, J.; Bencini, A.; Bianchi, A.; Doménech, A.; García-España, E. *J. Chem. Soc., Dalton Trans. 2.* 1992, 319–324.

105. Bernardo, M. A.; Parola, A. J.; Pina, F.; García-España, E.; Marcelino, V.; Luis, S. V.; Miravet, J. F.; *J. Chem. Soc., Dalton Trans.* 1995, 993–997.

106. Bencini, A.; Bianchi, A.; Burguete, M. I.; Dapporto, P.; Doménech, A.; García-España, E.; Luis, S. V.; Paoli, P.; Ramírez, J. A. *J. Chem. Soc., Perkin Trans. 2.* 1994, 569–577.

107. Manfrin, M. F.; Sabbatini, N.; Moggi, L.; Balzani, V.; Hosseini, M. W.; Lehn, J.-M. *J. Chem. Soc., Chem. Commun.* 1984, 555–556.

108. Manfrin, M. F.; Moggi, L.; Castelvetro, V.; Balzani, V.; Hosseini, M. W.; Lehn, J.-M. *J. Am. Chem. Soc.* 1985, *107,* 6888–6892.

109. Balzani, V.; Sabbatini, N.; Scandola, F. *Chem. Rev.* 1986, *86,* 319–337.

110. Beer, P. D.; Keefe, A. D. *J. Organomet. Chem.* 1989, *375,* C40.

111. Beer, P. D.; Gale P. A.; Chen, G. Z. *Coord. Chem. Rev.* 1999, *185–186,* 3–36.

112. Pina, F.; Bernardo, M. A.; García-España, E. *Eur. J. Inorg. Chem.* 2000, *10,* 2143–2157.

113. Beer, P. D.; Cadman, J. *Coord. Chem. Rev.* 2000, *205,* 131–155.

114. Amendola, V.; Fabbrizzi, L.; Mangano, C.; Pallavicini, P.; Poggi, A.; Taglietti, A. *Coord. Chem. Rev.* 2001, *219–221,* 821–837.

115. García-España, E.; Díaz, P.; Llinares, J. M.; Bianchi, A. *Coord. Chem. Rev.* 2006, *259,* 2952–2986.

116. Steed, J. W. *Chem. Soc. Rev.* 2009, *38,* 506–519.

117. Gale, P. A.; Gunnlaugsson, T., eds., *Themed Issue: Supramolecular Chemistry of Anionic Species. Chem. Soc. Rev.* 2010, *39,* 3581–4008.

118. Li, A.-F.; Wang, J.-H.; Wang, F.; Jiang, Y.-B. *Chem. Soc. Rev.* 2010, *39,* 3729–3745.

119. Galbraith, E.; James, T. D. *Chem. Soc. Rev.* 2010, *39,* 3831–3842.

120. Duke, R. M.; Veale, E. B.; Pfeffere, F. M.; Kruger, V.; Gunnlaugson, T. *Chem. Soc. Rev.* 2010, *39,* 3936–3953.

121. Anzenbacher Jr., P.; Lubal, P.; Bucek, P.; Palacios, M. A.; Kozelkova, M. E. *Chem. Soc. Rev.* 2010, *39,* 3954–3979.

122. Lodeiro, C.; Capelo, J. L.; Mejuto, J. C.; Oliveira, E.; Sanos, H. M.; Pedrás, B.; Nuñez, C. *Chem. Soc. Rev.* 2010, *39*, 2733–3336.
123. Martínez-Máñez, R.; Sancenón, F. *Chem. Rev.* 2003, *103*, 4419–4476.
124. Moragues, M. E.; Martínez-Máñez, R.; Sancenón, F. *Chem. Soc. Rev.* 2011, *40*, 2593–2643.
125. Bühlman, P.; Pretsch, E.; Bakker, E. *Chem. Rev.* 1998, *98*, 1593–1687.
126. Antonisse, M. M. G.; Reinhoudt, D. N. *Electroanalysis.* 1999, *11*, 1035–1048.
127. Park, C. H.; Simmons, H. E. *J. Am. Chem. Soc.* 1968, *90*, 2431–2432.
128. Graf, E.; Lehn J.-M. *J. Am. Chem. Soc.* 1976, *98*, 6403–6405.
129. Schmidtchen, F. P. *Angew. Chem. Int. Ed. Engl.* 1977, *16*, 720–721.
130. Kintzinger, J.-P.; Lehn, J.-M.; Kauffman, E.; Dye, J. L.; Popov, A. I. *J. Am. Chem. Soc.* 1983, *105*, 7549–7553.
131. Bianchi, A.; Bowman-James, K.; García-España, E., eds., 1997. *Supramolecular Chemistry of Anions*. Wiley-VCH, New York.
132. Miyaji, H.; Anzenbacher, P.; Sessler, J. L.; Bleasdale, E. R.; Gale, P. A. *Chem. Commun.* 1999, 1723–1724.
133. Miyaji, H.; Sato, W.; Sessler, J. L. *Angew. Chem. Int. Ed.* 2000, *39*, 1777–1780.
134. Furuta, H.; Maeda, H.; Osuka, A. *J. Am. Chem. Soc.* 2001, *123*, 6435–6436.
135. Gale, P. A.; Sessler, J. L.; Král, V.; Lynch, V. *J. Am. Chem. Soc.* 1996, *118*, 5140–5141.
136. Kim, S. K.; Yoon, J. *Chem. Commun.* 2002, 770–771.
137. Kim, H. J.; Kim, S. K.; Lee, J. Y.; Kim, J. S. *J. Org. Chem.* 2006, *71*, 6611–6614.
138. Lin, Z.; Chen, H. C.; Sun, S.-S.; Hsu, C.-P.; Chow, T. J. *Tetrahedron.* 2009, *65*, 5216–5221.
139. Bozdemir, Q. A.; Sozmen, F.; Buyukcadir, O.; Guliyev, R.; Cakmak, Y.; Akkaya, U. E. *Org. Lett.* 2010, *12*, 1400–1403.
140. Cametti, M.; Dalla Cort, A.; Mandolin, L.; Nissinen, M.; Rissanen, K. *New. J. Chem.* 2008, *32*, 1113–1116.
141. Cametti, M.; Rissanen, K. *Chem. Commun.* 2009, 2809–2829.
142. Beer, P. D.; Dent, S. W. *Chem. Commun.* 1998, 825–826.
143. Li, H.; Han, C.; Zhang, L. *J. Mater. Chem.* 2008, *18*, 4543–4548.
144. Chen Y.-M.; Cheng, T.-L.; Tseng, W.-L. *Analyst.* 2009, *134*, 2106–2122.
145. Delgado-Pinar, E.; Carbonell, E.; Alarcón, J.; García-España, E. Work in preparation.
146. Tomasulo, M.; Sortino, S.; White, A. P.; Raymo, F. M. *J. Org. Chem.* 2006, *71*, 744–753.
147. Ekmecki, Z.; Yilmaz, M. D.; Akkaya, E. U. *Org. Lett.* 2008, *10*, 461–464.
148. Saha, S.; Ghosh, A.; Mahato, P.; Mishra, S.; Mishra, S. K.; Suresh, E.; Das, S.; Das, A. *Org. Let.* 2010, *12*, 3406–3409.
149. Chung, S.-Y.; Nam, S.-W.; Lim, J.; Park, S.; Yoon, J. *Chem. Commun.* 2009, 2866–2868.
150. Tetilla, M. A.; Aragoni, M. C.; Arca, M.; Caltagirone, C.; Bazzicalupi, C.; Bencini, A.; Garau, A.; Isaia, F.; Laguna, A.; Lippolis, V.; Meli, V. *Chem. Commun.* DOI: 10.139/c0cc54500d.
151. Fabbrizzi, L.; Faravelli, I.; Francese, G.; Licchelli, M.; Perotti A.; Taglietti, A. *Chem. Commun.* 1998, 971–972.
152. Reyheller, K.; Kubik, S. *Org. Lett.* 2007, *9*, 5271–5274.
153. Delgado-Pinar, E.; Rotger, C.; Alarcón, J.; Costa, A.; García-España, E. Work in preparation.
154. Xie, H.; Yi, S.; Yang, X.; Wu, S. *New J. Chem.* 1999, *23*, 1105–1110.
155. Azenbacher, P.; Jursíková, K.; Sessler, J. *J. Am. Chem. Soc.* 2000, *122*, 9350–9351.9.
156. Piña, M. N.; Soberats, B.; Rotger, C.; Ballester, P.; Deyà, P. M.; Costa, A. *New J. Chem.* 2008, *32*, 1919–1923.
157. Comes, M.; Marcos, M. D.; Martínez-Máñez, R.; Sancenón. F.; Soto, J.; Villaescusa, L. A.; Amorós, P. *Chem. Commun.* 2008, 3639–3641.
158. Gunnlaugsson, T.; Davis, A. P.; O'Brien, J. E.; Glynn, M. *Org. Lett.* 2002, *4*, 2449–2452.

159. Yamaguchi, S.; Yoshimura, I.; Kohira, T.; Tamuru, S.; Hamachi, I. *J. Am. Chem. Soc.* 2005, *127*, 11,835–11,847.
160. Rudkevich, D. M. *Angew. Chem. Int. Ed.* 2004, *43*, 558–571.
161. Rudkevich, D. M.; Leontiev, A. V. *Aus. J. Chem.* 2004, *57*, 713–722.
162. Rudkevich, D. M. *Eur. J. Org. Chem.* 2007, *20*, 3255–3270.
163. Hagleitner, C.; Hierlemann, A.; Lange, D.; Kummer, A.; Kerness, N.; Brand, O.; Baltes, H. *Nature.* 2001, *414*, 293–296.
164. Jiménez-Cadena, G.; Riu, J.; Rius, F. X. *Analyst.* 2007, *132*, 1083–1099.
165. Katsuki, A.; Fukui, K. *Sens. Actuat. B.* 1998, *52*, 30–37.
166. Bakker, E.; Telting-Diaz, M. *Anal. Chem.* 2002, *74*, 2781–2800.
167. Lupan, O.; Chai, G.; Chow, L. *Microelectron. Eng.* 2008, *85*, 2220–2225.
168. Huang, H.; Lee, Y. C.; Tan, O. K.; Zhou, W.; Peng, N.; Zhang, Q. *Nanotechnology.* 2009, *20*, 115501 (5pp).
169. Renard, L.; Elhamzaoui, H.; Jousseaume, B.; Toupance, T.; Laurent, G.; Ribot, F.; Saadaoui, H.; Brötz, J.; Fuess, H.; Riedel, R.; Gurlo, A. *Chem. Commun.* 2011, *47*, 1464–1466.
170. Noh, J.-S.; Lee, J. M.; Lee, W. *Sensors.* 2011, *11*, 825–851.
171. Yang, F.; Taggart, D. K.; Penner, R. M. *Nano Lett.* 2009, *9*, 2177–2182.
172. Lee, Y. T.; Lee, J. M.; Kim, Y. J.; Joe, J. H.; Lee, W. *Nanotechnology.* 2010, *21*, 165503 (5pp).
173. Chopra, S.; McGuire, K.; Gothard, N.; Rao, A. M.; Pham, A. *Appl. Phys. Lett.* 2003, *83*, 2280–2282
174. Dybtsev, D. N.; Chun, H.; Yoon, S. H.; Kim, D.; Kim, K. *J. Am. Chem. Soc.* 2004, *126*, 32–33.
175. Esteban, J.; Ros-Lis, J. V.; Martinez-Máñez, R.; Marcos, M. D.; Moragues, M.; Soto, J.; Sancenón, F. *Angew. Chem. Int. Ed.* 2010, *49*, 4934–4937.
176. Wanna, Y.; Srisukhumbowornchai, N.; Tauntranont, A.; Wisitsoraat, A.; Thavarungkul, N.; Singjai, P. *J. Nanosci. Nanotechnol.* 2006, *6*, 3893–3896.
177. Trotta, F.; Cavalli, R.; Martina, K.; Biasizzo, M.; Vitillo, J.; Bordiga, S.; Vavia, P.; Ansari, K. *J. Incl. Phenom. Macrocycl. Chem.* 2011, *71*, 189–194.
178. García-España, E.; Gaviña, P.; Latorre, J.; Soriano, C.; Verdejo, B. *J. Am. Chem. Soc.* 2004, *126*, 5082–5083.
179. D'Alessandro, D. M.; Smit, B.; Long, J. R. *Angew. Chem. Int. Ed.* 2010, *49*, 6058–6082.
180. Oter, O.; Ertekin, K.; Derinkuyu, S. *Talanta.* 2008, *76*, 557–563.
181. Banerjee, R.; Phan, A.; Wang, B.; Knobler, C.; Furukawa, H.; O'Keeffe, M.; Yaghi, O. M. *Science.* 2008, *319*, 939–943.
182. Phan, A.; Doonan, C. J.; Uribe-Romo, F. J.; Knobler, C. B.; O'Keeffe, M.; Yaghi, O. M. *Acc. Chem. Res.* 2010, *43*, 58–67.
183. An, J.; Geib, S. J.; Rosi, N. L. *J. Am. Chem. Soc.* 2010, *132*, 38–39.
184. Star, A.; Han, T. R.; Joshi, V.; Gabriel, J. C. P.; Grüner, G. *Adv. Mater.* 2004, *16*, 2049–2052.
185. Bomchil, G.; Harris, N.; Leslie, M.; Tabony, J.; White, J. W.; Gamlen, P. H.; Thomas, R. K.; Trewern, T. D. *J. Chem. Soc. Faraday Trans.* 1979, *75*, 1535–1541.
186. Brina, R.; Collins, G. E.; Lee, P. A.; Armstrong, N. R. *Anal. Chem.* 1990, *62*, 2357–2365.
187. Qi, Z.; Honma, I.; Zhou, H. *Anal. Chem.* 2006, *78*, 1034–1041.
188. Zhang, T.; Nix, M. B.; Yoo, B. Y.; Deshusses, M. A.; Myung, N. V. *Electroanalysis.* 2006, *18*, 1153–1158.
189. Kang, Y.; Zyryanov, G. V.; Rudkevich, D. M. *Chem. Eur. J.* 2005, *11*, 1924–1932.
190. Deng, X.; Wang, F.; Chen, Z. *Talanta.* 2010, *82*, 1218–1224.
191. Kauffman, D. R.; Star, A. *Nano Lett.* 2007, *7*, 1863–1868.
192. Tonzetich, Z. J.; McQuade, L. E.; Lippard, S. J. *Inorg. Chem.* 2010, *49*, 6338–6348.
193. Wang, J. X.; Sun, X. W.; Yang, Y.; Wu, C. M. L. *Nanotechnology.* 2009, *20*, 465501.

194. Ohira, S.-I.; Wanigasekara, E.; Rudkevich, D. M.; Dasgupta, P. K. *Talanta*. 2009, *77*, 1814–1820.
195. McAlpine, M. C.; Ahmad, H.; Wang, D.; Heath, J. R. *Nat. Mater.* 2007, *6*, 379–384.
196. Yuliarto, B.; Zhou, H. S.; Yamada, T.; Honma, I.; Asai, K. *Chem. Phys. Chem.* 2004, *5*, 261–265.
197. Li, J.; Lu, Y.; Ye, Q.; Cinke, M.; Han, J.; Meyyappan, M. *Nano Lett.* 2003, *3*, 929–933.
198. Qi, P.; Vermesh, O.; Grecu, M.; Javey, A.; Wang, Q.; Dai, H.; Peng, S.; Chao, K. J. *Nano Lett.* 2003, *3*, 347–351.
199. Higgins, C.; Wencel, D.; Burke, C. S.; MacCraith, B. D.; McDonagh, C. *Analyst*. 2008, *133*, 241–247.
200. Elibol, O.; Morisette, D.; Akin, D.; Denton, J.; Bashir, R. *Appl. Phys. Lett.* 2003, *83*, 4613–4615.
201. Frycek, R.; Vladimir, M.; Martin, V.; Filip, V.; Jelinek, M.; Nahlik, J. *Sens. Actuators B: Chem.* 2004, *98*, 233–238.
202. Chien, F. S.-S.; Wang, C.-R.; Chan, Y.-L.; Lin, H.-L.; Chen, M.-H.; Wu, R.-J. *Sens. Actuators B: Chem.* 2010, *144*, 120–125.
203. Maruo, Y. Y.; Akaoka, K.; Nakamura, J. *Sens. Actuators B: Chem.* 2010, *143*, 487–493.
204. Das, S.; Chakraborty, S.; Parkash, O.; Kumar, D. S.; Bandyopadhyay, S.; Samudrala, S. K.; Sen, A.; Maiti, H. S. *Talanta*. 2008, *75*, 385–389.
205. Albrecht, M.; Schlupp, M.; Bargon, J.; van Koten, G. *Chem. Commun.* 2001, 1874–1875.
206. Star, A.; Joshi, V.; Skarupo, S.; Thomas, D.; Gabriel, J. C. P. *J. Phys. Chem. B.* 2006, *110*, 21014.
207. Sathiyamoorthi, R.; Chandrasekaran, R.; Mathanmohan, T.; Muralidharan, B.; Vasudevan, T. *Sens. Actuators B: Chem.* 2004, *99*, 336–339.

# 3 Organic and Biological Analytes

*Hans-Jörg Schneider*

## CONTENTS

## INTRODUCTION AND SOME PRINCIPLES

When chemists began preparing a large number of host compounds that could form supramolecular complexes, one of the most obvious and earliest applications was the use of such systems for analytical purposes.[1] Receptor compounds that react selectively have been prepared for practically all possible analytes as described in thousands of publications.[2] Also synthetic receptors for biogenic analytes, in contrast to natural biosensors, can always be equipped with suitable signaling elements; furthermore, they are usually more stable and can be adopted to any desired medium. As stated recently, natural biosensors comprising biomaterials have, with

the exception of the glucose monitor, almost entirely failed to achieve their potential as reagentless, real-time analytical devices.[3] In this chapter, we can only discuss
a few important principles ruling the choice of a practically useful synthetic host
system and highlight applications with selected examples—if possible from recent
papers in which one may find leading references. The chosen examples are usually promising for analysis in technical, environmental, biological, and medicinal
fields. The preferred examples are those suited for detection by optical or electrochemical methods and for analyses in aqueous media or in the gaseous state, with
fluorescence offering a particular advantage with respect to sensitivity.[4] The importance of host–guest complexes in water and the development of suitable receptor
compounds have been aptly reviewed.[5] Several techniques rely on the immobilization of supramolecular receptor compounds on surfaces and allow fast screening of
many analytes, for example, in high-throughput screening (HTS). Besides optical
detection, these physical methods comprise mass-sensitive detection (quartz crystal
microbalance, QCM); electrochemical techniques (see Chapters 4 and 5) such as
ion-sensitive electrodes (ISEs), which are not restricted to ions; photonic crystals
(PCs); surface plasmon resonance (SPR); and others. First, we will discuss some
methodological aspects, which then will be illustrated by examples from different
analyte classes.

## SENSITIVITY AND SELECTIVITY

The primary aim of an ideal supramolecular system is high sensitivity, which is
based on a large-complexation, free-energy $\Delta G_t$, and high selectivity, which is
defined as the ratio between equilibrium constants $K_{RX}$ and $K_{RY}$ for two analytes X
or Y and depends on the difference $\Delta\Delta G$ between the binding energies (Equation
3.1).[6] Both sensitivity or affinity and selectivity can be optimized by the most perfect
complementarity between host and guest molecules with respect to matching shape
and functionalities. Ideally, one would then expect a correlation between the affinity
$\Delta G_t$ and the selectivity, as the difference $\Delta G_{RY} - \Delta G_{RX}$ obviously becomes larger
with increasing values of $\Delta G$ (Equation 3.1).

$$K_{RX}/K_{RY} = \exp[(\Delta G_{RY} - \Delta G_{RX})/RT] \qquad (3.1)$$

A correlation between affinity and selectivity is, however, typical only for complexes
such as those with ionophores in which all existing interactions contribute the same
way to $\Delta G_t$ and to the difference $\Delta\Delta G = \Delta G_{RY} - \Delta G_{RX}$. More typical for the complexation of organic analytes are receptors in which one part of the host provides a
primary interaction site securing a high affinity, whereas a secondary site secures the
selectivity—usually at a different location than the one that ensures high sensitivity.
In addition, small variations of the high affinities achieved by interaction with the
primary binding site overshadow the usually smaller, yet sufficiently large for selectivity, differences at the secondary binding site.

　　A good sensitivity in the sense of correspondence between analyte concentration
and measured signal does not only depend on the affinity $\Delta G_t$. If the concentration of

an analyte exceeds that of the receptor and is so high that almost all guest molecules are bound, there will be no significant change in the signal from the complex. In that case, a too-high affinity can be a disadvantage. Ideally, the dissociation constant $K_d$ of a complex should be around the desired concentration range of the analyte.

## COMPARTMENTALIZATION FOR HIGHER SENSITIVITY

An often-overlooked way to enhance sensitivity is miniaturization of the container in which both analyte and sensor are located.[7] If all analyte molecules are bound to the available receptor units within the compartment, the same signal emerges from a smaller container with fewer analyte molecules than from a larger container; however, this occurs only as long as the affinity is large enough to bind nearly all the guest molecules.[8,9] This compartmentalization effect[10] can, with femtoliter containers, even allow single-cell and single-molecule detection;[11] small-size containers can be nanoparticles,[12] tiny vesicles, living cells, or even viruses. Another way to amplify sensitivity is to provide more binding sites within a receptor than available in the complementary analyte guest molecule.[9]

## SIGNAL AMPLIFICATION WITH MOLECULAR WIRES

A promising way to enhance the signal from a single binding event is based on electrochemical detection with a polyreceptor assembly. If, in a molecular wire, one single analyte binding produces a single resistive element in the wire, electrons flowing through the wire will experience disruption of the current.[13] For instance, with crown-ether macrocycles in a polythiophene, the binding of sodium ions to one or a few crown units leads to a conformational twisting of the chain, which may result in a decreased current. Again, the presence of many more binding sites within the wire than needed for a response at a single site will greatly enhance the sensitivity.

## REPORTER GROUPS FOR UV/VIS AND FLUORESCENCE MEASUREMENT

The formation of supramolecular complexes is frequently only measured by time-consuming and often less sensitive methods such as NMR spectroscopy. For practical applications, changes in UV or visible light absorbance are preferred, as are fluorescence changes that can result from photoinduced electron transfer (PET) or fluorescence resonance energy transfer (FRET). Sometimes, the pH or other bulk medium properties change, which might be used by optical indicators. In many cases, the receptor unit that is responsible for attaining affinity and selectivity will not exhibit spectroscopic signals, and one then attaches, for example, a fluorescent reporter group to the host. Chapter 2 presents many corresponding examples from inorganic chemistry. In view of practical applicability, we will also focus on examples based on UV/vis or fluorescence methods for organic analytes as other techniques, such as electrochemical detection or chromatography, are dealt with in other chapters.

## Indicator Displacement Assay

The reporter unit, which allows UV/vis or fluorescence detection, must not necessarily be attached covalently to the host compound. Instead, one can use an indicator complexed by the receptor, which is displaced via competition by the analyte and which, thereby, changes its signal. Ideally, the indicator should have an affinity between that of competing analytes.[14] The indicator displacement assay (IDA) was introduced already in 1967 for studying the binding of, for example, perchlorate to cyclodextrin with nitrophenolate as the indicator.[15] The IDA method, particularly with fluorescence indicators, has found widespread application and has been aptly described in several reviews.[16]

## Sensor Array Systems and Combinatorial Searches

The selectivity of a single receptor is often not high enough, particularly if it comes to rapidly recognizing many analytes within one sample. Nature shows how an assembly of slightly different receptors, for instance in the nose or on the tongue, allow a simultaneous and highly selective recognition of many analytes. With the availability of necessary computer programs such as for principal component analyses (PCAs) and of microengineering, great progress has been made in recent years to adopt these principles with whole arrays of receptors to overcome selectivity problems.[17] The differential sensing array method is based on a series of receptors, accessible often by combinatorial syntheses; each of the single receptors gives an often just slightly different signal upon complexation. The search for an optimal host compound can be greatly helped by the use of dynamic combinatorial libraries in which the best receptors are fished out from equilibrating host compounds by complexation with the target analyte.[18] For sensor arrays, the signals from the different receptors are subjected to PCA, for example, based on learning sets containing a known composition of analytes. This method is particularly appealing to the analysis of biopolymers, for which it becomes difficult to synthesize large enough host compounds with sufficient size and selective interactions (see also Chapter 16 on interaction with biopolymers). Array sensing is also the basis of modern genomics; with genetic fingerprinting based on oligonucleotide arrays; the DNA microarray is one of the most well-known applications.[19] It is equally instrumental in proteomics[20] and glycomics.[21] Multiplexed assays with arrays of multiple cells are now a standard technique for high-throughput screening of bioactive compounds.[22]

## Molecular Imprinting

A promising alternative to the often time-consuming synthesis of artificial host compounds is the use of molecularly imprinted polymers (MIPs) as recognition elements in sensors (see Chapter 5).[23] Such polymers are formed in the presence of a template molecule; removal of the template after cross-linking creates binding cavities prepared for noncovalent interactions for the target analyte molecule, which is used as a templating agent. This method is particularly appealing for the design of multielement sensing arrays, which collect a sufficient number of recognition elements possessing different binding affinities for analytes of interest.

## Thin-Film Techniques/Thin-Film and Field-Effect Transistors

Thin films incorporating a supramolecular receptor can, upon interaction with an analyte, change many physical properties. This has been used early,[24] for example, with organic thin-film transistors, which are based on the modulation of current through thin organic semiconducting films and include field-effect and electrochemical transistors. Electroactive films based, for example, on polythiophene layers are discussed in Chapter 5 on MIPs; the use of field-effect transistors (FETs) for sensing is explained in Chapter 9 on supramolecular electronics. Some reviews[25] are available on the topic, as are several books; however, they mostly deal with other applications such as displays.[26] Interesting applications of thin films for sensing involve mostly gaseous analytes. Thus, SPR can be used for measuring as shown, for example, with a calix[4]pyrrole Langmuir–Schäfer (LS) film for detection of the gaseous anesthetic agent sevoflurane.[27] Thin films can also be used in combination with high-frequency quartz microbalances as shown recently for the detection of the explosive triacetone triperoxide.[28]

## Covalent Bonding for Recognition

An alternative to supramolecular complex formation by noncovalent bonds is the use of covalent bonds for molecular recognition if such bond formations occur rapidly enough.[29] If they do not form in a reversible or dynamic manner, the selectivity will be dictated by the speed of bond formation, and the receptor is, of course, used up. Even though such methods do not belong, in the narrow sense, to supramolecular complexation, for practical reasons, they are illustrated in the present chapter with a few examples. The distinction between noncovalent and covalent bonding becomes ambiguous anyway, for instance, in the case of transition-metal complexes. Also, one must preorganize the reacting groups in the host compound similarly to the design of supramolecular complexes. The use of fast covalent bond formation is particularly important for analytes such as carbohydrates, which in their natural aqueous environment are notorious for very weak intermolecular interactions.

## SELECTED ANALYTES

### Environmental and Process Control

The detection of volatile organic compounds (VOCs) is of significance for controlling such analytes in industrial processes and in the environment. Calixarene hosts may be used in thin films for such analyses preferably with fluorescence detection.[30] With phosphorus-bridged cavitands such as **1** ($R = C_{11}H_{23}$, $R^1 = Br$, $R^2 = Ph$), one can, with the help of the mass-sensitive QCM, analyze alcohols in the gas phase, for example.[31] The binding of the cavitands is attributed to H-bonding interactions between the PO units and the alkyl- and arylammonium cations. The interaction strength is correlated with the proton affinity of the interacting species; dispersive interactions increase with the number of carbon atoms in the alcohol chains. Thus, with linear alcohols, the QCM frequency changes with 3000 ppm of the analyte from 50 Hz for methanol to about 400 Hz for amylalcohol.

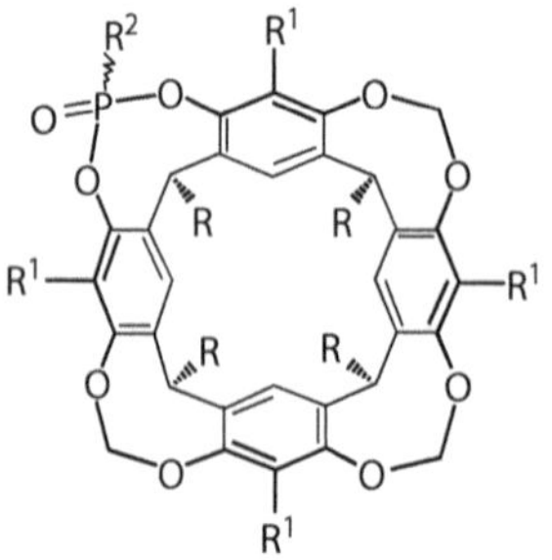

1 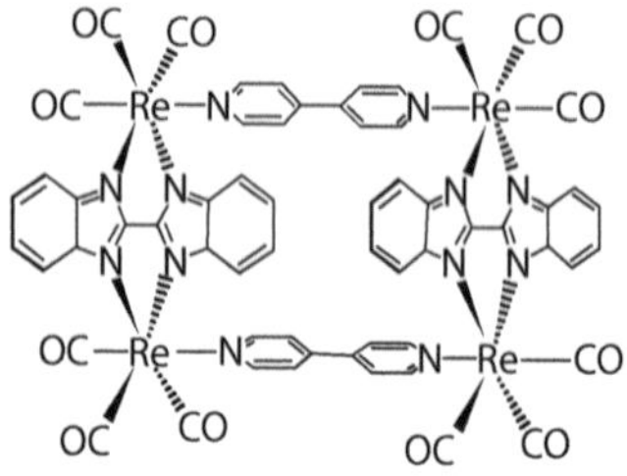                                                    2

Metallophthalocyanines with four bulky pentaphenylbenzene substituents can adsorb VOCs selectively into their cavities. Such macromolecular metal complexes were prepared by immersing polymer brush–modified QCMs into an aqueous solution of the sterically protected cobalt phthalocyanine. Anionic cobalt phthalocyanine was trapped in the polymer brushes and acted as a molecular receptor for the sensing of VOC molecules.[32]

Metallosupramolecular structures[33] can be designed for the relatively easy assembly of many host compounds and are also used increasingly as sensors.[34] VOCs can be sensed with self-assembled rectangular host **2** in the gas phase.[35] The emission of thin films containing the metallocomplex is quenched upon exposure to p-toluidine vapor, but the emission is enhanced and shifted to higher energy with tetrahydrofurane (THF) vapor, for example. The selectivity of the film-based mesoporous material was studied with QCM mass sensing, showing that aromatic guests with electron-withdrawing substituents diminish the affinity, whereas aromatic guests with electron-donating substituents enhance it. Affinities reach from approximately $K = 500$ [M$^{-1}$] for hexafluorobenzene to $K = 3000$ [M$^{-1}$] for toluene.

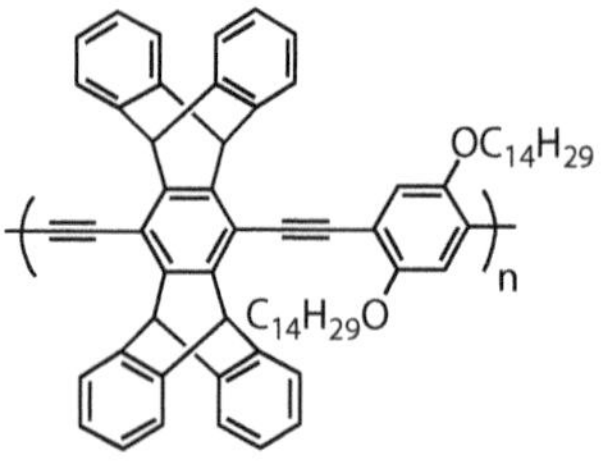

3 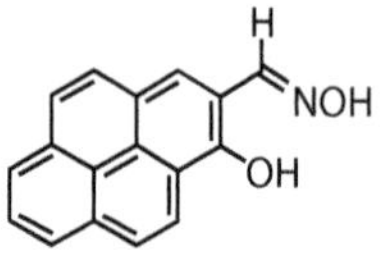                                                    4

Sensing of explosives has become an important target; in particular, one wants to detect traces of explosives leached from land mines. Conjugated polymers containing large electron-donating arenes hold promise for the detection of nitroarene-derived explosives such as trinitrotoluol (TNT).[36] A "sponge" for electron-poor analytes is based on fluorescence quenching by electron transfer with polymers such as **3**. Within seconds, thin films of this polymer display sizable quenching by exposure to 10 ppb of TNT in the gas phase. In this so-called iptycenes interchain, π–π interactions between the chains, which create self-quenching, are avoided, and pores are generated in which the solvent can be displaced by analytes. Light-emitting dendrimers with a bisfluorene-based chromophore core exhibit, by a distributed feedback laser, higher sensitivity than the change in photoluminescence against nitrobenzene-type explosives. Such high electron–affinity nitroaromatic analytes bind to the aromatic polymer surface, introducing nonradiative deactivation pathways that quench fluorescence by electron transfer. Such bisfluorene dendrimers are coated as thin films on silica and allow trace detection of analytes commonly found in explosive vapor.[37] Perylenes offer electron-accepting surfaces in contrast to systems as **3** and enable detection of volatile amines with high sensitivity if the perylenes are introduced as imides into cyclodextrins. These conjugates can be incorporated in a poly(vinylidenefluoride) membrane; gaseous amines are embedded in the cavities of permethyl-ß-cyclodextrins and induce fluorescence quenching of such perylene bisimide chromophores.[38] In solution, nitroaromatic explosives were sensed with a colorimetric assay using allosteric tetrathiafulvalene derivatives.[39] The overall binding constants vary in chloroform from $K = 3 \times 10^6$ [M$^{-1}$] with 1,3,5-trinitrobenzene and, for example, $K = 2 \times 10^4$ [M$^{-1}$] with 2,4,6-trinitrotoluene (TNT). The binding efficiency is not strictly related to the redox potential of the different tetrathiafulvalene derivatives.

In times of chemical warfare development and terrorist actions, the detection of nerve agents has become an important issue. In view of the high reactivity of activated phosphate esters, it is no surprise that fast covalent reactions have been most often used for detection. One naked-eye detection is based on fast phosphorylation of the hydroxy group in (dimethylamino-phenyl)ethanol that was coupled to an azo dye. The system exhibits a yellow band resulting from charge transfer, which decreases upon reaction with a nerve agent because the N,N-dimethyl-anilino function then has a drastically reduced donor capacity.[40] Similarly, the high nucleophilicity of the oxime group, for example, as in **4**, which is also the basis for antidotes presently used against poisoning by nerve gas, can be used for detection via fluorescence or color change; the reactions with **4** can be complete to 50% in about 6 seconds.[41] Imine derivatives of pyrene were incorporated in thin films and in fibers, leading to a subsequent fluorimetric response by similar fast reactions.[42]

## DRUGS, METABOLITES, NEUROTRANSMITTERS/CHIRAL DISCRIMINATION

Most bioactive compounds are chiral, and the enantioselective sensing of such analytes has been considerably advanced.[43] Often one applies derivatives of chiral 1,1′-bi-2-naphthol (BINOL) such as **5**. In chloroform, S-mandelate binds to **5** with $K = 5.8 \times 10^4$[M$^{-1}$], whereas the R-isomer binds with $K = 1 \times 10^4$ [M$^{-1}$]. With the Boc-protected alanine anions, the enantioselectivity amounts to K(S)-Ala)/K(R)-Ala) = 10.17. Fluorescence also shows, with ΔI(S(/ΔI (R) = 5.4, significant quenching differences.[44] Gels obtained from the copper complex of the BINOL and terpyridine conjugate **6** exhibit not only fluorimetric differences between phenylglycinol and other chiral amino alcohol enantiomers but interestingly also differences in the gel volume, which decreases with S-phenylglycinol more than with the R-isomer.[45] The BINOL derivative **7** exhibits significant enantioselectivity in the fluorescent recognition of various γ-hydroxycarboxylic acids; with phenyllactic acid, for instance, the difference in quenching leads to ΔI(S)/ΔI (R) = 11.[46]

5                                              6                                              7

The often-used bowl-shaped calixarenes and the related resorcarenes are interesting host compounds for many biogenic analytes,[47] including neurotransmitters.[48]
The p-sulfonatocalix[4]arene **8** has the advantage of bearing negative charges over
a large pH range; binding of the fluorescence indicator 2,3-diazabicyclo[2.2.2]oct-2-
ene leads to efficient quenching by up to 90%. If analytes such as carnitine or choline
are added, the fluorescence is regained.[49]

8                                                                          9

One methodology for sensing citrate that may be applied to related polar organic
anions is based on the formation of a ternary fluorophore host–anion complex with an
imidazolium-substituted calix[4]arene. Complexation of citrate induced p$K$a shifts of
an amino group attached to the fluorophore and, therefore, protonation of the amino
group, which switches off intramolecular PET and thus enhances the fluorescence
signal.[50] Alternatively, citrate can be selectively detected, even in  the presence of
malate or tartrate, using the triscationic host **9**, which complexes both the analyte
citrate and the indicator tris-anion carboxyfluorescein with a similar binding constant
of $5 \times 10^5$ [M$^{-1}$]. This is ideal for the IDA method and allows naked-eye detection by
the appearance of the quenched fluorescence after adding the analyte.[51]

The drug γ-hydroxybutyric acid (GHB/GABA) can be detected in millimolar
concentrations with a colorimetric sensor based on supramolecular host–guest complexes of fluorescent dyes with organic capsules from cucurbiturils. The basis is a
library combining four fluorescent tricyclic basic dyes as sensing guests and two
cucurbiturils (**10**, $n = 7,8$) as hosts, which bind the dyes with constants between $K = 3$
and $15 \times 10^6$ [M$^{-1}$].[52] Electrochemical sensing of acetylcholine (ACh) is possible with
a membrane electrode containing **10** ($n = 6$), which detects ACh with high selectivity
over choline and other ions such as Na$^+$, K$^+$, and NH$_4^+$.[53]

The isoquinoline alkaloid coptisine, which has many interesting pharmacological properties including antibacterial activity, also can be analyzed in urine and serum samples with the help of cucurbituril complexes with a detection limit of 0.012 ngm/L. The fluorescence intensity of the complex with CB[7] (**10**, $n = 7$) is enhanced 70-fold compared to that of coptisine itself.[54] The flat surface of the triphenylene-based ketal **11** and the side groups of the host secure effective binding of trimethylxanthine analytes by hydrogen bonding and stacking. Addition of caffeine to a solution of the host in $CH_2Cl_2$ results in an increased fluorescence signal with binding constants around $K = 4.5 \times 10^4$ [M$^{-1}$].[55]

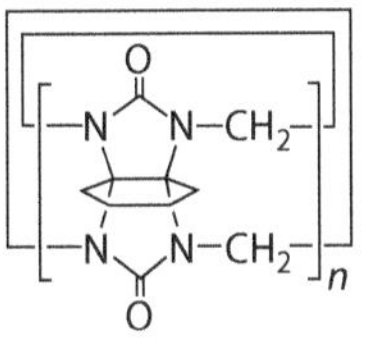

**10** 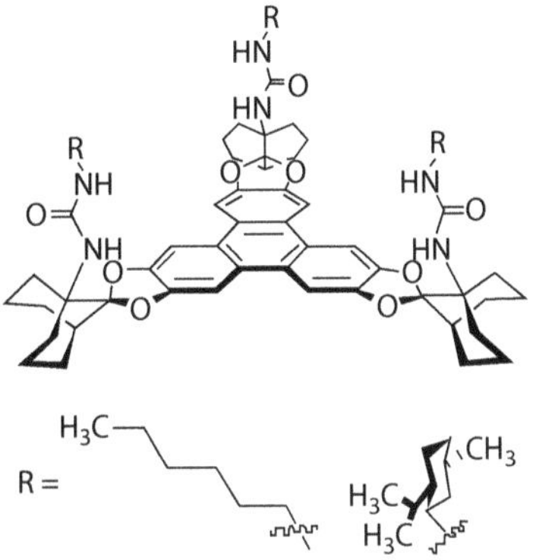                                              **11**

Chiral ammonium salts can be detected with an oligoresorcinol nonamer, which assumes a helical conformation with about five partially deprotonated resorcinol rings per turn. A preferred helicity induction of the oligoresorcinols was observed with the addition of chiral ammonium salts. The induced circular dichroism shows significant differences with different enantiomers.[56] With a chiral bisnaphthyl host, one observes in benzene as solvent with $10^{-3}$ M (S)-mandelic acid a more than 20-fold fluorescence enhancement and with (S)-hexahydromandelic acid a more than 80-fold enhancement and a remarkable enantioselectivity.[57]

Nonchiral fluorimetric sensing of electron-rich aryl derivatives is possible with ternary complexes in which aromatic π-donor–acceptor pairs occur inside the hydrophobic cavity of cucurbituril (CBn, Scheme 3.1). With 2,7-dimethyl diaza-pyrenium as an electron-poor dye and cucurbit[8]uril, one observes, in aqueous solution and with silica nanoparticles coated with the cucurbituril and the fluorescence dye,

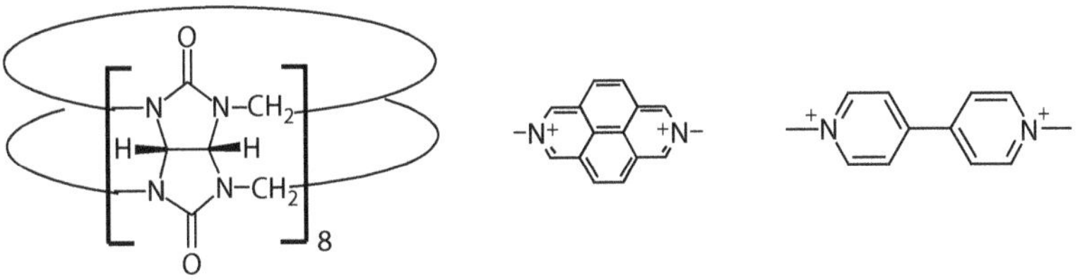

**SCHEME 3.1**   Cucurbit[8]uril and electron-poor dyes used in ternary complexes with dopamine, etc.

significant fluorescence quenching with about 1 mM catechol and dopamine as ana-lyte.[58] Similarly, methyl viologen was used in such ternary CB complexes for detec-tion of amino acids and peptides (see below).

Modified cyclodextrins have been often used for chiral sensing, frequently in combination with MIPs or at interfaces, for example, of thin films.[59] Self-assembled monolayers (SAMs) of some CD derivatives lead, with the QCM, to a chiral dis-crimination factor of 1.33 with the drug (methoxyphenyl)ethylamine as the analyte.[60] A macromolecular helicity inversion accompanied by a color change response was reported with a stereoregular (cis-transoidal), chromophoric poly phenylacetylene bearing an optically active α-cyclodextrin residue as a side group. The CD group seems to induce a predominantly one-handed helical conformation and an intense circular dichroism in the long absorption region of the conjugated polyene backbone. In dimethyl sulfoxide, one observes, for example with borneol enantiomers, which are known to complex with CD, quite different colors.[61]

Steroids such as bile acids can be detected by inclusion in cyclodextrin, which is equipped with fluorophores. A cyclodextrin-conjugated helical peptide bearing cou-marin and pyrene fluorophores in the oligopeptide side chains shows intramolecular FRET without quenching of two fluorophores.[62] The addition of hyodexoy cholic acid, for example, which binds in water with $K = 2 \times 10^5$ [M$^{-1}$] to the CD peptide, causes a drastic reduction of the fluorescence emission, likely by exclusion of the coumarin fluorophore from inside to outside the CD cavity. A MIP was used for QCM measurements with the often-abused drug methamphetamine; the sensor had significant selectivity for the D-isomer in human urine samples also.[63]

Self-assembly of labeled aptamer subunits in the presence of target analytes pro-vides a new method for optical or electrochemical detection. Figure 3.1 shows how

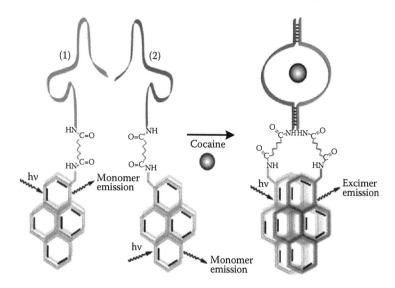

**FIGURE 3.1** Self-assembly of pyrene-labeled aptamer subunits allows detection of cocaine. (Freeman, R. et al., *Analyst*, 2009, *134*, 653–656. Reproduced by permission of The Royal Society of Chemistry.)

aptamer fragments, modified with pyrene units, lead to a supramolecular aptamer–analyte complex that allosterically stabilizes the formation of excimer supramolecular structure accompanied by a characteristic fluorescence emission. The thiolated aptamer subunit can be assembled on a gold electrode, which allows, after further modification, the detection of cocaine with a detection limit of $1 \times 10^{-5}$ [M].[64]

Fast and, if possible, covalent reactions are used for analytes, where the aim of rapid measurements without expensive instrumentation is complicated (see previous section). For example, the bisazo dye **12** was incorporated in thin layers of plasticized polyvinylchloride and undergoes an amphetamine reversible reaction to **13** upon which it changes color from blue to red. The sensitivity of the dye polymer layer is in the range from 0.3 to 30 mmol, with a 0.1-mmol detection limit.[65]

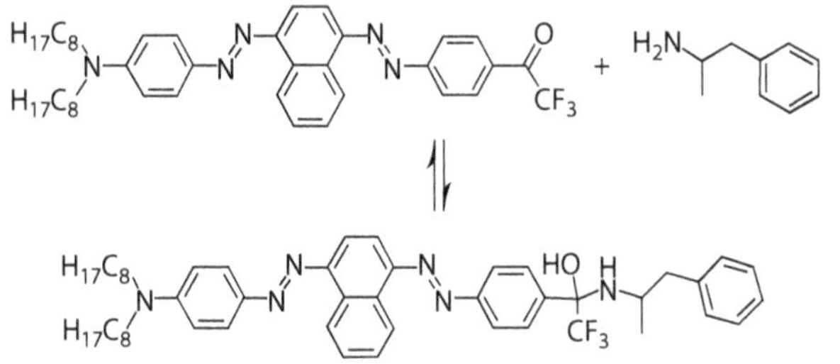

The most often used covalent indicator methods involve boronic acid derivatives. The dye **14** (Lucifer yellow derivatized with boronic acid) recognizes L-DOPA after reversible esterification and also by stacking between the aryl units as well as by ion pairing in **15**, leading to a drop in the fluorescence emission.[66]

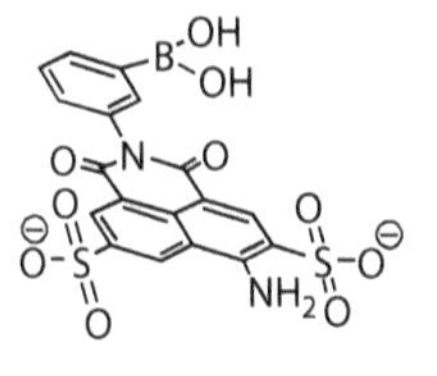

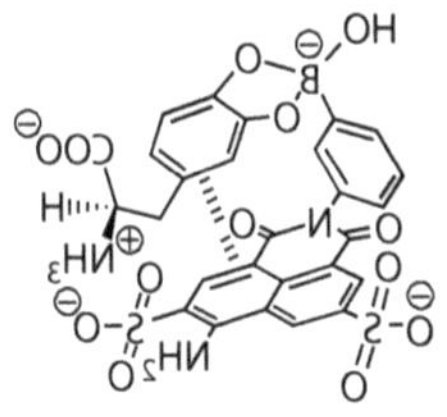

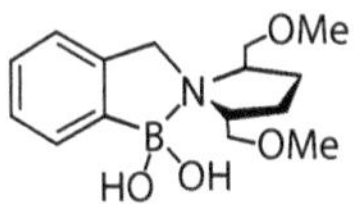

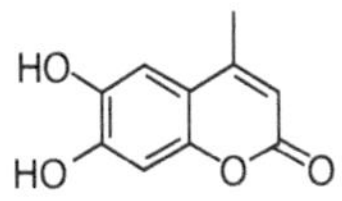

Chiral boronic acids such as **16** are very successfully used for the distinction of all kinds of chiral analytes bearing vicinal OH groups such as γ-hydrocarboxylates or vicinal diols, including sugars (see the following section). These undergo rapid esterification with the boronic acids[67] as exemplified with structure **15**. Indicators such as coumarin **17** exhibit an up to tenfold fluorescence enhancement if complexed with the boronic acid **16**. Indicator displacement by the addition of phenyl mandelic acid results in effective fluorescence quenching, much more so upon addition of the L-enantiomer than the D-enantiomer. With a receptor array containing guanidinium embedded in an aminoimidazoline group as well as two boronic acids and pyrocatechol violet as an indicator for UV/vis measurement, it was possible to distinguish different whiskeys, based essentially on the different content and concentration of gallic acids.[68] Additionally, these contain carboxylate 3,4,5-trihydroxy aryl units, which undergo reaction with the boronic acid.

## AMINO ACIDS AND PEPTIDES

Many artificial receptors for the sensing of protected amino acids have been described, but until now, there have been relatively few for amino acids in their natural form and in an aqueous environment. The exceptions are amino acids with additional charges in the side chain or reactive ones such as cysteine, in which covalent reactions can be used (see the following).[69] Electrochemical methods such as voltammetry at uncoated and modified electrodes and at immiscible liquid–liquid interfaces, potentiometry at polymer membrane electrodes, and electrochemical impedance spectroscopy open a way to analyze unprotected amino acids.[70] With an imprinted polymer derived from poly(o-phenylenediamine), phenylalanine (Phe) can be detected with a sensitivity of 0.5 mM.[71] Molecular imprinting with Phe as templating based on submicron sphere polymers from methacrylic acid and trimethylolpropane trimethacrylate as cross-linking agents is efficient for the recognition and transport of Phe.[72] Molecular SAMs on an electrode, adsorbed on a substrate with a highly ordered structure, allow serine detection down to 4 $\mu$M.[73]

## DETECTION OF NATURAL AMINO ACIDS WITHOUT PROTECTION

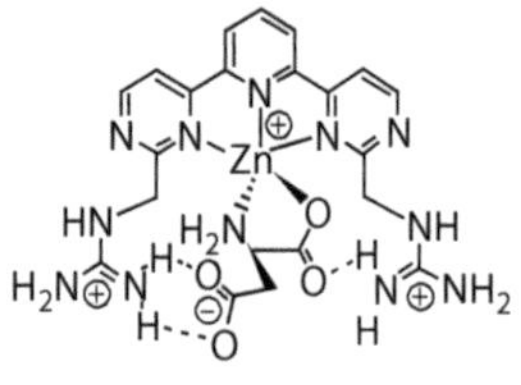

**18** 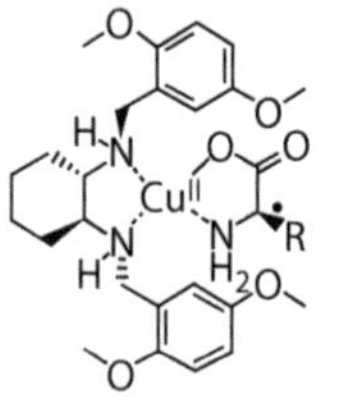                              **19** 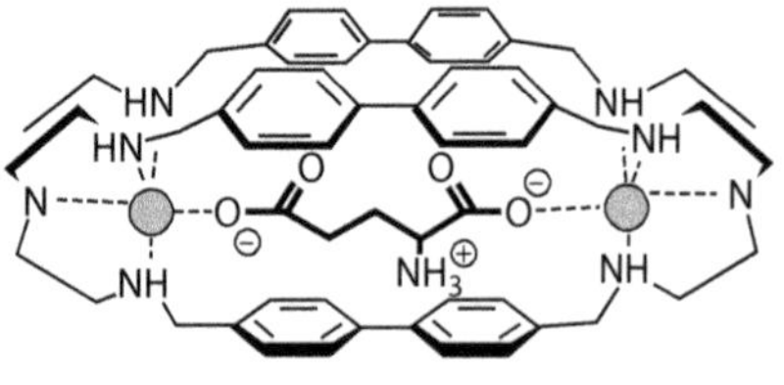                                      **20**

Amino acids that bind transition-metal ions by chelation with the carboxylate and the amine termini can be detected by optical signal changes upon complexation. Thus, the neurotransmitters aspartate and glutamate form a complex (**18**) with $K$ up to $10^5$ [$M^{-1}$] with terpyridine derivatives and zinc(II) salts. Pyrocatechol as an indicator shows, by displacement, a color change from deep blue to yellow.[74] Copper complexes derived from chiral *trans*-diaminocyclohexane like **19**, with the displaced indicator pyrocatechol violet, allow distinction of enantiomeric Val, Phe, Trp, or Leu with binding constants of up to $10^5$ [$M^{-1}$] and with selectivity ratios $K_D/K_L$ between 1.7 and 2.6.[75] The dicopper(II)octamine cage **20** can be used for the detection of L-glutamate; rhodamine is used here as an indicator. The fluorescence of rhodamine is fully restored by uptake of the L-glutamate. Other typical neurotransmitters (glycine, GABA) and related amino acids show only minor effects.[76] Cyclodextrins containing a metal binding site and a dansyl fluorophore, upon addition of D- or L-amino acids, show increases in fluorescence intensity, which depend on the amino acid used and, in some cases, on its absolute configuration.[77]

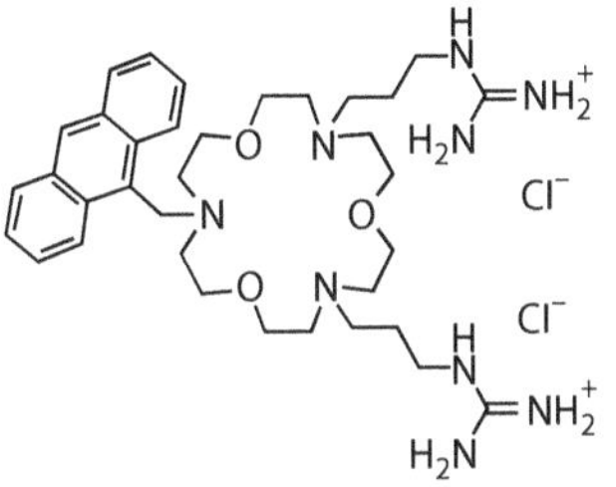

R' = H, R = Acylpyrrol

**21** 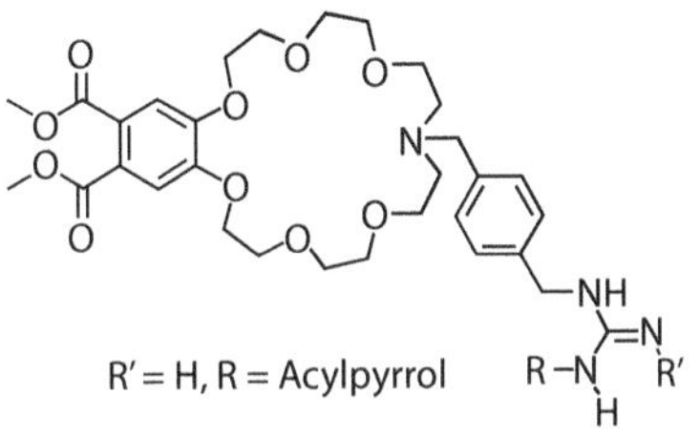                                                **22**

The combination of crown ethers with ammonium and/or guanidinium in receptors like **21** provides for binding by ion pairing in water with a selectivity determined by the placement of the charged sites.[78] Thus, receptor **21** preferably binds glycine, lysine, and 4-amino butyric acid, indicated by an increased emission. Host **22** shows, with fluorescence, emission selectivity toward GABA compared to similar amino acids and distinguishes GABA from its biological precursor, that is, the glutaminic acid.

Ternary complexes with cucurbituril in combination with the electron-poor dye methyl viologen shown in Scheme 3.1 are used for binding selectively electron-donating amino acid side chains such as those from tryptophan, phenylalanine, and tyrosine in contrast to the other biogenic 17 amino acids.[79] The mechanism of recognition with peptides also involves formation of ternary complexes held together by the combined action of electrostatic interactions of terminal peptide ammonium groups or free amino acids with carbonyl groups of cucurbituril as well as a stacking interaction with the included MV dication.[80] Peptide-based scaffolds functionalized with one, two, or three viologen groups show a 31- to 280-fold binding enhancement to cucurbituril.[81]

Detection of charged amino acids and peptides such as arginine effectively uses ion pairing. The presence of many counterions such as in cyclophane **23** leads to many salt bridges with complexation free energies around 5 kJ/Mol for each bridge.[82] Dipeptides and tripeptides bearing lysine or arginine residues are effectively complexed by p-sulfonatocalixarenes with binding constants increasing from $10^3$ to $10^5$ with peptides bearing one, two, or three basic amino acids.[83] The molecular tweezer **24** provides not only permanent charges at the anionic phosphonate groups but also an electron-rich interior and displays an affinity for lysine with $K = 5000$ [M$^{-1}$] in a neutral phosphate buffer.[84] Electrophoresis on a microchip relying on monitoring the difference in the ionic mobility of the receptor upon interaction with the ligand can be used for the complexes between acid-rich diketopiperazine receptors and basic tripeptides in aqueous solution.[85]

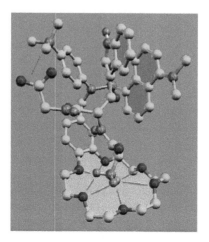

**23**                                          **24**

An ideal sensor for peptides would distinguish them according to length, sequence, and nature of the amino acids in the chain and would perform in aqueous media without limitation regarding the presence of ionic side groups. Host compounds such as **18–24** fulfill these tasks with some limitations. Receptor **25** illustrates how a combination of a crown ether for binding the $^+NH_3$ termini and a permanently charged $^+NMe_3$ group for ion pairing with the peptide $COO^-$ termini can secure primary association with length selectivity. Additional groups' RH can be mounted at suitable places along the receptor chain and allow selection with respect to the amino acid side group RG occurring at the complementary location on the peptide chain. If in **25**, the side group RG is, for example, an aromatic residue, there is a lipophilic or stacking interaction, which increases in the sequence Gly<Ala<Leu<<Phe<Trp, provided that the corresponding amino acid is in the right position within the peptide chain. At the same time, the dansyl residue as RG group exhibits a strong fluorescence signal if a peptide is bound because the quenching effect of the neighboring free crown ether oxygen atoms disappears. The binding constants with this receptor, which makes simultaneous use of ion pairing, hydrogen bonding, and lipophilic interactions (Figure 3.2), vary, for example, from 200 [$M^{-1}$] for Gly-Gly-Phe to 1600 [$M^{-1}$] for Gly-Phe-Gly, where the Phe unit is in the "right" place.[86]

**FIGURE 3.2**   Complex of 25 with Phe.

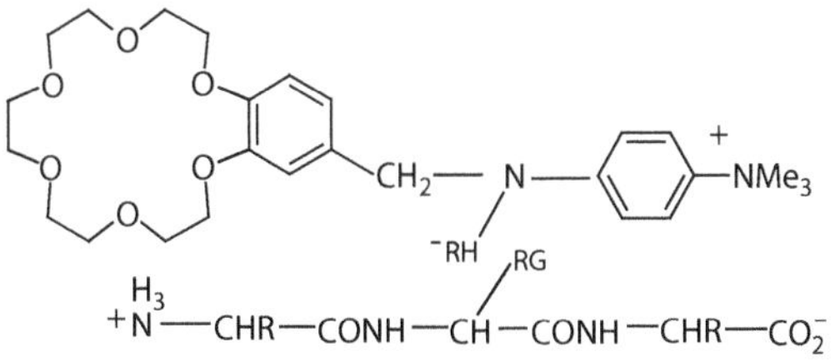

**25**

Receptor **26** again uses a crown ether moiety for hydrogen bonding with the $^+NH_3$ terminus of peptides and ion pairing, with pyridinium units attached to a porphyrin ring, which allows stacking with amino acid side groups. With Gly-Gly-Gly one observes lg $K$ = 3.4; with Gly-Gly-Phe lg = 4.4; with Gly-Gly-Gly-Gly lg $K$ = 5.0. The associations are visible in changes in the UV Soret band.[87] Receptor **27** uses a coordinative bond with the central $Zn^{2+}$ ion for superior affinity toward histidine-containing peptides and ion pairing with additional carboxyalkyloxy groups at the periphery of the porphyrin.[88] Host **27** is also a strong receptor for biogenic amines such as histamine and, upon complexation, exhibits a red shift in the Soret band of the porphyrin. Mixed monolayers of dioctadecyl glycylglycinamide amphiphiles incorporate aqueous dipeptides with binding constants greater than $10^3$ $M^{-1}$ at the air–water interface; the Fourier-Transform IR (FT-IR) spectra of the Langmuir-Blodgett films show characteristic changes. Ditopic receptors based on a covalently linked nucleobase such as cytosine to a metalloporphyrin can, with a PVC membrane–based ion-selective electrode, detect millimolar levels of nucleotides.[89]

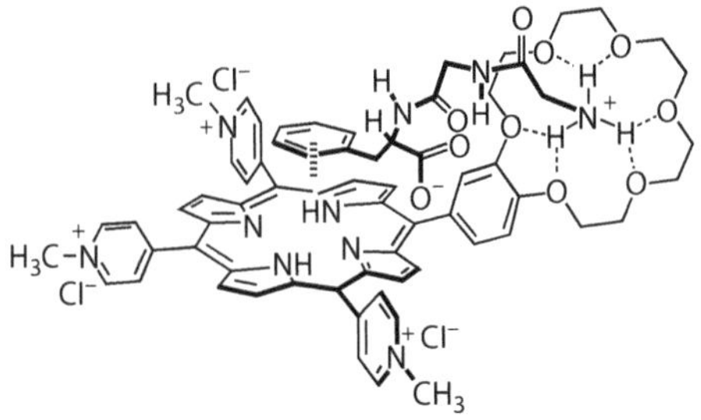

**26** 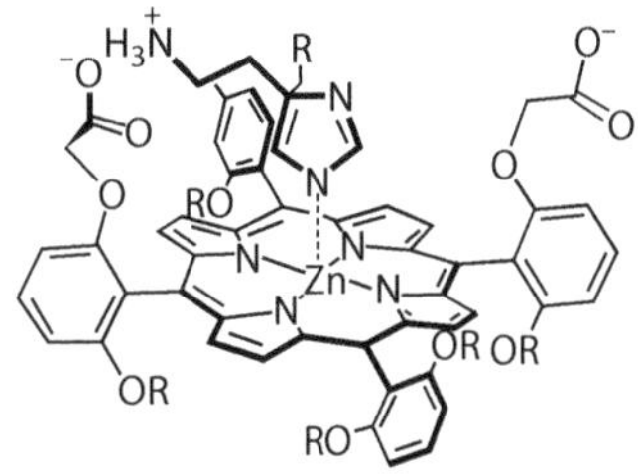            **27**

## COMPLEXATIONS OF PROTECTED AMINO ACIDS AND PEPTIDES

Many selective receptors have been reported for protected amino acids; N-Boc- or N-Bz derivatives allow and require the use of nonpolar solvents, suitable for the often-lipophilic large host compounds. The pseudopeptidic fluorescent receptor **28** shows, in dichloromethane via fluorescence emission, preferential binding for Z-protected aromatic (Phe) over aliphatic amino acids (Ala, Val) by a factor of 3–4.[90] Host **29** binds N-protected anionic amino acids in acetonitrile with fluorescence quenching by photoinduced charge transfer; with *t*-Boc alanine, the equilibrium constants are $K$ = 2300 [$M^{-1}$] for the *L*-enantiomer and $K$ = 23,900 [$M^{-1}$]

for the *D*-enantiomer, thus exhibiting the remarkable stereoselectivity of *KL/KD* = 10.4.[91] The 1,8-diheteroarylnaphthalene-derived compound **30** is an enantiose-lective fluorosensor for N-t-Boc-protected serine.[92] Large self-assembled metallo-coordination cages can accommodate as many as three amino acid residues of oligopeptides in a sequence-selective fashion along with visible color changes in the solutions, however, only after O-protection.[93] The peptide derivative Ac-Trp-Trp-Ala-NH$_2$ binds in water with $K > 10^6$ [M$^{-1}$]; sequences such as Ac-Trp-Ala-Trp-NH$_2$ or Ac-Ala-Trp-Trp-NH$_2$ are about 100 times weaker. N-acetyl methyl esters of peptides in chloroform sequence show selectivity with a host derived from Kemp's triacid.[94] A tweezer receptor incorporating a guanidinium head group and two peptide-derived side arms binds selectively protected peptides with up to 10$^5$ [M$^{-1}$] in DMSO/water.[95] In a library of related tweezers containing peptide arms, one host is able to selectively recognize the sequence Val-Val-Ile-Ala, which is the C-terminal tetrapeptide sequence of the amyloid-β protein.[96] Receptors based on cationic acylguanidinium with additional H-bond donor sites complex O-protected dipeptides such as Val-Val in water with binding constants of up to 50,000.[97] Calixarenes bridged with an oligopeptide chain bind N-acetyl-D-Ala-D-Ala similarly to vancomycin.[98]

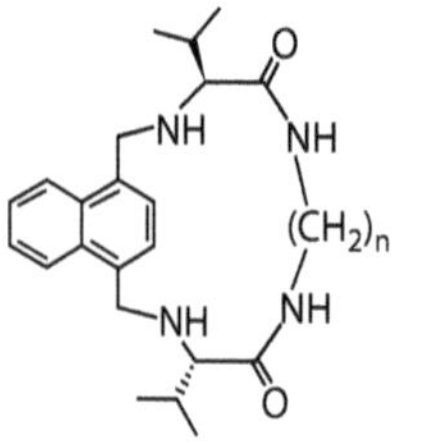

**28** 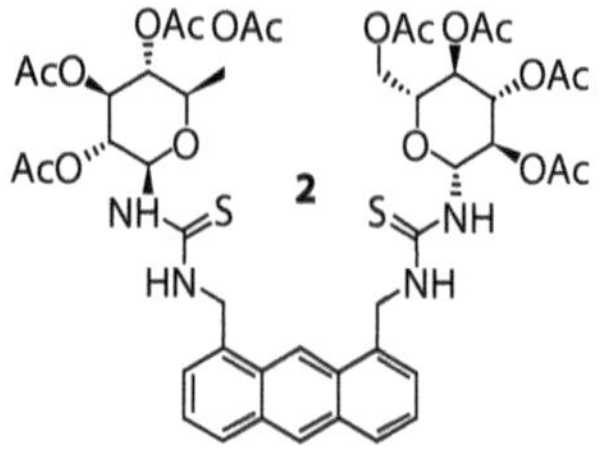                                    **29** 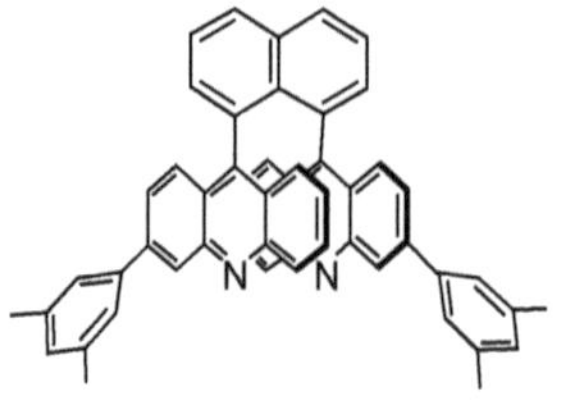                                    **30**

Amino acids with reactive side chains can be detected with the aid of fast cova-lent reactions. The fluorescein derivative **31** reacts with a significant fluorogenic response only with compounds such as cysteine that contain both a free thiol and amine function, yielding a thiazinane or thiazolidine heterocycle.[99] Depending on the structure of the amino thiols, electrostatic interactions occur between the NH$_3^+$ group with both the phenolate and the carboxylate moieties of the product, leading to different emission enhancement in the order Cys (41-fold) > 4 Hcy (29-fold) > 4 mercaptoethanol (6-fold) > 4 mercaptopropionate (1.5-fold), after com-plete reaction with excess thiol (1 mM compound 1:10.0 mM thiol). The assay is particularly promising as it is suitable for all N-terminal cysteine peptides and for homocysteine, which is a risk factor in many diseases. Homocysteine can be mea-sured in blood plasma based on the kinetically favored formation of the alpha-amino carbon-centered radical of this amino acid, which allows for the selective reduction of methyl viologen dication to its corresponding radical cation.[100] Electrochemical detection of aminothiols such as glutathione relies on the analytical signal gen-erated from oxidation of catechol-thiol adducts and can discriminate glutathione from cysteine and homocysteine.[101]

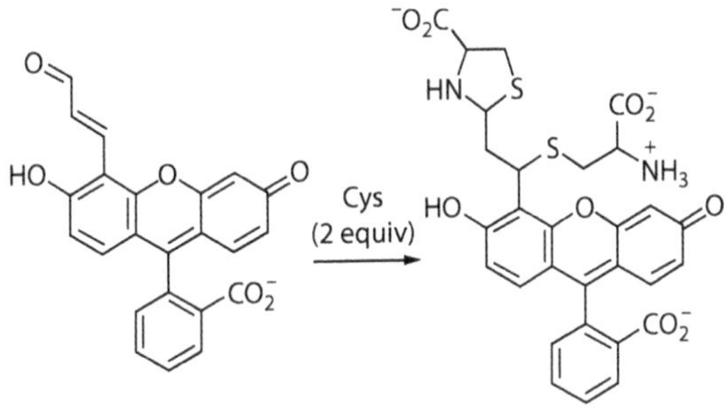

**31**

## NUCLEOTIDES AND NUCLEOSIDES

Azacrown ethers equipped with fluorescent moieties were found early to be efficient sensing units for nucleotides and related analytes bearing negative charges. Host **32** binds with significant fluorescence enhancement ATP and NADPH with $K = 3 \times 10^8$ [M$^{-1}$] as a result of ion pairing and stacking of the two acridine units with both the adenine and the nicotinamide moieties.[102] With many positively charged receptors, the affinity is a function of the number of ion pairs and decreases, for example, in the sequence ATP > ADP > AMP.[103] This is also seen with receptors based on cyclodextrins bearing up to seven amino substituents at the lower rim (host **33**); the approximately eight salt bridges between the fully protonated amino groups and the fully deprotonated phosphate residues lead to a theoretical association constant of approximately $10^{10}$ Mol$^{-1}$ for ATP$^{4-}$, with ATP$^{2-}$ to $10^6$ Mol$^{-1}$. Under practical conditions in presence of 0.1 M NaCl and with a 5 mM nucleotide, the values are, for example, $3 \times 10^6$ Mol$^{-1}$ for ATP and $1.2 \times 10^5$ Mol$^{-1}$ for AMP. There are relatively small differences for different nucleobases, with $0.4 \times 10^5$ Mol$^{-1}$ for GMP, $0.2 \times 10^5$ Mol$^{-1}$ for CMP, and $0.9 \times 10^5$ Mol$^{-1}$ for UMP. Both the small nucleobase selectivity and the fact that ribose phosphate has, with $8 \times 10^5$ Mol$^{-1}$, an even larger affinity than AMP are in line with the inclusion mode as shown in **33**; the placement of the nucleobase outside the CD cavity is corroborated by NMR analyses.[104]

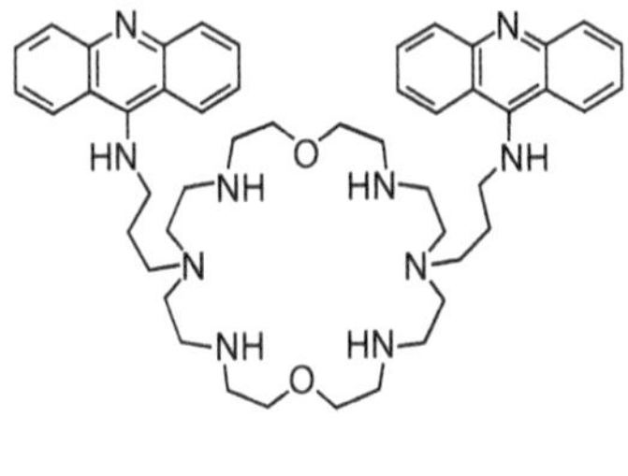

**32**

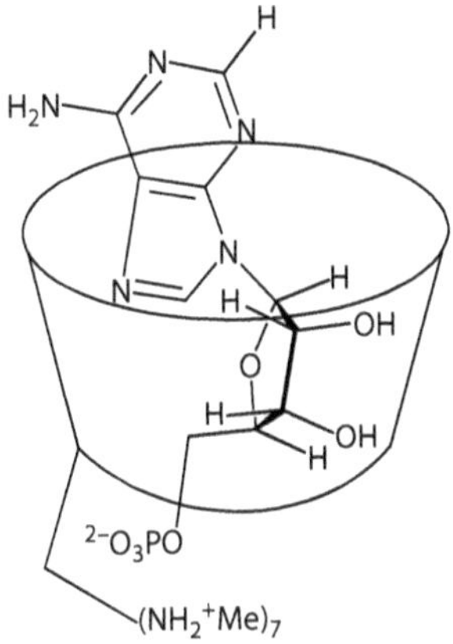

**33**

The porphyrin tweezer **34** is one of the few receptors that complex nucleotides equally as well as the electroneutral nucleosides because of the dominating stacking interaction in comparison to ion pairing. Thus, receptor **34**, exhibiting significant changes in the UV Soret band, shows a large affinity with, for example, cytidine, reaching lg $K$ = 4.7 in 0.3 M buffer and the same value with dCMP$^{2-}$. There are, however, no significant differences between different nucleobases.[105] The bisintercaland **35** exhibits higher affinities for guanosine derivatives than for other nucleotides with a decrease in the absorption and fluorescence spectra of cyclophane. The monophosphates AMP, GMP, and CMP are complexed with lg $K$ values of 4 ± 0.2, but the diphosphates show, with lg $K$ = 6.8 for GTP, a larger difference to, for example, ATP (lg $K$ 5.4).[106] Pyrenophanes rendered water soluble by attached polycationic or amphiphilic side chains complex ATP and GTP with $K = 1 \times 10^6$ [M$^{-1}$], CTP with $0.3 \times 10^6$ [M$^{-1}$], and UTP with $0.8 \times 10^6$ [M$^{-1}$]. The diphosphates are, as usual, bound approximately 500–1000 times less.[107] Similar results were obtained with anthraquinone or acridine-containing cyclophanes.[108]

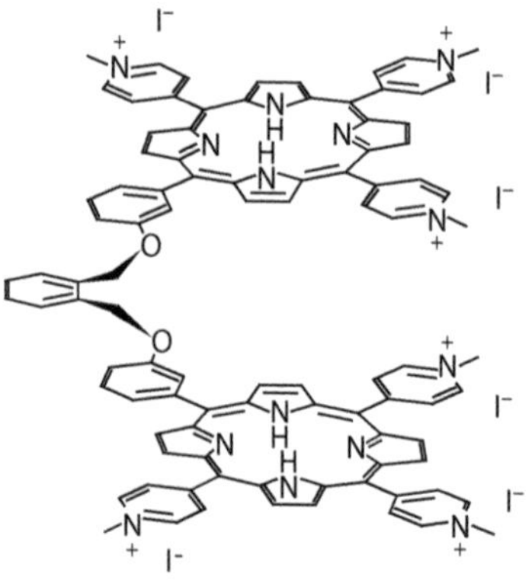

34

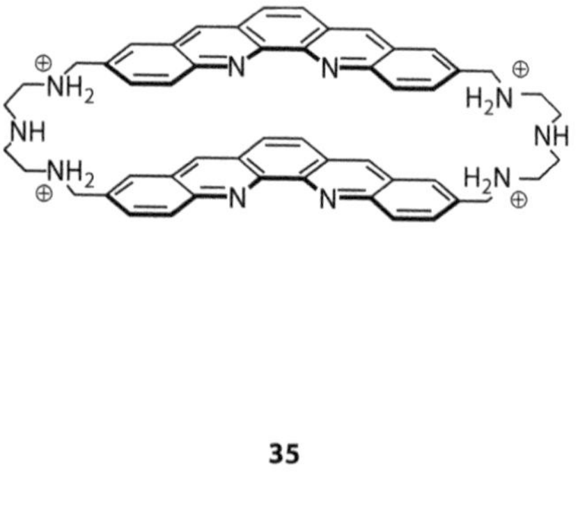

35

Many receptors show only small selectivity with respect to the nucleobase. A new acridine-based sensor is remarkably sensitive, exhibiting binding constants of $K = (2 \pm 0.7) \times 10^6$ [M$^{-1}$], with the same value for the triphosphates ADP, GDP, and UDP, however.[109] With host **36**, one finds preference for G with significant fluorescence quenching by the guanine base; the binding constants are for GTP, ATP, ADP, and AMP 87,000, 15,000, 610, and 120 M$^{-1}$, respectively.[110] The Cu(II)-dimetallic derivative of cryptate **37** exhibits selective recognition of GMP with respect to other nucleoside monophosphates, tentatively ascribed to the capability of GMP to match, with its phosphonate and enolate oxygen atoms, the distance between the two CuII ions inside the cryptate. The assay relies on the displacement of the indicator 6-carboxyfluorescein, which releases a yellow fluorescent emission.[111]

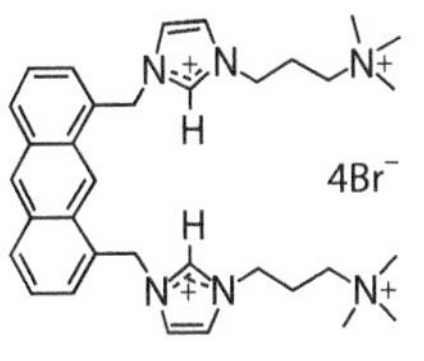

36 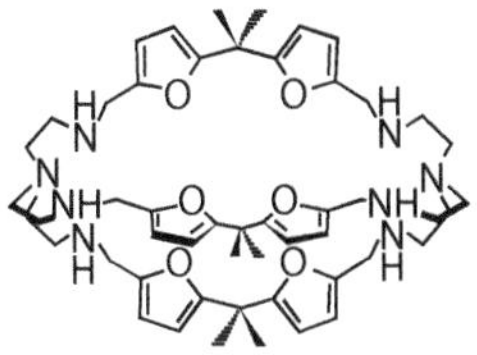          37 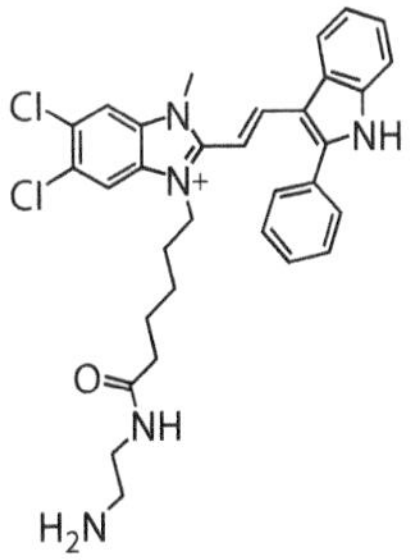          38

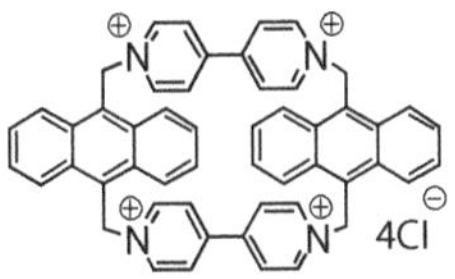

39 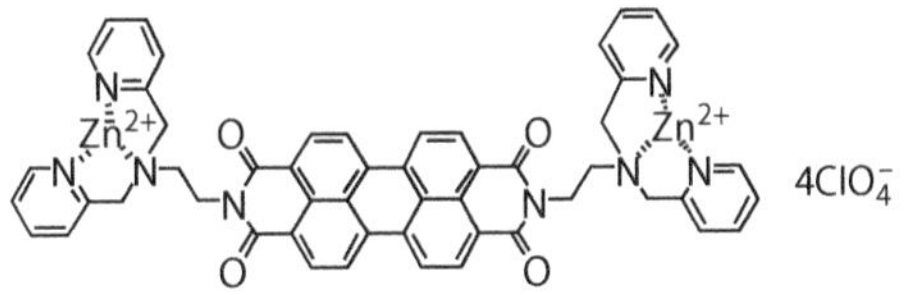          40

Host **38** performs as rather selective sensor for GTP with an 80-fold fluorescence increase; in contrast, only twofold (with ATP) smaller changes were observed for all other analytes. This indicates that the 2-hydroxyl group of GTP is crucial for the strong interaction.[112] The cyclophane **39** discriminates GTP from ATP and other nucleotides and nucleosides through an *on–off–on* fluorescence, using the fluorophore 8-hydroxy-1,3,6-pyrene trisulfonate for indicator displacement.[113] Host **40** is remarkably selective for UTP and UDP, displaying significant fluorescence enhancements, while a negligible fluorescence emission is promoted by CTP. In contrast, ATP, GTP, ADP, AMP, and UMP even induce some fluorescence decrease. The observed fluorescence enhancement is associated with UTP/UDP-promoted strengthening of the $Zn^{2+}$ coordination.[114] A recently described aminonaphthalimide imidazolium podand can be used for fluorescent imaging of ADP and ATP in living cells.[115]

The benzene-bridged tweezer-like host **41** with pyrene as an excimer and imidazolium as a phosphate receptor exhibits strong fluorescence only in the presence of ATP because of stacking of the encapsulated adenine residue. Other bases such as those from GTP, CTP, UTP, and TTP interact only from the outside with the already stabilized stacked pyrene–pyrene dimer, resulting in excimer fluorescence quenching.[116] A hairpin peptide **42** with 12 residues (acetyl-Arg-Trp-Val-Lys-Val-Asn-Gly-Orn-Trp-Ile-Lys) can detect ATP via fluorescence quenching of the tryptophan moiety with $K = 6 \times 10^3$ [M$^{-1}$]. In the presence of NaCl (0.2 M), the binding constant decreases eightfold.[117] A combinatorial search for an ATP selector with fluorescence emission, which is also based on peptide hairpins with 5-carboxyfluorescein appended as fluorophores to the ends of the peptide chain, yielded a Ser-Tyr-Ser derivative that binds ATP with $K = 3 \times 10^4$ [M$^{-1}$], whereas ADP and GTP lead to much smaller fluorescence emission.[118]

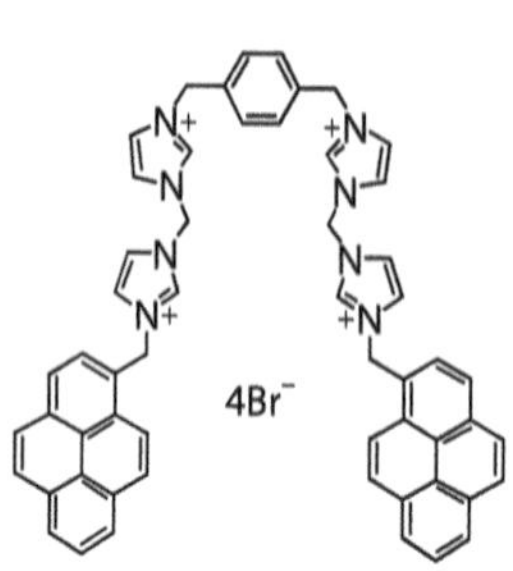

**41**

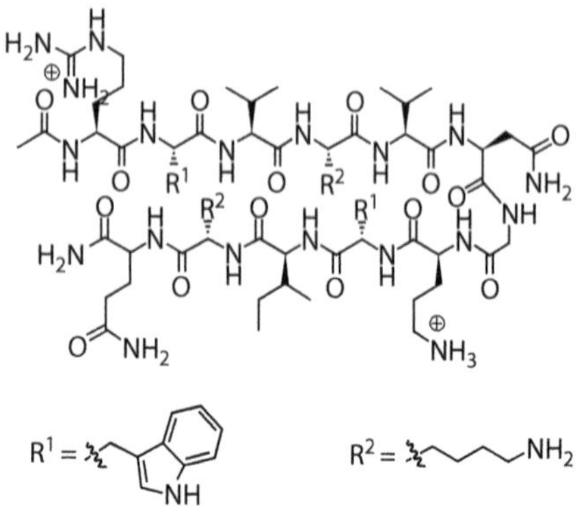

**42**

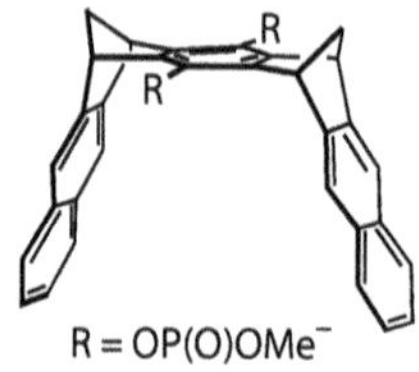

$R^1 = $

$R^2 = $

**43**

R = OP(O)OMe⁻

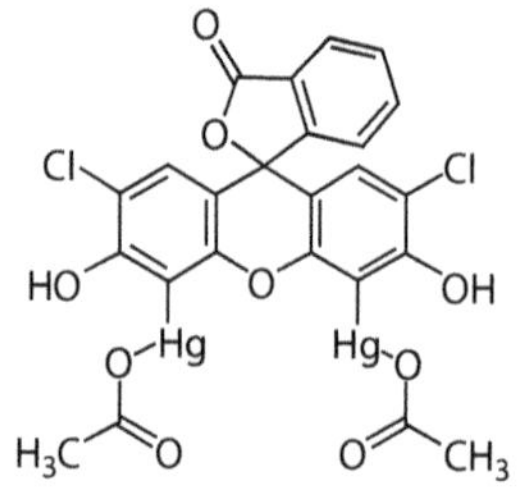

**44**

Recognition of NAD⁺ and related compounds is accomplished with the cleft **43**, exhibiting an association constant of 6500 [M⁻¹]; remarkably, complexation in methanol is weaker than in water, which demonstrates that van der Waals interactions such as the cation-π effect contribute more than the ion pairing with the anionic phosphates.[119] The fluorescein mercury derivative **44** recognizes NADH by two mercury metal ions as a result of metal–anion interaction with fluorescence changes. The receptor can detect a NADH concentration of approximately 1 μM with a large selectivity over other anions, including NAD⁺.[120] Imprinted polymer films in combination with QCM detection were found to discriminate adenine from structurally related analytes such as 2-aminopurine or guanine, with a detection limit of 5 nM adenine.[121]

## CARBOHYDRATES

In view of their notoriously weak intermolecular interactions, the detection of carbohydrates poses a particular challenge for the design of suitable receptors.[122] As the important hydrogen bonding suffers from the strongly competing bulk medium, extraction of saccharides out of water into a lipophilic phase such as chloroform was one of the early methods, mostly based on complexes with, for example, the resorcarene derivative **45**. Many studies with the aim of understanding the mechanism of carbohydrate complexation have been done in apolar solvents with the aid of saccharides, which were made soluble by substitution with long alkyl chains, for example, with receptors such as **46**. Characteristically binding constants with such saccharides drop from, for instance, $K = 10^5$ [$M^{-1}$] in chloroform by at least two orders of magnitude in chloroform solutions containing just 10% methanol.[123] In chloroform, even open-chain host compounds such as **47** bind, for example, n-octyl-b-D-glucopyranoside with the remarkably high constant of $K = 1.4 \times 10^5$ [$M^{-1}$].[124] Introduction of stronger anionic hydrogen bond acceptors in cyclophanes improves affinities, but they still work only with alkyl glycosides and in rather unpolar media.[125] With anionic sugars, one can, with cationic receptors such as **9**, achieve binding constants, which obviously are pH-dependent and reach from $6 \times 10^3$ with, for example, galacturonic acid to $25 \times 10^3$ with glucose-1-phosphate at pH 4, dropping to $13 \times 103$ at pH 7.4.[126]

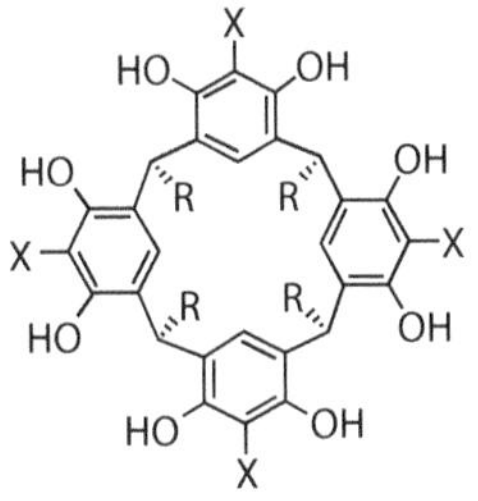

R = (CH$_2$)$_2$SO$_3$Na; X = H

**45**

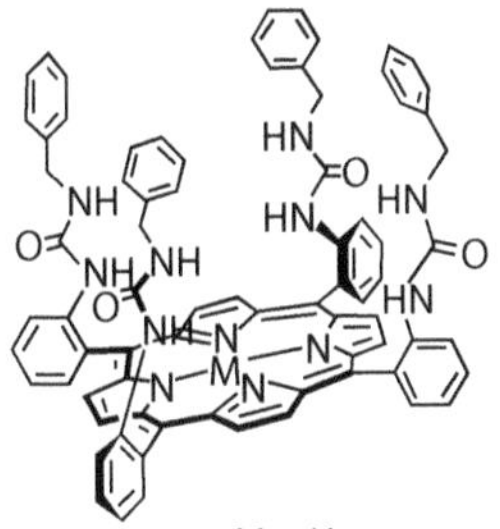

M = H$_2$

**46**

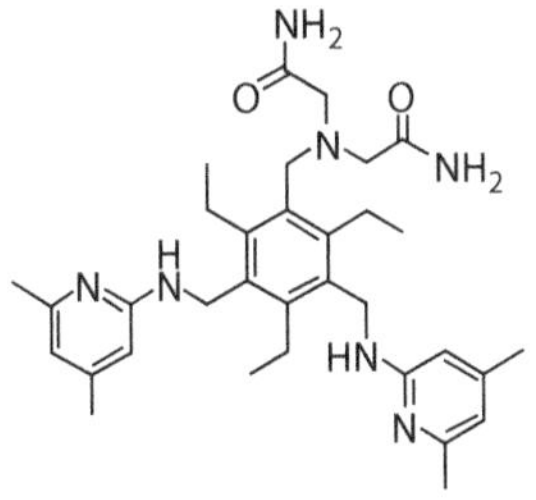

**47**

Relatively few receptors have been described until now for the efficient complexation of nonionic carbohydrates in aqueous media.[122] Binaphthyl-substituted macrocycles with a calixarene core bind in water/acetonitrile (1:1), for example, glucose with $K = 100$ [$M^{-1}$] and maltose with up to 230. Selectivity between disaccharides and trisaccharides is governed by the cavity size. Methyl red can be used here for a colorimetric IDA assay.[127] With a porphyrin core, significantly enhanced affinities can be achieved with natural sugars in pure water.[128] The porphyrin rings, which have the advantage of exhibiting analyte-sensitive UV absorptions, can be equipped with covalently linked UV/vis or fluorescence units for optical detection and can also be used in SPR methods.[129] Even relatively simple derivatives such as **48** with R = SO$_3^-$ show, for example, lg $K = 1.4$ for galactose and nearly the same value for maltotriose. Attached peptide chains such as R = CONH-Lys-Ala-AspOMe, OH, however, show for maltotriose the impressive value of lg $K = 4.85$, whereas other tripeptides lead to significantly smaller affinities. This indicates the importance of biomimetic interactions with complementary amino acid sequences. Bile acid-porphyrin conjugates **48** (R = bile acids linked by ethylenediamine) complex, for example, glucose even with lg $K = 5.7$ and lend themselves for fluorimetric recognition of saccharides typically present on malignant tumor cells.[130] In this respect, it is interesting that even small peptides can recognize, with enhanced fluorescence emission, sugars with relatively high affinity; for example, the pentapeptide WGDEY binds erythrose in water with $3.5 \times 10^4$ and galactose with $0.5 \times 10^4$ [$M^{-1}$]. Studies with a large combinatorial peptide library show that aromatic residues are necessary for stacking, and acidic residues are needed for hydrogen-bonding interactions.[131] Selection of disulfide-bridged macrocyclic peptides for saccharide complexation was achieved with a dynamic combinatorial library. The optimal macrocycles are able to bind the neurotransmitter N-acetyl neuraminic acid or the disaccharide trehalose in water at millimolar concentrations.[132]

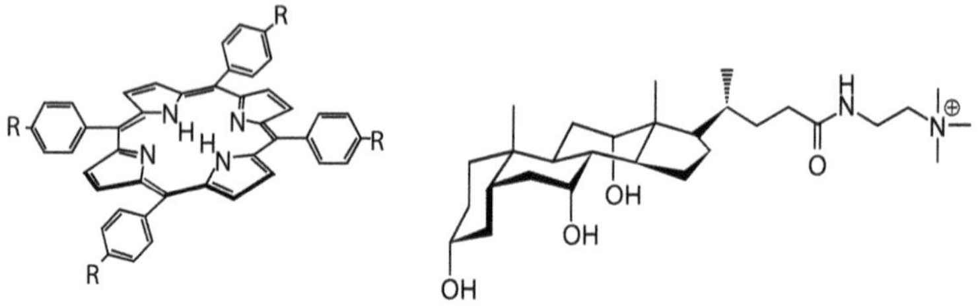

**48**

The tricyclic cage receptor **49** reduces competition of bulk water inside the cavity and binds 1-O-methyl-b-D-cellobiose with an association constant of approximately 900 [M$^{-1}$], which comes close to those of natural lectins. Also, the selectivity is similar to that of lectin; the affinity for lactose, for example, an epimer of cellobiose with a single axially oriented OH group, is with **49** almost two orders of magnitude lower.[133] Modified derivatives of **49** show enhanced affinities but still only up to $K = 60$ [M$^{-1}$] for glucose.[134]

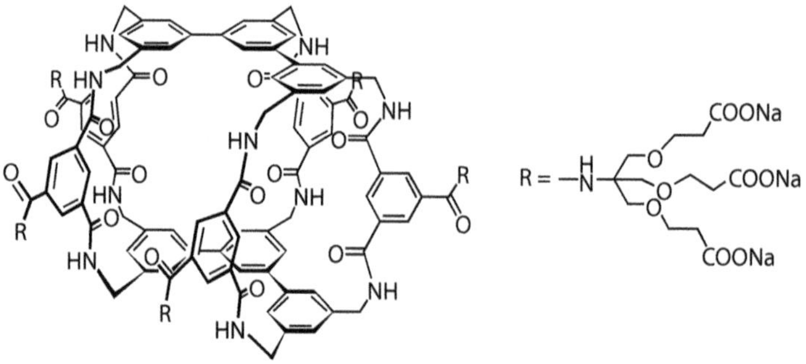

**49**

Host compounds such as **50** make use of the relatively strong affinity of lanthanide ions toward saccharides and display increased fluorescence with saccharides in millimolar concentrations. Also, lysophosphatidic acid, a biomarker for ovarian cancer, for example, is selectively detected by the europium complex via a fluorescence increase in human plasma samples.[135] Dinuclear copper complexes are also promising receptors for monosaccharides although in alkaline solution.[136] The apparent binding strength of complexes between unprotected disaccharides and the dinuclear Cu(II) host **51** ranges between $10^3$ and $10^4$ [M$^{-1}$]; the selectivities are not directly correlated with the X-ray–determined Cu–Cu distances.[137]

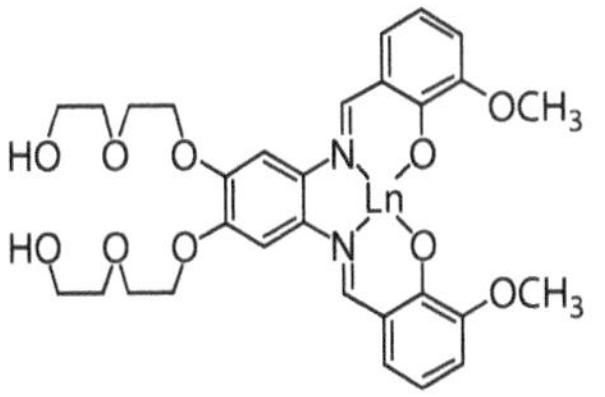

50

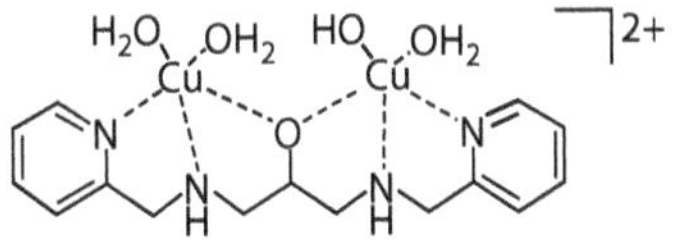

51

## BORONIC ESTER REACTIONS FOR CARBOHYDRATE ANALYSIS

Fast and reversible covalent reactions have been most often applied to the analyses of carbohydrates, invariably via boronic ester formation.[138] Receptors can be tailored to react particularly fast with glucose.[139] Such so-called boronolectins, also called synthetic lectins, are of obvious diagnostic interest for medicine.[140] Miniaturization of boronic ester–based sensing has been already reviewed.[141] The principle is illustrated with receptor **52**, which is suitable for the fluorimetric detection of glucose at physiological levels by PET.[142] The reaction of glucose leads to macrocycle **53**, which places the sugar unit above the anthracene face as monitored via NMR and fluorescence quenching.

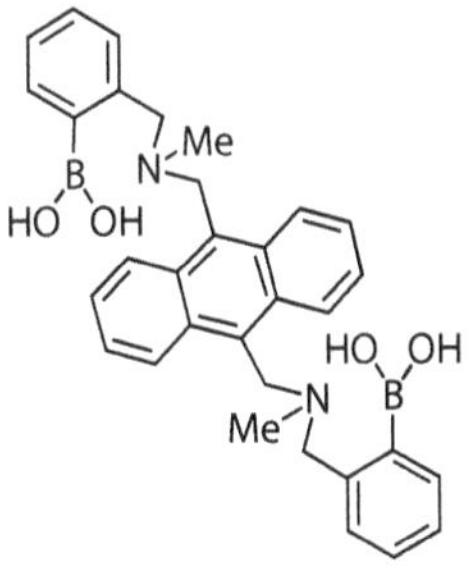

52

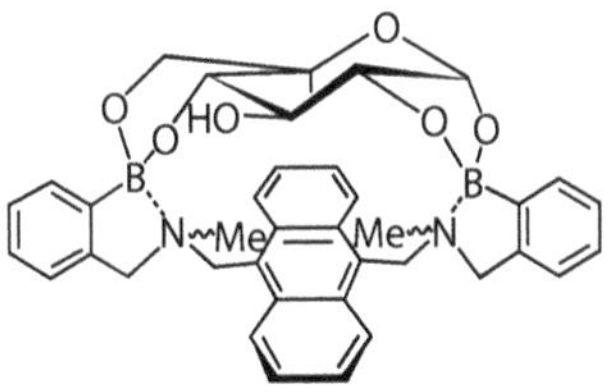

53

Resorcarenes decorated with boronic ester residues at the rim can visually distinguish structurally related saccharides, including glucose phosphates and amino and carboxylic acid sugars by different colors.[143] Related resorcarene-based boronic acids such as **54** exhibit a high colorimetric fructose selectivity. In contrast, at physiological levels, selective glucose monitoring can be achieved via fluorescence.[144] The boronic acid–functionalized rhodamine derivative **55** is selective for ribose, adenosine, nucleotides, nucleosides, and congeners.[145]

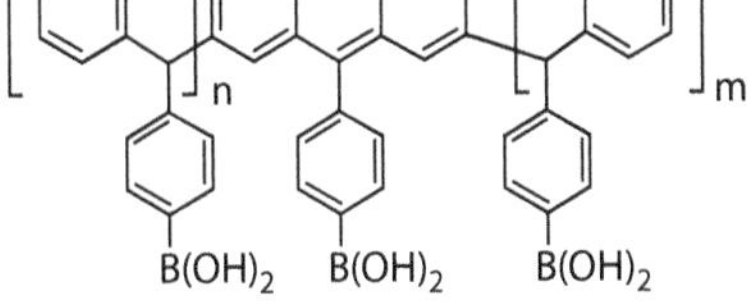

54

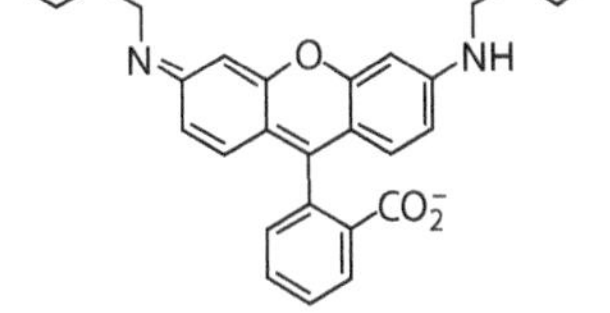

55

Solid-supported pentapeptide diboronic acids present hybrid systems such as **56**, which combine covalent interactions provided by the boronic acids with the noncovalent interactions delivered by the backbone and side chains of the peptides.[146] Measurements at pH 10.7 with covalently bound alizarine as a competitive indicator show considerable variation in carbohydrate reaction strengths with association constants in the range of 60–5300 $[M^{-1}]$. The strongest D-glucose binding ($K = 3.6 \times 10^3\,[M^{-1}]$) was achieved with the sequence N-Ac-BPA-BPA-Ala-Arg-Arg-AHA-, in which BPA is 4-borono-L-phenyl-alanine, and AHA is 6-aminohexanoic acid. With nucleotides, a sevenfold preference for CMP over AMP was shown with the same peptide sequence; other sequences exhibited enantioselectivity with an 8.4-fold binding preference for *L*-glucose over *D*-glucose.

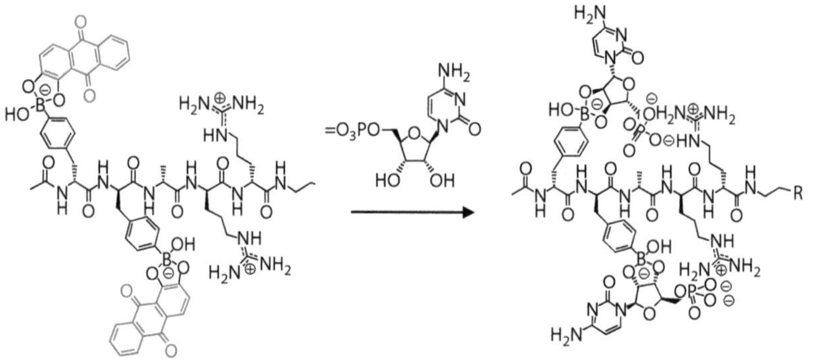

**56**

New sensors based on the fluorophore **57** exhibit increased emission; they also show improved selectivity. Thus, binding constants are for D-Glucose 1970, for D-Galactose 115, for D-Fructose 132, and for D-Mannose 19 [M-1].[147] Vesicles based on 2-(hexadecyloxy)-naphthalene-6-boronic acid can be used as fluorescent miniature detectors.[148] The incorporation of boronic acid–based sensors into nanoparticles has also been reported.[149] Photonic borax crystals in crystalline colloidal arrays may also be used for glucose sensing.[150] FRET between CdSe–ZnS quantum dots and fluorophore-labeled galactose or fluorophore-labeled dopamine linked to phenyl boronic acid–functionalized quantum dots was used as a competitive assay for optical detection of galactose, glucose, or dopamine.[151] Boronic acid derivatives were immobilized for sugar detection on the surface of FETs[152] or on piezoresistive microcantilevers.[153]

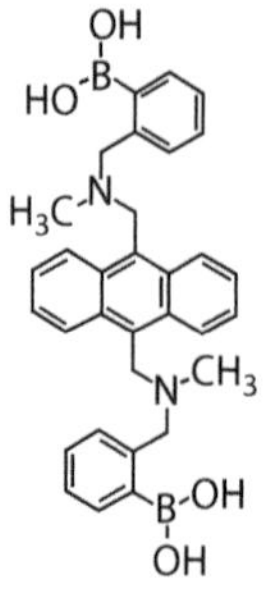

**57**

## Biopolymers

Interactions of proteins and other biopolymers with synthetic receptors are discussed in depth in Chapter 15. Enzyme assays as well as, for example, nucleic acid interactions are dealt with in numerous books and special reviews. DNA biochips[154] and, more recently, protein-based microarrays represent the commercially most important application of supramolecular complexation. Protein biochips require selective immobilization of a protein, most often by covalent reactions but also by usually more gentle so-called bioaffinity methods.[155] Besides strong avidin–biotin interactions with biotinylated proteins, one can make use of metal chelation; for example, proteins with a (His)6 tag at the C- or N-terminus can be immobilized via a nickel-chelated complex such as Ni-nitriloacetic acid. In the present section, we will only illustrate with a few examples how classical supramolecular complexations are applied to the analyses of large natural molecules. Often one encounters the same principles as those used for smaller analytes.

Heparin, which is a biopolymer with iduronic acid and glucosamine units, can be measured in serum down to nanomolar concentrations with a fluorescent sensor based on a 1,3,5-triphenyl ethynylbenzene core with side arms containing boronic acid and ammonium groups.[156]

A new concept for enzymatic activity assays relies on macrocycles such as **10**, which binds an enzyme substrate weakly but the product strongly. The macrocycle with a complexed fluorescence dye inside the cavity is used as a reporter unit. The weakly binding substrate cannot displace the dye from the complex, but the product as a strong competitor displaces the fluorescent dye from the macrocycle. As an example, dapoxyl was used as dye, which formed a strong complex with cucurbituril **10** ($n = 7$), which is up to 200 times more strongly fluorescent than the free dye. The activity of the amino acid decarboxylase can then be conveniently followed because the free (protonated) amines as product bind much more strongly to the receptor than the dye.[157]

A quite sensitive protein assay makes use of meso-tetraphenylporphyrin (TPP) and meso-tetraphenylporphyrin cobalt(II) (CoTPP) in the presence of the cyclodextrin derivative heptakis (2,6-di-O-$n$-octyl)-β-cyclodextrin. In the presence of the CD, a significant increase in TPP fluorescence is observed, but the increased fluorescence is quenched by CoTPP. The addition of bovine serum albumin (BSA) or human serum albumin (HSA) restores the fluorescence emission with detection limits of 0.32 μg/ml for BSA and 1.06 μg/ml for HSA, which can be measured this way in human serum.[158]

Tetra-anionic receptors based on calixarenes similar to **8**, which bear anionic head groups at the upper rim and nonpolar butoxy tails at the lower rim, behave as amphiphiles and can be incorporated into lipid monolayers. Such doped layers can be used for protein sensing at the air–water interface, as they attract peptides and proteins from the aqueous subphase. Addition of basic peptides into the aqueous subphase produces distinct additional expansions in pressure/area diagrams, for example, with tri-arginine even at $10^{-7}$ M concentration. Arginine-rich proteins such as cytochrome c can lead to pronounced shifts in their pressure/area diagrams even at the nanomolar level.[159]

A new assay of the protein concanavalin (ConA) is based on the formation of weakly fluorescent noncovalent complexes of ConA with various saccharide-appended porphyrins as pseudosubstrates. These macrocycles are highly aggregated in aqueous media, resulting in diminished fluorescence. The addition of ConA leads to deaggregation by formation of a noncovalent complex with the lectin accompanied by a strong increase in fluorescence. Subsequent addition of a natural substrate (e.g., D-mannose), which has a higher affinity for ConA, leads to the release of the weakly bound pseudosubstrate with concomitant reaggregation and quenching of fluorescence.[160]

Nucleic acids belong to the most important targets for molecular recognition, mostly in the context of drug design. For leading references, we mention here only one recent paper, which highlights the potential of multivalent association with spectroscopic probes. The guanidiniocarbonyl-pyrrole–pyrene **58** interacts strongly with dsDNA but only weakly with dsRNA. The heterocycles in the central part of the structure control the alignment of the molecules within different forms of dsDNA minor grooves by steric effects and hydrogen bonding. This leads to small differences in the position of the attached pyrene units within the groove, which is reflected in different fluorimetric responses. Such compounds also exhibit *in vitro* selectivity among various human tumor cell lines.[161]

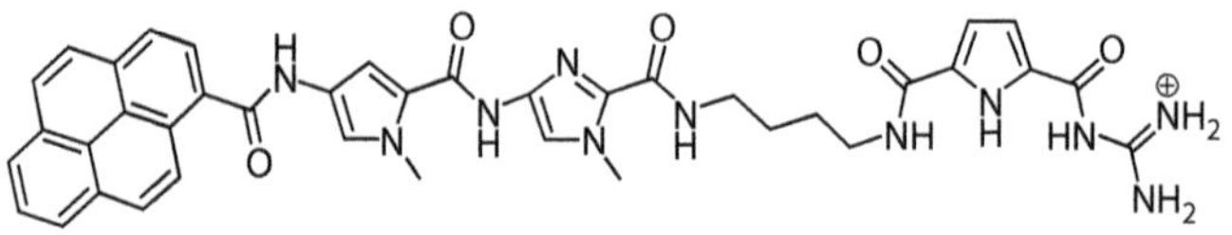

**58**

## OUTLOOK

The reader will notice that most publications discussed in this chapter stem from quite recent years. This demonstrates that the application of supramolecular chemistry for the analysis of compounds that are of environmental, industrial, biological, or medicinal interest is in very active development. The challenge is to design highly sensitive and selective host systems, which allow detection preferably by optical or electrochemical methods, for all kinds of analytes. These can replace more complicated separation techniques and will be of particular importance not only for environmental control but also for so-called bedside monitoring in medicine. The possibilities of modifying the selectivity of synthetic receptors are virtually endless; in comparison to enzymatic assays, for example, they have the advantage of higher stability.

## REFERENCES

1. Yoshio, M.; Noguchi, H.; Yoshio, M.; Ugamura, M.; Noguchi, H.; Nagamatsu, M. *Anal. Lett.* 1978, *11*, 281–286; Blasius, E.; Janzen, K. P. *Top. Curr. Chem.* 1981, *98*, 163–189.
2. Bell, J. W.; Hext, N. M. *Chem. Soc. Rev.* 2004, *33*, 589–598; Anslyn, E. V. *J. Org. Chem.* 2007, *72*, 687–699; Santana Rodriguez, J. J.; Halko, R.; Betancort Rodriguez, J. R.;

Aaron, J. J. *Analyt. Bioanalyt. Chem.* 2006, *385*, 525–545; Joyce, L. A.; Shabbir, S. H.; Anslyn, E. V. *Chem. Soc. Rev.* 2010, *39*, 3621–3632.

3. Lubin, A. A.; Plaxco, K. W. *Acc. Chem. Res.* 2010, *43*, 496–505.
4. Basabe-Desmonts, L.; Reinhoudt, D. N.; Crego-Calama, M. *Chem. Soc. Rev.* 2007, *36*, 993–1017.
5. Oshovsky, G. V.; Reinhoudt, D. N.; Verboom, W. *Angew. Chem. Int. Ed. Engl.* 2007, *46*, 2366–2393.
6. Schneider, H.-J.; Yatsimirsky, A. K. *Chem. Soc. Rev.* 2008, *37*, 263–277.
7. Kopelman, R.; Dourado, S. *Proc. SPIE Int. Soc. Opt. Eng.* 1996, *2836*, 2–11.
8. Schneider, H.-J.; Tianjun, L.; Lomadze, N. *Chem. Comm.* 2004, 2436–2437.
9. Schneider, H.-J. *Angew. Chem. Int. Ed. Engl.* 2009, *48*, 3924–3977.
10. Schneider, H.-J.; Kato, K. *Angew. Chem. Int. Ed. Engl.* 2007, *46*, 2694–2696.
11. Gorris, H. H.; Walt, D. R. *Angew. Chem. Int. Ed. Engl.* 2010, *49*, 3880–3895.
12. Lee, Y. E. K.; Kopelman, R. *Wiley Interdiscip. Rev., Nanomed.* 2009, *1*, 98–110.
13. Swager, T. M. *Acc. Chem. Res.* 1998, *31*, 201–207.
14. Fabbrizzi, L.; Licchelli, M.; Taglietti, A. *Dalton Trans.* 2003, 3471–3479; Fabbrizzi, L.; Poggi, A. *Chem. Soc. Rev.* 1995, *24*, 197–202.
15. Cramer, F.; Saenger, W.; Spatz, H.-C. *J. Am. Chem. Soc.* 1967, *89,* 14–20; Saenger, W. *Angew. Chem. Int. Ed. Engl.* 1980, *19*, 344–362.
16. Wiskur, S. L.; Ait-Haddou, H.; Lavigne, J. J.; Anslyn, E. V. *Acc. Chem. Res.* 2001, *34*, 963–972; Kitamura, M.; Shabbir, S. H.; Anslyn, E. V. *J. Org. Chem.* 2009, *74*, 4479–4489.
17. Anzenbacher, P.; Lubal, P.; Bucek, P.; Palacios, M. A.; Kozelkova, M. E. *Chem. Soc. Rev.* 2010, *39*, 3954–3979; Lavigne, J. J.; Anslyn, E. V. *Angew. Chem. Int. Ed. Engl.* 2001, *40*, 3119–3130; Wright, A. T.; Anslyn, E. V. *Chem. Soc. Rev.* 2006, *35*, 14–28; Ciosek, P.; Wroblewski, W. *Analyst.* 2007, *132*, 963–978.
18. Lehn, J. M. *Chem. Eur. J.* 1999, *5*, 2455–2463; Corbett, P. T.; Leclaire, J.; Vial, L.; West, K. R.; Wietor, J. L.; Sanders, J. K. M.; Otto, S. *Chem. Rev.* 2006, *106*, 3652–3711.
19. Christensen, C. B. V. *Talanta.* 2002, *56*, 289–299.
20. Schubert, P.; Hoffman, M. D.; Sniatynski, M. J.; Kast, J. *Anal. Bioanal. Chem.* 2006, *386*, 482–493; Panicker, R. C.; Chattopadhaya, S.; Yao, S. Q. *Analyt. Chim. Acta.* 2006, *556*, 69–79; Jander, G.; Barth, C. *Trends Plant Sci.* 2007, *12*, 203–210.
21. Liang, C. H.; Wu, C. Y. *Expert Rev. Proteomics.* 2009, *6*, 631–645; Jelinek, R.; Kolusheva, S. *Chem. Rev.* 2004, *104*, 5987–6015.
22. Beske, O. E.; Goldbard, S. *Drug Discov. Today.* 2002, *7*, S131–S135; Henderson, G.; Bradley, M. *Curr. Opin. Biotechnol.* 2007, *18*, 326–330.
23. Shimizu, K. D.; Stephenson, C. J. *Curr. Opin. Chem. Biol.* 2010, *14*, 743–750; Moreno-Bondi, M. C.; Navarro-Villoslada, F.; Benito-Pena, E.; Urraca, J. L. *Curr. Anal. Chem.* 2008, *4*, 316–340.
24. Liedberg, B.; Nylander, C.; Lundstrom, I. *Sensors Actuators.* 1983, *4*, 299–304.
25. Mabeck, J. T.; Malliaras, G. G. *Anal. Bioanal. Chem.* 2006, *384*, 343–353.
26. Li, F.; Nathan, A.; Wu, Y.; Ong, B.S. *Organic Thin Film Transistor Integration: A Hybrid Approach,* Wiley-VCH Weinheim, 2011; Bao, Z.; Locklin, J. *Organic Field-Effect Transistors,* CRC Press: Boca Raton, 2007.
27. Petralia, S. *Sensors Transducers J.* 2011, *5,* 115–122.
28. Lubczyk, D.; Siering, C.; Lorgen, J.; Shifrina, Z. B.; Müllen, M.; Waldvogel, S. R. *Sensors Actuators B Chem.* 2011, *143*, 561–566.
29. Mohr, G. J. *Anal. Bioanal. Chem.* 2006, *386*, 1201–1214.
30. Holloway, A. F.; Nabok, A.; Hashim, A. A.; Penders, J. *Sensors Transducers J.* 2010, *113*, 71–81.
31. Pinalli, R.; Suman, M.; Dalcanale, E. *Eur. J. Org. Chem.* 2004, 451–462; Melegari, M. et al. *Chem. Eur. J.* 2008, *14*, 5772–5779.

32. Kimura, M.; Sugawara, M.; Sato, S.; Fukawa, T.; Mihara, T. *Chem. Asian J.* 2010, *5*, 869–876.
33. Yoshizawa, M.; Klosterman, J. K.; Fujita, M. *Angew. Chem. Int. Ed. Engl.* 2009, *48*, 3418–3438; Klosterman, J. K.; Yamauchi, Y.; Fujita, M. *Chem. Soc. Rev.* 2009, *38*, 1714–1725; Dalgarno, S. J.; Power, N. P.; Atwood, J. L. *Coord. Chem. Rev.* 2008, *252*, 825–841; Holliday, B. J.; Mirkin, C. A. *Angew. Chem. Int. Ed. Engl.* 2001, *40*, 2022–2043; Northrop, B. H.; Yang, H. B.; Stang, P. J. *Chem. Commun.* 2008, 5896–5908; Würthner, F.; You, C. C.; Saha-Moller, C. R. *Chem. Soc. Rev.* 2004, *33*, 133–146; Amijs, C. H. M.; van Klink, G. P. M.; van Koten, G. *Dalton Trans.* 2006, 308–327.
34. Kumar, A.; Sun, S. S.; Lees, A. J. *Coord. Chem. Rev.* 2008, *252*, 922–939.
35. Benkstein, K. D.; Hupp, J. T.; Stern, C. L. *Angew. Chem. Int. Ed. Engl.* 2000, *39*, 2891–2893.
36. McQuade, D. T.; Pullen, A. E.; Swager, T. M. *Chem. Rev.* 2000, *100*, 2537–2574; Swager, T. M. *Acc. Chem. Res.* 2008, *41*, 1181–1189.
37. Richardson, S.; Barcena, H. S.; Turnbull, G. A.; Burn, P. L.; Samuel, I. D. W. *Appl. Phys. Lett.* 2009, *95*, 0633051–0633053.
38. Liu, Y; Wang, K. R.; Guo, D. S.; Jiang, B. P. *Adv. Funct. Mater.* 2009, *19*, 2230–2235.
39. Park, J. S.; Le Derf, F.; Bejger, C. M.; Lynch, V. M.; Sessler, J. L.; Nielsen, K. A.; Johnsen, C.; Jeppesen, J. O. *Chem. Eur. J.* 2010, *4, 6 16*, 848–854.
40. Costero, A. M.; Gil, S.; Parra, M.; Mancini, P. M. E.; Martinez-Manez, R.; Sancenon, F.; Royo, S. *Chem. Commun.* 2008, 6002–6004.
41. Dale, T. J.; Rebek, J. *Angew. Chem. Int. Ed. Engl.* 2009, *48*, 7850–7852.
42. Rathfon, J. M.; Al-Badri, Z. M.; Shunmugam, R.; Berry, S. M.; Pabba, S.; Keynton, R. S.; Cohn, R. W.; Tew, G. N. *Adv. Funct. Mater.* 2009, *19*, 689–695.
43. Ariga, K.; Richards, G. J.; Ishihara, S.; Izawa, H.; Hill, J. P. *Sensors.* 2010, *10*, 6796–6820.
44. Xu, X. D.; Chen, C. S.; Lu, B.; Cheng, S. X.; Zhang, X. Z.; Zhuo, R. X. *J. Phys. Chem. B.* 2010, *114*, 2365–2372.
45. Chen, X.; Huang, Z.; Chen, S. Y.; Li, K.; Yu, X. Q.; Pu, L. *J. Am. Chem. Soc.* 2010, *132*, 7297–7299; for another example of an enantioselective chemomechanical polymer, see Schneider, H.-J.; Kato, K. *Angew. Chem. Int. Ed. Engl.* 2007, *46*, 2694–2696.
46. Liu, H. L.; Peng, Q.; Wu, Y. D.; Chen, D.; Hou, X. L.; Sabat, M.; Pu, L. *Angew. Chem. Int. Ed. Engl.* 2010, *49*, 602–606; Yu, S.; Pu, L. *J. Am. Chem. Soc.* 2010, *132*, 17,698–17,700.
47. Sansone, F.; Baldini, L.; Casnati, A.; Ungaro, R. *New J. Chem.* 2010, *34*, 2715–2728.
48. Schneider, H.-J.; Güttes, D.; Schneider, U. *J. Am. Chem. Soc.* 1988, *110*, 6449–6454.
49. Bakirci, H.; Nau, W. M. *Adv. Func. Mater.* 2006, *16*, 237–242.
50. Koner, A. L.; Schatz, J.; Nau, W. M.; Pischel, U. *J. Org. Chem.* 2007, *72*, 3889–3895.
51. Schmuck, C.; Schwegmann, M. *Org. Biomol. Chem.* 2006, *4*, 836–838.
52. Baumes, L. A.; Sogo, M. B.; Montes-Navajas, P.; Corma, A.; Garcia, H. *Chem. Eur. J.* 2010, *16*, 4489–4495.
53. Zhao, J. Z.; Kim, H. J.; Oh, J.; Kim, S. Y.; Lee, J. W.; Sakamoto, S.; Yamaguchi, K.; Kim, K. *Angew. Chem. Int. Ed. Engl.* 2001, *40*, 4233–4235.
54. Li, C. F.; Du, L. M.; Wu, W. Y.; Sheng, A. Z. *Talanta.* 2009, *80*, 1939–1944; The use of cucurbiturils has been patented: United States Patent 6,365,734.
55. Siering, C.; Beermann, B.; Waldvogel, S. R. *Supramol. Chem.* 2006, *18*, 23–27.
56. Goto, H.; Furusho, Y.; Yashima, E. *Chem. Commun.* 2009, 1650–1652.
57. Li, Z. B.; Lin, J.; Sabat, M.; Hyacinth, M.; Pu, L. *J. Org. Chem.* 2007, *72*, 4905–4916.
58. Sindelar, V.; Cejas, M. A.; Raymo, F. M.; Chen, W. Z.; Parker, S. E.; Kaifer, A. E. *Chem. Eur. J.* 2005, *11*, 7054–7059.
59. Shahgaldian, P.; Pieles, U. *Sensors* 2006, *6*, 593–615.
60. Lu, M. L.; Zhang, W. G.; Zhang, S.; Fan, J.; Su, W. C.; Yin, X. *Chirality* 2010, *22*, 411–415.

61. Yashima, E.; Maeda, K.; Sato, O. *J. Am. Chem. Soc.* 2001, *123*, 8159–8160.
62. Hossain, M. A.; Mihara, H.; Ueno, A. *Bioorg. Med. Chem. Lett.* 2003, *13*, 4305–4308.
63. Arenas, L. F.; Ebarvia, B. S.; Sevilla, F. B. *Anal. Bioanal. Chem.* 2010, *397*, 3155–3158.
64. Freeman, R.; Li, Y.; Tel-Vered, R.; Sharon, E.; Elbaz, J.; Willner, I. *Analyst* 2009, *134*, 653–656.
65. Mohr, G. J.; Wenzel, M.; Lehmann, F.; Czerney, P. *Anal. Bioanal. Chem.* 2002, *374*, 399–402.
66. Coskun, A.; Akkaya, E. U. *Org. Lett.* 2004, *6*, 3107–3109.
67. Zhu, L.; Zhong, Z.; Anslyn, E.V. *J. Am. Chem. Soc.* 2005, *127*, 4260–4269.
68. Wiskur, S. L.; Anslyn, E. V. *J. Am. Chem. Soc.* 2001, *123*, 10,109–10,110.
69. Peczuh, M. W.; Hamilton, A. D. *Chem. Rev.* 2000, *100*, 2479–2493; Schneider, H.-J. *Adv. Supramol. Chem.* 2000, *6*, 185–216.
70. Herzog, G.; Arrigan, D. W. M. *Analyst* 2007, *132*, 615–632.
71. Peng, H.; Zhang, Y.; Zhang, J.; Xie, Q.; Nie, L.; Yao, S. *Analyst* 2001, *126*,189–164.
72. Morelli, I.; Chiono, V.; Vozzi, G.; Ciardelli, G.; Silvestri, D.; Giusti, P. *Sensors Actuators B Chem.* 2010, *150*, 394–401.
73. Huan, S. Y.; Shen, G. L.; Yu, R. Q. *Electroanalysis* 2004, *16*, 1019–1023.
74. Ait-Haddou, H.; Wiskur, S. L.; Lynch, V. M.; Anslyn, E. V. *J. Am. Chem. Soc.* 2001, *123*, 11,296–11,297.
75. Folmer-Andersen, J. F.; Lynch, V. M.; Anslyn, E. V. *J. Am. Chem. Soc.* 2005, *127*, 7986–7987.
76. Bonizzoni, M.; Fabbrizzi, L.; Piovani, G.; Taglietti, A. *Tetrahedron* 2004, *60*, 11,159–11,162.
77. Pagliari, S.; Corradini, R.; Galaverna, G.; Sforza, S.; Dossena, A.; Montalti, M.; Prodi, L.; Zaccheroni, N.; Marchelli, R. *Chem. Eur. J.* 2004, *10*, 2749–2758.
78. Späth, A.; König, B. *Tetrahedron 66*, 1859–1873.
79. Rajgariah, P.; Urbach, A. R. *J. Incl. Phenom. Macrocycl. Chem.* 2008, *62*, 251–254; Cong, H.; Tao, L. L.; Yu, Y. H. et al. *Asian J. Chem.* 2007, *19*, 961–964.
80. Heitmann, L. M.; Taylor, A. B.; Hart, P. J.; Urbach, A. R. *J. Am. Chem. Soc.* 2006, *128*, 12,574–12,581.
81. Reczek, J. J.; Kennedy, A. A.; Halbert, B. T.; Urbach, A. R. *J. Am. Chem. Soc.* 2009, *131*, 2408–2415.
82. Ngola, S. M.; Kearney, P. C.; Mecozzi, S.; Russell, K.; Dougherty, D. A. *J. Am. Chem. Soc.* 1999, *121*, 1192–1201.
83. Douteau-Guevel, N.; Perret, F.; Coleman, A. W.; Morel, J. P.; Morel-Desrosiers, N. *Perkin Trans. 2.* 2002, 524–532.
84. Fokkens, M.; Schrader, T.; Klarner, F. G. *J. Am. Chem. Soc.* 2005, *127*, 14,415–14,421.
85. Stettler, A. R.; Krattiger, P.; Wennemers, H.; Schwarz, M. A. *Electrophoresis* 2007, *28*, 1832–1838.
86. Hossain, M. A.; Schneider, H. J. *J. Am. Chem. Soc.* 1998, *120*, 11,208–11,209.
87. Sirish, M.; Schneider, H.-J. *Chem. Commun.* 1999, 907–908.
88. Mizutani, T.; Wada, K.; Kitagawa, S. *J. Am. Chem. Soc.* 1999, *121*, 11,425–11,431.
89. Kral, V.; Shishkanova, T. V.; Sessler, J. L.; Brown, C. T. *Org. Biol. Chem.* 2004, *2*, 1169–1175.
90. Burguete, M. I.; Galindo, F.; Luis, S. V.; Vigara, L. *J. Photochem. Photobiol. Chem.* 2010, *209*, 61–67.
91. Kim, Y. K.; Lee, H. N.; Singh, N. J.; Choi, H. J.; Xue, J. Y.; Kim, K. S.; Yoon, J.; Hyun, M. H. *J. Org. Chem.* 2008, *73*, 301–304; Ryu, D.; Park, E.; Kim, D. S. et al. *J. Am. Chem. Soc.* 2008, *130*, 2394–2395.
92. Mei, X. F.; Martin, R. M.; Wolf, C. *J. Org. Chem.* 2006, *71*, 2854–2861.
93. Tashiro, S.; Tominaga, M.; Kawano, M.; Therrien, B.; Ozeki, T.; Fujita, M. *J. Am. Chem. Soc.* 2005, *127*, 4546–4547.
94. Wehner, M.; Janssen, D.; Schäfer, G.; Schrader, T. *Eur. J. Org. Chem.* 2006, 138–153.

95. Jensen, K. B.; Braxmeier, T. M.; Demarcus, M.; Frey, J. G.; Kilburn, J. D. *Chem. Eur. J.* 2002, *8*, 1300–1309.
96. Shepherd, J.; Langley, G. J.; Herniman, J. M.; Kilburn, J. D. *Eur. J. Org. Chem.* 2007, 1345–1356.
97. Schmuck, C. *Coord. Chem. Rev.* 2006, *250*, 3053–3067.
98. Casnati, A.; Fabbi, M.; Pelizzi, N.; Pochini, A.; Sansone, F.; Ungaro, R.; DiModugno, E.; Tarzia, G. *Bioorg. Med. Chem. Lett.* 1996, *6*, 2699–2704.
99. Lim, S.; Escobedo, J. O.; Lowry, M.; Xu, X.; Strongin, R. *Chem. Comm.* 2010, *46*, 5707–5709.
100. Escobedo, J. O.; Wang, W. H.; Strongin, R. M. *Nat. Protocols* 2006, *1*, 2759–2762.
101. Pacsial-Ong, E. J.; McCarley, R. L.; Wang, W. H.; Strongin, R. M. *Anal. Chem.* 2006, *78*, 7577–7581.
102. Fenniri, H.; Hosseini, M. W.; Lehn, J. M. *Helv. Chim. Acta* 1997, *80*, 786–803.
103. Schneider, H.-J.; Blatter, T.; Palm, B.; Pfingstag, U.; Rüdiger, V.; Theis, I. *J. Am. Chem. Soc.* 1992, *114*, 7704–7708.
104. Eliseev, A. V.; Schneider, H.-J. *J. Am. Chem. Soc.* 1994, *116*, 6081–6088; Schwinte, P.; Darcy, R.; O'Keeffe, F. *Perkin Trans. 2.* 1998, 805–808.
105. Sirish, M.; Schneider, H.-J. *J. Am. Chem. Soc.* 2000, *122*, 5881–5882.
106. Baudoin, O.; Gonnet, F.; Teulade-Fichou, M. P.; Vigneron, J. P.; Tabet, J. C.; Lehn, J. M. *Chem. Eur. J.* 1999, *5*, 2762–2771.
107. Abe, H.; Mawatari, Y.; Teraoka, H.; Fujimoto, K.; Inouye, M. *J. Org. Chem.* 2004, *69*, 495–504.
108. Moreno-Coral, R.; Lara, K. O. *Supramol. Chem.* 2008, *20*, 427–435.
109. Ojida, A.; Miyahara, Y.; Wongkongkatep, A.; Tamaru, S.; Sada, K.; Hamachi, I. *Chem. Asian J.* 2006, *1*, 555–563.
110. Kwon, J. Y.; Singh, N. J.; Kim, H. N.; Kim, S. K.; Kim, K. S.; Yoon, J. Y. *J. Am. Chem. Soc.* 2004, *126*, 8892–8893.
111. Amendola, V.; Bergamaschi, G.; Buttafava, A.; Fabbrizzi, L.; Monzani, E. *J. Am. Chem. Soc.* 2010, *132*, 147–156.
112. Wang, S. L.; Chang, Y. T. *J. Am. Chem. Soc.* 2006, *128*, 10,380–10,381.
113. Neelakandan, P. P.; Hariharan, M.; Ramaiah, D. *J. Am. Chem. Soc.* 2006, *128*, 11, 334–11,335.
114. Chen, X. Q.; Jou, M. J.; Yoon, J. *Org. Lett.* 2009, *11*, 2181–2184.
115. Wang, D.; Zhang, X.; He, C.; Duan, C. *Org. Biomol. Chem.* 2010, *8*, 2923–2925.
116. Xu, Z.; Singh, N. J.; Lim, J.; Pan, J.; Kim, H. N.; Park, S.; Kim, K. S.; Yoon, J. *J. Am. Chem. Soc.* 2009, *131*, 15,528–15,533.
117. Butterfield, S. M.; Sweeney, M. M.; Waters, M. L. *J. Org. Chem.* 2005, *70*, 1105–1114.
118. Schneider, S. E.; O'Neil, S. N.; Anslyn, E. V. *J. Am. Chem. Soc.* 2000, *122*, 542–543.
119. Jasper, C.; Schrader, T.; Panitzky, J.; Klarner, F. G. *Angew. Chem. Int. Ed. Engl.* 2002, *41*, 1355–1358.
120. Jung, S. O.; Ahn, J. Y.; Kim, S.; Yi, S.; Kim, M. H.; Jang, H. H.; Seo, S. H.; Eom, M. S.; Kim, S. K.; Ryu, D. H.; Chang, S. K.; Han, M. S. *Tetrahedron Lett.* 51, 3775–3778.
121. Pietrzyk, A.; Suriyanarayanan, S.; Kutner, W.; Chitta, R.; Zandler, M. E.; D'Souza, F. *Biosens. Bioelectron.* 2010, *25,* 2522–2529.
122. Davis, A. P. *Org. Biomol. Chem.* 2009, *7*, 3629–3638; Davis, A. P. *Nature.* 464, 169–170; Kubik, S. *Angew. Chem. Int. Ed. Engl.* 2009, *48*, 1722–1725; Mazik, M. *Chem. Bio. Chem.* 2008, *9*, 1015–1017; Mazik, M. *Chem. Soc. Rev.* 2009, *38*, 935–956.
123. Lee, J. D.; Kim, Y. H.; Hong, J. I. *J. Org. Chem.* 2010, *75*, 7588–7595 and references cited therein.
124. Mazik, M.; Kuschel, M. *Chem. Eur. J.* 2008, *14*, 2405–2419.
125. Droz, A. S.; Neidlein, U.; Anderson, S.; Seiler, P.; Diederich, F. *Helv. Chim. Acta.* 2001, *84*, 2243–2289.

126. Schmuck, C.; Schwegmann, M. *Org. Lett.* 2005, *7*, 3517–3520.
127. Rusin, O.; Lang, K.; Kral, V. *Chem. Eur. J.* 2002, *8*, 655–663.
128. Rusin, O.; Hub, M.; Kral, V. *Mater. Sci. Eng. C.* 2001, *18*, 135–140.
129. Charvatova, J.; Rusin, O.; Kral, V.; Volka, K.; Matejka, P. *Sensors Actuators B Chem.* 2001, *76*, 366–372.
130. Kralova, J.; Koivukorpi, J.; Kejik, Z.; Pouckova, P.; Sievanen, E.; Kolehmainen, E.; Kral, V. *Org. Biomol. Chem.* 2008, *6*, 1548–1552.
131. Sugimoto, N.; Miyoshi, D.; Zou, J. *Chem. Commun.* 2000, 2295–2296.
132. Rauschenberg, M.; Bomke, S.; Karst, U.; Ravoo, B. J. Missing *Angew. Chem. Int. Ed. Engl.* 2010, *49*, 7340–7345.
133. Ferrand, Y.; Crump, M. P.; Davis, A. P. *Science* 2007, *318*, 619–622.
134. Barwell, N. P.; Crump, M. P.; Davis, A. P. *Angew. Chem., Int. Ed.* 2009, *48*, 7673–7676.
135. Alptürk, O.; Rusin, O.; Fakayode, S. O.; Wang, W. H.; Escobedo, J. O.; Warner, I. M.; Crowe, W. E.; Kral, V.; Pruet, J. M.; Strongin, R. M. *Proc. Natl. Acad. Sci.* 2006, *103*, 9756–9760.
136. Striegler, S. *Curr. Org. Chem.* 2003, *7*, 81–102; Striegler, S. *Curr. Org. Chem.* 2007, *11*, 1543–1565 and references cited therein.
137. Striegler, S.; Gichinga, M. G. *Chem. Commun.* 2008, 5930–5932.
138. Boeseken, J. *Adv. Carbohydr. Chem.* 1949, *4*, 189–210.
139. James, T. D.; Shinkai, S. *Top. Curr. Chem.* 2002, *160*, 218; James, T. D.; Phillips, M. D.; Shinkai, S. *Boronic Acids in Saccharide Recognition.* RSC: Cambridge, 2006.
140. Yan, J.; Fang, H.; Wang, B. H. *Med. Res. Rev.* 2005, *25*, 490–520.
141. Mader, H. S.; Wolfbeis, O. S. *Microchim. Acta* 2008, *162*, 1–34.
142. James, T. D.; Linnane, P.; Shinkai, S. *Chem. Commun.* 1996, 281–288.
143. Davis, C. J.; Lewis, P. T.; McCarroll, M. E.; Read, M. W.; Cueto, R.; Strongin, R. M. *Org. Lett.* 1999, *1*, 331–334.
144. Rusin, O.; Alpturk, O.; He, M.; Escobedo, J. O.; Jiang, S.; Dawan, F.; Lian, K.; McCarroll, M. E.; Warner, I. M.; Strongin, R. M. *J. Fluorescence* 2004, *14*, 611–615.
145. Jiang, S.; Escobedo, J. O.; Kim, K. K.; Alpturk, O.; Samoei, G. K.; Fakayode, S. O.; Warner, I. M.; Rusin, O.; Strongin, R. M. *J. Am. Chem. Soc.* 2006, *128*, 12,221–12,228.
146. Duggan, P. J.; Offermann, D. A. *Tetrahedron* 2009, *65,* 109–114.
147. Larkin, J. D.; Frimat, K. A.; Fyles, T. M.; Flower, S. E.; James, T. D. *New J. Chem.* 2010, *34*, 2922–2931.
148. Zhang, Y. J.; He, Z. F.; Li, G. W. *Talanta* 2010, *81*, 591–596.
149. Zenkl, G.; Klimant, I. *Microchim. Acta.* 2009, *166*, 123–131.
150. Cui, Q.; Muscatello, M. M. W.; Asher, S. A. *Analyst* 2009, *134*, 875–880.
151. Freeman, R.; Bahshi, L.; Finder, T.; Gill, R.; Willner, I. *Chem. Commun.* 2009, 764–766.
152. Matsumoto, A.; Sato, N.; Miyahara, Y. *Curr. Appl. Physics* 2009, *9*, E214–E217.
153. Baker, G. A.; Desikan, R.; Thundat, T. *Anal. Chem.* 2008, *80*, 4860–4865.
154. Sassolas, A.; Leca-Bouvier, B. D.; Blum, L. J. *Chem. Rev.* 2008, *108*, 109–139; Wang, J.; Uttamehandani, M.; Sun, H. Y.; Yao, S. Q. *Qsar Comb. Sci.* 2006, *25*, 1009–1019.
155. Rusmini, F.; Zhong, Z. Y.; Feijen, J. *Biomacromolecules* 2007, *8*, 1775–1789.
156. Wright, A. T.; Zhong, Z. L.; Anslyn, E. V. *Angew. Chem. Int. Ed. Engl.* 2005, *44*, 5679–5682.
157. Hennig, A.; Bakirci, H.; Nau, W. M. *Nat. Methods* 2007, *4*, 629–632; Bailey, D. M.; Hennig, A.; Uzunova, V. D.; Nau, W. M. *Chem. Eur. J.* 2008, *14*, 6069–6077.
158. Long, L. P.; Jin, M. Y.; Zhang, Y.; Yang, R. H.; Wang, K. M. *Analyst* 2008, *133*, 1201–1208.
159. Zadmard, R.; Schrader, T. *J. Am. Chem. Soc.* 2005, *127*, 904–915.
160. Rusin, O.; Kral, V.; Escobedo, J. O.; Strongin, R. M. *Org. Lett.* 2004, *6*, 1373–1376.
161. Gröger, K.; Baretic, D.; Piantanida, I.; Marjanovic, M.; Kralj, M.; Grabar, M.; Tomic, S.; Schmuck, C. *Org. Biomol. Chem.* 2011, *9*, 198–209.

# 4 Potentiometric Ion Sensors

## Host–Guest Supramolecular Chemistry in Ionophore-Based Ion-Selective Membranes

Ernö Lindner, Róbert E. Gyurcsányi,
and Ernö Pretsch

## CONTENTS

## INTRODUCTION

Ion-selective potentiometry looks back more than a century, being among the earliest instrumental methods of analysis. Its very beginning is marked by fundamental contributions on understanding the electrode potential of electrodes of the first and second kind as well as the potential difference arising at the interface of aqueous electrolytes of different concentrations and immiscible liquids. These studies had already been initiated by Nernst and Riesenfeld at the end of the 19th century.[1,2]

The first ion-selective electrode (ISE) developed was the pH-sensitive glass electrode,[3] and the ubiquitous importance of pH measurements made this discovery by Max Cremer an extremely favorable preamble to further research focusing on the development of electrodes selective to ions other than $H_3O^+$. The research continued predominantly along the line of developing solid-state ISE membranes, for example, $LaF_3$[4] and various precipitate-based electrodes. Solid-state membranes, however, very soon reached their limits in terms of the number of ionic analytes to be selectively detected. Nevertheless, a rather large variety of solid-state ISEs enabling the measurement of anions, for example, $CN^-$, $SCN^-$, $Cl^-$, $Br^-$, $I^-$, $F^-$, $S^{2-}$, and metal ions such as heavy metals, were developed. Anion-selective ISEs covered a major hiatus in the range of analytical techniques of that time, and aside from their use for direct ion determination, potentiometric indicator electrodes enabled instrumental endpoint detection and, consequently, the automation of practically all titrimetric methods.

Modern potentiometric sensors are passive membrane-based systems for measuring and monitoring ion activities and concentrations of neutral species.[5] They revolutionized analytical chemistry through their superb selectivity, allowing separation and sample preparation–free measurements of ion activities (concentrations) in complex matrices such as during short-term, *in vivo* measurements or closed-loop monitoring of blood electrolytes.[6–12] Potentiometric sensors can be made in extremely small sizes by photolithographic microfabrication to analyze samples of submicroliter volumes or for probing ion activities inside single cells.[13,14] Microfabricated sensors were used to follow ionic transients in ischemic porcine hearts[6,12] and as detectors in liquid chromatography.[15]

The research on potentiometric ion sensors implementing host–guest supramolecular chemistry was initiated in the late 1960s by the independent introduction of lipophilic complex–forming natural antibiotics by Simon et al.[16–18] and charged $Ca^{2+}$-selective ionophores by Ross[19] and Bloch et al.[20] The use of antibiotics was based on the discovery by Moore and Pressman in 1964 that some macrocyclic antibiotics induce ion transport in mitochondria.[21] The major representatives of the early natural ionophores included the depsipeptide valinomycin (**10**, Scheme 4.3), enniatin B, and macrotetrolides (cyclic tetrahydrofuranyl-carboxylic acid derivatives) such as nonactin, monactin, dinactin, trinactin, and tetranactin. All these compounds have in common that one side of the macrocycle is hydrophobic while the other, the ion-complexing side, is hydrophilic. Thus, the ion–ionophore complexes become sufficiently hydrophobic to permeate through lipid bilayer membranes, which explains the ion-carrier properties of such compounds.[22]

Very soon, however, it was realized that beyond using various naturally occurring ion carriers, synthetic chemistry is a more powerful tool to generate a wide range of versatile ionophores. The paper by Pedersen,[23] a DuPont scientist, on cyclic polyethers, coining the name "crown ethers," not only started the quest for new synthetic ionophores but also marked the beginning of host–guest chemistry. The nearly simultaneous introduction of highly plasticized polyvinyl chloride (PVC) membranes[20,24] provided an additional boost to the interest toward sensors based on host–guest supramolecular chemistry, which went through an explosive growth and provided the most successful class of ISEs. Nowadays, solid-state and ionophore-based ISEs enable the measurement of more than 60 analytes[25] with major applications in clinical analysis of body-fluid electrolytes, environmental analysis, and process chemistry. The understanding of the importance of ionic fluxes in the potentiometric response or, more precisely, the minimization of ionic fluxes across the ISE membranes shifted the focus of the research toward the development of novel measuring techniques with microfabricated sensors and subnanomolar detection limits.[8,26–29]

The potential of an ISE is a function of the free ionic activity (a thermodynamic property) in the sample. This unique property of ISEs has distinctive advantages in studying biological systems, in which the free ion activity is of utmost importance. The same property enables the use of ISEs as indicator electrodes in complexometric titrations, during which only the free ion activity changes. In fact, chemists still consider ISEs primarily as simple routine tools for direct potentiometry and potentiometric titrations, an opinion strongly supported by most analytical chemistry textbooks.

The connection of ion-selective potentiometry and supramolecular chemistry is twofold. As a result of progress in supramolecular and synthetic chemistry, new selective lipophilic complexing agents, are regularly implemented in ISEs, and novel potentiometric methods provide essential analytical tools for characterizing ion–ligand complex formation reactions. The purpose of this chapter is to highlight the unique advantages of modern ion-selective potentiometric characterization methods in fundamental research related to supramolecular chemistry and synthesis of selective complexing agents. The selectivity of ionophores can be more conveniently and accurately evaluated potentiometrically than by picrate extraction methods commonly employed by organic chemists.[30] The same applies to the quantitative assessment of complex formation constants and to the characterization of certain properties of polymeric materials, for example, the determination of diffusion coefficients in plasticized polymeric membranes.[31,32] Therefore, the focus of this chapter is the implementation of highly selective complexing agents for ion recognition and their characterization in solvent polymeric membranes. The particularity of these systems relies on the use of water-insoluble, lipophilic ionophores in hydrophobic polymeric membranes in which the ion–ionophore complex formation takes place. To obtain complex formation constants, diffusion coefficients, and stoichiometry data of practical relevance for ISE development, their values have to be determined in the polymeric membrane phase. Although these measurements could be performed more conveniently in other media such as methanol, because of solvent effects, the data can differ widely from those gained in the membrane.[33,34] Consequently, such data can be rather irrelevant for predicting the potentiometric responses of ISE membranes. To extend the range of selective complexing agents compatible with common hydrophobic membranes, nanotechnology seems to offer unique possibilities.[35]

## IONOPHORE-BASED MEMBRANES AND
## POTENTIOMETRIC CELL CONFIGURATIONS

The ion-sensing membranes described in this contribution are hydrophobic, plasti-cized polymeric films with glass-transition temperatures below the room tempera-ture. In these membranes, ionic conductivities and diffusion coefficients are similar to highly viscous liquids. Therefore, such membranes are also termed "liquid" or "solvent polymeric" membranes. These plasticized polymeric membranes are robust with advantageous mechanical properties for handling and processing. For example, the most commonly used PVC membranes consist of 66 wt.% plasticizer and 33 wt.% high molecular weight PVC. If these membranes are prepared without any further additive, they behave like low-capacity cation or anion exchangers[5,36] that contain either negatively (e.g., PVC)[37] or positively charged (e.g., polyurethane) func-tional groups (intrinsic sites) as a consequence of ionic impurities in the polymer and the plasticizer,[38] e.g., $-COO^-$, $-SO_3^-$, or $-NR_3^+$ groups. Membranes with negative ionic sites are permeable to cations, and those with positive ionic sites are perme-able to anions. Low site–density membranes are, however, not ideally permselective and contain a significant concentration of co-ions (i.e., ions of the same charge sign as that of the ionic sites in the membrane) extracted from the bathing electrolyte. To ensure adequate permselectivity for plasticized polymeric membranes, fixed or mobile ionic sites are embedded into the membrane. The externally added sites, with a few exceptions, are mobile sites, for example, tetraphenylborate or tetraalkylam-monium ions for establishing cationic and anionic permselectivity, respectively. They are incorporated into the membrane as salts with hydrophilic counterions and used in large excess to the intrinsic sites of the membrane. The site density of fixed or mobile ionic sites determines whether a membrane is permselective and allows ions of only one charge sign from the sample into the membrane.

The origins of potential generation and current control are related to the selective extraction of the analyte ion into the membrane. The free energy of transfer of ions is a function of the dielectric constants and the charges. The measured signal, that is, the potential difference between the two sides of the membrane, is a function of competi-tive energetics of the interactions between ions, solvent, and complex-forming species in each phase, for example, aqueous bathing electrolyte and lipophilic membrane. Hydrophobic liquid ion-exchanger membranes inherently prefer large, soft ions of low charge density. In such membranes, the order of preference among different ions of the same charge is called the Hofmeister series.[39] The selectivity of the sensors can be modified through the dielectric properties of the membrane and, most importantly, by compounding the membranes with selective hydrophobic complexing agents, that is, electrically neutral or charged ionophores.[18,19] Therefore, the invention of liquid mem-brane electrodes whose selectivity could be controlled by incorporating an ionophore, either of biological or synthetic origin, in the liquid membrane brought real versatility and launched a new era in the field of potentiometric sensors.[25,40]

The general layout of a conventional ISE cell assembly is shown in Figure 4.1a. The sensing membrane of ionophore-based ISEs is usually plasticized PVC in which dioctyl sebacate (DOS) or 2-nitrophenyl octyl ether (NPOE; 1:2 weight ratio) serves as a plasticizer. In these highly plasticized polymeric films (liquid membranes),

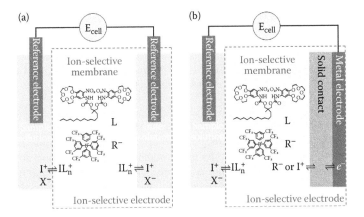

**FIGURE 4.1** Schematic representations of potentiometric cells comprising polymeric membrane based ISEs with (a) a liquid electrolyte and (b) a solid (electron or mixed conductor-based) contact. $I^+$ is the primary ion, L is the ionophore (**11**, Scheme 4.3), $R^-$ is a lipophilic mobile site (tetrakis[3,5-bis(trifluoromethyl)phenyl]borate), $IL_n^+$ is the ionophore–ion complex, and $e^-$ indicates electrons.

ion-exchange sites ($R^-$) and selective complexing agents (L) are trapped by their hydrophobic properties.

The electrochemical notations of the cell in Figure 4.1a and b are given in Schemes 4.1 and 4.2, respectively, in which phase boundaries are indicated by vertical lines and liquid–liquid interfaces by two parallel lines.

The overall cell voltage ($E_{cell}$) is measured between the two reference electrodes. It is composed of a number of interfacial potential differences. Consequently, $E_{cell}$ is the sum of these potential differences ($\varepsilon_1$ through $\varepsilon_8$ or $\varepsilon_1$ through $\varepsilon_7$ in Schemes 4.1 and 4.2, respectively). Among these, $\varepsilon_3$ and $\varepsilon_4$ are termed "liquid-junction" or "diffusion" potentials ($E_d$), and $\varepsilon_5$ and $\varepsilon_6$ are labeled as the membrane solution phase-boundary potentials ($E'_{PB}$ and $E''_{PB}$). Finally, the sum of the phase-boundary potentials ($\varepsilon_5$ and $\varepsilon_6$) is called the "membrane potential" ($E_M$). Because among all the potential differences only $E_d$ ($\varepsilon_4$) and $E_{PB}$ ($\varepsilon_5$) are directly influenced by the composition of the sample, the cell voltage is commonly given as follows:

$$E_{cell} = E^0 + E_d + E_{PB} \tag{4.1}$$

where $E^0 = \varepsilon_1 + \varepsilon_2 + \varepsilon_3 + \varepsilon_6 + \varepsilon_7 + \varepsilon_8$ holds for the cell of Scheme 4.1 and $E^0 = \varepsilon_1 + \varepsilon_2 + \varepsilon_3 + \varepsilon_6 + \varepsilon_7$ for that of Scheme 4.2. For the accurate determination of $E_{PB}$ from the measured $E_{cell}$, it is assumed that all potential terms in $E^0$ are constant, and one chooses a composition of the salt bridge electrolyte that assures constant, close to

$$Ag|AgCl|KCl||salt\ bridge||sample|membrane|I^+Cl^-|AgCl|Ag$$
$$\varepsilon_1 \quad \varepsilon_2 \quad \varepsilon_3 \qquad \varepsilon_4 \qquad \varepsilon_5 \qquad \varepsilon_6 \quad \varepsilon_7 \quad \varepsilon_8$$

**SCHEME 4.1** Electrochemical notation of the potentiometric cell in Figure 4.1a comprising a polymeric membrane based ISE with liquid electrolyte inner contact.

$$Ag|AgCl|KCl||salt\ bridge||sample|membrane|solid\ contact|Pt$$

$$\varepsilon_1 \qquad \varepsilon_2 \quad \varepsilon_3 \qquad\qquad \varepsilon_4 \qquad \varepsilon_5 \qquad\qquad \varepsilon_6 \qquad\qquad \varepsilon_7$$

**SCHEME 4.2** Electrochemical notations of the potentiometric cell in Figure 4.1b comprising a polymeric membrane based ISE with solid inner contact.

zero diffusion potentials. The selection of an adequate combination of the reference electrode and salt bridge electrolyte remains one of the most critical issues with respect to the precision and accuracy of the potentiometric methods. Besides the errors introduced by the unknown diffusion potential, the assumption of a constant $E^0$ is also a source of error in potentiometric analysis. Indeed, drifts and fluctuations in the potential difference at the membrane–solid contact interface ($\varepsilon_6$ in Scheme 4.2) impeded the spread of solid-contact electrodes in regular laboratory use for decades.[41–43] However, the motivations for solid-contact electrodes matching the performance of liquid-contact electrodes remained strong primarily because their manufacturing is claimed to be compatible with thin- and thick-film microfabrication.[8,27,43,44] The essential criteria for solid-contact electrodes showing reproducible and stable potentials at the membrane–solid contact interface ($\varepsilon_6$ in Scheme 4.2) are discussed elsewhere.[27,43]

More recently, it has been realized that a large part of the uncertainty in the potential difference at membrane–solid contact interface ($\varepsilon_6$) is related to the unintentional formation of a thin aqueous layer between the membrane and its solid contact. The transmembrane water transport was identified as the main source of such aqueous layers.[45,46] Fibbioli and coworkers suggested a simple protocol for tracing the presence of this undesirable aqueous film.[47] This aqueous layer test works best when the thickness of the aqueous film, sandwiched between the membrane and its solid contact, is only a few nanometers. To interpret the potential response of a novel solid-contact electrode, it is essential to prove that no aqueous layer is formed between the membrane and its solid contact during extended use.[27,43]

## THEORETICAL TREATMENT OF POTENTIAL RESPONSE

The phase-boundary potential ($E_{PB}$, Equation 4.2) is determined by the ion activities in the sample and the membrane.

$$E_{PB} = \frac{RT}{z_i F} \ln k_i + \frac{RT}{z_i F} \ln \frac{a_i(aq)}{a_i(m)}, \tag{4.2}$$

where $R$ is the gas constant, $T$ is the absolute temperature, $z_i$ is the charge of the ion (I), and $F$ is the Faraday constant; $a_i(aq)$ and $a_i(m)$ are ionic activities in the aqueous and membrane phases, respectively. The single ion partition coefficient, $k_i$, is defined by the standard free energies of the ions in the aqueous ($\tilde{\mu}_i^0(aq)$) and membrane ($\tilde{\mu}_i^0(m)$) phases:

$$RT \ln k_i = z_i F \varepsilon_i^0 = \tilde{\mu}_i^0(aq) - \tilde{\mu}_i^0(m). \tag{4.3}$$

The free energy change on the right-hand side of Equation 4.3 is the medium effect, that is, the work necessary to transfer an ion from water (an infinitely dilute solution) into the membrane. For a water-swollen synthetic ion-exchanger membrane, the medium effect is close to zero ($k_i = 1$). However, for lipophilic membranes containing plasticizers of low dielectric constant, $k_i$ is different from unity ($k_i \neq 1$). The sequence of the individual single ion partition coefficients in the ISE literature is termed the Hofmeister series based on the early works of Franz Hofmeister on the effect of cations and anions on the solubility of proteins.[39] The incorporation of an ionophore into the membrane phase confers a specific selectivity to the counterion exchange that forces a non-Hofmeister response, which can be predicted by considering the complexation equilibrium. In the presence of an ionophore (L), the ion exchange or partitioning of the counterion is accompanied by the formation of an association complex characterized by a complex formation constant in the membrane:

$$\beta_{IL} = \frac{a_{IL_n}(m)}{a_L^n(m)\, a_i(m)} \qquad (4.4)$$

To boost the permselectivity and control the ion-exchange capacity of ISE membranes, a salt, consisting of a hydrophilic cation and a highly hydrophobic (lipophilic) anion ($R^-$), is also incorporated into the membranes. The most generally used salts are tetraphenyl borate derivatives and are applied in excess to the intrinsic sites. The optimal $R_T^-/L_T$ concentration ratio (where $R_T^-$ is the total concentration of anionic sites and $L_T$ that of the ionophore) depends on the ion–ionophore complex stoichiometries and the charges of the primary and interfering ions. When lipophilic mobile sites are incorporated into ionophore-based membranes, besides influencing the membrane selectivity (Table 4.1),[48] they suppress the salt coextraction from the sample solution (thus improving the permselectivity), reduce the membrane resistance, and shorten the response time.[36,49]

---

## TABLE 4.1
### Optimum Anionic Site/Ionophore Concentration Ratios in Ion-Selective Membranes to Achieve the Most Favorable Selectivity for the Primary Ion (I) with Respect to the Interfering Ion (J)

| Charge of the Cation | | Ion–Ionophore Complex Stoichiometry: Ionophore/Ion (I or J) | | $R_T^-/L_T$ |
|---|---|---|---|---|
| Primary ($z_i$) | Interfering ($z_j$) | $n_i$ | $n_j$ | (mol ratio) |
| 2 | 2 | 1 | 2 | 1.41 |
| 2 | 2 | 2 | 3 | 0.77 |
| 2 | 2 | 3 | 4 | 0.54 |
| 2 | 1 | 1 | 1 | 1.62 |
| 2 | 1 | 2 | 2 | 0.73 |
| 2 | 1 | 3 | 3 | 0.46 |
| 1 | 1 | 1 | 2 | 0.71 |

Source: Eugster, R. et al., *Anal. Chem.* 1991, 63, 2285.

Considering the individual phase-boundary potentials on both sides of the membrane ($E'_{PB}$ and $E''_{PB}$), the membrane potential ($E_M$) can be expressed as follows:

$$E_M = \frac{RT}{z_i F} \ln \frac{a'_i(aq)a''_i(m)}{a'_i(m)a''_i(aq)} \qquad (4.5)$$

The phase-boundary potential arises from a charge separation of cations and anions across the interface because of their different standard free energies in the two phases. By combining Equations 4.1 and 4.2, $E_{cell}$ can be calculated (Equation 4.6) under the assumptions that the membrane (or its phase boundaries) is in chemical equilibrium with the aqueous sample and the diffusion potentials in the membrane are negligible. Experimental evidence supports the validity of these assumptions.

$$E_{cell} = E^0 + E_d + \frac{RT}{z_i F} \ln k_i + \frac{RT}{z_i F} \ln \frac{a_i(aq)}{a_i(m)} = E^{0'} + \varepsilon_i^0 + \frac{RT}{z_i F} \ln \frac{a_i(aq)}{a_i(m)} \qquad (4.6)$$

The challenge in calculating $E_{cell}$ for a particular membrane and experimental conditions lies in quantifying the concentration of the uncomplexed primary ion on both sides of the phase boundary. It has been shown that in dilute solutions $a_i(aq)$ can be considerably different from the ion activity in the bulk of the sample (if ion fluxes across the membrane are relevant).[50] Equation 4.6 reduces to the Nernst equation if $a_i(m)$ can be considered constant:

$$E_{cell} = E^{0''} + \frac{RT}{z_i F} \ln a_i(aq), \qquad (4.7)$$

where $E^{0''}$ contains all constant potential terms of the measuring cell.

## SELECTIVITY COEFFICIENTS AND THEIR RELATIONSHIP TO ION–IONOPHORE COMPLEX STABILITIES

The selectivity is probably the most important characteristic of an ion sensor.[51–54] Most often, it determines whether the quantitative assessment of a specific analyte in the target matrix is possible. It is especially critical in analytical tasks with strict precision and accuracy requirements, for example, in clinical applications where the concentrations of interfering components may fluctuate within their physiological concentration ranges.

Under ideal conditions, the response function of an ion-selective sensor can be described by Equation 4.7. Ideal conditions denote that the measured cell voltage is determined by the activity of a single ion (I), which in the aqueous phase boundary layer $a_i(aq)$ is equal to that in the bulk of the sample solution $a_i(I)$. The constant potential contribution in Equation 4.7 is unique for every ion measured because of the $\varepsilon_i^0$ term in Equation 4.6 containing the single-ion partition coefficient $k_i$. Therefore, the response function of an ion sensor in the presence of the primary ion (I) or the interfering ion (J) can be formulated as

$$E_{cell,i} = E_i^0 + \frac{RT}{z_i F} \ln a_i(I) \tag{4.8a}$$

$$E_{cell,j} = E_j^0 + \frac{RT}{z_j F} \ln a_j(J) \tag{4.8b}$$

Under regular conditions, the samples contain both primary and interfering ions, and consequently, both contribute to the measured $E_{cell}$. If the primary and interfering ions have the same charge ($z$), the contribution of the interfering ions to $E_{cell}$ can be taken into account with the selectivity coefficient $K_{i,j}^{pot}$.

$$E_{cell,I+J} = E_i^0 + \frac{RT}{z_i F} \ln\left( a_i(I) + \sum_J K_{i,j}^{pot} a_j(J) \right) \tag{4.9}$$

Equation 4.9 was first derived for $H_3O^+$-selective glass electrodes and is known as the Nikolsky equation. With somewhat more complex equations, the selectivity coefficients are also adequate for describing the response function in the presence of ions having different charges.[55,56]

The selectivity coefficient can be determined from measuring $E_{cell}$ in two solutions, that is, one containing only primary and the other only interfering ions (the separate solution method).[51]

$$\log K_{i,j}^{pot} = \frac{z_i F \left( E_{cell,i} - E_{cell,j} \right)}{2.303RT} + \log\left( \frac{a_i(I)}{a_j(J)^{z_i/z_j}} \right) \tag{4.10}$$

Insertion of Equations 4.8a and 4.8b into Equation 4.10 shows that the selectivity coefficients are mainly dependent on the $E^0$ values, which involve the single-ion partition coefficients:

$$\log K_{i,j}^{pot} = \frac{z_i F}{2.303RT} \left( E_j^0 - E_i^0 \right) \tag{4.11}$$

In the case of strongly discriminated ions (typically $\log K_{i,j}^{pot} < -4$), even minute amounts of primary ions leaching from the membrane into the aqueous phase might determine the phase boundary potential on the sample side and lead to biased selectivity coefficient values. Various methods have been suggested for eliminating such effects.[57]

The selectivity coefficients of an ionophore-free membrane with anionic or cationic sites are controlled by the single-ion partition coefficients:

$$K_{i,j}^{pot} = \frac{k_j}{k_i} \tag{4.12}$$

On the other hand, the selectivity coefficients of a membrane containing an ionophore are dominated by the stability constants ($\beta$) of the ion–ionophore complexes formed. In the simplest case, when an uncharged ionophore forms complexes of the same stoichiometry with the primary and interfering ions bearing equal charges, the selectivity coefficient is directly proportional to the ratio of the single-ion partition coefficients and the ratio of the corresponding stability constants:

$$K_{i,j}^{pot} = \frac{k_j \beta_{JL}}{k_i \beta_{IL}} \tag{4.13}$$

However, in more complex situations, besides the stability constants of the complexes of primary and interfering cations, the total concentrations of the ionophore ($L_T$) and the anionic sites ($R_T^-$), the charges of the primary and interfering ions ($z_i$ and $z_j$), and the complex stoichiometries ($n_i$ and $n_j$) influence the $K_{i,j}^{pot}$ values. This dependence can be utilized for optimizing the membrane composition for the best $K_{i,j}^{pot}$ value (Table 4.1).[48,58]

The optimized performance characteristics may be jeopardized when the membrane composition, for example, the $R_T^-/L_T$ ratio, changes owing to the dissolution or decomposition of membrane components.[59] These concentration changes can be particularly significant when microfabricated ISEs and microsphere optodes (optical sensors with the same constituents and complex equilibria as their ISE analogs) of a few micrometer overall dimensions (diameter and thickness) are exposed to large sample volumes or used in *in vivo* monitoring applications because the sensing membranes of such devices contain extremely small amounts of the components. These problems increase in membranes with reduced concentrations of ionophore and ion-exchanger sites, which are favored for measurements in highly diluted solutions.

## DETERMINATION OF CONCENTRATION AND DIFFUSION COEFFICIENTS OF IONOPHORES AND ION–IONOPHORE COMPLEXES IN LIQUID MEMBRANES

Related to the concerns of uncontrolled changes in the optimized membrane composition and their detrimental effect on the functionality of the sensors, the correlation between the lipophilicity of the ionophore and the lifetime of the ion-selective sensor was determined.[60,61] In addition, the chemical stability of various chromoionophores[62] and anionic additives was studied.[59,63,64] Later, chronoamperometric (CA) and chronopotentiometric (CP) methods were worked out to estimate the changes in membrane composition during extended use.[65,66]

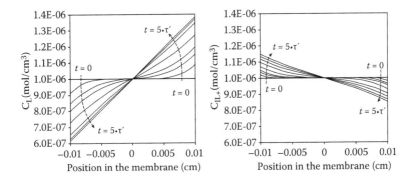

**FIGURE 4.2** Calculated traces of the concentration profiles across the cross section of an ionophore-loaded ion-selective membrane before and following galvanostatic polarization with 75 nA/cm$^2$ current density. Left: free ionophore (L). Right: ion–ionophore complex (IL$^+$). Parameters used for the calculations were $c_L^o = c_{IL^+}^o = 1$ mM, $d = 0.01$ cm, $D_L = 2 \cdot 10^{-8}$ cm$^2$/s, $t_+ = 0.873$, and $D_S = 0.507 \cdot 10^{-8}$ cm$^2$/s. $t_+$ is the transference number of cations in the membrane, and $D_S$ is the effective salt diffusion coefficient $D_S = \left(2D_{R^-}D_{IL^+}\right)/\left(D_{R^-} + D_{IL^+}\right)$. Individual traces were calculated for $t/\tau = 0, 0.1, 0.3, 0.5, 1, 2, 3, 4, 5$ ($\tau = d^2/(\pi^2 D_S) = 0.556$ h). (From Zook, J. M. et al., *J. Phys. Chem. B* 2008, 112, 2008; Zook, J. M. et al., *Electroanalysis* 2008, 20, 259. With permission.)

In CA and CP experiments, cations are forced into the membrane on its positively polarized side and are then complexed by the ionophore. As a result of this complexation reaction, the concentration profiles of the ionophore, the ion–ionophore complex, and the mobile sites in the membrane become tilted as shown in Figure 4.2.[67,68] When the applied current is sufficiently large, at some finite transition time $\tau$, the boundary concentrations of the free ionophore, the ion–ionophore complex, or both will reach zero on opposite sides of the membrane. At this transition time, a breakpoint emerges in the CA and CP transients. The equations used to calculate the CP breakpoints for the free ionophore ($\tau_{L,free}$) and the ion–ionophore complex ($\tau_{IL_n^{z+}}$) of 1:$n$ ion/ionophore stoichiometries and primary ions of charge $z$ are provided by Equations 4.14a and 4.14b. Equation 4.14b for breakpoint time of the ion–ionophore complex is more complicated because of the migration of charged species in current-polarized membranes. However, it becomes simpler if this migration effect is minimized by loading a background electrolyte such as ETH 500 (tetradodecylammonium tetrakis(4-chlorophenyl) borate) in excess into the membrane (Equation 4.14c).[31,32,69]

$$\tau_{L,free}^{1/2} = \frac{zFAC_{L,free}^o \sqrt{D_{L,free}\pi}}{2nI_{appl}} \tag{4.14a}$$

$$\tau_{IL_n^{z+}}^{1/2} = \frac{zFAC_{IL_n^{z+}}^o}{2I_{appl}} \sqrt{\frac{(z+1)\pi D_{IL_n^{z+}}\left(zD_{IL_n^{z+}} + D_{R^-}\right)}{zD_{R^-}}} \tag{4.14b}$$

$$\tau_{IL_n^{z+}}^{1/2} = \frac{zFAC_{IL_n^{z+}}^{0}\sqrt{D_{IL_n^{z+}}\pi}}{2I_{appl}} \tag{4.14c}$$

Here, $I_{appl}$ is the constant applied current in CP experiments; $A$ is the membrane surface area; and $D_{L,free}$, $D_{IL_n^{z+}}$, and $D_{R^-}$ are the diffusion coefficients of the free ionophore, ion–ionophore complex, and lipophilic anion, respectively. $C_{L,free}^{0}$ and $C_{IL_n^{z+}}^{0}$ are the average concentrations of the free and complexed ionophore, respectively, and can be calculated from the total ionophore ($C_{L_T}^{0}$) and lipophilic anion ($C_{R_T^-}^{0}$) concentrations in the membrane, where $C_{L,free}^{0} = C_{L_T}^{0} - \frac{n}{z}C_{R_T^-}^{0}$ and $C_{IL_n^{z+}}^{0} = \frac{C_{R_T^-}^{0}}{z}$. Because the transition times are proportional to the diffusion coefficients and the square of the concentrations of the free ionophore (Equation 4.14a) or the ion–ionophore complex (Equation 4.14b or 4.14c), they can be used to determine either of these parameters.

## DETERMINATION OF COMPLEX FORMATION CONSTANTS

The phase-boundary potential depends on the activity of the uncomplexed ions in the membrane phase. This provides an opportunity to determine complex formation constants in the membrane phase. However, the determination of a single phase boundary potential of a membrane with two phase-boundaries is not trivial because the membrane potential ($E_M$) is a function of the ion activities at both phase boundaries ($E_{PB}'$ and $E_{PB}''$) as it is seen in Equation 4.15:

$$E_{cell} = \text{const} + E_M = \text{const} + \frac{RT}{z_i F}\ln\frac{a_i(aq)'\ a_i(m)''}{a_i(m)'\ a_i(aq)''}, \tag{4.15}$$

where (const) contains all constant potential terms in the cell, $a_i(aq)$ and $a_i(m)$ are the ion activities in the aqueous and membrane phase at the phase boundary, respectively, and primes and double primes refer to the two sides of the membrane, that is, the interfaces of membrane–sample solution and membrane–inner solution, respectively. Obviously, the determination of a single phase-boundary potential requires experiments in which the influence of the other phase boundary on $E_{cell}$ can be eliminated.

So far, three different procedures have been successfully applied. The first two methods rely on measurements with two membranes of different composition but the same phase-boundary potential $a_i(m)''/a_i(aq)''$ on their inner side. This can be achieved by utilizing either a reference ionophore in the membrane or a lipophilic reference cation in the inner solution that poises the phase boundary potential on the inner solution side of the membrane. The outer phase-boundary potential depends on $a_i(m)'/a_i(aq)'$, which is measured with both membranes. The only difference between the two membranes is that one membrane is cast with the ionophore whereas the other without it. The difference between the two membrane potentials provides the information on $a_i(m)'$.

In the first method, one membrane contains the studied ionophore together with a reference ionophore, while the other membrane contains only the reference ionophore. As reference ionophore a pH sensitive lipophilic base such as **1** (Scheme 4.3,

Table 4.2) is used which only interacts with $H_3O^+$.[70] Using these membranes, the detection limit of the pH response is determined in the presence of the cation whose binding constant to the studied ionophore is measured. The difference in the detection limits of the two membranes provides information on $a_i(m)'$. This method was used previously to obtain formation constants in optode membranes.[71]

The second method makes use of a reference cation (e.g., tetraalkylammonium) that does not interact with the studied ionophore. In this protocol the selectivity coefficients of two membranes are determined for the ions of interest compared to the reference cation. One membrane contains the ionophore under study together with a cation exchanger while the other membrane contains only a cation exchanger. The difference in the selectivity coefficients of the two membranes is a direct measure of $a_i(m)'$.[72]

The most widely used third procedure is the so-called segmented sandwich membrane method.[73] It is based on determining the membrane potential of a sandwich membrane made by pressing together two membrane segments as shown in Figure 4.3. One membrane segment (MEM A) contains the ionophore and a lipophilic ion exchanger while the other membrane segment (MEM B) contains only the cation exchanger. The membrane segments are conditioned symmetrically in the aqueous solution of the ion of interest (most often, the primary ion typically at a concentration of 10 mM) for at least one day. Then, the individual membranes are mounted in a measuring cell and their potential is determined in the solution used for conditioning. The membrane potential of symmetrically bathed, fully conditioned membranes should be zero in both cases.

Next, the two membranes are blotted dry, pressed together to form a sandwich membrane, and rapidly mounted in the measuring cell where the membrane potential of the sandwich membrane is recorded. The assemblage of the sandwich membrane

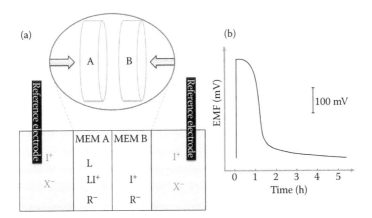

**FIGURE 4.3** Sandwich membrane arrangement for the determination of complex formation constants in plasticized polymeric membranes (a) and typical potential trace recorded with the sandwich membrane (b), which consists of two membrane segments combined together. One segment contains the lipophilic cation exchanger (R⁻, MEM B), while the other (MEM A) in addition to the cation exchanger contains also the ionophore (L). The membrane potential is measured between two reference electrodes dipped in aqueous solutions of the cation of interest (usually the primary ion, I⁺) on both sides of the sandwich membrane.

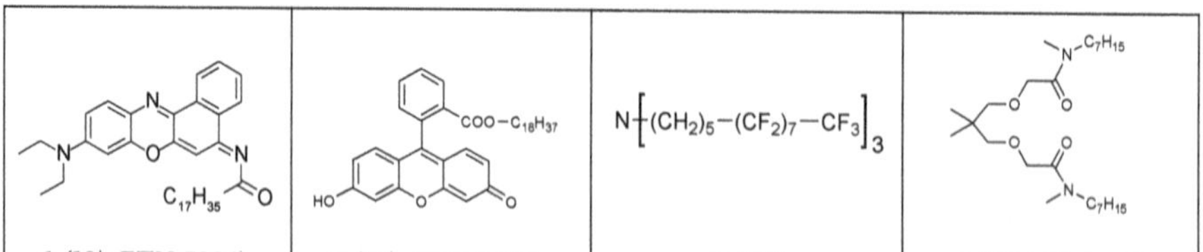

| | | | |
|---|---|---|---|
| **1** (H⁺; ETH 5294) | **2** (H⁺; ETH 7061) | **3** (H⁺) | **4** (Li⁺; ETH 149) |
| **5** (Li⁺) | **6** (Li⁺) | **7** (Na⁺; ETH 2120) | **8** (Na⁺) |
| **9** (Na⁺) | **10** (K⁺; valinomycin) | **11** (R = –C₁₂H₂₅K⁺; BME-44)  **12** (R = –O-CH₂-PVC) | **13** (K⁺) |
| **14** (K⁺) | **15** (K⁺) | **16** (Cs⁺) | **17** (Mg²⁺; ETH 1117) |
| **18** (Ca²⁺; ETH 1001) | **19** (Ca²⁺; ETH 129) | **20** (Ba²⁺) | **21** (Ag⁺) |

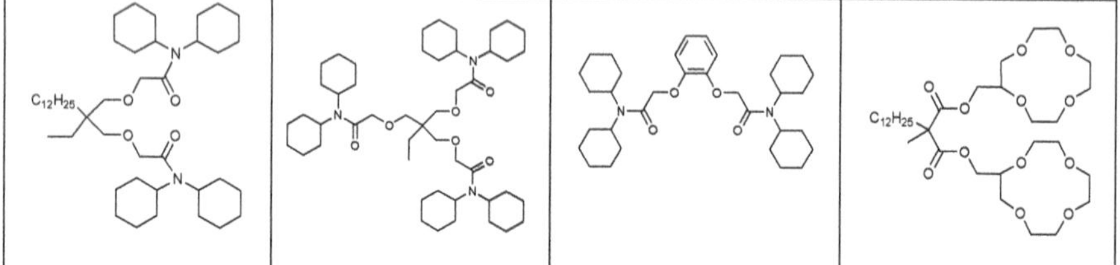

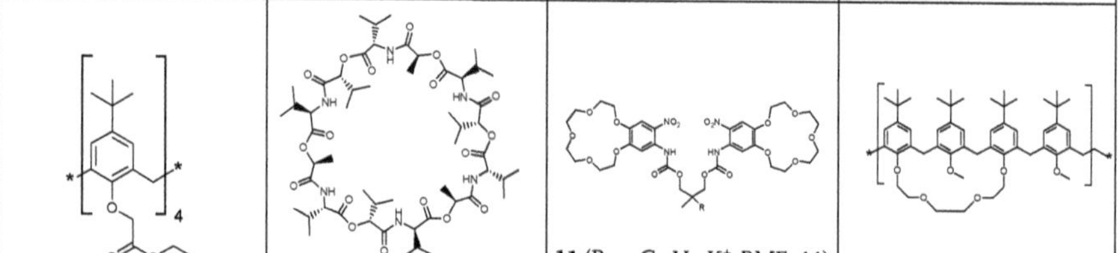

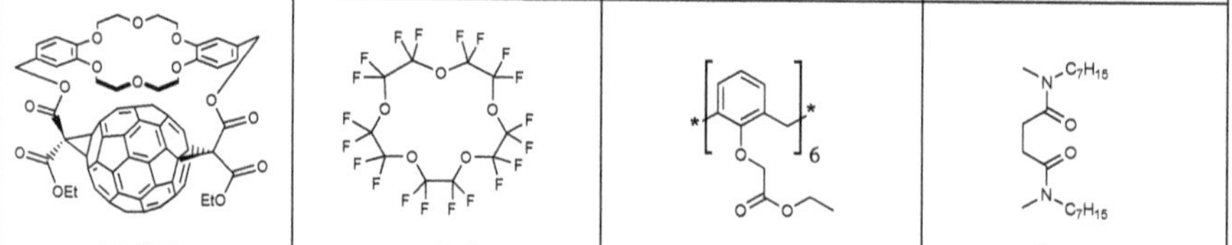

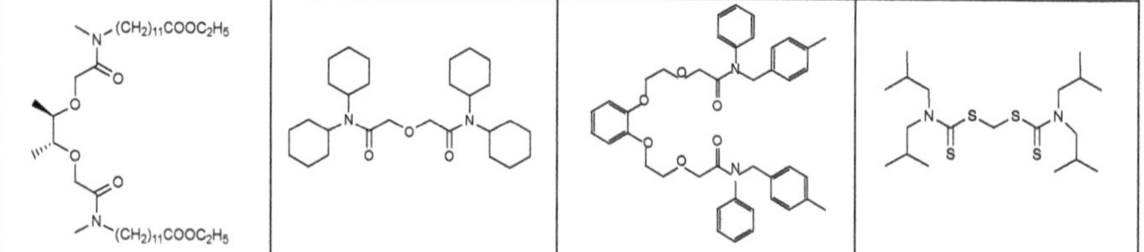

**SCHEME 4.3** Structures of a variety of cation- and anion-selective ionophores commonly used in ISE membranes.

and its incorporation into the electrochemical cell are the most critical steps of the procedure. To minimize the time of ionophore diffusion from the ionophore-loaded into the ionophore-free membrane before starting the membrane potential measurement is the key for accurate determinations of the formation constants with the sandwich membrane method. Typically, the time delay between contacting the

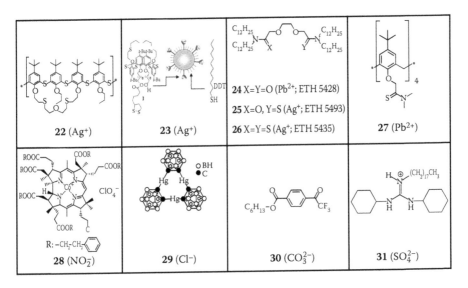

| | | $C_{12}H_{25}$ $C_{12}H_{25}$ N X Y N $C_{12}H_{25}$ $C_{12}H_{25}$ | |
|---|---|---|---|
| 22 (Ag⁺) | 23 (Ag⁺) | 24 X=Y=O (Pb²⁺; ETH 5428) <br> 25 X=O, Y=S (Ag⁺; ETH 5493) <br> 26 X=Y=S (Ag⁺; ETH 5435) | 27 (Pb²⁺) |
| 28 (NO₂⁻) | 29 (Cl⁻) | 30 (CO₃²⁻) | 31 (SO₄²⁻) |

**SCHEME 4.3 (Continued)**

two membranes and initiating the membrane potential measurement is less than 1 minute.

The membrane potential of the sandwich membrane separating two aqueous solutions of identical composition is determined solely by the ion activities in the two membrane segments:

$$E_M = \frac{RT}{z_i F} \frac{a_i(\text{org})''}{a_i(\text{org})'} \tag{4.16}$$

Because the free, uncomplexed ion activity in the membrane containing the ionophore is determined by the strength of the ion–ionophore complex, the complex formation constant ($\beta_{IL}$) can be calculated from the membrane potential of the sandwich membrane. Neglecting the ion pair formation in the membrane leads to the following equation:

$$\beta_{IL} = \left( L_T - \frac{n R_T^-}{z_i} \right)^{-n} \exp\left( \frac{E_M z_i F}{RT} \right), \tag{4.17}$$

where $L_T$ and $R_T^-$ are the total concentrations of the ionophore and ion exchanger in the membrane, respectively.

The complex formation constant can also be expressed by taking into account the ion pair formation in the membrane; however, the resulting relationship is considerably more complicated. As shown for a monovalent cation, the complex formation constant in this case is also a function of the ion pair formation constants $K_{IL_nR}$ and

$K_{IR}$. However, their contribution is moderate because the ion–ionophore complex formation constant varies with the square root of the ratio of the two ion pair formation constants:

$$\beta_{IL} = \left(L_T - nR_T^-\right)^{-n} \sqrt{\frac{K_{IL_nR}}{K_{IR}}} \exp\left(\frac{E_M F}{RT}\right). \qquad (4.18)$$

Using the previously described potentiometric methods, effective complex formation constants of more than 100 ionophores have been determined in the membrane phase for various ions. A selection of ionophore structures is shown in Scheme 4.3. The corresponding binding constants are summarized in Table 4.2. Because $E_{cell}$ measurements only yield free ion activities, the stoichiometry of the complex is required for calculating the binding constant. Thus, several measurements with different relative concentrations of the ionophore and ions (defined by the concentration of the lipophilic ion exchanger) are necessary[72,74] or else the stoichiometry must be assumed on the basis of other information. The situation might be complicated by the fact that a ligand may form different complexes with the same ion. For example, the widely used $Ca^{2+}$-selective ionophore **18** (ETH 1001, see Scheme 4.3 and Table 4.2) forms a 2:1 (ligand/ion) complex in different membrane phases. However, for a structurally similar compound (lacking the two methyl substituents on the ethylene glycol moiety), it was shown by $^{13}C$ NMR spectroscopy that it may likewise form a 1:1 complex in which the two ester groups at the end of the chain also coordinate the cation. Additionally, in methanol, it can also form a 1:1 complex without participation of the ester carbonyl groups.[75]

Considering the stoichiometry is also an important issue of ligand design. In the 2:1 complexes of a series of dioxaoctane diamides (such as **18**), $Ca^{2+}$ has eight coordinating sites. Based on the fact that $Mg^{2+}$ prefers an octahedric (sixfold) coordination sphere, a series of oxapentane diamides (such as **19**, ETH 129, see Scheme 4.3 and Table 4.2) were synthesized. Surprisingly, at first sight, they all showed high preference for $Ca^{2+}$, which turned out to be a result of the formation of 3:1 complexes with nine coordinating O atoms.[76]

The data in Table 4.2 show that effective complex-binding constants in lipophilic membrane phases can be extremely high, for example, $10^{15.9}$ for a 1:1 complex of **27** with $Pb^{2+}$.[72] Even values of up to $10^{27.6}$ were reported in the literature.[77] The direct measurement of such high binding constants is possible because the potentiometric signal is a function of the logarithm of the activity of the uncomplexed ion. It would not be possible to directly measure such high values if the signal, as is the case with most other methods, were linearly dependent on the concentration.

To obtain such high complex-formation constants, a lipophilic membrane phase with poor coordinating properties of its components is required. For example, the stability constant of the $K^+$ complex of the antibiotic valinomycin (**10**) is $10^{10.10}$ in bis(2-ethylhexyl) sebacate (DOS) and $10^{11.63}$ in NPOE as a membrane solvent. These high values observed in lipophilic membrane phases are in sharp contrast to those obtained in polar solvents of good coordinating properties ($10^{6.08}$ in ethanol,[78] $10^{4.48}$ in methanol,[79] and even below 1 in water[80]).

## TABLE 4.2
## Effective Formation Constants, $\log\beta_{IL}$, for Complexes of Lipophilic Hosts and Ionic Guests in Solvent Polymeric Membranes with Given Host–Guest Stoichiometry [in brackets][a]

| Ionophore | Effective Formation Constants, $\log\beta_{IL}$, of Host–Guest Complexes[a] |
|---|---|
| 1 | $H^+$ [1:1] 11.41 (DOS), 14.82 (NPOE); $Na^+$ [1:1] 2.61 (DOS); $K^+$ [1:1] 2.59 (DOS)[73] |
| 2 | $H^+$ [1:1] 18.04 (DOS), 20.00 (NPOE); $Na^+$ [1:1] 4.83 (DOS); $K^+$ [1:1] 4.44 (DOS)[73] |
| 3 | $H^+$ [1:1] 15.4 (PTFE)[86] |
| 4 | $Li^+$ [1:1] 7.90 (DOS), 10.71 (NPOE)[87] |
| 5 | $Li^+$ [1:1] 5.9, 7.0; $Na^+$ [1:1] 4.1, 5.2; $K^+$ [1:1] 3.1, 4.1 (DOS)[88] |
| 6 | $Li^+$ [1:1] 7.40 (BBPA),[89] 8.24 (DOS),[87] 6.7, 7.4 (DOS),[88] 10.40 (NPOE);[88] $Na^+$ [1:1] 5.75 (BBPA),[89] 4.5, 5.1 (DOS);[88] $K^+$ [1:1] 4.62 (BBPA),[89] 3.2, 2.8 (DOS)[88] |
| 7 | $Na^+$ [2:1] 8.76 (DOS), 10.91 (NPOE)[87] |
| 8 | $Na^+$ [1:1] 6.55 (DOS), 9.19 (NPOE)[87] |
| 9 | $Na^+$ [1:1] 7.69,[87] 7.60[83] (DOS), 10.27 (NPOE)[87] |
| 10 | $K^+$ [1:1] 10.10 (DOS),[87] 7.52 (DBP),[74] 11.63 (NPOE);[87] $Na^+$ [1:1] 4.4 (DBP);[74] $NH_4^+$ [1:1] 5.7 (DBP)[74] |
| 11 | $K^+$ [1:1] 7.84,[87] 7.75 (DOS),[90] 10.04 (NPOE);[87] $Li^+$ [1:1] 4.22 (DOS);[83] $Na^+$ [1:1] 6.07 (DOS)[83] |
| 12 | $K^+$ [1:1] 6.50; $Na^+$ [1:1] 4.63 (DOS)[90] |
| 13 | $K^+$ [1:1] 6.1; $Na^+$ [1:1] 3.7; $Rb^+$ [1:1] 5.7; $Cs^+$ [1:1] 4.0 (BBPA)[91] |
| 14 | $K^+$ [1:1] 5.4; $Li^+$ [1:1] 2.9; $Na^+$ [1:1] 4.0; $Cs^+$ [1:1] 3.3 (DOS)[92] |
| 15 | $K^+$ [1:1] 5.5; $Na^+$ [1:1] 1.7 (neat)[93] |
| 16 | $Cs^+$ [1:1] 8.74 (DOS)[94] |
| 17 | $Mg^{2+}$ [3:1] 9.72 (DOS), 13.84 (NPOE)[87] |
| 18 | $Ca^{2+}$ [2:1] 19.70 (DOS),[87] 24.54,[87] 14.0[95] (NPOE) |
| 19 | $Ca^{2+}$ [3:1] 25.5 (DOS),[87] 29.2,[87] 15.2[95] (NPOE) |
| 20 | $Ba^{2+}$ [2:1] 19.7; $H^+$ [1:2] 4.1; $Mg^{2+}$ [2:1] 12.0; $Ca^{2+}$ [2:1] 15.1 (NPOE)[96] |
| 21 | $Ag^+$ [2:1] 12.42,[83] 12.6 (DOS);[72] $H^+$ [1:1] < 3 (DOS);[72] $Na^+$ [2:1] 2.8 (DOS);[83] $K^+$ [1:1] < 3 (DOS);[72] $Mg^{2+}$ [1:1] < 3 (DOS);[72] $Ca^{2+}$ [1:1] < 3 (DOS);[72] $Pb^{2+}$ [1:1] < 3 (DOS);[72] $Cu^{2+}$ [1:1] < 3 (DOS);[72] $Cd^{2+}$ [1:1] < 3 (DOS)[72] |
| 22 | $Ag^+$ [1:1] 10.85, 11.31 (NPOE)[97] |
| 23 | $Ag^+$ [1:1] 8.5 (11.1 for the free ionophore); $K^+$ [1:1] 3.1 (2.7 for the free ionophore) (NPOE)[35] |
| 24 | $Pb^{2+}$ [2:1] 16.8; $H^+$ [1:1] 4.2; $Na^+$ [1:1] 5.4; $K^+$ [1:1] 4.6; $Ag^+$ [1:1] 6.4; $Mg^{2+}$ [1:1] 9.1; $Ca^{2+}$ [1:1] 14.3; $Cu^{2+}$ [2:1] 13.7; $Cd^{2+}$ [2:1] 16.5 (DOS)[72] |
| 25 | $Ag^+$ [2:1] 15.9; $H^+$ [1:1] 2.6; $Na^+$ [1:1] 3.7; $K^+$ [1:1] 3.3; $Mg^{2+}$ [2:1] 7.2; $Ca^{2+}$ [2:1] 10.5; $Pb^{2+}$ [2:1] 15.7; $Cu^{2+}$ [2:1] 13.5; $Cd^{2+}$ [2:1] 17.1 (DOS)[72] |
| 26 | $Ag^+$ [2:1] 19.0; $H^+$ [1:1] < 3; $Na^+$ [1:1] < 3; $K^+$ [1:1] < 3; $Mg^{2+}$ [1:1] < 3; $Ca^{2+}$ [1:1] < 3; $Pb^{2+}$ [2:1] 14.7; $Cu^{2+}$ [2:1] 14.8; $Cd^{2+}$ [2:1] 16.4 (DOS)[72] |
| 27 | $Pb^{2+}$ [1:1] 15.9 (DOS), 18.4 (NPOE); $H^+$ [1:1] < 3; $Na^+$ [1:1] 3.1; $K^+$ [1:1] < 3; $Mg^{2+}$ [1:1] < 3; $Ca^{2+}$ [1:1] < 3; $Cu^{2+}$ [1:1] 12.1; $Cd^{2+}$ [1:1] 10.0 (DOS)[72] |
| 28 | $NO_2^-$ [1:1] 10.58 (DOS), 10.59 (NPOE)[73] |

*(continued)*

**TABLE 4.2 (Continued)**
**Effective Formation Constants, $\log\beta_{\mathrm{IL}}$, for Complexes of Lipophilic Hosts and Ionic Guests in Solvent Polymeric Membranes with Given Host–Guest Stoichiometry [in brackets]**[a]

| Ionophore | Effective Formation Constants, $\log\beta_{\mathrm{IL}}$, of Host–Guest Complexes[a] |
|---|---|
| 29 | $Cl^-$ [2:1] 13.4 (DOS; $\log K_1 = 9.9$, $\log K_2 = 3.5$)[98] |
| 30 | $CO_3^{2-}$ [4:1] 12.8; $CH_3COO^-$ [2:1] 5.9; benzoate [2:1] 5.3; $Cl^-$ [2:1] 3.32 (DBP)[99] |
| 31 | $SO_4^{2-}$ [2:1] 16.8 (NPOE)[100] |

[a] PVC membranes were based on the following plasticizers: bis(butylpentyl) adipate (BBPA), bis(2-ethyl-hexyl) sebacate (DOS), dibutyl phthalate (DBP), bis(2-ethylhexyl) phthalate (DOP), and 2-nitrophenyl octyl ether (NPOE). Further membrane matrices were perfluoroperhydrophenanthrene (PTFE) or plasticizer-free (neat).

As shown previously, the potentiometric selectivity coefficient $K_{i,j}^{pot}$ is related to the complex formation constants of an ionophore with the respective ions (for two ions of the same charge forming sufficiently strong complexes, it is given by the ratio of the effective complex binding constants; see Equation 4.13). A strong complexation of the primary ion is, therefore, necessary for relevant sensors because extremely small selectivity coefficients are required for measuring traces of an ion in the presence of abundant interfering ions such as in environmental samples. With certain ionophore-based membranes, selectivity coefficients as low as $10^{-15}$ were achieved.

Fluorous phases are the least polar and least polarizable liquid phases known. Their use as a sensing membrane phase is very promising because, on one hand, matrices with weaker ion-binding properties exhibit both greater selectivity and wider measuring ranges.[81] On the other hand, interferences by lipophilic acids or bases, which make the direct application of ISEs in biological fluids difficult,[82] do not occur in such phases. For these reasons, the design and synthesis of fluorous hosts, that is, ionophores that can be dissolved in highly fluorinated liquids, is a very promising novel field of host–guest chemistry. In Scheme 4.3 and Table 4.2, **15** is the first example of a fluorous ionophore for cations other than $H_3O^+$.

All values shown in Table 4.2 are effective complex-binding constants; the influence of ion pairs is included.[83] However, the sandwich membrane method can also be used to obtain ion pair formation constants either by varying the concentration of a second lipophilic electrolyte[84] or by relying on a reference salt for which the absence of any ion pair formation can be shown.[85] The association constants between alkali metal cations and a tetraphenylborate derivative are in the order of $10^2$,[84] while those between derivatives of the benzyltrimethylammonium cation and chloride are about $10^4$.[85] Especially strong ion pairs with association constants in the order of $10^{20}$ are formed between tetrakis[3,5-bis(perfluorohexyl)phenyl]borate and alkali metal cations in fluorous phases.[81]

# OUTLOOK

## NANOSTRUCTURE-BASED ION-SELECTIVE POTENTIOMETRY

Ionophore–nanostructure conjugates have been shown recently to offer significant functional advantages over conventional ISE membrane formulations, opening new perspectives in the design of ISEs. There are two major factors limiting the applicability of polymeric ISE membranes: the gradual leaching of active components from the membrane into the sample solution and the coextraction of lipophilic sample components into the membrane. Both phenomena have detrimental effects on the lifetime and analytical performance of the respective ISEs. To prevent the leaching of active membrane components, efforts were made to covalently immobilize them in the membrane matrix.[90] However, this most often implied the use of functionalized polymers that could alter the selectivity of the ISE membrane. On the other hand, the conjugation of ionophores to inert Au nanoparticles of approximately 5-nm diameter was shown to provide a new route to immobilize the active components in the membrane as ionophore–Au nanoparticle conjugates (IP-AuNP). The silver selective ionophore conjugated to gold nanoparticles (**23**, Scheme 4.3) was found to be immobile in plasticized polymeric membranes.[35] The IP–AuNPs were synthesized by the spontaneous self-assembly of thiol and dithiolane bearing compounds on the surface of Au nanoparticles. This approach offers the opportunity to independently adjust the composition of the membrane matrix components and immobilize the ionophore; to tune the properties of the IP–AuNPs, most notably, their solubility by using mixed self-assembled monolayers in which the selective complexing agents are co-immobilized with thiol derivatives bearing various end functionalities; and to eliminate the light sensitivity of some solid-contact ISEs because IP–AuNP–loaded membranes are not transparent.

Although the basic inspiration for ionophore-based ISEs can be traced back to the ionophore-mediated ion transport through lipid bilayer membranes, the synthetic reconstruction of biological ion channels has received little attention. Biological ion channels are protein pores with amino acid sequences, providing rigorously spaced functionalities in the channel to induce selective recognition and passage for ions through the cell membrane. The selective ion transport through ion channels is key in the electrical activity and signaling of living cells, for example, transport through $K^+$ channels is at least $10^4$ times more effective for $K^+$ than for $Na^+$. While such membrane proteins and their suspending lipid bilayers are extremely fragile, the same functionality may be achieved by using chemically modified solid-state nanopores or nanopore arrays. An early potentiometric study showed that Au-impregnated track-etch membranes with charged nanopores rejected ions of the same charge sign and transported those of the opposite one.[101] However, selective solid-state ion channels have been introduced only very recently. In these ion channels, the effective diameter of 5-nm nanopores was further restricted to molecular dimensions by their subsequent modification with selective complexing agents. The ion-selective response further required the establishment of ionic sites within Au nanopores and adjusting their hydrophobicity by modifying their surface with a mixed self-assembled monolayer consisting of three different thiol (disulfide) derivatives in a proper ratio

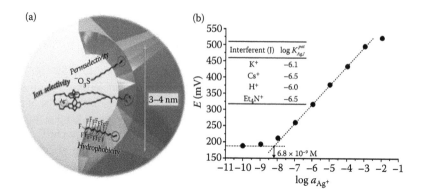

**FIGURE 4.4** (a) Schematic cross section of a chemically modified Au nanopore depicting the three different derivatives immobilized by Au–S bonds on the inner wall of the nano-pores, inducing the Ag+ selectivity of the respective nanoporous membrane. (b) Typical calibration curve and selectivity coefficients of the modified Au-nanopore–based ISEs. (From Jágerszki, G. et al., *Angew. Chem. Int. Ed.* 2010, *50*,1656–1659.)

(Figure 4.4).[102] The proof of principle was made by using a synthetic Ag+-selective thiacalixarene derivative bearing dithiolane moieties (see **23** in Scheme 4.3) with cation-exchanger sites generated with mercaptodecanesulfonate while the hydropho-bicity of the nanoporous membrane was ensured by a perfluorinated thiol derivative, thus taking advantage of the latest results, which showed the superiority of fluorous ISE membranes.[81]

Solid-state, ion channel–based, solvent-free membranes exhibited excellent selec-tivities, exceeding six orders of magnitude for a number of critical interfering cations and nanomolar detection limits (Figure 4.4). Because all membrane components are covalently attached to the nanoporous membrane phase, there are no restrictions in terms of their lipophilicity as in the case of conventional ISE membranes. Therefore, Au nanopores seem to offer a unique versatility to integrate ionophores and other components of widely different properties given that they possess thiol or disulfide functionalities. Thus, nanopore-based ISE membranes are expected to extend the range of applicable selective complexing agents beyond those with lipophilic charac-ter and to provide a new, robust, all-solid-state platform for ion sensing, separation, and characterization of host–guest interactions.

### FLUOROUS MEMBRANES

To eliminate interferences related to the coextraction of lipophilic anions and the uptake of electrically neutral lipophilic sample components from real samples into the mem-brane, as well as to membrane biofouling the use of fluorous membranes appears to be the most promising. The extremely poor solvation capacity of fluorous phases, which phase-separate both from aqueous solutions and from hydrocarbons, makes them resist the extraction of highly lipophilic components. As in such membranes the ion-pair formation constants are unprecedentedly large (in the order of $10^{20}$), even ionophore-free potentiometric sensors based on fluorous membranes doped with a fluorophilic

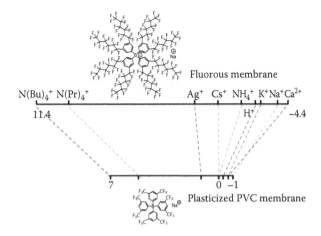

**FIGURE 4.5** Comparison of the logarithmic selectivity coefficients for several cations referenced to $Cs^+$ obtained with a membrane based on a fluorinated tetraphenylborate-based cation exchanger (see chemical structure) and a chloroparaffin/PVC membrane. (From Boswell, P., and Bühlmann, P. *J. Am. Chem. Soc.* 2005, 127, 8958. With permission.)

cation exchanger (a tetraphenylborate derivative) were shown to have a range of selectivities that exceeds the selectivity range of conventional polymeric membranes by eight orders of magnitude.[81] Thus, the comparative selectivity coefficients of cation exchanger–based plasticized PVC and perfluoroperhydrophenanthrene fluorous phases demonstrate that the latter is extremely well suited to discriminate between various cations because the differences between their lipophilicities are amplified (Figure 4.5).

The weakly coordinating components of such fluorous membranes are expected to increase the potentiometric selectivity of ionophore-based membranes by favoring a stronger binding between the ionophore and the primary ion in combination with a weak solvation of the interfering ions. However, on the downside, poor solvation makes fluorous membranes incompatible with commercially available ionophores and impedes their general applicability. Still, besides fluorous membranes based on cation[81] and anion exchangers,[86] successful examples of a pH-sensitive[86] and, recently, of a $Ag^+$-selective membrane[93] were reported.

## ACKNOWLEDGMENTS

We thank Dr. D. Wegmann for careful reading of the manuscript. This work was supported in part the National Institute of Health, National Heart, Lung and Blood Institute (NIH/NHLBI), # 1 RO1 HL079147-01, and the U.S. Army Medical Research and Material Command (USAMRMC) W81XWH-10-1-0358 grants, the Tennessee Technology Development Corporation (TTDC) "Infusensor" Tech Maturation Fund and by the financial support from the FedEx Institute of Technology (FIT) for the establishment of the Sensor Institute of the University of Memphis "SENSORIUM". REGy thanks for the support of the Hungarian Scientific Research Fund (OTKA) NF 69262 and the Social Renewal Operational Programme TÁMOP-4.2.1/B-09/1/KMR-2010-0002.

## REFERENCES

1. Nernst, W. *Z. Phys. Chem.* 1889, *4*, 129.
2. Nernst, W.; Riesenfeld, E. H. *Ann. Phys.* 1902, *8*, 600.
3. Cremer, M. *Z. Biol. Munich.* 1906, *47*, 562.
4. Frant, M. S.; Ross, J. W., Jr. *Science.* 1966, *154*, 1553.
5. Buck, R. P.; Lindner, E. *Acc. Chem. Res.* 1998, *31*, 257.
6. Cosofret, V. V.; Erdösy, M.; Johnson, T. A.; Buck, R. P.; Ash, R. B.; Neuman, M. R. *Anal. Chem.* 1995, *67*, 1647.
7. Lauks, I. R. *Acc. Chem. Res.* 1998, *37*, 317.
8. Lindner, E.; Buck, R. P. *Anal. Chem.* 2000, *72*, 336A.
9. Lindner, E.; Cosofret, V. V.; Ufer, S.; Johnson, T. A.; Ash, R. B.; Nagle, H. T.; Neuman, M. R.; Buck, R. P. *Fresenius J. Anal. Chem.* 1993, *346*, 584.
10. Lindner, E.; Cosofret, V. V.; Ufer, S.; Kusy, R. P.; Buck, R. P.; Ash, R. B.; Nagle, H. T. *J. Chem. Soc. Faraday Trans.* 1993, *89*, 361.
11. Uhlig, A.; Lindner, E.; Teutloff, C.; Schnackenberg, U.; Hintsche, R. *Anal. Chem.* 1997, *69*, 4032.
12. Buck, R. P.; Cosofret, V. V.; Lindner, E.; Ufer, S.; Madaras, M. B.; Johnson, T. A.; Ash, R. B.; Neuman, M. R. *Electroanalysis* 1995, *7*, 846.
13. Messerli, M. A.; Robinson, K. R.; Smith, P. J. S. In *Plant Electrophysiology: Theory and Methods*; Volkov, A. G., Ed.; Springer-Verlag: Berlin, Heidelberg, 2006, Sections 4.1.
14. Smith, P. J. S.; Sanger, R. H.; Messerli, M. A. In *Methods and New Frontiers in Neuroscience*; Michael, A. C., Ed.; CRC Press: Boca Raton, FL, 2007; Vol. 2007, Chapter 18, p. 373.
15. Isildak, I.; Covington, A. *Electroanalysis* 1993, *5*, 815.
16. Pioda, L. A. R.; Stankova, V.; Simon, W. *Anal. Lett.* 1969, *2*, 665.
17. Štefanac, Z.; Simon, W. *Chimia.* 1966, *20*, 436.
18. Štefanac, Z.; Simon, W. *Microchem. J.* 1967, *12*, 125.
19. Ross, J. W. *Science.* 1967, *156*, 1378.
20. Bloch, R.; Shatkay, A.; Saroff, H. A. *Biophys. J.* 1967, *7*, 865.
21. Moore, C.; Pressman, B. C. *Biochem. Biophys. Res. Comm.* 1964, *15*, 562.
22. Läuger, P. *Science.* 1972, *178*, 24.
23. Pedersen, C. J. *J. Am. Chem. Soc.* 1967, *89*, 2495.
24. Moody, G. J.; Oke, R. B.; Thomas, J. D. R. *Analyst* 1970, *95*, 910.
25. Bühlmann, P.; Pretsch, E.; Bakker, E. *Chem. Rev.* 1998, *98*, 1593.
26. Bakker, E.; Pretsch, E. *Anal. Chem.* 2002, *74*, 420A.
27. Lindner, E.; Umezawa, Y. *Pure Appl. Chem.* 2008, *80*, 85.
28. Sokalski, T.; Ceresa, A.; Fibbioli, M.; Zwickl, T.; Bakker, E.; Pretsch, E. *Anal. Chem.* 1999, *71*, 1210.
29. Sokalski, T.; Ceresa, A.; Zwickl, T.; Pretsch, E. *J. Am. Chem. Soc.* 1997, *119*, 11,347.
30. Kimura, K.; Maeda, T.; Shono, T. *Talanta* 1979, *26*, 945.
31. Bodor, S.; Zook, J. M.; Lindner, E.; Tóth, K.; Gyurcsányi, R. E. *Analyst* 2008, *133*, 635.
32. Zook, J.; Bodor, S.; Lindner, E.; Tóth, K.; Gyurcsányi, R. E. *Electroanalysis* 2009, *21*, 1923
33. Frensdorff, H. K. *J. Am. Chem. Soc.* 1971, *93*, 600.
34. Wipf, H. K.; Pioda, L. A. R.; Štefanac, Z.; Simon, W. *Helv. Chim. Acta* 1968, *51*, 377.
35. Jágerszki, G.; Grün, A.; Bitter, I.; Tóth, K.; Gyurcsányi, R. E. *Chem. Commun.* 2010, *46*, 607.
36. Lindner, E.; Gráf, E.; Niegreisz, Z.; Tóth, K.; Pungor, E.; Buck, R. P. *Anal. Chem.* 1988, *60*, 295.
37. van den Berg, A.; van der Waal, P. D.; Skowronska-Ptasinska, M.; Sudhölter, E. J. R.; Reinhoudt, D. N.; Bergveld, P. *Anal. Chem.* 1987, *58*, 2827.
38. Gyurcsányi, R. E.; Lindner, E. *Anal. Chem.* 2002, *74*, 4060.

39. Hofmeister, F. *Arch. Exp. Pathol. Phar.* 1888, *24*, 247.
40. Bakker, E.; Bühlmann, P.; Pretsch, E. *Chem. Rev.* 1997, *97*, 3083.
41. Bobacka, J. *Anal Chem.* 1999, *71*, 4932.
42. Bobacka, J.; Ivaska, A.; Lewenstam, A. *Chem Rev.* 2008, *108*, 329.
43. Lindner, E.; Gyurcsányi, R. E. *J. Solid State Electrochem.* 2009, 51.
44. Gyurcsányi, R. E.; Rangisetty, N.; Clifton, S.; Pendley, B. D.; Lindner, E. *Talanta* 2004, *63*, 89.
45. Lindfors, T.; Sundfors, F.; Höfler, L.; Gyurcsányi, R. E. *Electroanalysis* 2009, *21*, 1914.
46. Sundfors, F.; Lindfors, T.; Höfler, L.; Bereczki, R.; Gyurcsányi, R. E. *Anal. Chem.* 2009, *81*, 5925.
47. Fibbioli, M.; Morf, W. E.; Badertscher, M.; de Rooij, N. F.; Pretsch, E. *Electroanalysis* 2000, *12*, 1286.
48. Eugster, R.; Gehrig, P. M.; Morf, W. E.; Spichiger, U. E.; Simon, W. *Anal. Chem.* 1991, *63*, 2285.
49. Ammann, D.; Pretsch, E.; Simon, W.; Lindner, E.; Bezegh, A.; Pungor, E. *Anal. Chim. Acta* 1985, *171*, 119.
50. Gyurcsányi, R. E.; Pergel, E.; Nagy, R.; Kapui, I.; Lan, B. T. T.; Tóth, K.; Bitter, I.; Lindner, E. *Anal. Chem.* 2001, *73*, 2104.
51. Buck, R. P.; Lindner, E. *Pure Appl. Chem.* 1994, *66*, 2527.
52. Umezawa, Y. *Handbook of Ion-Selective Electrodes: Selectivity Coefficients*; CRC Press: Boca Raton, FL, 1990.
53. Umezawa, Y.; Bühlmann, P.; Umezawa, K.; Tohda, K.; Amemya, S. *Pure Appl. Chem.* 2000, *72*, 1851.
54. Umezawa, Y.; Umezawa, K.; Bühlmann, P.; Hamada, N.; Aoki, H.; Nakanishi, J.; Sato, M.; Xiao, K. P.; Nishimura, Y. Y. *Pure Appl. Chem.* 2002, *74*, 923.
55. Bakker, E. *J. Electroanal. Chem.* 2010, *639*, 1.
56. Bakker, E.; Meruva, R. K.; Pretsch, E.; Meyerhoff, M. E. *Anal. Chem.* 1994, *66*, 3021.
57. Bakker, E. *Anal. Chem.* 1997, *69*, 1061.
58. Bakker, E.; Bühlmann, P.; Pretsch, E. *Talanta* 2004, *63*, 3.
59. Langmaier, J.; Lindner, E. *Anal. Chim. Acta* 2005, *543*, 156.
60. Dinten, O.; Spichiger, U. E.; Chaniotakis, N.; Gehrig, P.; Rusterholz, B.; Morf, W. E.; Simon, W. *Anal. Chem.* 1991, *63*, 596.
61. Oesch, U.; Simon, W. *Anal. Chem.* 1980, *52*, 692.
62. Bakker, E.; Lerchi, M.; Rosatzin, T.; Rusterholz, B.; Simon, W. *Anal. Chim. Acta* 1993, *278*, 211.
63. Peper, S.; Telting-Diaz, M.; Almond, P.; Albrecht-Schmitt, T.; Bakker, E. *Anal. Chem.* 2002, *74*, 1327.
64. Rosatzin, T.; Bakker, E.; Suzuki, K.; Simon, W. *Anal. Chim. Acta* 1993, *280*, 197.
65. Pendley, B. D.; Gyurcsányi, R. E.; Buck, R. P.; Lindner, E. *Anal. Chem.* 2001, *73*, 4599.
66. Pendley, B. D.; Lindner, E. *Anal. Chem.* 1999, *71*, 3673.
67. Zook, J. M.; Buck, R. P.; Langmaier, J.; Lindner, E. *J. Phys. Chem. B.* 2008, *112*, 2008.
68. Zook, J. M.; Buck, R. P.; Gyurcsányi, R. E.; Lindner, E. *Electroanalysis* 2008, *20*, 259.
69. Bodor, S.; Zook, J. M.; Lindner, E.; Tóth, K.; Gyurcsányi, R. E. *J. Solid State Electrochem.* 2009, *13*, 171.
70. Bakker, E.; Pretsch, E. *Anal. Chem.* 1998, *70*, 295.
71. Bakker, E.; Willer, M.; Lerchi, M.; Seiler, K.; Pretsch, E. *Anal. Chem.* 1994, *66*, 516.
72. Ceresa, A.; Pretsch, E. *Anal. Chim. Acta* 1999, *395*, 41.
73. Qin, Y.; Bakker, E. *Talanta.* 2002, *58*, 909.
74. Shultz, M. M.; Stefanova, O. K.; Mokrov, S. B.; Mikhelson, K. N. *Anal. Chem.* 2002, *74*, 510.
75. Büchi, R.; Pretsch, E. *Helv. Chim. Acta* 1975, *58*, 1573.

76. Pretsch, E.; Ammann, D.; Osswald, H. F.; Güggi, M.; Simon, W. *Helv. Chim. Acta* 1980, *63*, 191.
77. Malinowska, E.; Górski, L.; Wojciechowska, D.; Reinoso-Garcia, M. M.; Verboom, W.; Reinhoudt, D. N. *New J. Chem.* 2003, *27*, 1440.
78. Möschler, H. J.; Weder, H.-G.; Schwyzer, R. *Helv. Chim. Acta* 1971, *54*, 1437.
79. Funck, T.; Eggers, F.; Grell, E. *Chimia.* 1972, *26*, 637.
80. Feinstein, M. B.; Felsenfeld, H. *P. Natl. Acad. Sci. USA* 1971, *68*, 2037.
81. Boswell, P.; Bühlmann, P. *J. Am. Chem. Soc.* 2005, *127*, 8958.
82. Bühlmann, P.; Hayakawa, M.; Ohshiro, T.; Amemiya, S.; Umezawa, Y. *Anal. Chem.* 2001, *73*, 3199.
83. Mi, Y. M.; Bakker, E. *Anal. Chem.* 1999, *71*, 5279.
84. Peshkova, M. A.; Korobeynikov, A. I.; Mikhelson, K. N. *Electrochim. Acta* 2008, *53*, 5819.
85. Egorov, V. V.; Lyaskovski, P. L.; Il'inchik, I. V.; Soroka, V. V.; Nazarov, V. A. *Electroanalysis.* 2009, *21*, 2061.
86. Boswell, P. G.; Szijjártó, C.; Jurisch, M.; Gladysz, J. A.; Rábai, J.; Bühlmann, P. *Anal. Chem.* 2008, *80*, 2084.
87. Qin, Y.; Mi, Y.; Bakker, E. *Anal. Chim. Acta* 2000, *421*, 207.
88. Bochenska, M.; Korczagin, I. *Polish J. Chem.* 2002, *76*, 601.
89. Mikhelson, K. N.; Bobacka, J.; Ivaska, A.; Lewenstam, A.; Bochenska, M. *Anal. Chem.* 2002, *74*, 518.
90. Bereczki, R.; Gyurcsányi, R. E.; Ágai, B.; Tóth, K. *Analyst* 2005, *130*, 63.
91. Bochenska, M.; Zielinska, A.; Pomecko, R.; Kravtsov, V. C.; Gdaniec, M. *Electroanalysis* 2003, *15*, 1307.
92. Bourgeois, J.-P.; Echegoyen, L.; Fibbioli, M.; Pretsch, E.; Diederich, F. *Angew. Chem. Int. Ed. Engl.* 1998, *37*, 2118.
93. Lai, C.-Z.; Reardon, M. E.; Boswell, P. G.; Bühlmann, P. *J. Fluorine Chem.* 2010, *131*, 42.
94. Radu, A.; Peper, S.; Gonczy, C.; Runde, W.; Diamond, D. *Electroanalysis* 2006, *18*, 1379.
95. Lee, M. H.; Yoo, C. L.; Lee, J. S.; Cho, I.-S.; Kim, B. H.; Cha, G. S.; Nam, H. *Anal. Chem.* 2002, *74*, 2603.
96. Peshkova, M. A.; Timofeeva, N. V.; Grekovich, A. L.; Korneev, S. M.; Mikhelson, K. N. *Electroanalysis* 2010, *22*, 2147.
97. Szigeti, Z.; Malon, A.; Vigassy, T.; Csokai, V.; Grün, A.; Wygladacz, K.; Ye, N.; Xu, C.; Chebny, V.; Bitter, I. *Anal. Chim. Acta* 2006, *572*, 1.
98. Ceresa, A.; Qin, Y.; Peper, S.; Bakker, E. *Anal. Chem.* 2003, *75*, 133.
99. Shultz, M. M.; Stefanova, O. K.; Mokrov, S. B.; Mikhelson, K. N. *Anal. Chem.* 2002, *74*, 510.
100. Koseoglu, S. S.; Lai, C.-Z.; Ferguson, C.; Bühlmann, P. *Electroanalysis* 2008, *20*, 331.
101. Nishizawa, M.; Menon, V. P.; Martin, C. R. *Science.* 1995, *268*, 700.
102. Jágerszki, G.; Takács, Á.; Bitter, I.; Gyurcsányi, R. E. *Angew. Chem. Int. Ed.* 2010, *50*, 1656–1659.

# 5 Supramolecular Self-Assembly Governed Molecularly Imprinted Polymers for Selective Chemical Sensing

*Agnieszka Pietrzyk-Le, Chandra K. C. Bikram,*
*Francis D'Souza, and Wlodzimierz Kutner*

## CONTENTS

## INTRODUCTION

Molecularly imprinted polymers (MIPs) are synthetic polymer materials with artificially controlled properties.[1] Molecular imprinting initially involves complexation in solution of a target compound or its close analogue by suitably selected functional monomers (FMs). These monomers are composed of polymerizing and complexing moieties. Next, the resulting complex is polymerized to yield MIP. At this stage, the imprinted compound plays the role of a template (T). Then, T is removed from the MIP, leaving discrete molecular cavities in it. In this form, MIP is used for a pre-defined application involving accumulation of the compound used for imprinting or its close analogue now as the target analyte. Proper design of interactions between the binding sites of T and the recognition sites of these molecular cavities in the polymer formed is most important in MIP preparation. The technology of molecular imprinting, involving interdisciplinary studies, has recently developed remarkably. This technology, combined with supramolecular, macromolecular, and analytical chemistry, greatly affects the synthesis of MIPs. These polymers, with properly adjusted flexibility, excellent mechanical and chemical stability, and biocompatibility, have been applied extensively as stationary phases for chromatographic separations,[2] such as chiral separations using SupelMIP,[3] as selective absorbents for solid phase microextraction,[4] as selective catalysts for pharmaceutically important reactions, for example,[5] to mimic the functioning of antibodies in immunoassays (biological receptor mimics);[6] and in devices for controlled drug release and drug monitoring.[7] Particularly, MIPs have found enormously wide application as recognition units in chemical sensors.[1]

The IUPAC defines both chemical sensors (or chemosensor)[8] and biochemical sensors (or biosensor)[9] as devices transforming information that originates from a chemical reaction of an analyte or from a physical property of the system investigated into an analytically useful signal, ranging from the concentration of a given sample component to a total composition analysis.[8] Both chemosensors and biosensors are two-component devices, combining a chemical or biochemical recognition unit, respectively, with a transduction unit.

The focus of this chapter is to compile and critically evaluate different supramolecular principles adopted to fabricate highly sensitive and selective chemosensors based on MIPs. The supramolecular concepts discussed herein provide better understanding of the mechanisms of both molecular imprinting and MIP functioning.

### SUPRAMOLECULAR CONCEPTS OF MOLECULAR IMPRINTING

Molecular recognition of MIPs is based on concepts of supramolecular chemistry involving complementarity of the guest (G) and host (H) molecules. Therefore, H discriminates between one G and another.[10] A binding cavity of H must feature the exact size and shape, proper orientation of recognition sites as well as electronic character such as polarity, electron donor or acceptor ability, hydrogen bond formation property, hardness, softness, etc., to match that of G. The affinity of an H for a particular G can readily be assessed by a stability constant, $K_s$, of the H–G complex formed in solution. For the host complexation of the guest at the 1:1 stoichiometry, this constant is H + G $\leftrightarrows$ H–G, $K_s$ = [H–G]/[H][G]. Chelating and macrocyclic effects are important in stabilizing H–G.

Selective H–G binding resembles that of a lock-and-key fit such as that between an enzyme and its substrate. This binding is provided by different individual weak interactions, which synergetically are intensified, that is, they are stronger than the sum of individual interactions in an additive action. The ratio of the stability constant for one complex to that for the other stands for selectivity of the devised supramolecular system. This selectivity results from intelligent manipulation of the lock-and-key analogy and complementarity models combined with a detailed knowledge of the interactions between H and G. Moreover, kinetic selectivity is evaluated from the rate of transformation of competing guests along a reaction path as in supramolecular (enzymatic) catalysis or guest sensing.

## DESIGN OF SUPRAMOLECULAR IMPRINTED POLYMER

The first step in MIP design, similar to host design, is the selection of T (Scheme 5.1) appropriate for the selected target compound, being an analyte in the case of MIP chemosensing. Most suitably, this analyte itself serves as T. However, a close analogue of this analyte is occasionally used. This is in case the analyte is not available in its genuine form or it is not appropriate for imprinting (see Section "Ditopic Interactions between Functional Monomers and a Template for Imprinting via Electropolymerization", below).

The second step is selection of FMs (Scheme 5.1) best suited for intelligent complexation of this T in solution. The complexing moieties of FM, that is, functional groups or other recognition sites, are responsible for either covalent or noncovalent linking of FMs to binding sites of T (Scheme 5.1a).[11]

On the one hand, interactions between the complexing moieties of FMs and the binding sites of T should be sufficiently strong in order to survive polymerization by preserving the geometry of the complex with bonds exactly arranged in space (Scheme 5.1b). On the other hand, however, they should be sufficiently weak to allow for releasing T from cavities of the resulting MIP. In other words, the geometry of the T–FM complex in solution, desirably, should be maintained in the MIP. That way, orientation of the recognition sites inside the imprinted molecular cavity and the size and shape of this cavity should selectively correspond to those of the T molecule but

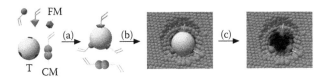

**SCHEME 5.1**   Schematic representation of the molecular imprinting principle: (FM) functional monomer, (CM) cross-linking monomer, (T) template molecule, (a) assembling of the prepolymerization complex in solution, (b) polymerization, (c) extraction of the template thus liberating the imprinted molecular cavity with recognition sites. (Adapted from Haupt, K., *Chem. Commun.*, 2003, 171–178. With permission.)

not to those of the molecule of the structurally or functionally similar interfering compound.

Molecular cavities of MIPs should be spaced somewhat apart from one another in order to minimize mutual repulsions. But they should be arranged in a way that allows all of them to interact simultaneously with T.[10,12] Moreover, analyte molecules should freely diffuse through MIPs, readily attaining these cavities.

MIPs with a predefined spatial arrangement of recognition sites in their imprinted molecular cavities can mimic, for example, enzymes featuring multirecognition sites. Enzyme binding is specific because it involves a combination of several electrostatic interactions, hydrogen bonds, hydrophobic interactions, and weak van der Waals bonds.

An MIP film used as a recognition unit of a chemosensor should be both sufficiently rigid and flexible in order to enable a complete release of T from the cavity under mild solution conditions (Scheme 5.1c) and provide full accessibility of the cavities for the analyte. Moreover, a cross-linking monomer (CM in Scheme 5.1) and/or a porogenic solvent are needed for the synthesis of MIP of a robust 3D geometry. Here, the porogenic or dispersion solvent is one of the factors responsible for exact arrangement of the recognition sites in the MIP network. Moreover, this solvent is responsible for both the macrostructure and microstructure of a cross-linked polymer and, that way, affects molecular recognition.[12] The amount of the porogenic solvent used affects the adsorption effectiveness of an MIP film. Obviously, this film should well adhere to the transducer surface.

Many different types of templates have successfully been imprinted,[13] and many different interactions have been utilized for that purpose. Those most often exploited include noncovalent (including stoichiometric noncovalent), coordinating, and covalent binding described in the following sections.

## NATURE OF SUPRAMOLECULAR INTERACTIONS

In principle, all intermolecular forces can be treated by quantum mechanics. All of them are encountered in molecular imprinting. In electrostatic interactions, ions, permanent dipoles, quadrupoles, etc., mutually interact by Coulombic forces. However, ion pairing in an aqueous environment is not exclusively electrostatic, but is also entropy driven; only in organic solvent solutions or in a gas phase do Coulombic interactions prevail.[14] Dipoles induced in atoms and molecules by electric fields of neighboring ions or permanent dipoles can interact by polarized forces. All interactions in solution involve polarization effects. Covalent or noncovalent bonds, including charge–transfer interactions, give rise to attractive interactions. Repulsive steric or exchange interactions balance the attractive interactions at very short distances.[15]

Forces leading to covalent bonds are of a short range. In molecules, they operate over the order of interatomic distances of 0.1–0.2 nm.[15] Their energy is lower the longer the bond is. The energy of ionic bonds is comparable to that of covalent bonds.[10] All these energies are, of course, extremely medium dependent.[14] Covalent bonds are sufficiently strong to allow for single-point T binding in MIPs, on the one hand, and sufficiently weak for reversible analyte recognition on the other (see "Covalent Template Binding in MIPs"). Noncovalent bonds are more effective in the reversible interaction of MIP with T (see Section "Noncovalent Template Binding in MIPs").

Ion–dipole interactions are encountered in many MIP preparation procedures. These interactions, as in those in complexes of cations of alkali metals or quarternized primary amines with crown ethers, have extensively been exploited.[16–19] Here, the ether oxygen atoms play the same electron-donating role as those of the polar water molecules where oxygen atoms' lone electron pairs are attracted to the cation positive charge.

Coordinating interactions are included in those of the ion–dipole nature. In the case of interactions of nonpolarizable metal cations with hard bases, the coordinating bonds are mostly electrostatic. However, coordinating (dative) bonds with a significant covalent component, as in $[Ru(bpy)_3]^{2+}$, are often encountered in supramolecular assemblies.[10]

Binding energy of ion pairs in aqueous solutions is dependent on the ionic strength of these solutions as is the distance between the ions, which differs for all noncovalent forces.[14]

Dipole–dipole interactions demonstrate, for instance, carbonyls. However, these interactions can be weak in solution.

Relatively strong and highly directional hydrogen bonds,[10] regarded as a particular kind of dipole–dipole interactions can be formed both in polar and nonpolar solvent solutions.[15] In these H-bridged bonds, hydrogen atoms can strongly interact with neighboring electronegative atoms by virtue of their tendency to become positively polarized and their uniquely small size.[15] Typically, length of the H···O bond is 0.25–0.28 nm; however, a hydrogen bond as long as 0.30 nm may also be significant, depending on the environment. Hydrogen bonds involving larger atoms, such as chlorine, are generally longer and may be weaker because of the lower electronegativity of the larger halide atom acceptor. They can occur intermolecularly and intramolecularly.[15] Macromolecular biological assemblies linking different segments inside the molecules, as in proteins or nucleic acids holding the double helix structure of DNA, are formed via hydrogen bonding. Their involvement in setting up a macromolecular structure is sometimes referred to as hydrogen-bond polymerization.[15] Importantly, the acidity of the C–H proton is increased in the presence of electronegative atoms located near the carbon atom, as in permanent dipoles such as C–H···O or C–H···N. However, the O–H···π hydrogen bonds are rather weak. Very high values of stability constants ($K_s = 10^3$–$10^7$ M$^{-1}$) of complexes of the templates bound by multiple hydrogen bonds[11] confirm that these bonds resemble a covalent bond during the imprinting (see Section "Noncovalent Template Binding in MIPs"). Importantly, interaction of MIPs with T via hydrogen bonds is reversible, and T removal is facile.

Among the cation–π interactions, the interaction of alkali and alkaline earth metal cations with the C=C double bonds is important in biological systems.[10]

The π–π stacking between electron-withdrawing and -donating aromatic rings is rather weak. These interactions are considered as electrostatic interactions between whole molecules. They are also engaged in molecular imprinting.[16] However, dispersion forces are more important than electrostatic interactions in π–π stacking in the case of individual pairs of atoms.[10] The π–π interactions participate in valuable contemporary molecular imprinting involving heteroarene FMs (such as pyrrole or thiophene).[16] Interactions between π-donors and π-acceptors have efficiently been used in imprinting.[20]

Charge-fluctuation forces, electrodynamics' forces, and (induced dipole)–(induced dipole) forces[15] are described as classical London–Eisenschitz dispersion forces with electron correlation.[14] Dipoles' fluctuation can originate either from repulsive or attractive interactions of a long (>10 nm) and short (~0.2 nm) range. These forces do not obey a simple power law with respect to interatomic or intermolecular distance. Instead, they decrease with the sixth and the twelfth, respectively, power of this distance. The attraction is additive with respect to every other bond in the molecule contributing to the overall interaction energy.[10] In addition to bringing molecules together, dispersion forces also tend to mutually align or orient molecules. However, this orientation effect is weak and affected by the neighboring molecules introducing nonadditive interactions. Dispersion forces contribute to the van der Waals interactions. The distinction of van der Waals interactions from high-order electrostatic interactions, for example, dispersive forces, is hardly justified. They are not restricted to polarization of an electron cloud.

These van der Waals interactions arise from the polarization of an electron cloud by an adjacent nucleus located in close proximity, leading to weak electrostatic attractions.[10] These interactions govern many important phenomena such as adhesion, surface tension, physical adsorption, or wetting. They are responsible for "properties of gases, liquids and their thin films, the strengths of solids, the flocculation of particles in liquids, and the structures of condensed macromolecules, such as proteins and polymers."[15] Moreover, organic guest molecules like toluene are incorporated within molecular cavities of a host macrocycle, forming inclusion complexes via van der Waals interactions.[10]

While the hydrophobic effect is rarely used to prepare MIPs, the imprinted cavities formed can bind analytes in water by virtue of this effect. The hydrophobic effect consists in exclusion from a solution of polar solvents, particularly water, of large molecules or those that are weakly solvated (e.g., via hydrogen bonds or dipolar interaction).[10] For instance, cyclodextrins (CDs) bind organic guest compounds in water by a hydrophobic effect. Hydrophobic effects can be enthalpy or entropy driven.[14]

## REMOVAL OF IMPRINTED TEMPLATE

Exhaustive T removal is crucial for efficient MIP operation. In this removal, the force of interaction between an MIP and T is decisive. It dictates T molecules' emptying of the imprinted cavities, which that way become ready for interaction with analyte molecules. For this T removal, several different procedures are developed. These include microwave-assisted extraction,[21,22] cyclic electrochemical oxidation and reduction of T in MIP,[23] electrochemical overoxidation of MIP,[24] supercritical fluid desorption,[22,25] and most commonly used, extraction with a solvent that strongly interacts with MIP resulting in its swelling.[26]

Noticeably, the most efficient, especially if using trifluoroacetic acid or formic acid, is microwave-assisted extraction.[26] This microwave treatment shortens the time of T removal and decreases the solvent amount used as compared to that consumed in the most common extraction procedures such as those that involve exposing template-loaded MIPs to the extracting solvent followed by Soxhlet extraction.[21] That way, for example, β-estradiol used as T was removed from the MIP film prepared at the mole ratio of T to methacrylic acid (MAA) to ethylene glycol dimethacrylate (EGDMA) as 1:30:150 with 2,2′-azobis(2-methylisobutyronitrile) (AIBN) used as the initiator.[21] Then, β-estradiol was determined under flow-injection analysis (FIA) conditions with the fluorimetric transduction of the detection signal.[21] The dynamic linear concentration range for β-estradiol extended from 4 to 80 µg L$^{-1}$ with the limit of detection (LOD) of 1.2 µg L$^{-1}$. A disadvantage of the microwave-assisted extraction is, however, that the MIP may degrade and, therefore, lose its selectivity.

An electrochemical removal of T, for example, by linear potential cycling through its oxidized and reduced states inside MIP or overoxidation of an MIP, is even faster than that microwave assisted. For example, the (paracetamol T)-loaded polypyrrole-based MIP film was extracted with the 0.1 M KCl, 0.05 M phosphate buffer (pH = 7.0) solution by potential cycling between −0.60 and +1.0 V at 100 mV s$^{-1}$.[23]

In another example, a polypyrrole MIP film loaded with sulfamethoxazole as T was electrochemically overoxidized in order to remove this T.[24] For that, the potential was 20 times linearly cycled in a range of 0.80 to 1.20 V at 50 mV s$^{-1}$ in 0.1 M NaOH. For sulfamethoxazole, the dynamic linear concentration range was 0.025 to 0.75 mM and LOD was 35.9 µM.

Extraction with a strong base solution is most useful for removal from MIP of, for example, a protonated amine used as T.[17,19,27] This extraction is very efficient because the amine becomes deprotonated in this solution and, therefore, loses its charge, thus allowing for dissociation of the inclusion complex of this amine, for example.

None of the T extraction techniques can completely eliminate T bleeding after the extraction, not even desorption with the use of a supercritical fluid,[22,25] for example, carbon dioxide, as the extracting solvent,[28] or extraction with a solvent that strongly interacts with the MIP causing its swelling.[26]

Solvation of recognition sites of the MIP cavity with an extracting solvent leads to cavity swelling and efficient removal of the template.[11] The original shape is then restored if T binds again (see Section "Covalent Template Binding in MIPs", and Scheme 5.2).

## COVALENT TEMPLATE BINDING IN MIPS

Weak covalent bonds can bind T to boronic diesters or Schiff bases for example. Recognition of 1,2- and 1,3-diols in carbohydrates, and saccharides in particular, by boronic acid groups is important in supramolecular chemistry. The saccharide chemosensors based on boronic acid receptors are sensitive and selective with respect to any chosen saccharide. These receptors form five- or six-membered covalent rings of cyclic boronate esters. In these rings, the O−B−O bond angle is compressed to ~108° as compared to that of boronic acid itself, which is ~120°. This compression is a result of the boron atom hybridization change from sp$^3$ to sp$^2$. Favorably with respect to reversibility of the chemosensor operation, these rings can readily be ruptured and the original sp$^3$ hybridization of the boron atom restored by solution alkalization. This alkalization leads to the formation of a new covalent B−OH bond between the boron atom and the hydroxy group taken from the basic solution.

For instance, two FM molecules of 4-vinylbenzeneboronic acid can selectively recognize one T molecule of phenyl-α-D-mannopyranoside (Scheme 5.2a). The covalently bound system with two boronic acid binding sites is cross-linked with

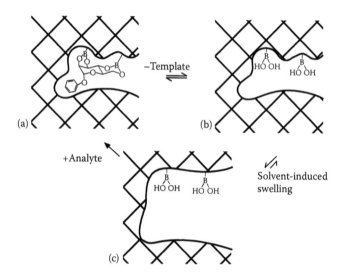

**SCHEME 5.2**   4-Vinylbenzeneboronic acid template (a) can be removed from MIP with water or methanol, leaving empty imprinted cavities (b). Solvation of the boronic acid functional groups in these cavities leads to their swelling (c). If the target analyte binds, the original shape of the cavity is restored. (Adapted from Wulff, G., *Chem. Rev.*, 2002, 102, 1–27. With permission.)

the excess of EGDMA.[11] Then, up to 95% of T is removed with the water–methanol mixture (Scheme 5.2b). Upon analyte binding, the original cavity shape is restored (Scheme 5.2c). Most often, recognition sites of boronic acid have been used in combination with other recognition sites for chemosensor preparation. For instance, monoaza-18-crown-6 ether (Scheme 5.3) served as the other site for binding of the amine group of D-glucosamine hydrochloride in the fluorescent photoinduced electron transfer (PET) chemosensor.[29] In this chemosensor, the anthracene fluorophore participates in PET involving the nitrogen atom of the azacrown ether moiety. This

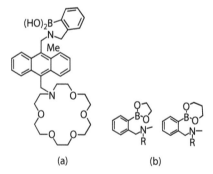

**SCHEME 5.3**   Structural formula of (a) the functional monomer for a ditopic fluorescence chemosensor for determination of D-glucosamine hydrochloride and (b) five- and six-member rings of boronic ester moieties with diol-containing compounds; R = -CH$_3$. (From Hall, D.G., *Boronic Acids Preparation, Applications in Organic Synthesis and Medicine*, Wiley-VCH: Weinheim, 2005, pp. 441–479.)

moiety can recover fluorescence by hydrogen bonding of the quarternized amine group. Actually, the fluorescence is recovered as a result of two chemical inputs, that is, interactions of anthracene with the quarternized amine group and with the diol group. The complex stability constant is, $K_s \approx 18$ M$^{-1}$, in the 33.2% (v/v) ethanol–water buffer of triethanolamine (pH = 7.18). Moreover, the anthracene substituent acts as a rigid spacer between the two recognition sites with the appropriate space to accommodate for the D-glucose guest (Schemes 5.3b and 5.10).[30]

Recently, Au nanoparticles (AuNPs) were functionalized both with thioaniline electropolymerizable sites (**1** in Scheme 5.4) and (mercaptophenyl)boronic acid recognition sites (**2** in Scheme 5.4).[31] Then, these AuNPs were used as FMs for binding of an antibiotic template, such as neomycin, kanamycin, or streptomycin, all bearing diol groups. The electropolymerization of the thioaniline-functionalized AuNPs in the presence of one of the antibiotics onto Au surfaces yields bisaniline–cross-linked AuNP composites. Mercaptoethanesulfonic acid (**3** in Scheme 5.4), which additionally modifies these AuNPs, stabilizes them, thus preventing precipitation. Because of the wave coupling between the localized plasmon of the AuNPs and the surface plasmon associated with the surface of the Au substrate, the response of the surface plasmon resonance (SPR) spectroscopy is amplified, enabling determination of small molecules like the antibiotics. The LOD for neomycin, kanamycin, and streptomycin was very low, equaling $(2.00 \pm 0.21)$ pM, $(1.00 \pm 0.10)$ pM, and $(200 \pm 30)$ fM, respectively.[31]

Another example of covalent imprinting might be a nanosize matrix of C$_{60}$ fullerene, bifunctionalized with boronic acid, used for regioselective and chiroselective imprinting of saccharides as templates (Scheme 5.5).[32] In this matrix, only single

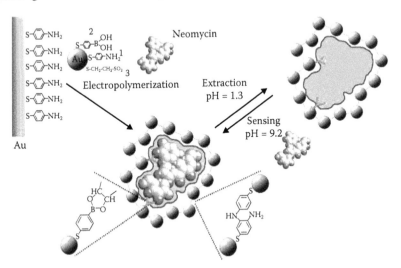

**SCHEME 5.4**  Imprinting of molecular recognition sites for determination of antibiotics (for example, neomycin) through deposition by electropolymerization of a bisaniline–cross-linked AuNP composite on the surface of an Au substrate; 1-thioaniline, 2-(mercaptophenyl) boronic acid, 3-mercaptoethanesulfonic acid. (Adapted from Frasconi, M. et al., *Anal. Chem.*, 2010, 82, 2512–2519. With permission.)

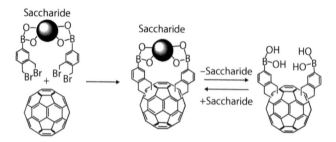

**SCHEME 5.5** Regioselective and chiroselective addition of two 3,4-bis(bromomethyl)-phenylboronic acid groups to $C_{60}$ fullerene. (Adapted from Ishi-i, T. et al., *Tetrahedron*, 1999, 55, 3883–3892. With permission.)

binding sites are formed; therefore, generating of many different binding sites is avoided. These chiroselective MIPs were used to test the D- versus L-form selective binding of the target saccharide analyte molecules.[32]

## NONCOVALENT TEMPLATE BINDING IN MIPS

Out of noncovalent binding, electrostatic interactions and hydrogen bonds are often employed for molecular imprinting.[11]

In order to reach high MIP selectivity, a nearly fourfold molar excess of FM is required for noncovalent imprinting. In effect, binding sites impressed in MIP are not only located inside the imprinted cavities but also distributed all throughout the polymer. For instance, MAA, often used as FM for free-radical polymerization, distributes its carboxy groups over the entire MIP film (Scheme 5.6a).[11,33] However, only a limited number of molecules of this monomer are engaged in covalent binding of T molecules (Scheme 5.2). Moreover, the T-emptied cavities are swelled as a result of solvation of their carboxy recognition sites (Scheme 5.6b). However, to differentiate, the cavities

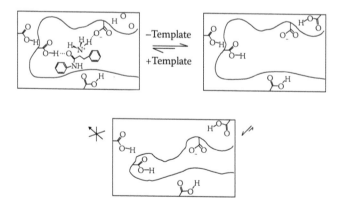

**SCHEME 5.6** Molecular imprinting by noncovalent bonding of L-phenylalanine anilide as a template and MAA forming a recognition site. (Adapted from Wulff, G., *Chem. Rev.*, 2002, 102, 1–27; Sellergren, B. et al., *J. Am. Chem. Soc.*, 1988, 110, 5853–5860. With permission.)

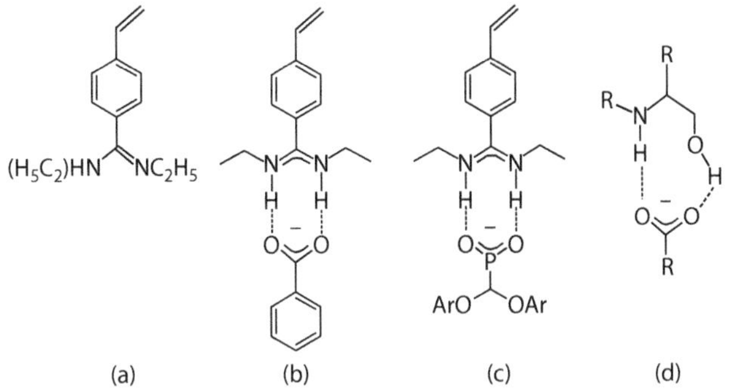

**SCHEME 5.7**   Structural formula of (a) *N,N'*-diethyl-4-vinylbenzamidine, (b) amidine–benzene–carboxylic acid complex, and (c) amidine–phosphate complex. (d) Example of charge-enforced hydrogen bonds in the complex of 2-α-*O*-methyl-*N*-acetylneuraminic acid and rhesus rotavirus hemagglutinin. (Adapted from Wulff, G., *Chem. Rev.*, 2002, 102, 1–27; Strikovsky, A.G. et al., *J. Am. Chem. Soc.*, 2000, 122, 6295–6296; and Mazik, M., *Chem. Soc. Rev.*, 2009, 38, 935–956. With permission.)

shrink to their original volumes during covalent T rebinding (Scheme 5.2). This is unlike in the case of noncovalent binding, which results in merely ~15% rebinding (Scheme 5.6c).

Strong stoichiometric noncovalent bonds can, for instance, be formed between the amidine group of FM (Scheme 5.7a) and the carboxy (Schemes 5.7b and 5.7d) or phosphate group of T (Scheme 5.7c).[34,35]

## NONCOVALENT INCLUSION COMPLEX FORMATION IN MIPs

Noncovalent inclusion complexes, or H–G complexes, most often involve macrocyclic hosts such as CDs, calixarenes, carcerands, cryptands, cucurbiturils, or crown ethers.[36]

Conformationally mobile hosts, such as 18-crown-6, can rapidly adjust to the locally imposed topological conditions of inclusion complex formation or dissociation.

Inclusion complexes of CDs are formed by virtue of interaction of their hydrophobic inner molecular cavities with different hydrophobic guest molecules. The hydrophilic outer surface of the CD molecule stabilizes the resulting inclusion complex in a polar solvent. Advantageously, CDs can be chemically polymerized at ambient temperature without any initiator but with the use of, for example, glutardialdehyde or toluene-2,4-diisocyanate (TDI) as CM. For MIP preparation, β-CD can be used as FM to interact with T in addition to serving as CM to form a 3D polymer of the β-CD (β-CDP) matrix of decreased flexibility of the β-CD moieties. Then, there is neither a problem of the presence of nonspecific binding sites nor that of collapse/distortion of the cavity during the subsequent T removal. Thus, the binding molecular cavity of β-CD becomes highly selective.[37]

CDs featuring cation exchange sites in addition to their inclusion sites, such as carboxymethylated CDs, are available for ion exchange at pH values exceeding p$K_a$

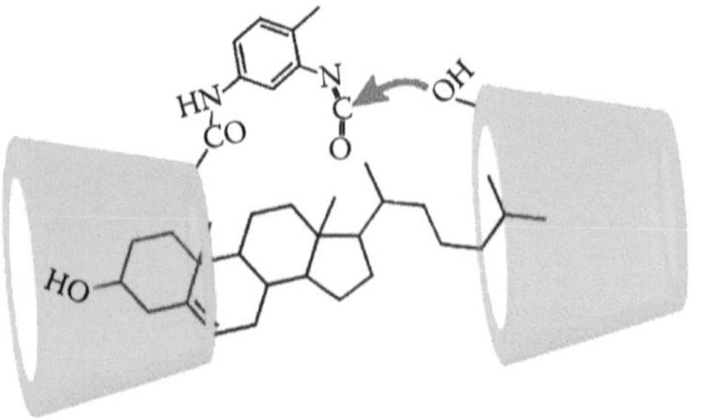

**SCHEME 5.8**   Proposed mechanism for molecular imprinting of cholesterol by using β-CD as the functional monomer and toluene-2,4-diisocyanate as the cross-linking monomer. (Adapted from Hishiya, T. et al., *J. Am. Chem. Soc.*, 2002, 124, 570–575. With permission.)

≅ 4.6 of their carboxy groups.[38,39] Binding properties of films of the carboxymethyl-ated β-CD polymer (β-CDPA) with respect to a range of tricyclic antidepressants, such as azepine and phenothiazine, were demonstrated by using simultaneous cyclic voltammetry and piezoelectric microgravimetry at an electrochemical quartz crystal microbalance.[38] A simple two-step procedure of determination of these drugs comprises a preconcentration step followed by the differential pulse voltammetry quantification. As expected, the accumulating properties of the β-CDPA film exceeded those of the β-CDP film.

Cavities of two or three β-CDs can simultaneously accommodate large guest molecules such as steroids. Scheme 5.8 presents a β-CD dimer, which provides its two β-CD cavities for binding of one cholesterol molecule.[40] Bridging together two β-CD molecules by TDI used as CM is enormously accelerated by the presence of T. Moreover, secondary -OH groups, located on the larger rim of the β-CD toruses, which are otherwise hardly reactive, eagerly participate in this bridging (Scheme 5.8).[40]

A thin β-CDP film imprinted with the common cough suppressant dextromethorphan was used as the recognition unit in the SPR chemosensor. The LOD was 0.035 μM dextromethorphan, and the dynamic linear concentration range varied from 0.035 μM to 6.00 mM.[41] TDI was used, as in the case of the chemosensor for determination of cholesterol described above, for condensation of the β–CD molecules.

## COORDINATING TEMPLATE BINDING IN MIPS

Excess of recognition sites is unnecessary for definite T coordination in the course of polymerization. This issue is addressed here.

### COORDINATING OF BORONIC ACID AND ITS ESTERS IN MIPS

For instance, an imidazole, phosphate, or citrate moiety can be bound by a coordinating moiety, for example, boronic acid, to form a binary complex (Scheme 5.9).[11,42]

Notably, the dissociation state of functional groups might, at the same solution acidity, be different in the polymer and in solution because of different $pK_a$ values of these groups in these two environments.[11] For example, $pK_a = 10.6$ characterizes

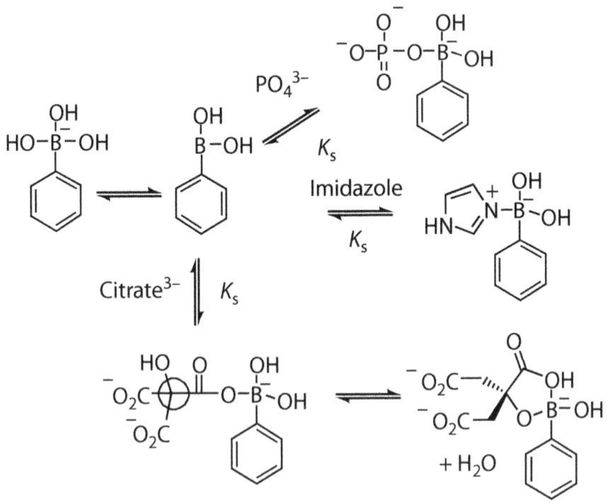

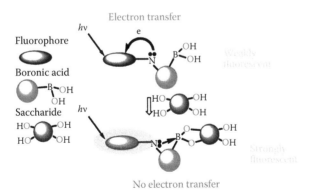

**SCHEME 5.9** Combined Lewis acid-base and complex equilibria of phenylboronic acid. (Adapted from Wulff, G., *Chem. Rev.*, 2002, 102, 1–27; Bosch, L. I. et al., *Tetrahedron*, 2004, 60, 11175–11190. With permission.)

amidine groups in solution. However, $pK_a$ of polymeric amidine is lower, remaining in the range of 8.62 to 9.02.[11]

The boron–nitrogen covalent bond is weak; its strength is comparable to that of a typical hydrogen bond.[43] The (boronate ester)–saccharide complex more strongly interacts intramolecularly with the nitrogen atom of the amine group than the boronic acid does. Upon binding of saccharide to the boronic acid group, acidity of the boron atom increases and formation of the B–N bond is promoted.[43] Therefore, binding of a saccharide results in a significant increase in the fluorescence intensity (Schemes 5.3 and 5.10).

**SCHEME 5.10** Modular design strategy in a mechanism of photoinduced electron transfer (PET). Formation of a boron-nitrogen bond upon binding of boronic acid with diol-containing compounds causes the fluorescence change. (Adapted from Franzen, S. et al., *J. Phys. Chem. B*, 2003, 107, 12942–12948; James, T.D. et al., *Chem Commun.*, 1996, 281–288. With permission.)

## Coordinating via Metal–Porphyrin Complex Formation in MIPs

Transition-metal atoms readily coordinate nitrogen atoms. Accordingly, a central transition-metal atom in metalloprotoporphyrins can easily coordinate axially a nitrogen atom of a heteroaromatic ring.

For example, a Zn atom of zinc(II) protoporphyrin (ZnPP) may coordinate a nitrogen atom of the purine moiety of a nucleic base such as 9-ethyladenine (9EA)[45] (Scheme 5.11), of the quinoline moiety of cinchonidine[46] (Scheme 5.12), or of the imidazol ring of histamine[47] (not shown). One of the ways of recognizing hemoglobin uses coordination of its Fe central metal atom by porphyrin.[48]

Metalloprotoporphyrins were used as luminescent FMs.[45–47] Additionally, MAA was used as a second FM and EGDMA as CM. The free-radical polymerization initiator was 2,2′-azobis(2,4-dimethylvarelonitrile) (ADMVN). In the fluorescence chemosensors using these monomers, polymer exposure to the imprinted compound quenches the steady-state fluorescence of the ground bulk polymer.

Interaction of 9EA with Zn(II) porphyrin (Scheme 5.11) resulted in quenching of MIP luminescence at 605 nm (excited at 423 nm). The polymer was sensitive to 9EA in the concentration range of 0.01 to 0.1 mM. However, the test solution has already become saturated with 9EA at a concentration of 0.15 mM.[45]

Excited at 404 nm, the luminescence at 604 nm of MIP containing Zn(II) porphyrin was significantly quenched upon binding of (–)-cinchonidine, even in the low concentration range of 0.01 to 2 mM. However, this quenching was negligible if the polymer was exposed to (+)-cinchonine. That way, the presence of binding sites highly complementary to (–)-cinchonidine in MIP was confirmed (Scheme 5.12).[46]

The stability constant for the complex of zinc(II)-[7,12-diethenyl-3,8,13,17-tetramethyl-21$H$,23$H$-porphine-2,18-dipropanoato(4-)-$\kappa N$21,$\kappa N$22,$\kappa N$23,$\kappa N$24] with histamine in the concentration range of 0.1 to 1 mM was $K_{MIP} = 4500$ M$^{-1}$, and it was $K_{MIP} = 270$ M$^{-1}$ in the range of 4 to 10 mM.[47] For the lower histamine concentration, the fluorescence intensity of MIP decreased with the increase in the histamine concentration. The dynamic linear concentration range was 0.1 to 4 mM histamine associated with a 40% decrease in the emission intensity at 583 nm. However, the

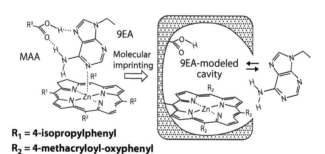

$R_1$ = 4-isopropylphenyl
$R_2$ = 4-methacryloyl-oxyphenyl

**SCHEME 5.11**  Representation of molecular imprinting of 9-ethyladenine (9EA) using 5,10,15-tris(4-isopropylphenyl)-20-(4-methacryloyl-oxyphenyl)porphyrin zinc(II) and MAA as FMs. (Adapted from Matsui, J. et al., *J. Am. Chem. Soc.*, 2000, 122, 5218–5219. With permission.)

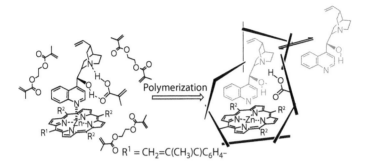

**SCHEME 5.12**   Schematic representation of molecular imprinting of (–)-cinchonidine using vinyl-substituted Zn(II) porphyrin and MAA as FMs. Ethylene glycol dimethacrylate (EGDMA) was used as the cross-linking monomer and 2,2'-azobis(2,4-dimethylvarelonitrile) (ADMVN) as the polymerization initiator. (Adapted from Takeuchi, T. et al., *Anal. Chem.*, 2001, 73, 3869–3874. With permission.)

fluorescence intensity no longer decreased for histamine concentration exceeding 4 mM, indicating that ZnPP-based recognition sites have become saturated.

Double-metal chelation has been employed for the imprinting of 8-hydroxy-2'-deoxyguanosine (8-OHdG) with the use of two FMs, namely, methacryloylaminoantipyrine-Fe(III) [MAAP-Fe(III)] and methacryloylamidohistidine-Pt(II) [MAH-Pt(II)] (Scheme 5.13), and fabrication of a selective piezoelectric microgravimetry (PM), i.e., acoustic, chemosensor, using a quartz crystal microbalance (QCM).[49] This (double-metal)–chelate monomer system was more effective than a (single-metal)–chelate

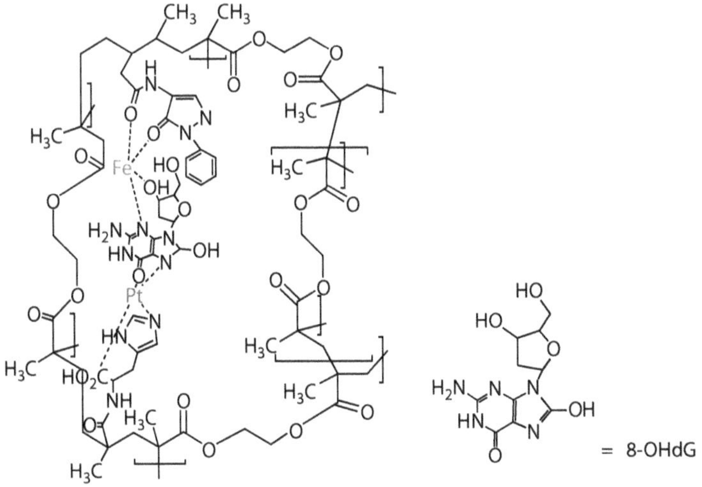

**SCHEME 5.13**   Representation of templating of 8-hydroxy-2'-deoxyguanosine (8-OHdG) with the MAH-Pt(II)–MAAP-Fe(III) double-metal chelating functional monomer for piezoelectric microgravimetry chemosensor based on QCM. (Adapted from Say, R. et al., *Anal. Chim. Acta*, 2009, 640, 82–86. With permission.)

monomer of MAH-Pt(II) system. For the (double-metal)–chelating system, LOD and the dynamic linear concentration range was 0.0075 and 0.01–3.50 µM, respectively.[49] The stability constant was $K_{MIP} = 1.54 \times 10^5$ M$^{-1}$ for the complex of 8-OHdG and MAH-Pt-8-OHdG-MAAP-Fe formed inside the MIP thin film.[49]

MAH-Pt(II) was also used as a metal-chelating FM for coordination of guanosine templates of DNA (Scheme 5.13). For that, nanoshell chemosensors were prepared by polymerizing thiolated methacryloylamidocysteine attached to CdS quantum dots (QDs).[50] The binding affinity of the guanosine-imprinted nanocrystals, investigated by the Langmuir and Scatchard methods, discriminated between a guanosine nucleotide ($K_s = 4.841 \times 10^6$ M$^{-1}$) and a free guanine base ($K_s = 0.894 \times 10^6$ M$^{-1}$). Additionally, the guanosine T was more favored for complexing of single-stranded rather than double-stranded DNA.[50]

## BINDING OF TEMPLATES IN MIP FILMS PREPARED BY ELECTROPOLYMERIZATION

The MIP films described above were prepared by free-radical polymerization induced by heat or UV light. However, this polymerization encounters several limitations. These include low or no control over the film thickness (in case of spin coating or surface casting), no possibility of use of volatile solvents (in case of thermally induced polymerization) and decomposition of light-sensitive templates during polymerization induced by UV light.

Despite these disadvantages, several biologically important amines were successfully imprinted by virtue of this polymerization. For instance, a melamine T was imprinted with MAA and EGDMA as FM and CM, respectively (Scheme 5.14).[51] The resulting MIP film was applied as the melamine recognition unit in a potentiometric chemosensor. For protonated melamine, the dynamic linear concentration range extended from 0.15 µM to 0.01 M and LOD was 6.0 µM.[51]

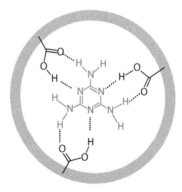

**SCHEME 5.14** Representation of the MIP cavity with melamine imprinted by free-radical polymerization. (Adapted from Liang, R. et al., *Sens. Actuators B*, 2009, 141, 544–550. With permission.)

If MIP is made of an insulating material, its application as a recognition unit of an electrochemical sensor may be difficult because of the lack of a direct path for electron conduction between its recognition sites and the electrode substrate. Novel hybrid materials combining MIPs with conducting polymers feature a network of molecular wires connecting the recognition sites of an MIP film to the electrode surface.[41]

Alternately, highly electrically and thermally conducting multiwall carbon nanotubes (MWCNTs) were used to transfer charge in the MIP composite applied as a recognition unit of a potentiometric chemosensor. For that, a copolymer of FMs such as MAA and trimethylolpropane trimethacrylate (TMPTMA), used for imprinting of dopamine, was cross-linked with vinyl groups modifying the surface of MWCNTs. A glassy carbon electrode coated with this MIP film was selective with respect to dopamine in the presence of ascorbic acid. The dynamic linear concentration range was 50 µM to 0.2 mM dopamine.[52]

Electrochemical polymerization is a frequently selected procedure among those available for fabrication of MIP thin films directly onto a transducer surface.[41,53] In this procedure, no polymerization initiator such as EGDMA, neither UV light, nor heat is needed. A polymer film is deposited directly onto an electrode surface. This film adheres well to a (roughened) electrode surface. Film thickness is governed by the amount of charge transferred. Surface morphology is controlled by the selection of a suitable porogenic solvent and supporting electrolyte. Solvent swelling and inclusion of ions of a supporting electrolyte tune the rigidity and porosity of the film.

For electropolymerization, electroactive FMs containing electropolymerizing moieties such as pyrrole,[23,24,54–58] phenol,[59] 1,2-aminophenol,[60] aminophenyl boronic acid,[61] 1,2-aminothiophenol,[62] aniline,[41,63] 1,2-phenylenediamine,[64–67] mercaptobenzimidazole,[68,69] metalloporphyrin,[70] or thiophene,[16–19,71] are most often used.

## FORMATION OF HYDROGEN BONDS BETWEEN FUNCTIONAL MONOMERS AND TEMPLATE FOR IMPRINTING VIA ELECTROPOLYMERIZATION

In this section, MIPs with templates imprinted using hydrogen bonds and deposited by electropolymerization are described.

For example, T forming hydrogen bonds with a positively charged polymer can selectively be imprinted in a polypyrrole film. For instance, a paracetamol-imprinted polypyrrole film (Scheme 5.15) deposited onto a pencil-graphite electrode served as the recognizing element of a selective chemosensor.[23] For paracetamol, the dynamic linear concentration range and LOD was 5 µM to 5 mM and 0.79 µM (S/N = 3), respectively. The response of this chemosensor was stable and reproducible.[23]

Moreover, caffeine,[72] glycoproteins,[73] and sulfamethoxazole were imprinted in polypyrrole.[24] The probability of hydrogen bond formation is higher for overoxidized polypyrrole.[24] Molecularly imprinted overoxidized polypyrrole was enantioselective with respect to L-glutamic acid.[54]

Dopamine was molecularly imprinted using a 1,2-aminophenol electroactive FM.[67] Hydrogen bonds formed between the dopamine T and the resulting MIP film were then ruptured by washing the film with 0.5 M $H_2SO_4$ to remove the imprinted T. Linear concentration range of the calibration plot for dopamine determination was 0.02 to 0.25 µM with LOD of 1.98 nM dopamine at S/N = 3.[67]

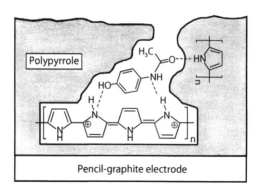

**SCHEME 5.15**   Representation of a pencil-graphite electrode coated with the paracetamol-imprinted polypyrrole film. (Adapted from Özcan, L., Sahin, Y., *Sens. Actuators B*, 2007, 127, 362–369. With permission.)

In another MIP chemosensor approach to the dopamine determination, iron *tetra*(1,2-aminophenyl)porphyrin was used as FM for dopamine imprinting.[74] Amino groups of porphyrins formed hydrogen bonds with hydroxy groups of dopamine. A carbon-fiber microelectrode was coated with the resulting dopamine-selective MIP film to fabricate an electrochemical microsensor for determination of dopamine.

Several different signal transduction schemes, such as those involving spectroscopic, electrochemical, or PM techniques with MIP-aided sensing, have already been implemented.[1,75–77] The combination of molecular imprinting and PM transduction is one of the most attractive routes to enhance both selectivity and sensitivity of a PM (acoustic) sensor.[78] Toward this end, a PM shear-thickness-mode bulk-acoustic-wave resonator is a suitable transducer, offering LOD at a single-nanogram mass level.

## DITOPIC INTERACTIONS BETWEEN FUNCTIONAL MONOMERS AND TEMPLATE FOR IMPRINTING VIA ELECTROPOLYMERIZATION

Recently, thiophene derivatives gained particular attention as electroactive FMs.[16–19] They are insoluble in aqueous solutions, and the resulting MIP films are highly conductive and quite stable both chemically and mechanically.

Chemically recognizing moieties of electroactive FMs should be positioned spatially around the molecular cavity, formed to function as a permanent memory element for imprinted T. Imprinting of T in the polythiophene film is facilitated, as in the case of polypyrrole, by formation of hydrogen bonds between the heteroatoms of T and the atoms of recognition sites of thiophene. Additionally, $\pi$–$\pi$ interactions between the conducting polymer and T make the cavity more selective. The presence of $\pi$ electrons enhances the stabilizing dispersion interactions between the aromatic moiety of FM and T. This stabilization is enhanced if a $\pi$-delocalized bonding is involved, for example, in the $\pi$-conducting polymer chain. Presumably, a T molecular dipole induced, depending on the T structure, may increase electrostatic interactions between T and FM in their complex.[79] There are several examples of derivatives of bithiophene useful for imprinting, by electropolymerization, of biogenic amines, such as melamine,[16] histamine,[17] adenine,[18] or dopamine.[19] In this

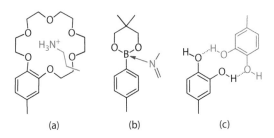

(a)                    (b)                    (c)

**SCHEME 5.16**   Representation of interactions of electroactive FMs with the imprinted templates by formation of (a) inclusion complex, (b) coordination bond, and (c) hydrogen bonds.

imprinting, primary protonated amine groups of the amines were complexed by a benzo-18-crown-6 moiety of benzo-[18-crown-6]-bis(2,2′-bithien-5-yl)methane used as FM **4** (Schemes 5.16a and 5.17). The imine nitrogen atom of the imidazole or purine moiety of histamine and adenine, respectively, was coordinated by the boron atom of the dioxaborinane substituent in the [4-(5,5-dimethyl-1,3,2-dioxaborinane-2-yl)-phenyl]-bis(2,2′-bithien-5-yl)methane FM **5** (Schemes 5.16b and 5.17). Diol groups of dopamine were immobilized by hydrogen bonds with 3,4-dihydroxyphenyl substituent of the *meso*-(3,4-dihydroxyphenyl)-bis(2,2′-bithien-5-yl)methane FM **6** (Schemes 5.16c and 5.18).[80,81] Sensitivity and selectivity of the MIP film was largely increased by cross-linking its FMs with the 2,2′-bis(2,2′-bithiophene-5-yl)]-3,3′-bithianaphthene electroactive CM **7** and by the presence of a porogenic ionic liquid, trihexyl(tetradecyl)phosphonium tris(pentafluoroethyl)-trifluorophosphate **8**, in the prepolymerization solution (Scheme 5.18).[16]

If T is electroactive in the potential range of electropolymerization of FM, products of its electrode reaction may be imprinted instead of the genuine T itself. Moreover, these products may adsorb and block the electrode surface. Therefore, the electrode reaction of T should be eliminated. Toward that, at least two imprinting procedures are available. In one, an electroinactive analogue is used as T, instead.[75–77,82] In another, a barrier underlayer is first deposited,[17,19] which prevents electrochemical transformation of T. Conducting polymer films have widely been used as such

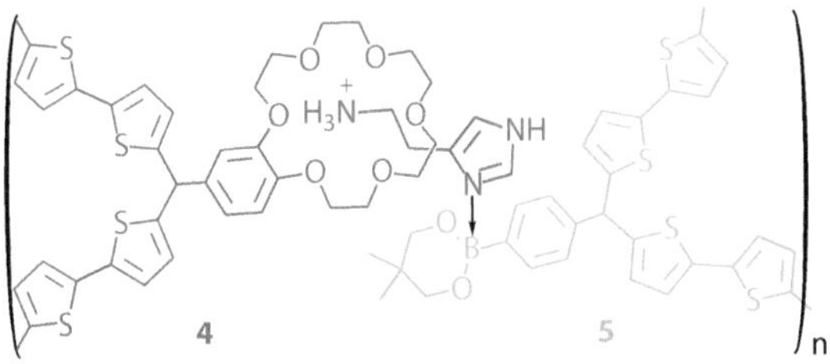

**SCHEME 5.17**   Schematic structural formula of the histamine-imprinted MIP film prepared by electropolymerization of benzo-[18-crown-6]-bis(2,2′-bithien-5-yl)methane **4** and [4-(5,5-dimethyl-1,3,2-dioxaborinane-2-yl)-phenyl]-bis(2,2′-bithien-5-yl)methane **5** in the presence of monoprotonated histamine (in red). (Adapted from Pietrzyk, A. et al., *Anal. Chem.*, 2009, 81, 2633–2643. With permission.)

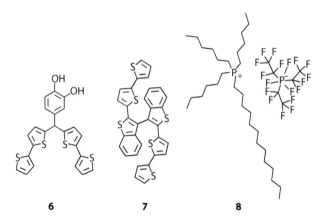

**SCHEME 5.18**  Structural formulae of the electroactive *meso*-(3,4-dihydroxyphenyl)-bis(2,2′-bitien-5-yl)methane **6** FM, the electroactive 2,2′-bis(2,2′-bithiophene-5-yl)]-3,3′-bithianaphthene **7** CM, and the trihexyl(tetradecyl)phosphonium tris(pentafluoroethyl)-trifluorophosphate **8** ionic liquid.

underlayers. Moreover, they were applied as protective coatings and buffering layers on the metal and semiconductor electrodes not only for proper operation of chemical sensors but also in electrocatalysis,[83] batteries,[84] and electrochromic devices.[85,86] Polybithiophene films, prepared by electropolymerization of 2,2′-bithiophene, have quite extensively been used as protecting films, owing to their advantages with respect to other conjugated polymer films.[86] These advantages include higher chemical and electrochemical stability. The polymer electroactivity can be controlled by prolonged potential cycling. This cycling improves the operation stability of the polymer.[87] The growth and morphology of the polybithiophene film can easily be incurred by suitable adjustment of electropolymerization conditions.

Note that (biogenic amine)-templated MIP films, used as recognition units of PM chemosensors, were underlayered with the polybithiophene barrier films in order to prevent electrooxidation of these amines.[17–19,27] For that, 10-MHz AT-cut shear-thickness-mode bulk-acoustic-wave quartz crystal resonators with Pt film electrodes were effectively used as signal transducers (Scheme 5.19).

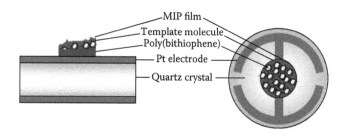

**SCHEME 5.19**  Piezomicrogravimetric (acoustic) chemosensor assembly - a bilayer arrangement of the chemosensor film deposited by electropolymerization onto a bulk-acoustic-wave thickness-shear-mode quartz crystal resonator.

In another application, a poly(3-methylthiophene) barrier layer prevented the electrocatalytic oxidation of ascorbic acid from competing with electrooxidation of dopamine.[88,89]

## CONCLUSIONS

A stable complex of the T molecule and the MIP cavity involves a covalent, nonco-valent, or coordinating bond resembling those between H and G in a supramolecu-lar H–G complex. This is how supramolecular chemistry concepts permeate ideas of molecular imprinting. Advanced selective and reversible binding of T in the man-made recognition materials successfully competes with that of the natural receptor materials. The combination of an MIP recognition unit and an appropriate signal transducer allows for preparation of a selective chemosensor.

Both free-radical and electrochemical polymerization are effectively used for MIP preparation. Because of its inherent advantages, the latter is becoming more and more frequently used. Advanced synthetic methods for preparation of MIPs led to better control of macromolecular architecture of the polymer films.

## PROSPECTIVE RESEARCH ON MIP CHEMOSENSORS

Future studies on MIP chemosensors may largely involve artificial supramolecular hosts such as calixarenes, carcerands, cryptands, cucurbiturils, and CDs, or self-assembled constructs of supramolecular complexes formed in solution.

Improving the morphology of MIPs in such a way that the imprinted compounds have better access to the imprinted cavities remains challenging. Future studies will most likely focus on this issue.

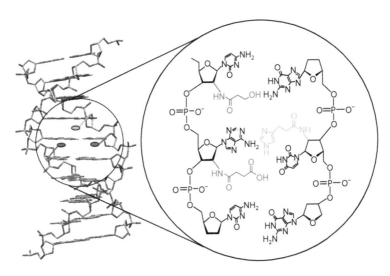

**SCHEME 5.20** Active sites in the groove of the DNA duplex. (Adapted from Catry, M.A., Madder, A., *Molecules*, 2007, 12, 114–129. With permission.)

Supramolecular recognition structures involving MIPs perceivably be more widely used in the future for selective drug delivery and implantable material systems. For instance, it is likely that engagement of oligonucleotides with different functional groups incorporated in the duplex groove (Scheme 5.20) will open new areas in mimicking biological recognition systems.[90]

## ACKNOWLEDGMENTS

WK and APL thank the European Regional Development Fund (Project ERDF, POIG.01.01.02-00-008/08 2007-2013) and the Ministry of Science and Higher Education of Poland (Project Iuventus Plus, IP 2010 031770), respectively, for their financial support. FD is thankful to the National Science Foundation (Grant 0804015 and 1110942) and NSF-EPSCoR for their financial support.

## REFERENCES

1. Suriyanarayanan, S.; Cywinski, P. J.; Moro, A. J.; Mohr, G. J.; Kutner, W. In *Molecular Imprinting Technology and Applications*, Haupt, K., ed.; *Topics in Current Chemistry*, Springer-Verlag, Berlin, Heidelberg, DOI: 10.1007/128_2010_92.
2. Barahona, F.; Turiel, E.; Cormack, P. A. G.; Martín-Esteban, A. *J. Polym. Sci., A: Polym. Chem.* 2010, *48*, 1058–1066.
3. Widstrand, C.; Bergström, S.; Wihlborg, A.-K.; Trinh, A. *Reporter Eur.* 2008, *31*, 11–13.
4. Spietelun, A.; Pilarczyk, M.; Kloskowski, A.; Namiesnik, J. *Chem. Soc. Rev.* 2010, *39*, 4524–4537.
5. Song, H.; Ozkan, U. S. *J. Mol. Catal. A: Chem.* 2010, *318*, 21–29.
6. Urusov, A. E.; Zherdev, A. V.; Dzantiev, B. B. *Appl. Biochem. Microbiol.* 2010, *46*, 253–266.
7. Alvarez-Lorenzo, C.; Yañez, F.; Concheiro, A. *J. Drug Deliv. Sci. Technol.* 2010, *20*, 237–248.
8. Hulanicki, A.; Glab, S.; Ingman, F. *Pure Appl. Chem.* 1991, *63*, 1247–1250.
9. Theavenot, D. R.; Toth, T. K.; Durst, R. A.; Wilson, G. S. *Pure Appl. Chem.* 1999, *71*, 2333–2348.
10. Steed, J. W.; Atwood, J. L. *Supramolecular chemistry*, John Wiley & Sons, Ltd: Baffins Lane, Chichester, 2000, pp. 1–33.
11. Wulff, G. *Chem. Rev.* 2002, *102*, 1–27.
12. Yoshizako, K.; Hosoya, K.; Iwakoshi, Y.; Kimata, K.; Tanaka, N. *Anal. Chem.* 1998, *70*, 386–389.
13. Haupt, K. *Chem. Commun.* 2003, 171–178.
14. Schneider, H.-J. *Angew. Chem., Int. Ed.* 2009, *48*, 3924–3977.
15. Israelachvili, J. N. *Intermolecular and Surface Forces*, Academic Press Limited: London, 1992, pp. 1–136.
16. Pietrzyk, A.; Kutner, W.; Chitta, R.; Zandler, M. E.; D'Souza, F.; Sannicolo, F.; Mussini, P. R. *Anal. Chem.* 2009, *81*, 10,061–10,070.
17. Pietrzyk, A.; Suriyanarayanan, S.; Kutner, W.; Chitta, R.; D'Souza, F. *Anal. Chem.* 2009, *81*, 2633–2643.
18. Pietrzyk, A.; Suriyanarayanan, S.; Kutner, W.; Chitta, R.; Zandler, M. E.; D'Souza, F. *Biosens. Bioelectron* 2010, *25*, 2522–2529.
19. Pietrzyk, A.; Suriyanarayanan, S.; Kutner, W.; Maligaspe, E.; Zandler, M. E.; D'Souza, F. *Bioelectrochemistry* 2010, *80*, 62–72.

20. Riskin, M.; Tel-Vered, R.; Bourenko, T.; Granot, E.; Willner, I. *J. Am. Chem. Soc.* 2008, *130*, 9726–9733.
21. Bravo, J. C.; Fernandez, P.; Durand, J. S. *Analyst* 2005, *130*, 1404–1409.
22. Ellwanger, A.; Berggren, C.; Bayoudh, S.; Crecenzi, C.; Karlsson, L.; Owens, P. K.; Ensing, K.; Cormack, P.; Sherringtonc, D.; Sellergren, B. *Analyst* 2001, *126*, 784–792.
23. Özcan, L.; Sahin, Y. *Sens. Actuators B.* 2007, *127*, 362–369.
24. Özkorucuklu, S. P.; Şahin, Y.; Alsancak, G. *Sensors.* 2008, *8*, 8463–8478.
25. Ricanyova, J.; Gadzala-Kopciuch, R.; Reiffova, K.; Buszewski, B. *Crit. Rev. Anal. Chem.* 2009, *39*, 13–31.
26. Andersson, L. I.; Hardenborg, E.; Sandberg-Stall, M.; Moller, K.; Henriksson, J.; Bramsby-Sjostrom, I.; Olsson, L.-I.; Abdel-Rehim, M. *Anal. Chim. Acta.* 2004, *526*, 147–154.
27. Heinze, J. In *Topics in Current Chemistry (Electrochemistry IV)*; Steckhan, E., ed., Springer-Verlag: Berlin, Heidelberg, 1990; Vol. 152, pp. 1–47.
28. Vandenburg, H. J.; Clifford, A. A.; Bartle, K. D.; Carroll, J.; Newton, I.; Garden, L. M.; Dean, J. R.; Costley, C. T. *Analyst* 1997, *122*, 101R–116R.
29. Cooper, C. R.; James, T. D. *J. Chem. Soc., Perkin Trans. 1.* 2000, 963–969.
30. Hall, D. G. In *Boronic Acids Preparation, Applications in Organic Synthesis and Medicine*; Hall, D. G., ed., Wiley-VCH: Weinheim, 2005, pp. 441–479.
31. Frasconi, M.; Tel-Vered, R.; Riskin, M.; Willner, I. *Anal. Chem.* 2010, *82*, 2512–2519.
32. Ishi-i, T.; Iguchi, R.; Shinkai, S. *Tetrahedron* 1999, *55*, 3883–3892.
33. Sellergren, B.; Lepisto, M.; Mosbach, K. *J. Am. Chem. Soc.* 1988, *110*, 5853–5860.
34. Strikovsky, A. G.; Kasper, D.; Grun, M.; Green, B. S.; Hradil, J.; Wulff, G. *J. Am. Chem. Soc.* 2000, *122*, 6295–6296.
35. Mazik, M. *Chem. Soc. Rev.* 2009, *38*, 935–956.
36. Lehn, J.-M. In *Supramolecular Chemistry*, VCH: Weinheim, 1995.
37. Roche, P. J. R.; Ng, S. M.; Narayanaswamya, R.; Goddarda, N.; Pag, K. M. *Sens. Actuators, B.* 2009, *139*, 22–29.
38. Ferancová, A.; Korgová, E.; Buzinkaiová, T.; Kutner, W.; Stepánek, I.; Labuda, J. *Anal. Chim. Acta.* 2001, *447*, 47–54.
39. Ferancová, A.; Labuda, J.; Kutner, W. *Electroanalysis* 2001, *13*, 1417–1423.
40. Hishiya, T.; Asanuma, H.; Komiyama, M. *J. Am. Chem. Soc.* 2002, *124*, 570–575.
41. Lakshmi, D.; Bossi, A.; Whitcombe, A. J.; Chianella, I.; Fowler, S. A.; Subrahmanyam, S.; Piletska, E. V.; Piletsky, S. A. *Anal. Chem.* 2009, *81*, 3576–3584.
42. Bosch, L. I.; Fyles, T. M.; James, T. D. *Tetrahedron* 2004, *60*, 11,175–11,190.
43. Franzen, S.; Ni, W.; Wang, B. *J. Phys. Chem. B.* 2003, *107*, 12,942–12,948.
44. James, T. D.; Linnane, P.; Shinkai, S. *Chem. Commun.* 1996, 281–288.
45. Matsui, J.; Higashi, M.; Takeuchi, T. *J. Am. Chem. Soc.* 2000, *122*, 5218–5219.
46. Takeuchi, T.; Mukawa, T.; Matsui, J.; Higashi, M.; Shimizu, K. D. *Anal. Chem.* 2001, *73*, 3869–3874.
47. Tong, A.; Dong, H.; Li, L. *Anal. Chim. Acta.* 2002, *466*, 31–37.
48. Chakraborti, A. S. *Mol. Cell. Biochem.* 2003, *253*, 49–54.
49. Say, R.; Gültekin, A.; Özcan, A. A.; Denizli, A.; Ersöz, A. *Anal. Chim. Acta.* 2009, *640*, 82–86.
50. Diltemiz, S. E.; Say, R.; Buyuktiryaki, S.; Hur, D.; Denizli, A.; Ersoz, A. *Talanta* 2008, *75*, 890–896.
51. Liang, R.; Zhang, R.; Qin, W. *Sens. Actuators, B.* 2009, *141*, 544–550.
52. Kan, X.; Zhao, Y.; Geng, Z.; Wang, Z.; Zhu, J.-J. *J. Phys. Chem. C.* 2008, *112*, 4849–4854.
53. Öpik, A.; Menaker, A.; Reut, J.; Syritski, V. *Proc. Est. Acad. Sci.* 2009, *58*, 3–11.
54. Deore, B.; Chen, Z.; Nagaoka, T. *Anal. Chem.* 2000, *72*, 3989–3994.
55. Ebarvia, B. S.; Cabanillab, S.; Sevilla III, F. *Talanta* 2005, *66*, 145–152.
56. Albano, D. R.; Sevilla III, F. *Sens. Actuators, B.* 2007, *121*, 129–134.

57. Syritski, V.; Reut, J.; Menaker, A.; Gyurcsanyi, R. E.; Öpik, A. *Electrochim. Acta.* 2008, *53*, 2729–2736.
58. Ramanavicius, A.; Ramanavicienė, A.; Malinauskas, A. *Electrochim. Acta.* 2006, *51*, 6025–6037.
59. Panasyuk, T. L.; Mirsky, V. M.; Piletsky, S. A.; Wolfbeis, O. S. *Anal. Chem.* 1999, *71*, 4609–4613.
60. Li, J.; Zhao, J.; Wei, X. *Sens. Actuators, B.* 2009, *140*, 663–669.
61. Deore, B.; Freund, M. S. *Analyst* 2003, *128*, 803–806.
62. Zhang, J.; Wang, Y.-Q.; Lv, R.-H.; Xu, L. *Electrochim. Acta.* 2010, *55*, 4039–4044.
63. Sreenivasan, K. *J. Mater. Sci.* 2007, *42*, 7575–7578.
64. Malitesta, C.; Losito, I.; Zambonin, P. G. *Anal. Chem.* 1999, *71*, 1366–1370.
65. Weetall, H. H.; Rogers, K. R. *Talanta.* 2004, *62*, 329–335.
66. Feng, L.; Liu, Y.-J.; Tan, Y.-Y.; Hu, J.-M. *Biosens. Bioelectron.* 2004, *19*, 1513–1519.
67. Liu, Y.; Song, Q.-J.; Wang, L. *Microchem. J.* 2009, *91*, 222–226.
68. Aghaei, A.; Hosseini, M. R. M.; Najafi, M. *Electrochim. Acta.* 2010, *55*, 1503–1508.
69. Gong, J.-L.; Gong, F.-C.; Zeng, G.-M.; Shen, G.-L.; Yu, R.-Q. *Talanta* 2003, *61*, 447–453.
70. Mazzotta, E.; Malitesta, C. *Sens. Actuators, B.* 2010, *148*, 186–194.
71. Apodaca, D. C.; Pernites, R. B.; Ponnapati, R. R.; Del Mundo, F. R.; Advincula, R. C. *ACS Appl. Mater. Interfaces.* 2010, DOI: 10.1021/am100805y.
72. Ebarvia, B. S.; Cabanilla, S.; Sevilla, S. F. *Talanta* 2005, *66*, 145–152.
73. Ramanaviciene, A.; Ramanavicius, A. *Biosens. Bioelectron.* 2004, *20*, 1076–1082.
74. Gómez-Caballero, A.; Ugarte, A.; Sánchez-Ortega, A.; Unceta, N.; Goicolea, M. A.; Barrio, R. J. *J. Electroanal. Chem.* 2010, *638*, 246–253.
75. Kriz, D.; Mosbach, K. *Anal. Chim. Acta.* 1995, *300*, 71–75.
76. Pizzariello, A.; Stredansky, M.; Stredanska, S.; Miertus, S. *Sens. Actuators, B.* 2001, *76*, 286–294.
77. Yasuo, Y.; Ryo, O.; Chiaki, I.; Kiyotaka, S. *Sens. Actuators, B.* 2001, *73*, 49–53.
78. Marx, K. A. *Biomacromolecules.* 2003, *4*, 1099–1120.
79. Dance, I. In *Encyclopedia of Supramolecular Chemistry*, Marcel Dekker, Inc., New York, 2004, pp. 1076–1092.
80. Foti, M. C.; DiLabio, G. A.; Ingold, K. U. *J. Am. Chem. Soc.* 2003, *125*, 14,642–14,647.
81. Foti, M. C.; Barclay, L. R. C.; Ingold, K. U. *J. Am. Chem. Soc.* 2002, *124*, 12,881–12,888.
82. Kröger, S.; Turner, A. P. F.; Mosbach, K.; Haupt, K. *Anal. Chem.* 1999, *71*, 3698–3702.
83. Ozcan, L.; Sahin, M.; Sahin, Y. *Sensors.* 2008, *8*, 5792–5805.
84. Biancardo, M.; Krebs, F. C. *Sol. Energy Mater. Sol. Cells.* 2008, *92*, 353–356.
85. Rammelt, U.; Hebestreit, N.; Fikus, A.; Plieth, W. *Electrochim. Acta.* 2001, *46*, 2363–2371.
86. Aeiyach, S.; Bazzaoui, E. A.; Lacaze, P. C. *J. Electroanal. Chem.* 1997, *434*, 153–162.
87. Wang, G.; Wong, T. K. S.; Hu, X. *Appl. Phys. A.* 2000, *71*, 117–120.
88. Dayton, M. A.; Ewing, A. G.; Wightman, R. M. *Anal. Chem.* 1980, *52*, 2392–2396.
89. Gonon, F. G.; Navarre, F.; Buda, M. J. *Anal. Chem.* 1984, *56*, 573–575.
90. Catry, M. A.; Madder, A. *Molecules.* 2007, *12*, 114–129.

# 6 Supramolecular Chromatography

*Volker Schurig*

## CONTENTS

## INTRODUCTION

"Column chromatography is a physical method of separation in which the components to be separated are distributed between two phases, one of which is stationary (the stationary phase) while the other (the mobile phase) moves in a definite direction."[1,2] Depending on the nature of the mobile phase (a gas, a supercritical fluid, a liquid, an electrolyte), four methods can be differentiated: gas chromatography (GC), supercritical fluid chromatography, liquid chromatography (LC), and electrochromatography (EC). Chromatography (from the Greek *chroma* = color, *graphein* = to write) was first described in 1903 by Mikhail Semenovich Tswett (in Russian: *tswett* = color).[3] This powerful separation method, originally rejected by Richard M. Willstätter, gained general acceptance only after its rediscovery in 1931 by Kuhn et al., who separated four xanthophylls by LC on calcium carbonate.[4]

The chromatographic retention factor $k$ is defined as the ratio of the time a component B spends in the stationary phase $t' = t - t_0$ ($t$ = total retention time) to the time it spends in the mobile phase $t_0$, that is, $k = t'/t_0 = (t - t_0)/t_0$. The retention factor $k$ is equal to the product of the distribution (partition) constant $K_c$ (the ratio of the stationary to mobile phase concentrations of B) and the phase ratio $\varphi$ (the ratio of the stationary to mobile phase volumes), that is, $k = K_c \varphi$ (universal equation of chromatography).[5] The temperature dependence of $k$ is derived from the Gibbs–Helmholtz equation: $RT \ln K_c = -\Delta G° = -\Delta H° + T\Delta S°$. It follows that $\ln k = \ln K_c + \ln \varphi = -\Delta H°/RT + \Delta S°/R + \ln \varphi$. The plot of $\ln k$ vs. $1/T$ (van't Hoff plot) is linear, yielding the standard enthalpy $\Delta H°$ and the standard entropy $\Delta S°$ of the transfer of the component B from the mobile to the stationary phase when ideal conditions prevail ($T$ = absolute temperature in Kelvin, $R$ = gas constant).

In chromatography, a new momentum is gained when an additive A is incorporated into the stationary phase and exhibits a chemical affinity toward the component B. By this strategy, *chemical selectivity* is introduced, in addition to the conventional physical chromatographic process of partitioning, as the result of a fast and reversible complexation process taking place in the stationary phase, that is, A + B = AB ($K$), where $K$ is the thermodynamic complexation (association) constant of A and B in the stationary phase. The first example of highly selective π-complexation chromatography utilized silver(I) nitrate added to polyethylene glycol for the GC separation of alkenes from alkanes.[6] Argentation chromatography was later extended to selective alkene (C2–C7) separations on rhodium(I) and rhodium(II) coordination compounds accompanied by the determination of relative π-complexation constants $K_{rel}$ by the retention-increment approach $R'$ (Figure 6.1).[7]

In the retention-increment approach of thermodynamic quantitation, the component B is both chromatographed on a reference column devoid of the additive A and on a complexation column containing the activity $a$ of A in an inert matrix, for example, in the solvent S.[7,8] If complexation of A and B takes place, a retention-increment $R'$ is defined, which is determined by the product of the complexation constant $K$ and the activity $a$ (or concentration $c$ at high dilution) of A in S (Figure 6.1). For two components $B_1$ and $B_2$ competing for A, the physical separation factor $\alpha^{phys} = k_{o2}/k_{o1}$ should be differentiated from the chemical separation factor $\alpha^{chem} = R'_2/R'_1 = K_2/K_1$ arising from molecular association in the stationary phase (Figure 6.1).

A special case concerns the separation of enantiomers (optical isomers) on a chiral (nonracemic) additive A, a method called *chiral chromatography*.[9,10] Because

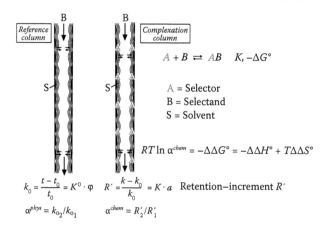

$$k_0 = \frac{t - t_0}{t_0} = K^0 \cdot \varphi \qquad R' = \frac{k - k_0}{k_0} = K \cdot a \quad \text{Retention–increment } R'$$

$$\alpha^{phys} = k_{o2}/k_{o1} \qquad \alpha^{chem} = R'_2/R'_1$$

**FIGURE 6.1** Principle of complexation GC. (Left) Reference column containing the pure solvent S. (Right) Reactor (complexation) column containing the selector A (additive) in the solvent S. $k_0$ is the retention factor of the selectand B on the reference column. $k$ is the retention factor of the selectand B on the complexation column. $K_c$ is the distribution constant of B in the reference column, and β is the phase ratio. $K$ is the complexation constant between A and B in the complexation column, and $a$ is the activity of A in S. (From Schurig, V., *J. Chromatogr. A*, 2002, 965, 315–356; Schurig, V., *J. Chromatogr. A*, 2009, 1216, 1723–1736. With permission.)

enantiomers cannot be differentiated in an achiral environment, $\alpha^{phys}$ is equal to unity. Thermodynamic data for chirality recognition are determined by the Gibbs–Helmholtz equation (Figure 6.1). For a complexation equilibrium A + B = AB, enthalpy/entropy compensation takes place at the isoselective temperature $T_{iso}$ because the stronger bonded complex is more ordered. At $T_{iso} = \Delta\Delta H°/\Delta\Delta S°$, the enantiomers $B_1$ and $B_2$ cannot be separated ($\alpha^{chem} = 1$, $\Delta\Delta G° = 0$), and an inversion of the elution order commences below $T_{iso}$ (by enthalpy control) and above $T_{iso}$ (by entropy control).[10,11] This phenomenon was first observed by enantioselective GC in a hydrogen-bonding association system[12] and in a metal complexation system.[13] The retention-increment method is based on a differentiation between the physical partition equilibrium of B between the mobile and stationary phases and the chemical equilibrium of A and B in the stationary phase. The validity of the method has been amply demonstrated,[14,15] and it will be employed for the thermodynamic quantitation of chromatographic [60]fullerene/perchlorinated biphenyl and [60]fullerene/cyclodextrin equilibria, respectively (Chapter 4). A three-phase approach considering also the separate partitioning of the component B between the mobile phase and the additive A in the stationary phase has also been advanced.[16]

In LC, the additive A is usually linked to silica particles via refined spacer chemistry. Although nonselective contributions to overall retention are usually present,[10] they are often not considered when the effect of the temperature on retention and selectivity $\alpha$ for competing components $B_1$ and $B_2$ is determined according to the equation $\ln \alpha = \ln(k_2/k_1) = -\Delta\Delta G/RT = -\Delta\Delta H/RT + \Delta\Delta S/R$.[17] After having established the quantitation of chromatographic selectivity, different types of supramolecular equilibria A + B = AB will now be considered.

The terms "selector" (SO) and "selectand" (SA) were introduced by Mikeš into chromatography in analogy to Ashby's operator–operand terminology[18] to avoid ambiguities that may arise from the use of terms such as solvent–solute, ligand–substrate, and host–guest in highly selective chromatographic partitioning systems, whereby the selector can be present as a stationary phase or as an additive to the liquid phase.[19] This terminology is adopted here, and it applies also to various supramolecular interactions such as antibody–antigen, receptor–substrate/inhibitor, and metal ion–ligand.

## LINK BETWEEN SUPRAMOLECULAR CHEMISTRY AND SEPARATION SCIENCE

The link between supramolecular chemistry and chromatography has been pointed out in the pioneering work of Bianco et al. in 1997.[20] The term "supramolecular chromatography" has recently been used in the system [60]fullerene (as selector) and cyclodextrin congeners $CD_n$ (as selectands; Chapter 4) because the observed selectivity pattern did not follow the conventional elution order according to molecular weight differences but was governed by a remarkable shape selectivity for medium ring sizes of cyclodextrin congeners.[21] The term supramolecular chromatography has also occasionally been mentioned in the literature.[22–24] Supramolecular metal complexes were characterized and purified by gel-permeation chromatography,[25] and a theoretical study simulated size exclusion chromatography for the characterization of a supramolecular complex.[26] The presence of supramolecular effects involved in the enantioseparation of the racemic drug clenbuterol by reversed-phase LC in the presence of a chiral (nonracemic) helical nickel(II) chelate has also been studied.[27] These representative examples show that *supramolecular interactions*[28] are omnipresent in selective chromatography because molecular interactions leading to a retention-increment $R'$ are the result of fast and reversible molecular associations via noncovalent interactions (i.e., van der Waals, ion pair, dipole–dipole, metal coordination, hydrogen bonding, π–π, cation–π, intercalation, charge-transfer, etc.) between selectors and selectands. Thus, affinity chromatography,[29,30] inverse GC,[31,32] electron donor–acceptor (EDA) or charge-transfer (CT) chromatography,[33] biopolymer chromatography,[34] peptide chromatography,[35,36] and enantioselective LC based on a variety of different chiral stationary phases (CSPs) such as modified cyclodextrins, amylose and cellulose derivatives, macrocyclic antibiotics, ion exchangers, proteins, antibodies, crown ethers, synthetic donors or acceptors, metal ligand exchangers,[9,10,37,38] molecularly imprinted polymers (MIPs),[39] and nanoscale materials[40] can be considered *supramolecular chromatographic systems.*

A remarkable case is that in which unusual size selectivity is combined with enantioselectivity via refined supramolecular complexation equilibria. A vivid example is provided by the selectand/selector system of chiral 2-methyl-substituted cyclic ethers and manganese(II) *bis*[(3-heptafluorobutyryl)-(*1R*)-camphorate] as studied by complexation GC (Figure 6.2).[41] The complexation selectivity of the cyclic ethers on

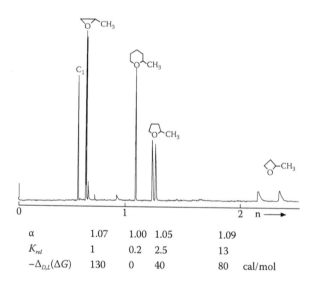

| | | | | | |
|---|---|---|---|---|---|
| α | 1.07 | 1.00 | 1.05 | 1.09 | |
| $K_{rel}$ | 1 | 0.2 | 2.5 | 13 | |
| $-\Delta_{D,L}(\Delta G)$ | 130 | 0 | 40 | 80 | cal/mol |

**FIGURE 6.2** Influence of ring size of chiral 2-methyl-substituted cyclic ethers on the gas-chromatographic enantioseparation factor α, $K_{rel}$ (first eluted enantiomer) and enantioselectivity $-\Delta_{D,L}(\Delta G)$ at 60°C. A 160 m × 0.4 mm i.d. stainless-steel capillary column coated with enantiomerically pure 0.05 *m* manganese(II) *bis*[(3-heptafluorobutyryl)-(*1R*)-camphorate] in squalane (because of the short elution times, the preferential enantioseparation of methyl-oxirane is not readily visible). (From Schurig, V., In *Chromatographic Separations Based on Molecular Recognition*, Jinno, K. Ed., Chapter 7, Wiley-VCH, New York, 1997, 371–418. With permission.)

Mn(II), $K_{rel}$, differed by a factor of 65, and it was accompanied by an inverse elution order for [4]–[6] ring sizes: [3]oxirane < [6]oxane < [5]oxolane < [4]oxetane. Yet the enantioselectivity $-\Delta_{D,L}\Delta G$ (determined by the retention-increment approach) followed the trend: [3]oxirane > [4]oxetane > [5]oxolane >> [6]oxane at a given $K_{rel}$, irrespective of being small or large.[41]

Supramolecular π-acceptor/π-donor CT interactions have widely been used in the early development of enantioselective LC. Following pioneering studies on the use of columns of silica acid impregnated with 2,4,6-trinitrophenol (picric acid) or 2,4,7-trinitrofluorenone for adsorption chromatography of mixtures of aromatic hydrocarbons,[42] Klemm and Read extended this type of molecular complexation chromatography to enantioseparations also.[43,44] The chiral CT selector (R)-(-)-2-(2,4,5,7-tetranitro-9-fluorenylideneamino-oxy)-propionic acid ((R)-(-)-TAPA) (Figure 6.3), a strong π electron acceptor introduced by Newman and Lednicer for the enantioseparation of hexahelicene via crystallization,[46] was impregnated on silicic acid and then packed into a 41 × 4.5 cm glass column. Racemic 1-naphthyl 2-butyl ether and racemic 9,10-dihydro[5]helicene were partially resolved by recycle LC. The colored elution zone could be directly visualized because of on-column CT complex formation. Enantiomers of varying degrees of specific rotations were identified by polarimetry.[43]

In the early stages of LC development in 1976, Mikeš and Gil-Av employed high-efficiency homemade chromatographic equipment for trials of enantioseparation in a supramolecular π–π selector/selectand system. Both enantiomers of TAPA (Figure 6.3) were employed and either impregnated on silica gel *in situ* or attached ionically or bonded covalently to aminopropyl silica gel. The CSPs were subsequently packed into a column.[19,47] The racemic carbohelicenes ([6]helicene–[14]helicene) were rapidly and completely resolved by high-performance LC (HPLC) on TAPA impregnated on silica gel (Figure 6.4),[47] and partially resolved on binaphthyl-2,2′diyl hydrogen phosphate linked to silica gel.[48]

The mutual structural recognition of host and guest is the basic principle of nature's transport and self-regulating systems. In order to form a 1:1 complex, a host (selector) and guest (selectand) must possess a complementary stereoelectronic arrangement of binding sites and steric barriers.[49] In the spirit of this pioneering concept, Cram et al. rationally designed chiral macrocyclic polyethers (crown ethers) capable of enantioselectively complexing racemic alkylammonium cations. The chiral $C_2$-symmetric cavitand di-binaphthylo-22-crown-6, containing pendant

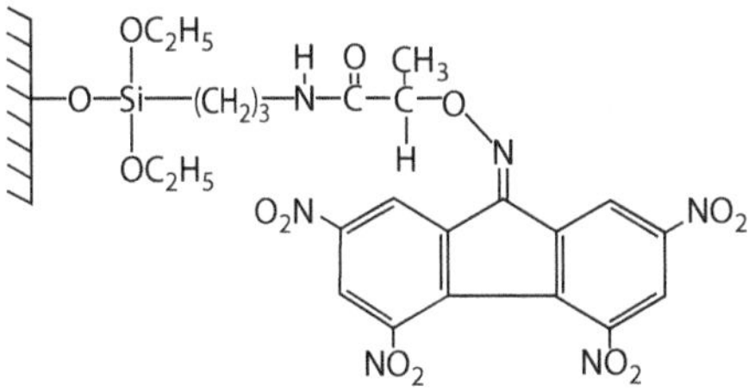

FIGURE 6.3    Structure of TAPA bonded to silica gel. (From Diack, M. et al., *J. Chromatogr.*, 1993, 639, 129–140. With permission.)

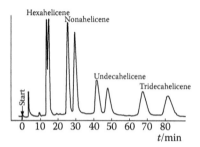

**FIGURE 6.4** Simultaneous enantioseparation of racemic hexa-, nona-, undeca-, and trideca-helicenes on an *in situ* prepared (S)-(+)-TAPA column (20 × 0.23 cm (i.d.)). (From Mikeš, F. et al., *Chem. Commun.* 1976, 99–100. With permission.)

atropisomeric $(R,R)$-1,1′binaphthyl moieties (Figure 6.5), was covalently attached to highly cross-linked polystyrene-divinylbenzene beads that were subsequently packed into an LC column.[50] Various ammonium perchlorate (or hexafluorophosphate) salts of free α-amino acids or α-amino acid methyl esters, respectively, were analytically and preparatively enantioseparated, and the $-\Delta\Delta G$ data of chiral recognition at 0°C were reported.[50] The changes in the $-\Delta\Delta G$ values between the diastereomeric association complexes correlated roughly with the steric requirements of the amino acid residue group, whereby larger R groups provided larger $-\Delta\Delta G$ values.[50]

The selector ideally combined the proper ring size of the central 22-crown-6 moiety for accommodating the substituted ammonium ion, whereas the atropisomeric binaphthyl rings, which are stable toward rotation, act as crowded chiral barriers. The rigid structure of the host prevented conformational changes common to

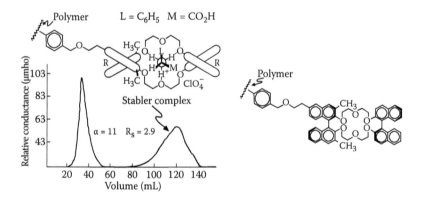

**FIGURE 6.5** Right: Structure of the polymer-linked chiral crown ether bis-$((R,R)$-1,1′-binaphthyl)-22-crown-6. Left: Enantioseparation of phenylglycinate perchlorate salt on the chiral crown ether. Column: 60 × 0.75 cm (i.d.), flow rate 0.36–2.0 mL/min. Eluent: $CHCl_3$/$CH_3CN$). Insert: diastereomeric association complex (L = large, M = medium). (From Sogah, G. D. Y., Cram, D. J., *J. Amer. Chem. Soc.*, 1976, 98, 3038–3041; Sogah, G. D. Y., Cram, D. J., *J. Amer. Chem. Soc.*, 1979, 101, 3035–3042. With permission.)

crown ethers upon complexation of guests (induced fit). The absolute configuration of guests was correlated with their elution order via molecular modelling studies. The two-phase chromatographic data could be correlated with data obtained in solution. The complexes are stabilized by multiple hydrogen bonds formed between the ammonium ion and the three ether oxygens and by ion-dipole forces between the ammonium ion and the ether oxygens. Dipole–dipole interactions and hydrophobic and π–π bonding (for aromatic α-amino acids) may also be present.[50]

In the following chapters, a restriction toward characteristic examples of supramolecular chromatographic applications was necessary. For chromatographic separations involving cyclophanes, rotaxanes, catenanes, dendrimers, helical polymers, and MIPs, the original literature should be consulted. Also naturally occurring building blocks (peptides, proteins, antibodies, nucleic acids, and carbohydrates) are omitted. As cases in point, supramolecular chromatographic interactions in the realm of fullerenes, carbon nanotubes, and synthetic receptors will be considered in detail.

## SUPRAMOLECULAR CHROMATOGRAPHIC SEPARATIONS OF FULLERENES ON VARIOUS STATIONARY PHASES

The third stable allotrope of carbon, that is, fullerenes, escaped the attention of chemists for a long time until they were predicted theoretically in 1970 by Osawa in a treatise about superaromaticity.[51] Their existence was later confirmed mass spectrometrically by Kroto et al.[52] Fullerenes ("bucky balls") were named for their resemblance to the architect Buckminster Fuller's geodesic domes, which have interlocking 5- and 6-membered circles on the surface of a partial sphere. [60]Fullerene ($C_{60}$) and [70]fullerene ($C_{70}$) were recently discovered in extraterrestrial space.[53]

The molecular recognition of fullerene congeners on various selectors by LC has been reviewed.[54,55] The isolation and purification of fullerenes and their derivatives were, at first, performed on conventional, solid, HPLC stationary phases such as neutral alumina and silica gels, on a number of reversed stationary phases such as octadecylsilica (ODS) and multipodal phenyl-bonded phases, on gel permeation phases, and on graphite.[54] Yet for more selective recognition leading to large separation factors between $C_{60}$ and $C_{70}$ ($\alpha > 2$), a supramolecular approach was first employed by Hawkins et al.[56] As a strong π-acceptor selector for the π-donor aromatic fullerene selectands, ionically bonded N-(3,5-dinitrobenzoyl)-phenylglycine (DNBPG) to silica (Figure 6.6, top) developed by Welch for enantioseparations[57] was used (Figure 6.7). In contrast to the condensed aromatic selectand anthracene, an unprecedented increase in retention of $C_{60}$ and $C_{70}$ with increasing column temperature was observed by Pirkle and Welch[58] on a column (250 × 4.6 mm i.d.) containing the covalently bonded π-acidic selector DNBPG to silica (Figure 6.6, top). Van't Hoff plots revealed the rare case of positive enthalpy and entropy values, that is, the gain in entropy is accompanied by an endothermic adsorption. From a practical viewpoint, the preferable elevated separation temperature decreased the retention of the fullerene-solvent benzene and, at the same time, increased the solubility and the retention of $C_{60}$ and $C_{70}$, and it produced narrower band shapes.[58]

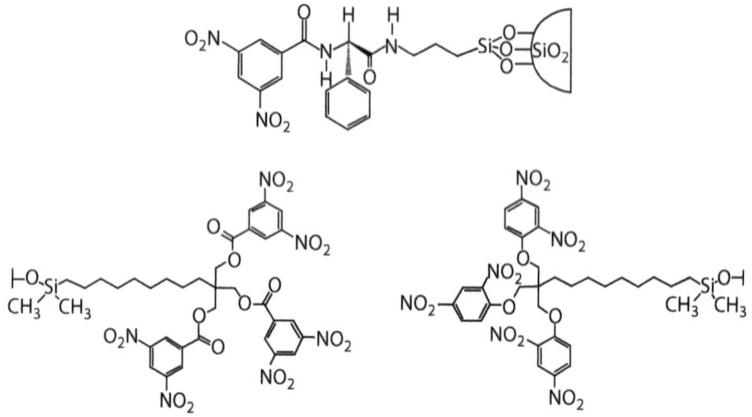

**FIGURE 6.6** Top: Structure of the selector DNBPG. (From Welch, C. J., *J. Chromatogr. A*, 1994, 666, 3–26; Pirkle, W. H., Welch, C. J., *J. Org. Chem.*, 1991, 56, 6973–6974. With permission.) Bottom: Structure of silica-bound tripodal 3,5-dinitrobenzoate ester and 2,4-dinitrophenyl ether stationary phases. (From Welch, C. J., Pirkle, W. H., *J. Chromatogr.*, 1992, 609, 89–101. With permission.)

Although the large-scale purification and separation of fullerenes, being only sparingly soluble in organic solvents, were considered to be beyond the scope of conventional chromatography, novel silica-bound tripodal 3,5-dinitrobenzoate ester and 2,4-dinitrophenyl ether stationary phases (Figure 6.6, bottom) containing three π-acidic functionalities for simultaneous multipoint supramolecular interaction with fullerenes were developed. They provided the highest retention and separation factor for the $C_{60}/C_{70}$ mixture.[59] The so-called "Buckyclutcher I" stationary phase (Regis) consists of the tripodal arrangement of three 2,4-dinitrophenyl groups in a π-acidic semicavity array covalently attached through a $C_{10}$ spacer link on 5-μm silica gel,[59] whereas "Cosmosil Buckyprep" represents a commercial 3-(1-pyrenyl)propyl group bonded silica column (250 × 4.6 mm i.d.) for preparative-scale separation of fullerenes in toluene solution.[60,61] A remarkable separation of 15 fullerene congeners in the range of $C_{60}$–$C_{96}$, including enantiomerically enriched specimens, by HPLC on

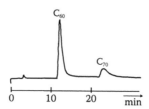

**FIGURE 6.7** HPLC separation of $C_{60}$ (12.2 minutes) and $C_{70}$ (23.5 minutes) on a Pirkle-type (from Welch, C. J., *J. Chromatogr. A*, 1994, 666, 3–26) ionically DNBPG containing column (250 × 10 mm i.d.) eluted with 5.0 mL/min *n*-hexane and detected at 280 nm, α = 2.25 (r.t.). (From Hawkins, J. M. et al., *J. Org. Chem.*, 1990, 55, 6250–6252. With permission.)

a COSMOSIL 5PBB column in 45 minutes with chlorobenzene as eluent has been described recently.[62]

Cox et al. found that the chromatographic parameters of $C_{60}/C_{70}$ on a π-acidic dinitroanilinopropyl (DNAP) silica gel support closely resembled certain planar aromatic counterparts with characteristic molecular footprints when eluted with a mobile-phase gradient of n-hexane (100%) to n-hexane/dichloromethane (1:1), that is, $C_{60}$ and planar triphenylene coelute, whereas the molecular footprint of $C_{70}$ lies between that of benzo[α]pyrene and coronene.[63]

As a strong π-electron acceptor, (R)-(-)-2-(2,4,5,7-tetranitro-9-fluorenylideneamino-oxy)propionic acid (TAPA) (Figure 6.3) forms supramolecular charge transfer complexes with appropriate π-donor molecules. TAPA was used to separate $C_{60}$ and $C_{70}$ with a separation factor α up to 2.6, and the same peculiar increase in retention with increasing temperature previously observed with π-acidic DNBPG was found and interpreted as arising from solute solvation at low temperatures.[45] The same authors mentioned that $C_{60}$ and $C_{70}$ can behave as both π-electron donor and π-electron acceptor (the electron affinity of $C_{60}$ is 2.6 eV). Large separation factors for $C_{60}/C_{70}$ (up to α = 4.8) in the preferred solvent toluene were observed on silica gel supports containing immobilized porphyrins (metalated and nonmetaled) via supramolecular π–π interactions.[64] Jinno et al. described copper-phthalocyanine stationary phases chemically linked to aminopropyl silica with residual aminopropyl groups being end-capped with either short butanoyl or long n-decanoyl groups.[65] No striking difference was observed by comparison with an ODS (Develosil ODS-5) phase (α = 1.9 to 2.3), but an effect of π–π interactions of fullerenes with the planar conjugated selector was nevertheless evident.[65] Shape selectivity may operate in the HPLC separation of $C_{60}$ and $C_{70}$ on native γ-cyclodextrin chemically bonded to silica. $C_{70}$ was much more strongly retained than $C_{60}$ on this stationary phase, whereas no separation of the two fullerenes occurred on the corresponding unmodified silica, indicating that the separation is a result of the selective supramolecular interaction with the cycloamylose selector.[66]

Dihedral-symmetric fullerenes such as $C_{76}$–$D_2$, $C_{78}$–$D_3$, $C_{80}$–$D_2$, $C_{80}$–$D_3$, and $C_{84}$–$D_2$ and $C_2$–symmetric fullerenes such as $C_{82}$–$C_2$ (three congeners) are chiral and occur as pairs of enantiomers.[67] As judged from the minuscule difference in the shapes of fullerene enantiomers and because of the lack of proper functionalities, the direct enantioseparation of chiral carbon cage molecules into enantiomers represents an extremely difficult task. Previous experiments to enantioseparate fullerenes on silica-bonded TAPA (Figure 6.3) failed because of the incomplete separation of $C_{76}$ and $C_{78}$ on this phase (α = 1.07).[68] However, the extraordinary challenge has been met by Yamamoto et al. who described the first direct resolution of $C_{76}$–$D_2$ enantiomers by chiral HPLC.[69] The authors used amylose tris(3,5-dimethylphenylcarbamate) (Chiralpak AD) as a chiral selector developed by the Okamoto group,[70] and they succeeded in the resolution of $C_{76}$–$D_2$ by recycle HPLC (Figure 6.8). The collected enantiomeric fractions were identified via an on-column chiral CD detector (Figure 6.8, left) and by their opposite CD spectra (Figure 6.8, right). The enantiomeric purity of $C_{76}$–$D_2$ was claimed to be higher than that of a sample previously obtained by kinetic resolution of racemic $C_{76}$–$D_2$ by asymmetric osmylation.[71]

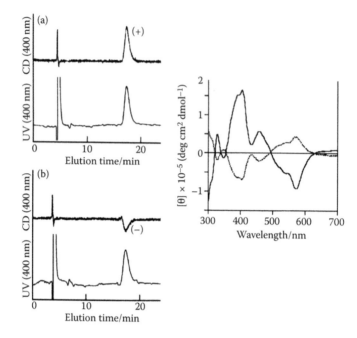

**FIGURE 6.8** Left: HPLC chromatogram of the first (a) and the second (b) eluted enantiomers isolated after recycled enantioseparation of $C_{76}$–$D_2$. Column: chemically bonded Chiralpak AD. Eluent: $n$-hexane-chloroform (80:20). Flow rate: 1.0 mL/min UV-CD-dual detector (Jasco). Right: CD spectra of $C_{76}$–$D_2$ (solid line: first eluted enantiomer; dotted line: second eluted enantiomer). (From Yamamoto, C. et al., *Chem. Commun.*, 2001, 925–926. With permission.)

Achiral $C_{60}$ is rendered chiral by proper substitution. The enantiomers of the *cis* and *trans* photo cycloadducts of $C_{60}$ and 3-methylcyclohexanone (Figure 6.9a) were separated by HPLC on the CSP Whelk O1[57] (Figure 6.9b), and the mirror image CD spectra of the isolated enantiomers were reported (Figure 6.9c).[72] The synthetic Whelk-O-type selector possesses π-acceptor and π-donor properties as well as a hydrogen-bonding site ideally suited for enantioselective supramolecular interaction with the $C_{60}$ photo adduct. The inherently chiral [1,9]methanocarboxylic acid diethylamide of $C_{70}$ (Figure 6.10, left) was enantioseparated on the Whelk-O selector, and the mirror image CD spectra of the isolated enantiomers were recorded.[73] The $C_{70}$ derivative is rendered chiral by a stereogenic bridging cyclopropane carbon. Whereas the corresponding $C_{60}$ derivative has a plane of symmetry through the cyclopropane ring, the $C_{70}$ derivative forms stable enantiomers because of the asymmetry of the $C_{70}$ core as shown by molecular modelling (Figure 6.10). Thus, chirality recognition is solely the result of a minute asymmetry of the curvature on the $C_{70}$ surface, that is, the presence of a bulge on one side of the fullerene core, which the Whelk-O selector is able to recognize. This result is a clear manifestation for refined *enantioselective supramolecular recognition*.

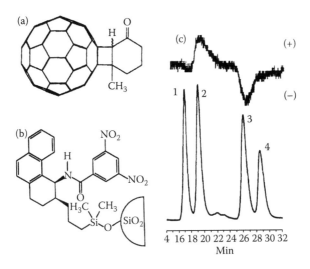

**FIGURE 6.9** (a) Structure of chiral *cis*- and *trans*-fused $C_{60}$-enone [2+2] photocycloadducts. (b) Structure of the CSP Whelk-O1. (From Welch, C. J. *J. Chromatogr. A*, 1994, 666, 3–26.) (c) Enantioseparation of *cis*- and *trans*-$C_{60}$-enone [2+2] photocycloadducts by HPLC on the Whelk-O1 selector. Peak assignment: 1 and 4, *cis* enantiomers ($\alpha$ = 1.90); 2 and 3, *trans*-enantiomers ($\alpha$ = 1.45). Above the peaks: the polarimetric response. (From Wilson, S. R. et al., *J. Org. Chem.*, 1993, 58, 6548–6549. With permission.)

Diederich et al. previously reported the chromatographic separation of several diastereomeric chiral $C_{70}$ *bis*-adducts formed by addition of chiral bromomalonate enolates to $C_{70}$ by HPLC on conventional porous spherical silica.[74] The enantioseparation of a number of *bis*-, *tris*-, and *hexakis*-adducts of $C_{60}$ with achiral addends combined with an inherent chiral addition pattern was achieved by HPLC using the chiral Whelk-O1 stationary phase (Figure 6.9b).[75] The largest enantioseparation

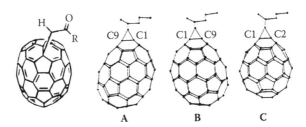

**FIGURE 6.10** Left: Structure of $C_{70}$-[1,9]methanocarboxylic acid diethylamide (R = NEt$_2$). Right: Computer models of the $C_{70}$-[1,9]methanocarboxylic acid ethyl ester enantiomers **A** and **B** (no plane of symmetry, C1 and C9 are nonequivalent) and the $C_{60}$ analogue **C** (plane of symmetry, C1 and C2 are equivalent). (From Wang, Y. et al., *J. Org. Chem.*, 1996, 61, 5198–5199. With permission.)

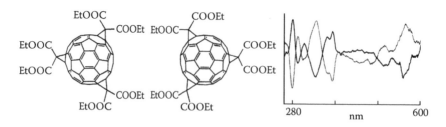

**FIGURE 6.11**   Structure of the $C_3$-symmetric *tris*-adduct $C_{60}[C(COOEt)_2]_3$ and mirror image CD spectra (dotted line: first eluted enantiomer; solid line: second eluted enantiomer). (From Gross, B. et al., *Chem. Commun.*, 1997, 1117–1118. With permission.)

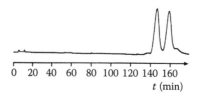

**FIGURE 6.12**   Enantioseparation of the *hexakis*-adduct $C_{60}[C(COOEt)_2]_6$ on the CSP TAPA bonded to aminopropyl silica gel by micro HPLC ($\alpha = 1.1$). Packed fused-silica capillary column ($200 \times 0.25$ mm i.d.), µL/min acetonitrile/water (75:25, v/v), room temperature. (From Gross, B. et al., *J. Chromatogr. A*, 1997, 791, 65–69. With permission.)

was observed with the $C_1$-chiral *hexakis*-adduct $C_{60}[C(COOEt)_2]_6$, which is a side product in the synthesis of the corresponding $T_h$-symmetrical *hexakis*-adduct of $C_{60}$ of unknown structure. The isolated enantiomers of the chiral $C_{60}$ adducts gave rise to mirror image CD curves (Figure 6.11). The enantioseparation of the chiral *tris*-adduct $C_{60}[C(COOEt)_2]_3$ and *hexakis*-adduct $C_{60}[C(COOEt)_2]_6$ (Figure 6.12) has also been described on the CSP TAPA ((*R*)-(-)-2-(2,4,5,7-tetranitro-9-fluorenylideneaminooxy)propionic acid) bonded to silica gel (Figure 6.3) by micro HPLC.[76] A systematic study on the enantiomeric separation of various [60]fullerene *bis*-adducts on polysaccharide HPLC columns (Chiralcel OD and Chiralpak AD)[70] have been advanced.[77]

## CHROMATOGRAPHIC FULLERENE-CONTAINING STATIONARY PHASES

The reciprocal principle of molecular recognition by changing the roles of selector and selectand in chromatography[78,79] has been discussed by Saito et al. in the supramolecular system fullerenes/polycyclic aromatic hydrocarbons (PAHs).[80] By identifying a PAH that undergoes a highly selective interaction with a fullerene stationary phase, this PAH should be a useful stationary phase selector in its own right for fullerene recognition by virtue of preferential π–π interaction and pronounced size and shape selectivity. Consequently, a chemically bonded $C_{60}$ silica phase was

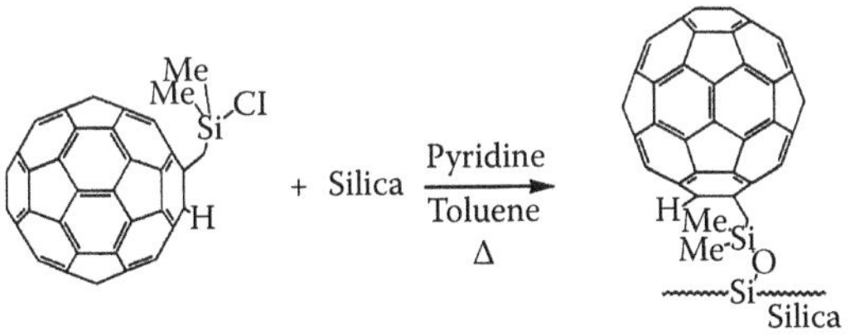

**FIGURE 6.13** Synthetic scheme for the $C_{60}$ bonded silica stationary phase. (From Saito, Y. et al., *J. High Resolut. Chromatogr.*, 1995, 18, 569–572. With permission.)

synthesized as a stationary phase for LC, and its retention behavior was evaluated for various PAHs.[54,80] The chemically bonded $C_{60}$ phase (Figure 6.13) showed preferential interaction with triphenylene and perylene, which possess partial structures similar to that of $C_{60}$. Interestingly, larger retention factors are observed for nonplanar vs. planar PAH molecules of comparable molecular size. Thus, effective molecular recognition occurs between the soccer ball $C_{60}$ selector and nonplanar selectands with a similar molecular curvature of the fullerene.[80] The bonded $C_{60}$ phase was also used for fullerene self-recognition, that is, for the congener separation of $C_{60}$ and $C_{70}$.[80] As compared to an ODS reversed phase devoid of aromatic moieties,[54] the $C_{60}$ phase entailed high retention factors for $C_{60}$ and $C_{70}$ and an enhanced separation factor ($\alpha = 2.9$).

Stalling et al. linked a $C_{60}/C_{70}$ mixture both to silica particles by the reaction $Si-NH_2 + C_{60/70}HBr \rightarrow Si-NH-C_{60/70}H$ and to polystyrene/divinylbenzene (PSDVB) beads via the reaction $PSDVB + C_{60/70} + AlCl_3/CS_2 \rightarrow PSDVB-C_{60/70}H$. LC-microcolumns, packed with these materials, retained PAHs in the order of increasing ring number, whereby nitrosubstitutents increased retention. Chlorinated dibenzo-*p*-dioxins and chlorinated dibenzofurans were more strongly retained than their parent counterparts, clearly indicating selective supramolecular $\pi$–$\pi$ interactions.[81] Planar polychlorinated biphenyls (PCBs) devoid of *ortho*-chlorine substituents showed an increased retention that further increased with higher chlorine substitution. Surface-linked-$C_{60/70}$-polystyrene divinylbenzene beads were used as chromatographic material for the enrichment of coplanar PCBs.[82] (3-Aminopropyl) silyl-bonded silica was employed to prepare a $C_{60}$-based stationary phase, which was utilized for the separation of haloaromatics by LC.[83]

Fullerenes have also been used as stationary phases in GC. A glass capillary column (15 m × 0.29 mm i.d.) was coated with pure $C_{60}$ (film thickness 0.1 μm) using a high-pressure static method.[84] The thermostable solid $C_{60}$ phase behaved like the saturated hydrocarbon squalane in its van der Waals interaction with *n*-alkanes. $C_{60}$ linked to poly(dimethylsiloxane) (Figure 6.14a) was coated on a 10 m × 0.25 mm i.d. fused silica column and used for the selective separation of isomeric hexachlorobiphenyls differing in *ortho*-chlorine substitution pattern.[85] Thermodynamic parameters were obtained using the retention-increment method $R'$ (Chapter 1), and it was shown that the retention behavior of PCBs depends strongly on the degree of *ortho*-chlorine substitution, which, in turn, is linked with nonplanarity (and reduced toxic potential by inserting into the aromatic hydrocarbon receptor; Figure 6.14b).

(a)

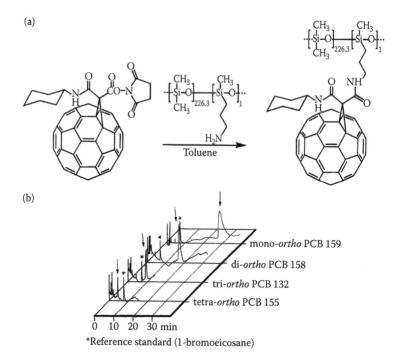

(b)

mono-*ortho* PCB 159

di-*ortho* PCB 158

tri-*ortho* PCB 132

tetra-*ortho* PCB 155

0  10  20  30 min
*Reference standard (1-bromoeicosane)

**FIGURE 6.14**    (a) Synthesis of $C_{60}$ linked to aminopropyl-poly(dimethylsiloxane). (b) Gas chromatographic elution order of hexachlorobiphenyl congeners with a different degree of *ortho*-substitution (10 m × 0.25 mm i.d. fused silica column coated the $C_{60}$ phase, 190°C, carrier: 0.6 bar (at gauge) helium, electron capture detection). (From Glausch, A. et al., *J. Chromatogr. A*, 1998, 809, 252–257. With permission.)

The deviation from planarity of PCBs strongly decreased the supramolecular interaction with $C_{60}$.[85]

Polysiloxanes containing bonded $C_{60}$-pyrrolidine were used as thermally stable stationary phases between 80°C and 360°C in capillary GC.[86] $C_{60}$ has also been attached to poly(dimethylsiloxane) via a side-on azidopropyl group and used for the gas chromatographic selective separation of PAHs, phthalic acid diesters, and aromatic positional isomers.[87] Mixed stationary phases composed of $C_{60}$ and dibenzo-24-crown-8 or β-cyclodextrin were investigated by GC.[88] Fullerene-impregnated ionic liquid stationary phases showed dual selectivity characteristics in GC.[89]

Saito et al. extended the molecular recognition principle operating in the LC selector system fullerene/PAHs to the system fullerene/calixarene.[90] The selective nonchromatographic complexation of $C_{60}$ and $C_{70}$ with calix[6]arenes and calix[8]arenes has previously been used as a versatile purification tool.[91,92] On the silica bonded $C_{60}$ phase (Figure 6.13) packed into a 15 cm × 0.32 mm i.d. fused silica capillary column, ᵗBu-calix[4]arene, ᵗBu-calix[6]arene, and ᵗBu-calix[8]arene were semiquantitatively separated by micro column LC with toluene/methanol (40:60 v/v) as eluent.[90] The elution sequence corresponded to that found on an ODS reversed phase; however, the separation factors α for the congeners were higher on the $C_{60}$ phase. A preferential fit between the 'Bu-calix[8]arene (cavity diameter 10 Å) and $C_{60}$ (diameter < 10 Å) based on $\pi-\pi$ interactions and multipoint interaction between the 'Bu-groups and $C_{60}$ were proposed.[90] Indeed, Bianco et al. observed a striking selectivity between 'Bu-calix[6]arene and 'Bu-calix[8]arene ($\alpha \sim 10$) on a [60]-fullereopyrrolidine-modified stationary phase (Figure 6.15, left) with the eluent $CH_2Cl_2$/2-propanol (99.5/0.5, v/v), whereas the congeners only coeluted with the eluent toluene.[20]

The strong effect of alcoholic modifiers in the eluent on the retention factor of 'Bu-calix[8]arene has been ascribed to its influence on the intramolecular hydrogen bond system required for complexation with $C_{60}$. On uncoated silica gel, the calix-arenes almost coelute, and their elution order is inverted as compared to the $C_{60}$ phase. Bianco et al. thus provided unequivocal evidence for their claim of the formation of supramolecular complexes and their provision that $C_{60}$-based HPLC materials have practical implications in both separation science and supramolecular chemistry (Chapter 2).[20] Also, on the $C_{60}$-fulleropyrrolidine-based silica four helically shaped and side-chain modified tyrosine nonapeptides possessing hydrophobic cavities were selectively separated.[20] The most tightly bound peptide carried two ferrocene moieties at the periphery of a hydrophobic binding cavity complementary in size to $C_{60}$. It was concluded that the shape and functional group selectivity of the fullerene stationary phase originates from the ability of the curved surface of $C_{60}$ to simultaneously establish $\pi-\pi$ and hydrophobic interactions.[20] The silica-grafted $C_{60}$-pyrrolidine was also tested as a solid singlet oxygen sensitizer resembling free $C_{60}$ to produce singlet oxygen in solution with a high quantum yield.[93] Gasparrini

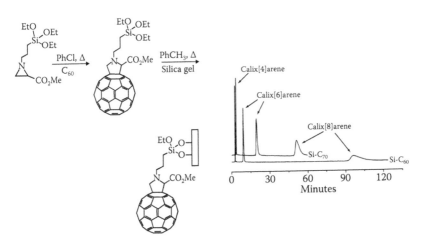

**FIGURE 6.15** Left: Synthetic pathway to a 6,6-$C_{60}$-pyrrolidine derivative by aziridine ring opening via 1,3-dipolar addition to $C_{60}$ and smoothly grafting the fullerene-containing triethoxysilane to silica microparticles, leading to a chemically homogeneous material with defined structure and high chromatographic efficiency. (From Bianco, A. et al., *J. Amer. Chem. Soc.*, 1997, 119, 7550–7554. With permission.) Right: Separation of 'Bu-calix[4]arene, 'Bu-calix[6]arene, and 'Bu-calix[8]arene on immobilized $C_{60}$ and $C_{70}$ by HPLC. A 250 × 1.8 mm i.d. stainless steel column packed with $C_{60}$-pyrrolidine-based silica (Hypersil, 5 μm). Eluent: 0.3 mL/min of dichloromethane/2-propanol (99.5:0.5 v/v), 25°C. (From Gasparrini, F. et al., *Tetrahedron*, 2001, 57, 6997–7002. With permission.)

et al. extended the investigations by immobilizing a $C_{70}$-pyrrolidine derivative on high-surface spherical silica microparticles.[94] Thermodynamic parameters for complex formation between the $^t$Bu-calix[6]arene/$^t$Bu-calix[8]arene selectands and the $C_{60}/C_{70}$ selectors were obtained for the two different column eluents dichloromethane (plus 0.5% 2-propanol) and toluene (plus 10% 2-propanol) at the temperature range 30°C–80°C (with 10°C intervals) by using the equations detailed in Chapter 1. Supramolecular complexation was always enthalpy-driven and entropy-opposed with the larger $^t$Bu-calix[8]arene forming the tighter complexes with $C_{60}$.[94]

In conclusion, Cancelliere et al. stated that "the basic idea for the use of fullerenes in the modification of HPLC supports was that they would have greatly facilitated evaluation of the relative affinities of potential hosts for the immobilized fullerene from the analysis of retention data; moreover, they would have enabled the use of solvents (as chromatographic eluents), in which fullerenes were not soluble, in the study of the host–guest chemistry of fullerenes."[95] It should be stressed that selectivity data obtained in a heterogeneous chromatographic system accompanied with a certain spacer chemistry may not necessarily agree with selectivity data obtained in the homogeneous system. Complexation enthalpies of a series of hydrogen-bonded host–guest complexes have been determined via temperature-dependent HPLC retention data, whereby one of the complementary supramolecular components was present in solution or was bonded to silica.[96] A very good agreement within the experimental error between the enthalpies of formation in dichloromethane solution and the enthalpies calculated from chromatographic retention data has been observed for hydrogen-bonded host–guest complexes involving derivatives of 2,6-diamidopyridine, adenine, and 1,8-naphthyridine (dissolved or bonded host) and various cyclic imides and thymine derivatives (dissolved guests).[96] A similar agreement between solution and solid phase enthalpies was obtained in host–guest systems in which complexation is a result of a combination of hydrogen bonding and π-stacking.[97]

A supramolecular selectand/selector system par excellence is represented by fullerenes ($C_n$, $n$ = number of carbon atoms) and cyclodextrins ($CD_n$, $n$ = number of D-glucose building blocks). The supramolecular recognition phenomenon between fullerenes and cyclodextrins is well established. γ-Cyclodextrin ($CD_8$) forms water-soluble bicapped 2:1 association complexes with [60]fullerene ($C_{60}$).[98–101] β-Cyclodextrin ($CD_7$) may also undergo complexation with $C_{60}$.[102,103] Bianco et al. used, for the first time, a $C_{60}$-bonded phase (Figure 6.15, left) in an aqueous/organic medium for the high-efficiency HPLC separation of α-, β-, and γ-cyclodextrins ($CD_6$–$CD_8$).[20] The trace shows an increased retention with an increased molecular weight of $CD_n$ (Figure 6.16). Because of an oversight, this early pioneering achievement unfortunately escaped the attention of Bogdanski et al. when referring to the chromatographic reciprocal principle between fullerenes $C_n$ and cyclodextrins $CD_n$.[21]

With the advent of the enzymatic access to large-ring cyclodextrins,[104] the separation of the congeners $CD_n$ represented a considerable analytical challenge. Koizumi et al. separated the mixture of $CD_6$–$CD_{85}$, obtained by the action of cyclodextrin glycosyltransferase from *Bacillus macerans*, on synthetic amylose by high-performance anion-exchange chromatography (HPAEC) with pulsed amperometric detection using a 25 cm × 4 mm i.d. Dionex CarboPac PA-100 column,[105] whereas Bogdanski et al. separated the mixture of $CD_6$–$CD_{21}$ by LC with electrospray ionization mass

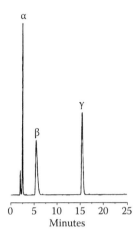

**FIGURE 6.16** Separation of $\alpha$-, $\beta$-, and $\gamma$-cyclodextrins ($CD_6$–$CD_8$) on immobilized $C_{60}$ by HPLC. A 250 × 1.8 mm i.d. stainless steel column packed with $C_{60}$-pyrrolidine-based silica (Hypersil, 5 µm). Eluent: 0.25 mL/min water/methanol/tetrahydrofuran (80:10:10 for 3 minutes and then linear gradient to 40:30:30 in 25 minutes, 25°C. (From Bianco, A. et al., *J. Amer. Chem. Soc.*, 1997, 119, 7550–7554. With permission.)

spectrometric (LC/ESI-MS) detection using a 25 cm × 4 mm i.d. LiChrospher $NH_2$ column.[106]

With one exception ($CD_9$ on HPAEC), the congeners $CD_n$ were eluted from the stationary phases according to the degree of polymerization $n$. However, a highly unusual selectivity pattern has been observed for the separation of $CD_6$–$CD_{15}$ congeners on a $C_{60}$ selector anchored to silica particles[21] that strongly deviated from the usual linear correlation of retention factor and molecular weight. Thus, whereas $CD_6$ and $CD_7$ exhibited a low retention and were eluted within 5 minutes, $CD_8$–$CD_{10}$ were eluted only after a long retention gap of about 15 minutes with the elution order $CD_8 < CD_{10} < CD_9$. The higher $CD_{11}$–$CD_{15}$ congeners again exhibited a low retention and were eluted together with $CD_6$ in 5 minutes (Figure 6.17). The drop of retention between $CD_{10}$ and $CD_{11}$ of 15 minutes was unprecedented. The unusual elution order has been detected with LC/ESI-MS of sodium ion adducts of $CD_6$–$CD_{15}$.[21] The results were confirmed for $CD_6$–$CD_{25}$ by employing an LC-evaporative light scattering detection system.[21] The observed selectivity changes that do not follow a molecular weight dependency of retention were considered as the result of supramolecular chromatography.[21]

The retention-increment method $R'$ (Chapter 1) has been employed to quantify the unusual selectivity. Two stationary phases, one containing the spacer alone and one containing the $C_{60}$ selector covalently bonded to the spacer, were prepared (Figure 6.18).[21] Thus, Nucleosil (5 µm, 100 Å) was reacted in toluene with 3-glycidoxypropyltrimethoxysilane to yield 3-glycidoxypropyl-silica. The epoxide was reacted with *bis*(1-aminohexyl)malonate to the silica-bonded spacer without a selector, and the same epoxide was reacted with *bis*(1-aminohexyloxycarbonyl)malonate-dihydro[60]fullerene.[107] The striking selectivity change of $CD_n$ occurred only on the $C_{60}$ selector but not on the silica-bonded spacer devoid of the $C_{60}$ selector. Because of the

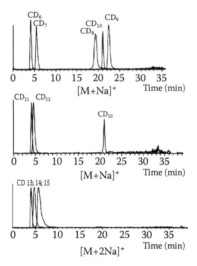

**FIGURE 6.17** Overlaid LC/ESI-MS traces of the selective separation of $CD_6$–$CD_{15}$ sodium ion adducts on $C_{60}$ by HPLC. 250 × 4 mm i.d. stainless steel column packed with silica-bonded $C_{60}$-stationary phase, 0.5 mL/min water to acetonitrile/tetrahydrofuran gradient elution, 25°C. (From Bogdanski, A. et al., *New J. Chem.*, 2010, 34, 693–698. With permission.)

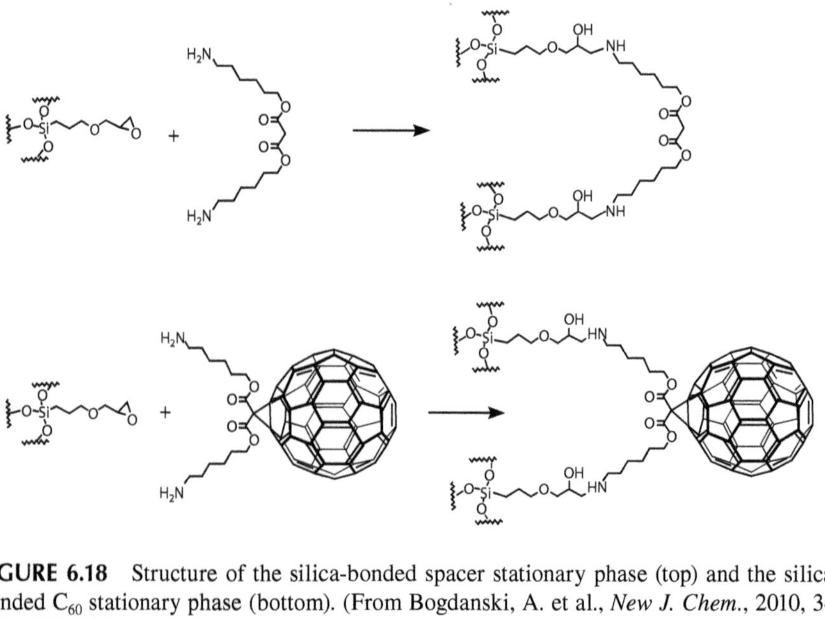

**FIGURE 6.18** Structure of the silica-bonded spacer stationary phase (top) and the silica-bonded $C_{60}$ stationary phase (bottom). (From Bogdanski, A. et al., *New J. Chem.*, 2010, 34, 693–698. With permission.)

required use of an LC elution gradient affecting the selectivity changes, only apparent relative complexation constants (given in brackets and normalized to $CD_6$) were reported, that is, $CD_6$ (1.0), $CD_7$ (4.5), $CD_8$ (39.0), $CD_9$ (46.0), $CD_{10}$ (42.0), $CD_{11}$ (1.0), and $CD_{12}$ (2.0). The data highlight the remarkable stability differences for the $CD_6$–$CD_{12}$ congeners in their supramolecular interaction with $C_{60}$.[20]

$C_{60}$ has also been attached to fused silica tubing via a 3-aminopropyltrimethoxysilane spacer and employed in open-tubular capillary EC (o-CEC) for the electrophoretic separation of plant phenols.[108] The $C_{60}$-based stationary phase possessed both hydrophilic and hydrophobic interaction sites and provided a positively charged capillary surface required for the electro-osmotic flow. The study involved a comparison between the bare fused silica surface, the 3-aminopropylsilylated surface, and the fullerene-bonded phase.[108]

## CARBON NANOTUBES

Carbon nanotubes (CNTs) consist of one or more graphene sheets rolled up into a seamless cylinder in nanoscale dimensions that comprise hexagon-rich $sp^2$ carbon on the wall and a few pentagons at the caps with increasing $sp^3$ character as tube diameters become smaller. They exist as single-walled nanotubes formed of a single graphene sheet (SWCNTs) or multiple-walled nanotubes (MWCNTs) with additional graphene tubes around the core of an SWCNT. Also open-ended structures or those capped with half-spheres resembling fullerenes are known. Complexation with potential selectands can occur on the outside surface, on the curved graphene planes, in the interstices between tubes that form bundles, and in the inner hollow cavity when they are open-ended. Because of their curved surface, CNTs are expected to show a stronger affinity for hydrophobic molecules as compared to planar graphite.[109] CNTs are known to have high thermal and mechanical stability. However, both of these carbon nanoparticles exhibit limited dispersibility in aqueous and organic solvents, which can be a source of difficulty for packing chromatographic columns with such material.[110]

Kartsova and Makarov first discussed the propensity of CNTs as selective stationary phases in GC despite their insolubility in organic solvents.[88,111] Li and Yuan used powdered MWCNT particles filled into a glass column (50 × 0.3 cm i.d.) as an improved stationary phase for the gas–solid chromatographic separation of aromatic hydrocarbons, alkanes, and halocarbons.[112] The CNT phase had improved properties as compared to activated charcoal and graphitized carbon black packing materials; however, the column plate numbers of the packed column were low.[112] The first application of self-assembled CNTs in long wide-bore (500 µm i.d.) capillaries for high-resolution gas–solid chromatography was reported by Saridara and Mitra.[113] A film (1–50 µm) of MWCNT was formed on an inner stainless steel capillary surface by chemical vapor deposition of ethylene used as a carbon source with preformed nanocrystals of iron serving as the catalyst for surface immobilization of MWCNT.[113] In a follow-up report, thermally highly stable self-assembled SWCNTs were deposited as film (300 nm) on a silica-lined steel capillary column.[114] Evaluation of McReynolds constants established that the SWCNT represented a nonpolar stationary phase. Selective π–π interactions are probably absent because benzene eluted before $n$-hexane.[114] Very fast GC on SWCNT stationary phases in microfabricated channels is suitable for field analysis applications with portable microinstrumentation.[115] Peak overlapping and peak tailing represented a problem that was partially alleviated by temperature programming.

SWCNTs were also chemically linked end-to-end to a fused-silica capillary column with a high surface area.[116] Gas chromatographic investigations were performed either on SWCNTs alone or in combination with ionic liquids (ILs).[116] Unfortunately, the strong adsorption ability of SWCNTs led to low efficiencies and peak tailing. A combination of a SWCNT with a chiral IL showed that the gas chromatographic enantioseparation was improved in the presence of the achiral CNTs.[117] True supramolecular interactions involving CNTs in GC remain to be established.

SWCNTs were also incorporated into an organic polymeric precursor containing vinylbenzyl chloride and ethylene dimethacrylate for the synthesis of a monolithic stationary phase for micro-HPLC and capillary electrochromatography (CEC). Incorporation of CNTs into the monolithic network increased the chromatographic retention of small organic selectands in reversed-phase HPLC, owing to its hydrophobic characteristics.[118] SWCNTs were also bonded to 3-aminopropyl silica gel and used as a selective stationary phase for the separation of PAHs by HPLC.[119]

MWCNT silica nanoparticle composite materials have been obtained by coupling aminopropyltriethoxysilane to a functionalized CNT-COOH through a carboxamide bond and formation of silica nanobeads in reverse micelles in a water-in-oil microemulsion.[120] For the coupling of shortened SWCNTs to HPLC silica microparticles (5 μm), the CNTs with oxidized termini and side walls were activated by reaction with thionyl chloride and then exposed to 3-aminopropyl-silica gel.[121] MWCNTs carrying protonated primary amino groups around their side walls were mixed with mesoporous 3-aminopropylsilica (5 μm). Both on the single-walled CNT packing material fixed by amide bonds and on the MWCNT packing material held together by hydrogen bonds between ammonium fragments of selector and amino groups of silica, the affinities of nitroaromatics, positional $o$-, $m$-, $p$-terphenyl, and small peptides were studied by HPLC through monitoring the retention characteristics under variable experimental conditions.[121] Kwon and Park packed MWCNTs into stainless-steel columns (150 × 1 mm i.d.) and used them as a stationary phase for reversed-phase LC.[122] They applied linear solvation energy relationships[123] to compare the CNT stationary phase to carbon-coated zirconia and octadecylsiloxane-bonded silica and thus showed similarity only between the carbon-based materials.[122] In agreement with the observation in GC (*vide supra*), $\pi$–$\pi$ and $\pi$–$n$ selector/selectand interactions are not pronounced with CNTs in comparison to graphitic carbon because of shielding of $\pi$ electrons by strongly adsorbed solvent molecules.

SWCNTs exhibit a pronounced adsorption ability for nanoscale interactions, whereas cellulose *tris*-phenylcarbamates possess high enantiorecognition ability in HPLC. Indeed, the mixed stationary-phase cellulose 2,3-*bis*-phenylcarbamate-6-SWCNT afforded high enantioseparating factors α because of a promoting effect of achiral CNT on chiral recognition.[124] For the challenging use of inherently chiral helical CNTs for chromatographic enantioseparations, they must first be obtained in an enantiomerically pure form.

## SYNTHETIC SUPRAMOLECULAR $C_n$-SYMMETRIC RECEPTORS

Following the pioneering example of Cram's chiral crown ether (Chapter 2),[50] other fully synthetic supramolecular receptors were developed as enantioselective selectors in LC. Supported by rational experiments, Pirkle et al. concluded that the use of $C_n$-symmetric selectors in CSP design allows for a more efficient use of geometric space, reducing nonselective interactions and making use of a greater surface density of binding sites–although $C_n$-symmetry of selectands or selectors in chromatography is *a priori* not a prerequisite to obtain higher enantioselectivities.[125] On chiral selectors possessing $C_n$ or $D_n$ axes ($n \neq 1$), the molecular association by a selectand can take place equally from either face of the molecule and, thus, produces the same complex.

Gasparrini et al. linked a synthetic $C_3$-symmetric basket-like receptor derived from $O$-allyl protected tyrosine and 1,3,5-trimercaptobenzene to γ-mercaptopropyl silica gel microparticles via free radical addition of the thiols groups to the allylic double bonds (Figure 6.19, left).[126] The chiral macrocyclic receptor developed by Erickson et al.[127] is unique in that it has a concave, rigid binding site and an array of alternating hydrogen bond donors and acceptors located around the periphery of the fully synthetic macrocycle (Figure 6.19, left). The CSP is readily available in both enantiomeric forms, and it shows high enantioselectivity for racemic $N$-Boc-$N'$-methylamides of α-amino acids (Figure 6.19, right). The receptor can be used to determine binding properties with several racemic substrates such as acylated amines and tripeptides.[126]

Gasparrini et al. also linked two $C_2$-symmetric double-armed molecules derived from identical tetra-amide subunits [obtained from $(R,R)$-1,2-diaminocyclohexane (DACH), trimesic acid, and isophthalic acid derivatives] and connected to $N$-(4-allyloxybenzoylated)-$(R,R)$-2,3-diaminopyrrolidine (Figure 6.20) to γ-mercaptopropyl

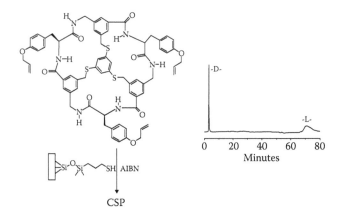

**FIGURE 6.19** Left: Synthesis of the $C_3$-symmetric receptor containing tyrosine and its immobilization on 3-mercaptopropyl silica gel. Right: HPLC enantioseparation of $N$-Boc-$D,L$-threonine $N'$-methylamide on the $C_3$-symmetric $O$-allyl protected tyrosyl macrocyclic receptor (α = 43, $-\Delta\Delta G$ = 2.2 kcal/mol). Column: 250 × 2 mm i.d. Eluent: 1% MeOH in $CH_2Cl_2$. (From Gasparrini, F. et al., *J. Org. Chem.*, 1995, 60, 4314–4315. With permission.)

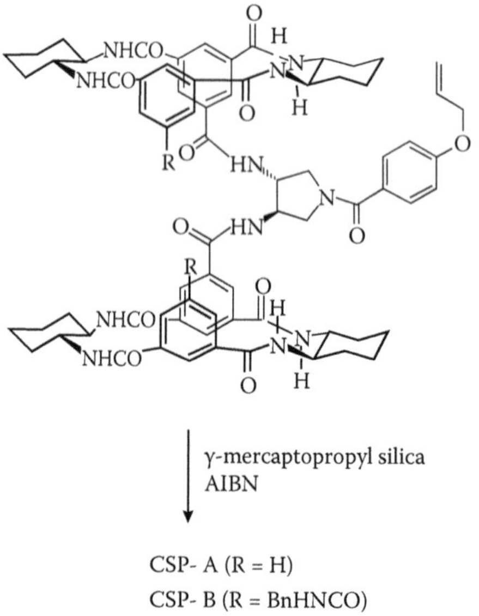

**FIGURE 6.20**   Synthesis of the $C_2$-symmetric receptor containing tetra-amide subunits and its immobilization on 3-mercaptopropyl silica gel. (From Gasparrini, F. et al., *J. Org. Chem.*, 1997, 62, 8221–8224. With permission.)

silica gel microparticles.[128] The macrocyclic receptors developed by Wennemers et al.[129] were used for the enantioseparation of *N*-3,5-dinitrobenzoyl(DNB)-α-amino acid *n*-hexylamides and the separation of the eight stereoisomers of the tripeptide acetyl-Pro-Val-Gln(γ-trityl)-propylamide. The multiple binding sites between host and guest induced large energetic variations in binding and in stereochemical discrimination.[128] Variable temperature chromatographic runs were performed to obtain thermodynamic data of stereochemical recognition. Van't Hoffs plots (see Chapter 1) revealed unusual behavior of the two CSPs toward the eight tripeptides: Retention of the loosely bound tripeptides was entropy driven, that is, the retention increased with temperature, and that of the strongly retained tripeptides was enthalpy driven.[128] The chromatographic approach employs a small amount of a host on which a multitude of potential guests can be analyzed. This represents an advantage as compared to traditional one-phase approaches for obtaining relative binding constants.

Gasparrini et al. synthesized a $C_2$-symmetric 18-membered tetramide macrocyclic minireceptor with an $A_2B_2$-type structural pattern by assembling 5-allyloxy-isophthaloyl chloride (A) and (*R,R*)-1,2-diphenylethylenediamine in a single step (Figure 6.21).[130] The chiral macrocycle preferentially binds the *L*-enantiomers of 3,5-DNP-α-amino acid *n*-hexylamides with enantioselectivity values up to −$\Delta\Delta G$ = 3.0 kcal/mol at 25°C, whereas an immobilized minireceptor analogue prepared from (*R,R*)-1,2-diaminocyclohexane (DACH) showed only modest enantioselectivities.[130] Molecular mechanics calculations revealed a low-energy structure of the minireceptor with the four peripheral phenyl groups responsible for the high affinity

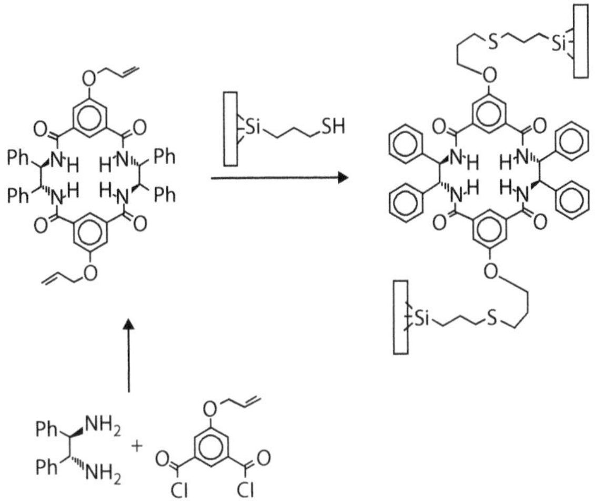

**FIGURE 6.21** Synthesis of the $C_2$-symmetric minireceptor with an $A_2B_2$-type structure and its immobilization on 3-mercaptopropyl silica gel. (From Gasparrini, F. et al., *Org. Lett.*, 2002, 4, 3993–3996. With permission.)

toward phenylacyl *L*-α-amino acid alkylamides.[130] All the synthetic chiral receptors described in this chapter exhibit very high enantioselectivities that are rather uncommon to many traditional CSPs.[37,38]

## CHALLENGES

The recognition of subtle structural differences of guests (selectands) puts an extreme selectivity requirement on a potential host (selector). One of the most difficult tasks in separation science constitutes the separation of isotopomers. Highly refined recognition modes between selectors and selectands are required for this challenge. Aided by the extreme separation power of high-resolution capillary complexation GC, the deuterated ethenes $C_2H_{4-n}D_n$ were completely resolved on a rhodium(I) coordination compound[41,131] and the α-deuterated tetrahydrofurans $C_4H_{4(\beta)}H_{4-n(\alpha)}D_{n(\alpha)}$ were resolved on a cobalt(II) coordination compound.[132] As might have been expected, the minuscule structural differences between *cis/trans*-1,2-dideuteroethene and that between the *meso/racem*-diastereomeric pair of 2,5-dideuterotetrahydrofuran and of the enantiomers of the *racem*-form could not further be distinguished on the nonracemic metal chelates carrying a chiral 3-acylated (*1R*)-camphor moiety.

Whereas the differentiation of enantiomers, owing their chirality entirely to isotopic substitution, is common to enzymes, their liquid chromatographic enantioseparation on CSPs represents a considerable challenge, which was independently met in 1997. One example is provided by Pirkle and Gan who observed an unprecedented isotope effect in the π-π face-to-edge interaction during the enantioseparation of racemic pivalamide of α,α'-phenyl-(phenyl-$d_5$)-methylamine and its *p,p'*-dibromo derivative on the selector (*3S,4R*)-Whelk-O1[57,72] (Figure 6.9b).[133]

Dynamic nuclear magnetic resonance (NMR) studies revealed that the protonated aromatic group is more strongly held in the binding cleft of the chiral selector than is the deuterated aromatic group.[133] Kimata et al. resolved racemic phenyl(phenyl-$d_5$) methanol using the CSP cellulose tribenzoate coated on silica with 2-propanol/$n$-hexane (5/95, v/v) as an eluent. The selector showed preferential retention of ($R$)-(−)-phenyl(phenyl-$d_5$)methanol compared to the ($S$)-(+)-enantiomer ($\alpha$ = 1.008 after extensive recycling).[134] Attempts to gas-chromatographically resolve racemic *tert*-butyl(*tert*-butyl-$d_9$)methanol have failed thus far, although the enantiomers could readily be distinguished by $^1$H-NMR spectroscopy in the presence of the chiral lanthanide shift reagent europium(III) *tris*[3-(heptafluorobutanoyl)-(*1R*)-camphor-ate].[135] It can be concluded that chromatographic resolution of racemic compounds of the type $R_1R_2C^*HD$ will hardly ever be possible.

Another challenge in separation science consists of the enantioseparation of unfunctionalized chiral alkanes because enantiorecognition in the presence of a chiral selector would be based solely on weak van der Waals forces, representing an extreme case of supramolecular noncovalent interactions. Indeed, all seven chiral C7–C8 alkanes, including ethylmethylpropylmethane (3-methylhexane), were resolved by GC on various modified cyclodextrin derivatives.[136] An inclusion mechanism is indicated as the chiral recognition phenomenon was totally lost when the corresponding linear dextrins (acyclodextrins) were used as CSPs. The smallest racemic (nonisotopically labeled) allene, that is, 2,3-pentadiene, has also been resolved on a cyclodextrin selector.[136]

Parity violation energy differences (PVEDs) have so far never been experimentally observed in chiral molecules. The enantiomers of the smallest chiral five-atomic molecules, that is, bromochlorofluoromethane C*HFClBr and chlorofluoroiodo-methane C*HClFI, represent (heavy) target molecules for PVED measurements.[137] Via supramolecular chemistry, Constante-Crassous et al. determined the absolute configuration of C*HFClBr as ($R$)-(−) or ($S$)-(+) from the molecular dynamics simulation of its enantioselective complexation by (−)-cryptophane-C (Figure 6.22).[138]

Access to the enantiomers of the tri(hetero)halogenomethanes relies on the decarboxylation (with retention of configuration) of the corresponding preresolved tri(hetero)halogenocarboxylic acids. The direct preparative GC enantioseparation has not been achieved yet; however, the quantitative analytical enantioseparation

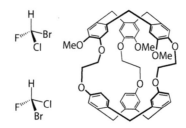

**FIGURE 6.22**  Structures of ($R$)- and ($S$)-C*HFClBr and absolute configuration of (−)-cryp-tophane-C. (From Costante-Crassous, J. et al., *J. Amer. Chem. Soc.*, 1997, 119, 3818–3823. With permission.)

of C\*HFClBr and C\*HFClI was achieved on the CSP Chirasil-γ-Dex (octakis(3-*O*-butanoyl-2,6-di-*O*-*n*-pentyl)-γ-cyclodextrin linked to poly(dimethylsiloxane)).[139,140] The following thermodynamic data of enantioselectivity (see Chapter 1) were obtained: C\*HFClBr: $\Delta\Delta H$ = −0.46 kJ/mol; $\Delta\Delta S$ = −1.37 J/kmol; $T_{iso}$ = 336 K, and C\*HFClI: $\Delta\Delta H$ = −1.52 kJ/mol; $\Delta\Delta S$ = −4.57 J/kmol; $T_{iso}$ = 333 K. Thus, the enantioselectivity difference on the CSP as expressed by thermodynamic data is threefold for the iodo compound. The identical and low inversion temperature $T_{iso}$ (~60°C) infers the same inclusion mechanism and a strong entropic contribution to chiral recognition.[140] All of these challenges of chromatographic resolutions undoubtedly require highly selective supramolecular interactions.

The study of supramolecular selector/selectand complexation equilibria can also be extended to partitioning systems whereby the selector is added to the mobile phase[141] or when two different selectors are present in the stationary and mobile phase.[142] Supramolecular chromatography may be regarded as a part of supramolecular analytical chemistry.[143]

## REFERENCES

1. Ettre, L. S. In International Union of Pure and Applied Chemistry (IUPAC), Nomenclature for Chromatography (Recommendation), *Pure Appl. Chem.* 1993, *65(4)*, 819–872.
2. Meyers, R. L.; Gehrke, C. W., eds. *Chromatography—A Science of Discovery*, John Wiley & Sons, Inc., Hoboken, New Jersey, 2010.
3. Senchenkova, E. M.; Michael Tswett. The Creator of Chromatography, *Scientific Council on Adsorption and Chromatography, Russ. Acad. Sci.* 2003.
4. Kuhn, R.; Winterstein, A.; Lederer, E. *Hoppe-Seyler's Z. Physiolog. Chem.* 1931, *197*, 141–160.
5. Poole, C. F. *The Essence of Chromatography*, Elsevier, Amsterdam, 2003.
6. Bradford, B. W.; Harvey, D.; Chalkley, D. E. *J. Inst. Petrol. London.* 1955, *41*, 80–91.
7. Schurig, V. *Inorg. Chem.* 1986, *25*, 945–949.
8. Schurig, V. *J. Chromatogr. A.* 2002, *965*, 315–356; *ibid.* 2009, *1216*, 1723–1736.
9. Stalcup, A. M. *Annu. Rev. Anal. Chem.* 2010, *3*, 341–363.
10. Lämmerhofer, M. *J. Chromatogr. A.* 2010, *1217*, 814–856.
11. Schurig, V. *Chirality.* 2005, *17*, S205–S226.
12. Watabe, K.; Charles, R.; Gil-Av, E. *Angew. Chem. Int. Ed.* 1989, *28*, 192–194.
13. Schurig, V.; Ossig, J.; Link, R. *Angew. Chem. Int. Ed.* 1989, *28*, 194–196.
14. Špánik, I.; Krupčik, J.; Schurig, V. *J. Chromatogr. A.* 1999, *843*, 123–128.
15. Schurig, V.; Schmidt, R. *J. Chromatogr. A.* 2003, *1000*, 311–324.
16. Lantz, A. W.; Pino, V.; Anderson, J. L.; Armstrong, D. W. *J. Chromatogr. A.* 2006, *1115*, 217–224.
17. Gebreyohannes, K. G.; McGuffin, V. L. *J. Chromatogr. A.* 2010, *1217*, 5901–5912.
18. Ashby, W. R. *An Introduction to Cybernetics*, Chapman & Hall Ltd., London, 1956, p. 10.
19. Mikeš, F. *The Resolution of Chiral Compounds by Modern Liquid Chromatography*, doctoral thesis, The Weizmann Institute of Science, Rehovot, Israel, 1975.
20. Bianco, A.; Gasparrini, F.; Maggini, M.; Misiti, D.; Polese, A.; Prato, M.; Scorrano, G.; Toniolo, C.; Villani, C. *J. Amer. Chem. Soc.* 1997, *119*, 7550–7554.
21. Bogdanski, A.; Wistuba D.; Larsen, K. L.; Hartnagel, U.; Hirsch, A.; Schurig V. *New J. Chem.* 2010, *34*, 693–698.
22. Peng, J.; Ruonong, F. *Fenxi Huaxue, Chinese J. Anal. Chem.* 1995, *23*, 104-110. http://www.chemyq.com/expert/ep134/1331191_845A0.htm.

23. Li, L. S.; Wang, S. W.; Huang, L. F.; Liu, M. F. *Acta Chim. Sinica.* 2008, *66*, 63–72.
24. Rozou, S.; Michaleas, S.; Antoniadou-Vyza, E. *J. Chromatogr. A.* 2005, *1087*, 86–94.
25. Graves, C. R.; Merlau, M. L.; Morris, G. A.; Sun, S.-S.; Nguyen, S. T.; Hupp, J. T. *Inorg. Chem.* 2004, *43*, 2013–2017.
26. Lou, X.; Zhu, Q.; Lei, Z.; van Dongen, J. L. J.; Meijer, E. W. *J. Chromatogr. A.* 2004, *1029*, 67–75.
27. Bazylak, G.; Aboul-Enein, H. Y. *Chirality.* 1999, *11*, 387–393.
28. Schneider, H.-J. *Angew. Chem. Int. Ed. Engl.* 2009, *48*, 3924–3977.
29. Bailon, P.; Ehrlich, G. K.; Fung, W.-J.; Berthold, W. *An Overview of Affinity Chromatography,* Humana Press, New York, 2000.
30. Lowe, C. R. In *Chromatography—A Science of Discovery,* Meyers, R. L.; Gehrke, C. W., eds., John Wiley & Sons, Inc., Hoboken, New Jersey, 2010, 141–150.
31. Thielmann, F. *J. Chromatogr. A.* 2004, *1037*, 115–123.
32. Katsanos, N. A.; Karaiskakis, G. *Time-resolved Inverse Gas Chromatography and its Applications,* HNB Publishing, New York, 2004.
33. Porath, J. *J. Chromatogr.* 1978, *159*, 13–24.
34. Huber, C. G. In *Encyclopedia Anal. Chem.*, Meyers, R. A., ed., John Wiley & Sons, Ltd., Chichester, 2000, 11,250–11,278.
35. Huber, C. G.; Schley, C.; Delmone, N. In *Proteomics & Peptidomics, Techn. Develop. Driving Biol.,* Chapter 2: Capillary HPLC for Proteomic and Peptidomic Analysis, *Comprehensive Anal. Chem.*, Elsevier, Amsterdam, 2005, 1–83.
36. Carta, G.; Jungbauer, A. *Protein Chromatography. Process Development and Scale-Up*, Wiley-VCH, Weinheim, 2010.
37. Allenmark, S.; Schurig, V. *J. Mater. Chem.* 1997, *7*, 1955–1963.
38. Okamoto, Y.; Ikai, T. *Chem. Soc. Rev.* 2008, *37*, 2593–2608.
39. Wulff, G. *Angew. Chem. Int. Ed. Engl.* 1995, *34*, 1812–1832.
40. Zhang, J.; Albelda, M. T.; Liu, Y.; Canary, J. W. *Chirality.* 2005, *17*, 404–418.
41. Schurig, V. In *Chromatographic Separations Based on Molecular Recognition*; Jinno, K., ed., Chapter 7; Wiley-VCH, New York; 1997, 371–418.
42. Klemm, L. H.; Reed, D.; Lind, C. D. *J. Org. Chem.* 1957, *22*, 739–743.
43. Klemm, L. H.; Reed, D. *J. Chromatogr.* 1960, *3*, 364–368.
44. Klemm, L. H.; Desai, K. B.; Spooner Jr., J. R. *J. Chromatogr.* 1964, *14*, 300–302.
45. Diack, M.; Compton, R. N.; Guiochon, G. *J. Chromatogr.* 1993, *639*, 129–140.
46. Newman, M. S.; Lednicer, D. *J. Amer. Chem. Soc.* 1956, *78*, 4765–4770.
47. Mikeš, F.; Boshart, G.; Gil-Av, E. *Chem. Commun.* 1976, 99–100; Mikeš, F.; Boshart, G.; Gil-Av, E. *J. Chromatogr.* 1976, *122*, 205–221.
48. Mikeš, F.; Boshart, G.; *J. Chem. Commun.* 1978, 173–174.
49. Cram, D. J.; Cram, J. M. *Acc. Chem. Res.* 1978, *11*, 8–14.
50. Sogah, G. D. Y.; Cram, D. J. *J. Amer. Chem. Soc.* 1976, *98*, 3038–3041; *ibid.* 1979, *101*, 3035–3042.
51. Osawa, E. *Kagaku.* 1970, *25*, 854–863.
52. Kroto, H. W.; Heath, J. R.; O'Brien, S. C.; Curl, R. F.; Smalley, R. E. *Nature* 1985, *318*, 162–163.
53. http://www.jpl.nasa.gov/news/news.cfm?release=2010-243.
54. Jinno, K. In *Chromatographic Separations Based on Molecular Recognition*; Jinno, K., ed., Chapter 3; Wiley-VCH: New York; 1997, 147–237.
55. Baena, J. R.; Gallego, M.; Valcárcel, M. *Trends Anal. Chem.* 2002, *21*, 187–198.
56. Hawkins, J. M.; Lewis, T. A.; Loren, S. D.; Meyer, A.; Heath, J. R.; Shibato, Y.; Saykally, R. J. *J. Org. Chem.* 1990, *55*, 6250–6252.
57. Welch, C. J. *J. Chromatogr. A.* 1994, *666*, 3–26.
58. Pirkle, W. H.; Welch, C. J. *J. Org. Chem.* 1991, *56*, 6973–6974.
59. Welch, C. J.; Pirkle, W. H. *J. Chromatogr.* 1992, *609*, 89–101.

60. http://www.echemshop.com/pages.php?pageid=16.
61. Hennrich, F.; Thesis, University of Karlsruhe, Germany, 2000, Table 2.4, p. 15.
62. Kawauchi, T.; Kitaura, A.; Kawauchi, M.; Takeichi, T.; Kumaki, J.; Iida, H.; Yashima, E. *J. Amer. Chem. Soc.* 2010, *132*, 12,191–12,193.
63. Cox, D. M.; Behal, S.; Disko, M.; Gorun, S. M.; Greaney, M.; Hsu, C. S.; Kollin, E. B.; Millar, J.; Robbins, J.; Robbins, W.; Sherwood, R. D.; Tindall, P. *J. Amer. Chem. Soc.* 1991, *113*, 2940–2944.
64. Kibbey, C. E.; Savina, M. R.; Parseghian, B. K.; Francis, A. H.; Meyerhoff, M. E. *Anal. Chem.* 1993, *65*, 3717–3719.
65. Jinno, K.; Kohrikawa, C.; Saito, Y.; Haiginaka, J.; Saito, Y.; Mifune, M. *J. Microcol. Sep.* 1996, *8*, 13–21.
66. Cabrera, K.; Wieland, G.; Schäfer, M. *J. Chromatogr. A.* 1993, *644*, 396–399.
67. Thilgen, C.; Diederich, F. *Chem. Rev.* 2006, *106*, 5049–5135.
68. Gross, B., doctoral thesis, University of Tübingen, 1998, p. 12.
69. Yamamoto, C.; Hayashi T.; Okamoto, Y.; Ohkubo, S.; Kato, T. *Chem. Commun.* 2001, 925–926.
70. Okamoto, Y.; Yashima, E. *Angew. Chem. Int. Ed.* 1998, *37*, 1020–1043.
71. Hawkins, J. M.; Meyer, A. *Science.* 1993, *260*, 1918–1920.
72. Wilson, S. R.; Wu, Y.; Kaprinidis, N. A.; Schuster, D. I.; Welch, C. J. *J. Org. Chem.* 1993, *58*, 6548–6549.
73. Wang, Y.; Schuster, D. I.; Wilson, S. R.; Welch, C. J. *J. Org. Chem.* 1996, *61*, 5198–5199.
74. Herrmann, A.; Rüttimann, M.; Thilgen, C.; Diederich, F. *Helv. Chim. Acta.* 1995, *78*, 1673–1704.
75. Gross, B.; Schurig, V.; Lamparth, I.; Herzog, A.; Djojo, F.; Hirsch, A. *Chem. Commun.* 1997, 1117–1118.
76. Gross, B.; Schurig, V.; Lamparth, I.; Hirsch, A. *J. Chromatogr. A.* 1997, *791*, 65–69.
77. Nakamura, Y.; O-kawa, K.; Nishimura, T.; Yashima, E.; Nishimura, J. *J. Org. Chem.* 2003, *68*, 3251–3257.
78. Pirkle, W. H.; House, D. W.; Finn, J. M. *J. Chromotogr.* 1980, *192*, 143–158.
79. Pirkle, W. H.; Däppen, R. *J. Chromatogr.* 1987, *404*, 107–115.
80. Saito, Y.; Ohta, H.; Terasaki, H.; Katoh, Y.; Nagashima, H.; Jinno, K.; Itoh, K. *J. High Resolut. Chromatogr.* 1995, *18*, 569–572.
81. Stalling, D. L.; Guo, C.; Kuo, K. C.; Saim, S. *J. Microcol. Sep.* 1993, *5*, 223–235.
82. Stalling, D. L.; Guo, C. Y.; Saim, S. *J. Chromatogr. Sci.* 1993, *31*, 265–278.
83. Chang, C.-S.; Wen, C.-H.; Den, T.-G. *Huaxue (Chemistry).* 1996, *54*, 11–21; *Chem Abstr.* 1997, *126*, 139272d.
84. Golovnya, R. V.; Terenina, M. B.; Ruchkina, E. L.; Karnatsevich, V. L. *Mendeleev Commun.* 1993, *3*, 231–233.
85. Glausch, A.; Hirsch, A.; Lamparth, I.; Schurig, V. *J. Chromatogr. A.* 1998, *809*, 252–257.
86. Ye, H.-Y.; Zeng, Z.-R.; Liu, Y.; Wu, C.-Y.; Chen, Y.-Y. *Chinese J. Anal. Chem.* 1999, *27*, 276–280.
87. Fang, P.-F.; Zeng, Z.-R.; Fan, J.-H.; Chen, Y.-Y. *J. Chromatogr. A.* 2000, *867*, 177–185.
88. Kartsova, L. A.; Makarov, A. A. *J. Anal. Chem. (translated from Russ.).* 2004, *59*, 724–729.
89. Tran, C. D.; Challa, S. *Analyst* 2008, *133*, 455–464.
90. Saito, Y.; Ohta, H.; Terasaki, H.; Katoh, Y.; Nagashima, H.; Jinno, K.; Itoh, K.; Trengove, R. D.; Harrowfield, J.; Li, S. F. Y. *J. High Resolut. Chrom.* 1996, *19*, 475–477.
91. Atwood, J. L.; Koutsantonis, G. A.; Raston, C. L. *Nature* 1994, *368*, 229–231.
92. Suzuki, T.; Nakashima, K.; Shinkai, S. *Tetrahedr. Lett.* 1995, *36*, 249–252.
93. Bonchio, M.; Maggini, M.; Menna, E.; Scorrano, G.; Garlaschelli, L.; Giacometti, A.; Paolucci, F.; Gasparrini, F.; Misiti, D.; Villani, C. *Proc. Electrochem. Soc.* 1999, *99*, 220–224.

94. Gasparrini, F.; Misiti, D.; Della Negra, F.; Maggini, M.; Scorrano, G.; Villani, C. *Tetrahedron* 2001, *57*, 6997–7002.

95. Cancelliere, G.; D'Acquarica, I.; Gasparrini, F.; Maggini, M.; Misiti, D.; Villani, C. *J. Sep. Sci.* 2006, *29*, 770–781.

96. Zimmerman, S. C.; Kwan, W.-S. *Angew. Chem. Int. Ed. Engl.* 1995, *34*, 2404–2406.

97. Zimmerman, S. C.; Saionz, K. W. *J. Amer. Chem. Soc.* 1995, *117*, 1175–1176.

98. Andersson, T.; Nilsson, K.; Sundahl, M.; Westman, G.; Wennerström, O. *J. Chem. Soc., Chem. Comm.* 1992, 604–606.

99. Priyadarsini, K. I.; Mohan, H.; Tyagi, A. K.; Mittal, J. P. *J. Phys. Chem.* 1994, *98*, 4756–4759.

100. Kanazawa, K.; Nakanishi, H.; Ishizuka, Y.; Nakamura, T.; Matsumoto, M. *Fullerene Sci. Techn.* 1994, *2*, 189–194.

101. Yoshida, Z.; Takekuma, H.; Matsubara, Y. *Angew. Chem. Int. Ed. Engl.* 1994, *33*, 1597–1599.

102. Samal, S.; Geckeler, K. E. *J. Chem. Soc., Chem. Comm.* 2000, 1101–1102.

103. Yu, Y.; Shi, Z.; Zhao, Y.; Ma, Y.; Xue, M.; Ge, J. *Supramol. Chem.* 2008, *20*, 295–299.

104. Larsen, K. L. *J. Inclusion Phenom. Macrocycl. Chem.* 2002, *43*, 1–13.

105. Koizumi, K.; Sanbe, H.; Kubota, Y.; Terada, Y.; Takaha, T. *J. Chromatogr. A.* 1999, *852*, 407–416.

106. Bogdanski, A.; Larsen, K. L.; Bischoff, D.; Ruderisch, A.; Jung, G.; Süßmuth, R.; Schurig, V. *Proceed. 12th International Cyclodextrin Symposium*, Montpellier, France, May 16–19, 2004, 171.

107. Braun, M.; Hartnagel, U.; Ravanelli, E.; Schade, B.; Böttcher, C.; Vostrowsky, O.; Hirsch, A. *Eur. J. Org. Chem.* 2004, 1983–2001.

108. Shiue, C. C.; Lin, S. Y.; Liu, C. Y. *J. Chinese Chem. Soc.* 2001, *48*, 1029–1034.

109. Speltini, A.; Merli, D.; Quartarone, E.; Profumo, A. *J. Chromatogr. A.* 2010, *1217*, 2918–2924.

110. Valcárcel, M.; Cárdenas, S.; Simonet, B. M.; Moliner-Martínez, Y.; Lucena, R. *Trends Anal. Chem.* 2008, *27*, 34–43.

111. Kartsova, L. A.; Makarov, A. A. *Russ. J. Appl. Chem.* 2002, *75*, 1725–1731.

112. Li, Q.; Yuan, D. *J. Chromatogr. A.* 2003, *1003*, 203–209.

113. Saridara, C.; Mitra, S. *Anal. Chem.* 2005, *77*, 7094–7097.

114. Karwa, M.; Mitra, S. *Anal. Chem.* 2006, *78*, 2064–2070.

115. Stadermann, M.; McBrady, A. D.; Dick, B.; Reid, V. R.; Noy, A.; Synovec R. E.; Bakajin, O. *Anal. Chem.* 2006, *78*, 5639–5644.

116. Yuan, L.-M.; Ren, C.-X.; Li, L.; Ai, P.; Yan, Z.-H.; Zi, M.; Li, Z.-Y. *Anal. Chem.* 2006, *78*, 6384–6390.

117. Zhao, L.; Ai, P.; Duan, A. H.; Yuan, L. M. *Anal. Bioanal. Chem.* 2011, *399*, 143–147.

118. Li, Y.; Chen, Y.; Xiang, R.; Ciuparu, D.; Pfefferle, L. D.; Horvath, C.; Wilkins, J. A. *Anal. Chem.* 2005, *77*, 1398–1406.

119. Chang, Y. X.; Zhou, L. L.; Li, G. X.; Li, L.; Yuan, L. M. *J. Liq. Chromatogr. & Rel. Technol.* 2007, *30*, 2953–2958.

120. Bottini, M.; Tautz, L.; Huynh, H.; Monosov, E.; Bottini, N.; Dawson, M. I.; Belucci, S.; Mustelin, T. *Chem. Commun.* 2005, 758–760.

121. Menna, E.; Negra, F. D.; Prato, M.; Tagmatarchis, N.; Ciogli, A.; Gasparrini, F.; Misiti, D.; Villani, C. *Carbon* 2006, *44*, 1609–1613.

122. Kwon, S. H.; Park, J. H. *J. Sep. Sci.* 2006, *29*, 945–952.

123. Abraham, M. H.; Roses, M.; Poole, C. F.; Poole, S. K. *J. Phys. Org. Chem.* 1997, *10*, 358–368.

124. Chang, Y.-X.; Ren, C.-X.; Ruan, Q.; Yuan, L.-M. *Chem. Res. Chinese Univ.* 2007, *23*, 646–649.

125. Pirkle, W. H.; Liu, Y.; Welch, C. J. *Enantiomer* 1998, *3*, 477–483.

126. Gasparrini, F.; Misiti, D.; Villani, C.; Borchardt, A.; Burger, M. T.; Still, W. C. *J. Org. Chem.* 1995, *60*, 4314–4315.

127. Erickson, S. D.; Simon, J.; Still, W. C. *J. Org. Chem.* 1993, *58*, 1305–1308.

128. Gasparrini, F.; Misiti, D.; Still, W. C.; Villani, C.; Wennemers, H. *J. Org. Chem.* 1997, *62*, 8221–8224.

129. Wennemers, H.; Yoon, S. S.; Still, W. C. *J. Org. Chem.* 1995, *60*, 1108–1109.

130. Gasparrini, F.; Misiti, D.; Pierini, M.; Villani, C. *Org. Lett.* 2002, *4*, 3993–3996.

131. Schurig, V. *Angew. Chem. Int. Ed. Engl.* 1976, *15*, 304.

132. Schurig, V.; Wistuba, D. *Angew. Chem. Int. Ed. Engl.* 1983, *22*, 772–773.

133. Pirkle, W.; Gan, K. Z. *Tetrahedr. Asymm.* 1997, *8*, 811–814.

134. Kimata, K.; Hosoya, K.; Araki, T.; Tanaka, N. *Anal. Chem.* 1997, *69*, 2610–2612.

135. Schurig, V. In *Houben-Weyl, Methods of Organic Chemistry, Volume E21a: Stereoselective Synthesis, Part A.* 3.1.4.5. Thieme Stuttgart, New York, 1995, p. 167.

136. Sicoli, G.; Kreidler, D.; Czesla, H.; Hopf, H.; Schurig, V. *Chirality* 2009, *21*, 183–198.

137. Darquié, B. et al., *Chirality* 2010, *22*, 870–884.

138. Costante-Crassous, J.; Marrone, T. J.; Briggs, J. M.; McCammon, J. A.; Collet, A. *J. Amer. Chem. Soc.* 1997, *119*, 3818–3823.

139. Grosenick, H.; Schurig, V.; Costante, J.; Collet, A. *Tetrahedr. Asymm.* 1995, *6*, 87–88.

140. Jiang, Z.; Crassous, J.; Schurig, V. *Chirality* 2005, *17*, 488–493.

141. Kalchenko, O. I.; Cherenok, S. O.; Solovyov, A. V.; Kalchenko, V. I. *Chromatographia.* 2009, *70*, 717–721.

142. Jakubetz, H.; Juza, M.; Schurig, V. *Electrophoresis* 1998, *19*, 738–744.

143. Anslyn, E. V. *J. Org. Chem.* 2007, *72*, 687–699.

# 7 Industrial and Environmental Applications
## Separation and Purification

*Hans-Jörg Schneider*

## CONTENTS

The potential of molecular recognition led early on to investigations of supramolecular complexation for the purpose of separation and purification. Environmental control is a particular challenge for supramolecular chemistry. Many metal ions such as mercury, thallium, chromium, cobalt, and nickel salts as well as anions such as chromates, phosphates, nitrates, and arsenates, are pollutants in wastewater, soil, and industrial emissions. Their toxicity, enhanced by accumulation in living systems, has led to many attempts to eliminate them also by supramolecular complexation. Purification of industrial and natural products also calls for application of new technologies, which allow the production, for example, of drugs with high purity and the recycling of valuable and potentially toxic compounds. Separations with the help of supramolecular complexation were traditionally aiming at analytical applications;[1] corresponding chromatographic methods are dealt with in detail in Chapter 6. Several other separation techniques also rely on selective intermolecular interactions but can only be mentioned here. Metal-organic frameworks (MOFs), which are hybrid porous materials, are presented already in the introduction of this book and can be used, in particular, for gas separation and storage of hydrogen and methane, for example.[2] Incorporation of additional functionalities into MOF backbones of zinc oxide and phenylene units leads, for example, to 400% better selectivity for carbon dioxide over carbon monoxide.[3]

**159**

In the following, promising applications will be discussed first according to the different separation techniques such as extraction-, membrane-, and exchange resin-based methods. In view of their particular role related to the use of nuclear power and radiation techniques, two separate sections will be devoted to removal and purification of actinide and cesium salts, and a final section will cover organic pollutants. As a result of the still rather new orientation of supramolecular chemistry toward practical uses, only a few applications are already playing an important role on an industrial scale; the increasing number of patents issued in the field, however, illustrates the great promise of supramolecular separation techniques with respect to intelligent solutions for many industrial, environmental, and medicinal problems.

## EXTRACTION TECHNIQUES

Often one uses extraction from aqueous media with supramolecular ligands dissolved in non-water-soluble organic solvents.[4] This method was applied early for removal of radioactive strontium;[5] extraction methods for cesium and actinide ions will be discussed in separate sections to follow. Uranium(IV) was extracted with chloroform solutions of a crown hydroxamic acid **1** with high selectivity vs. cerium, thorium, and lanthanide among many other metal ions; the binding constant of the $UO_2^{2-}$ cation was at least $10^{19}$ higher than with all other metals ions.[6] Open-chain crown ether analogs with terminal quinolyl groups showed an extractability with chloroform, which is significantly higher for $Th^{4+}$ than for $UO_2^{2+}$ ions.[7] Extraction of uranium from a nitric acid medium with supercritical carbon dioxide containing various crown ethers showed that the extraction efficiency among the 18-crown-6 series is influenced by the cavity size and by the basicity of ether oxygen atoms; the best efficiency was found with the ditertiarybutyldicyclohexano-18-crown-6.[8] The coextracted anion has a strong effect on the cation selectivity; this has been shown for affinities in calix complexes with ammonium substrates and with anions such as tosylate, chloride, acetate, trifluoroacetate, and picrate and is a result of different ion pairing contributions and allosteric effects on calix conformations.[9]

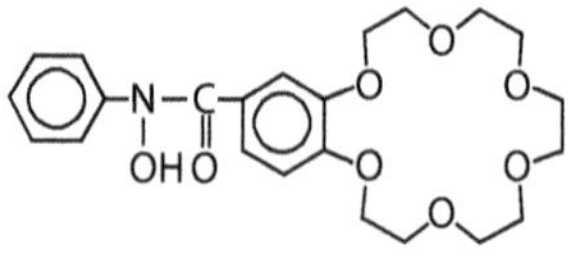

1 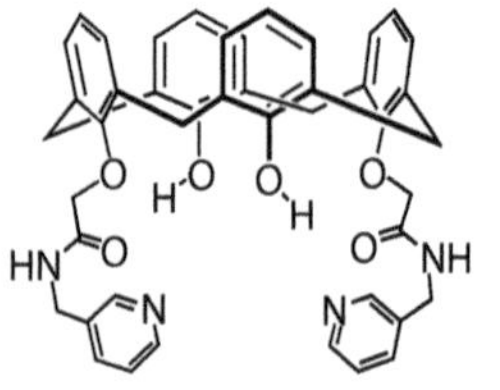                                                                                     2

Extraction is a relatively easy method for the removal of toxic compounds. For instance, 0.1 mM chromate or dichromate in water is treated with dichloromethane solutions containing the calixarene **2** and showed, for example, a 70% extraction of bichromate at pH 1.5, decreasing to 10% at pH 4.5; similar results were obtained with arsenate and phosphate.[10] The problem for technical applications of extraction methods is the cost of the ligands and of the solvents, which both should be recycled as often as possible.

## MEMBRANES AND IMMOBILIZATION OF HOST COMPOUNDS

Transport of such ions across liquid membranes combines extraction and stripping in a one-step process and also has been used as a model for membrane separations.[11] Bulk liquid membranes (BLMs) using tetra aza-14-crown-4 and oleic acid as a cooperative ion carrier have shown, for example, with an aqueous source phase containing only 0.1 mM of Cu(II), Cd(II), and Pb(II) ions as equimolar mixtures, a competitive transport with selectivity in the order Pb(II) >> Cu(II) > Cd(II).[12] The transport of uranium(VI) salts has been studied with a BLM with different lipophilic crown ethers as synergistic agents and thenoyltrifluoroacetone in chloroform; an anionic surfactant such as sodium dodececyl sulfate (SDS) in the receiving phase increases the efficiency. With dibenzocrowns the transport percentage decreased in the order of DC18C6 > DB18C6 > DB24C8 > DB21C7 > DB15C5; the dicyclohexyl-18-crown-6 was found to be particularly efficient.[13]

Another way to immobilize benzocrown ethers, for example, is to impregnate them by soaking them onto activated carbon cloth; an adsorption decrease in the order Cr(III) > Co(II) > Ni(II) ions was observed.[14] Nanofiltration with the help of water-soluble p-sulfonated calix[4]arene is another way to separate chromate anions from the less-retained chloride, sulfate, or nitrate anions.[15] A dicyclohexyl crown on an aminofunctionalized mesoporous molecular sieve can be used for separation and purification of molybdenum(VI) from fission products of uranium.[16]

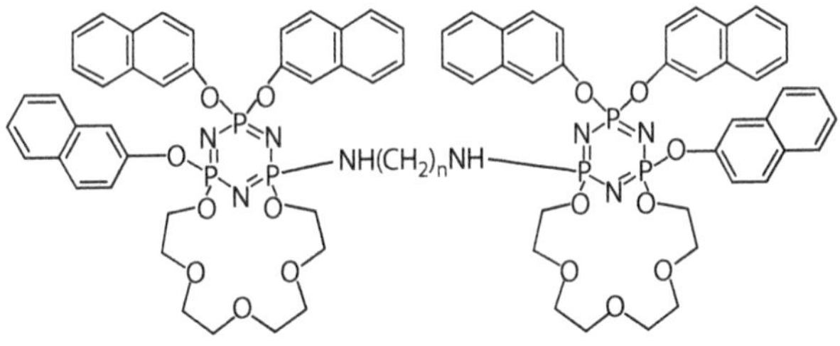

3

In contrast to BLMs and supported liquid membranes, polymer inclusion membranes (PIMs) are much more stable, easier to operate, and more economic as they save expensive receptor material. Such PIMs have been used for decades in the form of polymer membrane ion-selective electrodes (see Chapter 4) but have only been recently applied for separations. They are usually based on polyvinylchloride (PVC) or cellulose triacetate with added plasticizers for increasing flux and membrane flexibility.

Supramolecular applications rely on added carriers such as crown ethers and calixarenes, which secure transport of the complexes with the analyte across the membrane.[17] Lipophilic side chains in the receptors can secure retention in hydrophobic membranes and inhibit leakage into the water phase. Crown ethers were used from the beginning for separation of alkali and earth alkali-metal ions; for treating industrial waste, contaminated waters, and particularly nuclear waste, PIMs suitable for treating radioactive and heavy-metal ions, including cesium, were increasingly developed.[18] Basic carriers such as tri-*n*-octylamine or ammonium salts form ion pairs with a metal anion complex from the aqueous phase and can be used for extraction of heavy and precious metals, including the very toxic Cr(VI) salts in water.[19] A large variety of acids can also be used as carriers in which the proton is exchanged by the metal ion; countertransport of protons is the driving force for the transport based on a pH difference between the source and the receiving medium. These PMIs allow the extraction of heavy metals such as Pb(II), Cd(II), Zn(II), Cu(II), Ag(I), and Hg(II). As one of many examples,[17] we show a recent one with new references.[20] Here the bis-lariat ether **3** in carrier in a cellulose triacetate-based PIM exhibited a selectivity order Pb(II) > Zn(II) > Cd(II); for Pb/Cd a high separation factor of 183 was found. The increase in the linker length causes a decrease in the Pb(II) transport rate while an increase was observed for Cd(II) transport.[20] Generally, the transport efficiency is a function of size matching between the crown cavity and metal ion and, here, the linker length. With lariat ethers, the length of the side chain plays an important role. For example, a PIM containing sym-(alkyl) dibenzo-16-crown-5-oxyacetic acid as a carrier showed an excellent selectivity for Na$^+$ transport with no Li$^+$, K$^+$, Rb$^+$, and Cs$^+$ ions detectable in the receiving phase. Here, optimal flux was found only with a $C_9H_{19}-$ alkyl chain, which is attached geminally to the functional side arm of the lariat ether.[21] The bis(pyridylmethyl) amine units of the resorcinarene-based carrier **4** provides for coordination to transition-metal ions such as Fe(III), Cu(II), and Zn(II). At the same time, the metallated complex also allows transport of anions such as the ReO$_4^-$ ion because of the positive charges.[22] Selectivity varied as a function of carrier, plasticizer, and anion in the receiving phase. The tetrairon complex **4xFe$_4$** exhibits, for example, a remarkable selectivity with ReO$_4^-$/NO$_3^-$ = 7.5:1.0, which holds promise for separation of technate TcO$_4^-$ salts. The tetracopper carrier **4xCu$_4$** shows for NO$_3^-$/SeO$_4^-$ a selectivity of even 26:1; halides can also be separated, however, with smaller permeability than observed for oxoanions.

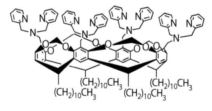

**4**

## EXCHANGE RESINS

Even isotope enrichment can be achieved with suitable host compounds. In 1976, chromatography with a resin-bound dibenzo[18]crown-6 was shown to enable enrichment of $^{44}$Ca.[23] The benzo 2.2.2 cryptand was later used for enrichment of $^{44}$Ca and $^{40}$Ca, which is of interest for labeling purposes, for example.[24] With a benzo 18-crown-6-ether resin, one found for $^{42}$Ca, $^{43}$Ca, $^{44}$Ca, $^{48}$Ca, and $^{40}$Ca the expected mass dependence of calcium isotope fractionations with the heavier calcium isotopes eluted at the front.[25] The depletion of the radioactive zinc isotope $^{64}$Zn in the light water of nuclear power plant tubings may also be achieved with crown ether resins.[26] Immobilization of vanillin-thiosemicarbazone, which is an efficient chelating agent for toxic metal ions, was achieved on a calix[4] arene polymer backbone **5** and allowed preconcentration factors of 117, 90, and 105 for Cr(VI), As(III), and Tl(I) ions, respectively, with detection limits around 6 μg/L.[27] A calix[4]arene-nitrile-loaded resin exhibited large retention of heavy metals such as Cu(II), Co(II), Cd(II), Ni(II), Hg(II), and Pb(II) and of the dichromate anion.[28] Unreacted azo dyes in wastewater from textile industries can be removed with a *p-tert*-butyl-calix[4]arene-based silica resin **6** (for related applications, see Chapter 17).[29]

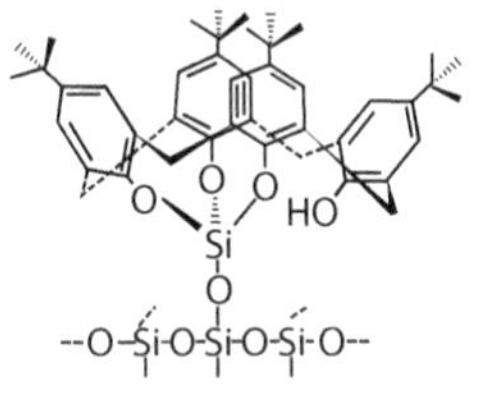

## ACTINIDE/LANTHANIDE SEPARATIONS FROM NUCLEAR WASTE

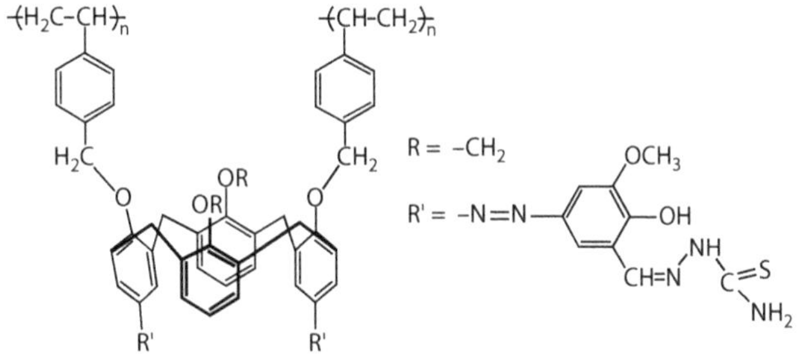

Besides uranium, plutonium, and lanthanides, nuclear waste contains high levels of the minor actinides Np, Am, and Cm, which pose severe problems in view of their long-lasting radioactivity of up to $t_{1/2} = 10^3$ to $10^4$ years. In order to minimize environmental risks and to reduce storage demand in repositories, much effort has been paid toward efficient separation between An(III) and Ln(III) salts, which is rather difficult because of their very similar ionic radii and hardnesses. If the long-lived actinides are removed, they can be transmuted to nonradioactive or short-lived nuclides. Many publications, technical reports, and an excellent review[30] provide insight into the field. For this reason, and because such complexations are essentially in the realm of coordination chemistry, we will only discuss some typical examples here. There are practical limitations as the solutions issuing from nuclear fuel reprocessing plants contain much nitric acid, which poses limits because of the high acidity and possible oxidations of the metal ions. Because of protonation, classical nitrogen or carboxylate donor sites are usually less appropriate; as N-donors, one therefore uses less basic ligands such as pyridine, pyrimidine, pyrazine, and triazine. A quantitative relationship between the efficiency of corresponding extractants and the gas-phase basicity of suitable model compounds has been proposed recently.[31]

Both An(III) and Ln(III) cations are hard Lewis acids, which, for efficient total complexation, call for ionic, charge density–controlled binding with harder, oxygen-containing ligands. However, Ln(III) cations are slightly harder than An(III) ions, so softer ligand atoms such as Cl, S, or N allow better selectivity. As a result of reduced basicity, triazine-pyridine ligands such as in **7**[32] exhibit the best ratio between metal concentration in the organic and the aqueous phase ($D_{Am} > 100$) among the N-donor systems and, at the same time, the best separation factor ($S_{M1/M2} > 100$), which characterizes the complex extractability between two metals M1 and M2. The malonamide derivative **8**[33] essentially binds the metal ions by the hard oxygen donor atoms and, therefore, shows low selectivity between An(III) and Ln(III) salts. The ligand **9**[34] exhibits quite large selectivity between An(III) and Ln(III) because of the soft sulfur donors but suffers from oxidation in the nitric acid medium. Tripodal ligands such as **10** can make use of chelating simultaneously with tree bidentate donor sites and, therefore, show enhanced affinity with moderate selectivity.[35]

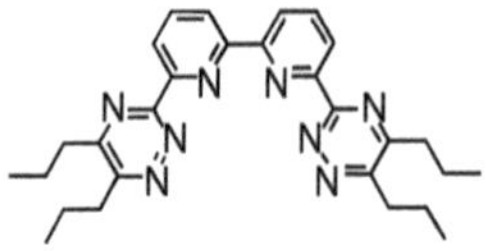

7

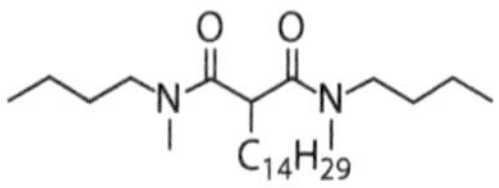

8

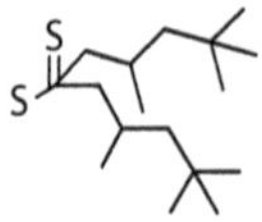

9

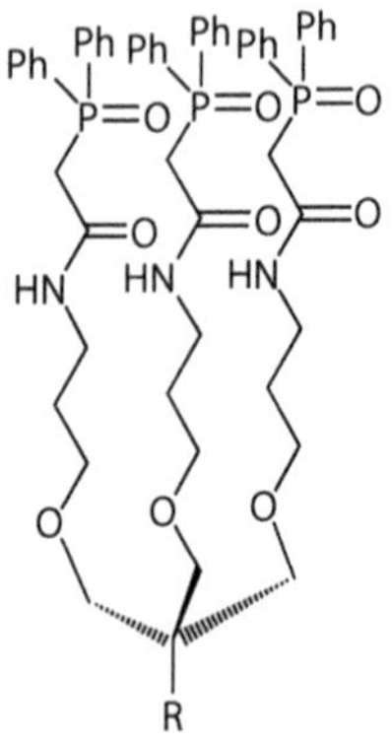

10 (R = C$_2$H$_5$)

The ligands discussed above lack the feature of preorganization, which is characteristic for supramolecular chemistry, but nevertheless perform relatively well, especially in view of their lower cost. Preorganized binding sites for the purpose of such metal ion complexation have been most often realized with calixarenes, which allow the attachment of suitable donor functions both at the upper rim and at the lower bottom side.[36] They still have sufficient flexibility for adopting an optimal contact between metal and donor atoms and, therefore, are usually better extractants than cavitands, which are too rigid.[30] As a rule, larger calixarenes **11** ($m = 6$ or 8) perform better than the smaller ones such as the calix[4]arenes **11**, $m = 4$ because of their higher flexibility. Particularly promising results were obtained with the phosphine oxide–containing ligand **11a**.[37] The upper-rim derivatized ligand **11b** even exhibits a significant size selectivity within the lanthanides with a remarkable selectivity of $S_{La/Yb}$ ~730.[38] Figure 7.1 illustrates that the extraction efficiency with the calix **11c** as measured by $D$ is essentially determined by the fit to the calixarene donors rather than independent of the medium.[38]

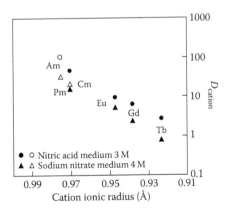

**11a** (m = 8; $R_1$ = $CH_2CH_2$–PO(Ph)$_2$ , $R_2$ = H)
**11b** (m = 4; $R_1$ = $C_5H_{11}$, $R_2$ = –NHCO–$CH_2$–PO(Ph)$_2$)
**11c** (R = $C_5H_{12}$, $R_2$ = –NHCO–$CH_2$–PO(Ph)$_2$)
**11d** ($R_1$ = $CH_3$ or $C_2H_5$, $R_2$ = –NHCO–$CH_2$–PO(Ph)$_2$)

**12**

Calix[4]arenes with smaller alkyl substituents at the bottom (e.g., **11d**) exhibit an increased flexibility, as smaller alkyl groups can pass the annulus, not keeping it in the symmetric cone conformation. In consequence, **11d** with $R_1$ = Me in alternate positions 1 and 3 and Et in positions 2 and 4 shows affinity with $D_{Am}$ ~570 with the selectivity $S_{Am/Eu}$ ~5. Small changes by placing $R_1$ = Me in positions 1 and 2 and Et in positions 3 and 4 improve the selectivity to $S_{Am/Eu}$ ~9 with almost the same value of $D_{Am}$ ~440.[39] The picolinamide derivative **12** is an example of a bottom-substituted calixarene, which, because of the low nitrogen basicity, with $S_{Am/Eu}$ = 3.87 still has

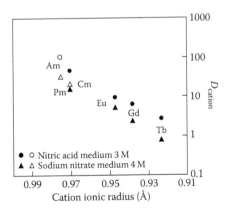

**FIGURE 7.1** Extraction of trivalent lanthanides and actinides by calixarene **11c** from two different media, [NaNO3] = 4 M/[HNO3] = 0.01 M or [HNO3] = 3 M. Organic phase: chloroform containing [**11c**]$_{init}$ = $10^{-3}$ M. Aqueous phase: [Ln3+]$_{init}$ = $10^{-6}$ M. Promethium and actinides were used at trace levels. Phase ratio 1:1. $T$ = 25°C. Filled symbols: lanthanides; open symbols: actinides. (From Delmau, L. H. et al., *Chem. Commun.* 1998, 1627–1628. With permission.)

an efficient selectivity. The distribution coefficients decrease with increasing nitric acid concentration and reach, for example, $D_{Am} = 4.58$, all at $[H+] = 5 \times 10^{-2}$ mol/L. A 70:30 mixture of o-nitrophenyl hexyl ether and 1,2-dichloroethane was used for the extraction, and the addition of Br-Cosan ([cesium commo-3,3-cobalta-bis(8,9,12-tribromo-1,2-dicarba-closo-dodecarborane)ate(1-)]) is necessary in order to secure sufficient extraction, which is likely a result of the exchange of the nitrate anion by this more lipophilic anion.

## CESIUM AND TECHNETIUM FROM NUCLEAR WASTE

Effluents from irradiated nuclear fuel reprocessing installations contain cesium-137 as one of the most noxious fission products with a half-life of 30 years and a high-energy gamma irradiation. Extraction of cesium with supramolecular ligands has therefore received particular attention with many ensuing patents. Sodium and cesium have very similar properties, so it is extremely difficult to selectively extract the cesium in the nuclear effluents, particularly with a Cs concentration usually even below the micromolar range and a high Na concentration of about 4 M. In addition, $^{137}$Cs is a valuable $\gamma$–radiation source, which may be used, for example, in food preservation or sterilization of medical accessories; therefore, Cs isolation from nuclear waste can generate significant value.

The separation of cesium with membrane-bound calixarenes has already been described in a patent from 1984.[40,41] Since then, calixarenes have been almost exclusively used, mostly in combination with attached crown ether moieties. Calix[4] arene-bis(benzo crown-6) and calix[4]arene-bis(napthocrown-6) were found to be superior extractants for cesium in solvents, which exhibited decreased efficiency in the order nitrobenzene > dichloroethane > chloroform > decanol, carbon tetrachloride, n-hexane, or toluene.[18] Generally, 1,3-alternate calix[4]arene-crown-6-ethers such as **13** possess high Cs selectivity over Na, reaching values of $S_{Cs/Na} > 10.000$.[42] The efficiency of these combination hosts is a result of the dual $\pi$-interactions by the aromatic rings and the lone electron pairs of the ether oxygen atoms and is used on a large scale in the so-called caustic-side solvent extraction process for cesium removal from nuclear waste.[43] The tetrabenzo-24-crown-8 **14** as a model was used to systematically study anion partitioning and ion pairing for Cs salt extraction in 1,2-dichloroethane.[44] The dissociation constants measured by conductometry were shown to be consistent with the extraction efficiency. The Cs$^+$ ion distribution ratios were essentially dependent on the nature of the anion as predicted by the Hofmeister series; different anions led to a 10-order-of-magnitude variation in the extractability of the anions, although differences in ion pairing in the complexes are relatively small.

Acidic residues in **13** such as R = COOH, R = CONHSO$_2$-R' (R' = CF$_3$, PhNO$_2$, Ph, or Me) can participate in metal ion complexation cooperatively with the crown ether unit and the aromatic rings of the calix[4]arene upon deprotonation of the carboxy or amide groups; the Cs extraction constants are proportional to their acidities.[45] Ionic liquids, consisting, for example, of 1-alkyl-3-methylimidazolium and [NTf2] and bis(trifluoromethylsulfonyl)imide, can be used at room temperature to extract Cs from aqueous media with the help of calix crown ethers such as bis(2-propyloxy) calix[4]crown-6. The extraction efficiency decreases with the increasing alkyl chain length of the ionic liquid imidazolium component. It also decreases with an increasing $NaNO_3$ concentration, reaching a maximum of $D_{Cs} > 1000$ at $[NaNO_3] = 0$ M.[46]

Another dangerous component of nuclear waste is the long-lived technetium (half-life time: $2 \times 10^5$ years), present as $TcO_4^-$ salt. At the same time, it is an important metal in nuclear medicine, so recovery leads to a high-value product. The calix[4] arene bis(crown-6) **13** is able to coextract cesium and pertechnetate ions from acidic or basic aqueous to organic solvents.[47] Extraction of $TcO_4^-$ salt from an alkaline solution is also possible with the bis-4,4′(5′)[(tert-butyl)cyclohexano]-18-crown-6.[48]

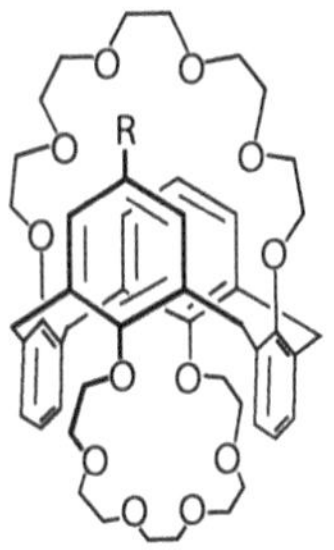

**13** 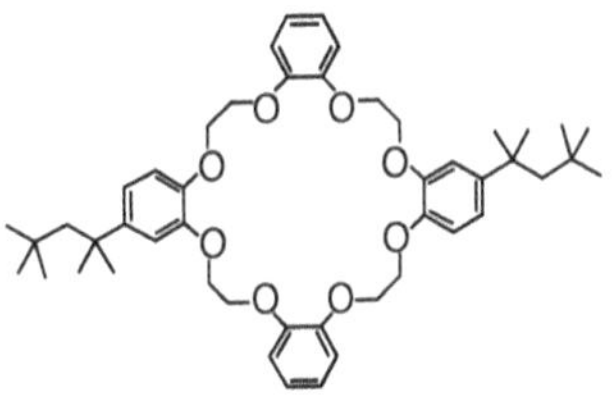                    **14** 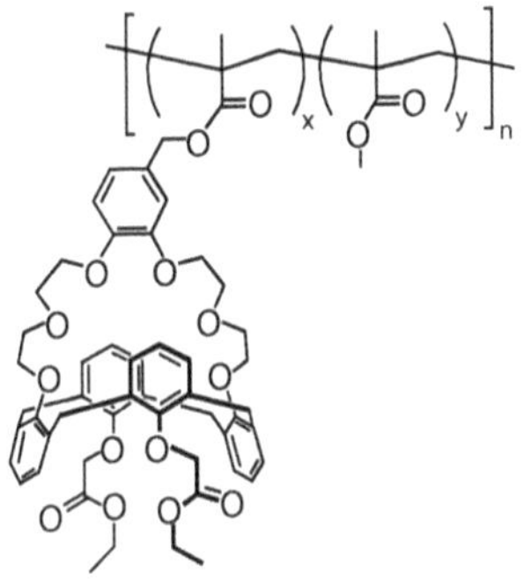                    **15**

The methacrylate-functionalized benzocrown-6-calix[4]arene **15**, as exchange resin, exhibits a greater selectivity for Cs over K picrate and was also found to be effective in extracting cesium nitrate from aqueous media in the presence of various other inorganic ions.[49] Another way to immobilize a cesium receptor is based on a macroporous silica-based calixcrown (1,3-[(2,4-diethyl-heptylethoxy) oxy]-2,4-crown-6-calix[4]arene); this polymeric composite showed high selectivity in 3.0 M $HNO_3$ solution for Cs against metal ions such as Na(I), K(I), Cs(I), Rb(I), Sr(II), and Ba(II).[50] Nanofiltration of the complexes offers a way to separate substrate ions according to the size of their supramolecular complexes. Introduction of hydroxy, carboxy, sulfato, or diethanolamino groups at the para position of the phenolic ring and/or on the benzo-ether moieties in **13** leads to water-soluble Cs receptors, which can be used for nanofiltration membranes. Cesium retention over sodium reached up to 90%, depending on the chosen calix and the $NaNO_3$ concentration.[51]

PIMs derived from plasticized cellulose triacetate were made containing, for example, di-benzo-18-crown-6 or di-benzo-21-crown-7 as crown ether carriers. The best selectivity and transport for Cs was found with 2-nitrophenyloctylether plasticizer and di-tert-butylbenzo-18-crown-6 as a carrier.[52]

## ORGANIC SUBSTANCES/POLLUTANTS

Many organic compounds such as polycyclic aromatic hydrocarbons (PAHs), polychlorinated biphenyls (PSBs), dioxins, aminoaryls, and benzopyrenes are pollutants that create an increasing demand for appropriate removal technologies. Many of these compounds are especially dangerous as they persist for a long time in the environment because of their low water solubility and their tendency to be adsorbed into soil organic matter and to accumulate in living systems. Traditional technologies such as the use of activated carbon, amphiphilic surfactants, cosolvents, or ultrafiltration often fail to remove such toxic contaminants down to parts per billion (ppb) levels and also pose problems with the desired recycling of such materials, particularly receptors. Supramolecular complexation of usually hydrophobic organic compounds offers an increasingly studied more sophisticated way to meet these challenges. The host compounds that are nearly always used for complexation of hydrophobic substances are cyclodextrins (CDs), which offer a lipophilic cavity of variable size and water solubility.[53] In addition, CDs are generally nontoxic—most are FDA-approved and are less expensive than many other host compounds. The most often used ß-CD accommodates substances of phenyl or, better, naphthyl size within the cavity; its bulk price is about $5/kg or less. The better water-soluble 2-hydroxypropyl-β-CD, however, costs about $300/kg. For CD-enhanced flushing, one can also use less expensive technical-grade CDs in liquid form.

A systematic investigation of sorption–desorption and transport processes of the herbicide norflurazon in soils revealed that with lower ß-CD concentration most of the CD molecules are adsorbed on soil particles, providing a coating that acts as a bridge between the herbicide and the soil surface, thus retarding the mobility of the herbicide. At higher CD concentrations, most of the CD is in the aqueous phase, and the herbicide is complexed in the CD solution.[54] Soil polluted by phenanthrene and pyrene was decontaminated with the help of, for example, a 10 mM solution of methyl-ß-CD; the hydrophobic pollutants were then extracted with an oil, and the remaining CD solution could be reused. Alternatively, filtration with an oil-impregnated membrane was found to be more economical.[55] Polychlorinated dibenzo-p-dioxins and polychlorinated dibenzofurans can be removed from soil and water with CDs and particularly well with hydroxypropyl-β-CD.[56]

Pesticides such as DDT and other chlorinated pollutants could be effectively cleaned from aqueous solutions in the typical pollution concentration range of 0.060– 0.270 μg/mL with CD-functionalized mesoporous silica adsorbents.[57] Ceramic membranes impregnated with (CD-Si) cross-linked CD polymers with siloxane bridges were shown to remove PAHs by up to 99% and several pesticides by up to 43% from water.[58] Figure 7.2 illustrates the high efficiency of the CD-loaded membranes over a nonimpregnated $Al_2O_3$ filter, particularly at lower flow rates.

Polymeric nanospheres based on a CD reaction with diisocyanates in miniemulsion efficiently absorb aromatic organic molecules such as toluene and, to a lesser degree, phenol. Regeneration of the absorbing particles is possible by treatment with organic solvents.[59] The separation of aromatic/aliphatic and isomeric mixtures, which is an important issue in the petrochemical industry, is possible with cross-linked polymers based on polyethylene glycol dimethacrylate and acrylated CDs.[60]

A solid channel-type assembly of crystalline γ-CD where CD molecules stacked in a head-to-head or head-to-tail manner can function as an effective adsorbent to remove chlorinated aromatic compounds, including PCBs from oil used for electric insulation. Most of the adsorbed chlorinated aromatics can be recovered from

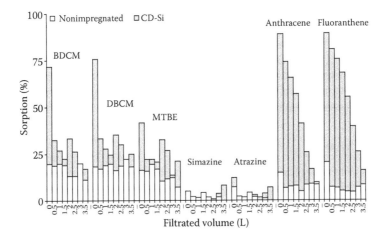

**FIGURE 7.2** Retention of organic pollutants during continuous filtration by nonimpregnated $Al_2O_3$ filters and by $Al_2O_3$ impregnated with ß-CD polymerized through siloxane bridges (CD-Si) as a function of filtrated volume. BDCM: bromodichloromethane; DBCM: dibromochloromethane; MTBE: methyl-tert-butyl ether. (Reprinted from *Water Research, 41*, Allabashi, R., Arkas, M., Hormann, G., Tsiourvas, D., 476–486, Copyright 2007, with permission from Elsevier.)

the solid CD material by washing it with n-hexane.[61] CD-containing polymers can act as nanosponges and allow water purification, for example, again with the possibility of regaining the CD material.[62] The ingenious combination of a CD and a functionalized carbon nanotube (f-CNT) within a cross-linked nanoporous polymer **16** is able to bind model organic species such as *p*-nitrophenol by as much as 99% from a 10 mg/L spiked water sample; in contrast, activated carbon removed only 47%. Trichloroethylene in 10 mg/L samples was removed to the detection limit of <0.01 ppb, compared to only 55% with activated carbon.[63] ß-CD grafted onto the surface of multiwalled carbon nanotubes exhibits a high adsorption in the removal of PCBs from aqueous solution.[64]

An interesting application of CD is the simultaneous removal of heavy metals and low-polarity organic compounds from contaminated soil. Flushing of soil with a CD solution could greatly enhance the simultaneous desorption and elution of phenanthrene as a model organic compound, for example, cadmium.[65] Similar results were obtained in soils contaminated by PCBs and by Cd, Cr, Cu, Mn, Ni, Pb, or Zn salts by washing them with solutions containing CD and ethylenediaminetetraacetic acid (EDTA).[66] Glutamic acid-β-CD and ethylene-diamine-β-CD were also shown to remove organics such as anthracene and, for example, cadmium salts from soil.[67] Alkylimidazolium- and pyridinium-based CD -ionic liquid polyurethanes can also simultaneously absorb organic and inorganic pollutants from water.[68]

## CONCLUSIONS

Supramolecular complexations offer new ways to separate problematic environmental compounds in a rationally designed manner. The until now often prohibitively high costs for large-scale applications are expected to become a lesser problem as synthetic pathways are becoming much improved. New technologies such as the use of stable PIMs allow more cost-effective separations—with respect to solvents as well. Dynamic supramolecular libraries offer an until now, barely explored way to find suitable receptors for a large variety of target compounds, which are typically present in waste matrices or in industrial mixtures. It is hoped that supramolecular chemistry can help as a remedy for many chemistry-inflicted problems and contribute significantly toward green technology. Several supramolecular systems are already in practical use, particularly for the management of radioactive waste. The possibility of recycling and purifying high-value compounds with the help of supramolecular complexes is a promising contribution for sustainable development.

## REFERENCES

1. Blasius, E.; Janzen, K. P. *Top. Current Chem.* 1981, *98*, 163–189.
2. Liu, B.; Sun, C. Y.; Chen, G. J. *Chem. Eng. Sci.* 2011, *66*, 3012–3019.
3. Deng, H. X.; Doonan, C. J.; Furukawa, H.; Ferreira, R. B.; Towne, J.; Knobler, C. B.; Wang, B.; Yaghi, O. M. *Science* 2010, *327*, 846–850.
4. Cox, B. G.; Schneider, H. *Coordination and Transport Properties of Macrocyclic Compounds in Solution*; Elsevier, New York, 1992.
5. Blasius, E.; Klein, W.; Schon, U. *J. Radioanal. Nuclear Chem.* 1985, *89*, 389–398.
6. Agrawal, Y. K.; Shah, G.; Vora, S. B. *J. Radioanal. Nuclear Chem.* 2006, *270*, 453–459.
7. Guo, Z. J.; Wu, W. S.; Shao, D. D.; Tan, M. Y. *J. Radioanal. Nuclear Chem.* 2003, *258*, 199–203.
8. Rao, A.; Kumar, P.; Ramakumar, K. L. *Radiochim. Acta* 2010, *98*, 403–412.
9. Arduini, A.; Giorgi, G.; Pochini, A.; Secchi, A.; Ugozzoli, F. *J. Org. Chem.* 2001, *66*, 8302–8308.
10. Bayrakci, M; Ertul, S.; Yilmaz, M. *Tetrahedron* 2009, *65*, 7963–7968.
11. Walkowlak, W.; Kozlowski, C. A. *Desalination* 2009, *240*, 186–197; Tomar, J.; Awasthy, A.; Sharma, U. *Desalination.* 2008, *232*, 102–109.
12. Kazemi, S. Y.; Hamidi, A. S. *J. Chem. Eng. Data* 2010, *56*, 222–229.
13. Shamsipur, M.; Davarkhah, R.; Khanchi, A. R. *Separ. Purif. Tech.* 2010, *71*, 63–69.
14. Duman, O.; Ayranci, E. *J. Hazard. Mater.* 2010, *176*, 231–238.
15. Tabakci, M.; Erdemir, S.; Yilmaz, M. *Supramol. Chem.* 2008, *20*, 587–591.
16. Ganjali, M. R.; Babaei, L. H.; Norouzi, P.; Pourjavid, M. R.; Badiei, A.; Saberyan, K.; Maragheh, M. G.; Salavati-Niasari, M.; Ziarani, G. M. *Anal. Lett.* 2005, *38*, 1813–1821.

17. Nghiem, L. D.; Mornane, P.; Potter, I. D.; Perera, J. M.; Cattrall, R. W.; Kolev, S. D. *J. Membr. Sci.* 2006, *281*, 7–41.

18. Mohapatra, P. K.; Lakshmi, D. S.; Bhattacharyya, A.; Manchanda, V. K. *J. Hazard. Mater.* 2009, *169*, 472–479.

19. Konczyk, J.; Kozlowski, C.; Walkowiak, W. *Desalination* 2010, *263*, 211–216, and references cited therein.

20. Konczyk, J.; Kozlowski, C.; Walkowiak, W. *Desalination* 2009, *263*, 211–216.

21. Walkowiak, W.; Bartsch, R. A.; Kozlowski, C.; Gega, J.; Charewicz, W. A.; Amiri-Eliasi, B. *J. Radioanal. Nucl. Chem.* 2000, *246*, 643–650.

22. Gardner, J. S.; Peterson, Q. P.; Walker, J. O.; Jensen, B. D.; Adhikary, B.; Harrison, R. G.; Lamb, J. D. *J. Membr. Sci.* 2006, *277*, 165–176.

23. Jepson, B. E.; Dewitt, R. *J. Inorg. Nucl. Chem.* 1976, *38*, 1175–1177.

24. Heumann, K. G.; Schiefer, H. P. *Angew. Chem. Int. Ed. Engl.* 1980, *19*, 406–407.

25. Fujii, Y.; Nomura, M.; Kaneshiki, T.; Sakuma, Y.; Suzuki, T.; Umehara, S.; Kishimoto, *Environm. Health Studies.* 2011, *46*, 233–241.

26. Tan, Y. Y.; Kan, D. L.; Ding, X. C.; Nomura, M. S.; Fujii, Y. K. *J. Nucl. Sci. Tech.* 2008, *45*, 1078–1083.

27. Jain, V. K.; Pandya, R. A.; Pillai, S. G.; Shrivastav, P. S.; Agrawal, Y. K. *Separ. Sci. Technol.* 2006, *41*, 123–147.

28. Tabakci, M.; Ersoz, M.; Yilmaz, M. *J. Macromol. Sci. Pure Appl. Chem.* 2006, *A43*, 57–69.

29. Kamboh, M. A.; Solangi, I. B.; Sherazi, S. T. H.; Memon, S. *Desalination* 2011, *268*, 83–89.

30. Dam, H. H.; Reinhoudt, D. N.; Verboom, W. *Chem. Soc. Rev.* 2007, *36*, 367–377.

31. Galletta, M.; Baldini, L.; Sansone, F.; Ugozzoli, F.; Ungaro, R.; Casnati, A.; Mariani, M. *Dalton Trans.* 2010, *39*, 2546–2553.

32. Drew, M. G. B.; Foreman, M.; Hill, C.; Hudson, M. J.; Madic, C. *Inorg. Chem. Commun.* 2005, *8*, 239–241.

33. Chan, G. Y. S.; Drew, M. G. B.; Hudson, M. J.; Iveson, P. B.; Liljenzin, J. O.; Skalberg, M.; Spjuth, L.; Madic, C. *Dalton Trans.* 1997, 649–660.

34. Jensen, M. P.; Bond, A. H. *J. Amer. Chem. Soc.* 2002, *124*, 9870–9877.

35. Reinoso-Garcia, M. M.; Janczewski, D.; Reinhoudt, D. N.; Verboom, W.; Malinowska, E.; Pietrzak, M.; Hill, C.; Baca, J.; Gruner, B.; Selucky, P.; Gruttner, C. *New J. Chem.* 2006, *30*, 1480–1492.

36. Arnaud-Neu, F.; Schwing-Weill, M.-J.; Dozol, J.-F. In *Calixarenes 2001*, Asfari, Z. et al., eds., Kluwer Academic Publishers, Dordrecht, 2001.

37. Malone, J. F.; Marrs, D. J.; McKervey, M. A.; Ohagan, P.; Thompson, N.; Walker, A.; Arnaud-Neu, F.; Mauprivez, O.; Schwingweill, M. J.; Dozol, J. F.; Rouquette, H.; Simon, N. *Chem. Commun.* 1995, 2151–2153.

38. Delmau, L. H.; Simon, N.; Schwing-Weill, M. J.; Arnaud-Neu, F.; Dozol, J. F.; Eymard, S.; Tournois, B.; Böhmer, V.; Gruttner, C.; Musigmann, C.; Tunayar, A. *Chem. Commun.* 1998, 1627–1628.

39. Matthews, S. E.; Saadioui, M.; Böhmer, V.; Barboso, S.; Arnaud-Neu, F.; Schwing-Weill, M. J.; Carrera, A. G.; Dozol, J. F. *J. Prakt. Chem.* 1999, *341*, 264–273.

40. Izatt, R. M.; Christensen, J. J.; Hawkins, R. T. *United States Patent* 4477377 (1984).

41. Blasius, E.; Nilles, K. H. *Radiochim. Acta.* 1984, *35*, 173–182; Blasius, E.; Nilles, K. H. *Radiochim. Acta* 1984, *36*, 207–214.

42. Casnati, A.; Sansone, F.; Dozol, J. F.; Rouquette, H.; Arnaud-Neu, F.; Byrne, D.; Fuangswasdi, S.; Schwing-Weill, M. J.; Ungaro, R. *J. Inclus. Phenom. Macrocycl. Chem.* 2001, *41*, 193–200; Casnati, A.; Pochini, A.; Ungaro, R.; Ugozzoli, F.; Arnaud, F.; Fanni, S.; Schwing, M. J.; Egberink, R. J. M.; Dejong, F.; Reinhoudt, D. N. *J. Amer. Chem. Soc.* 1995, *117*, 2767–2777.

43. Harmon, B. W.; Ensor, D. D.; Delmau, L. H.; Moyer, B. A. *Solv. Extraction Ion Exchange* 2007, *25*, 373–388.
44. Levitskaia, T. G.; Maya, L.; Van Berkel, G. J.; Moyer, B. A. *Inorg. Chem.* 2007, *46*, 261–272.
45. Talanov, V. S.; Talanova, G. G.; Gorbunova, M. G.; Bartsch, R. A. *Perkin Trans. 2.* 2002, 209–215.
46. Xu, C.; Yuan, L.; Shen, X.; Zhai, M. *Dalton Trans.* 2010, *39*, 3897–3902.
47. Grunder, M.; Dozol, J. F.; Asfari, Z.; Vicens, J. *J. Radioanal. Nucl. Chem.* 1999, *241*, 59–67.
48. Leonard, R. A.; Conner, C.; Liberatore, M. W.; Bonnesen, P. V.; Presley, D. J.; Moyer, B. A.; Lumetta, G. J. *Separ. Sci. Technol.* 1999, *34*, 1043–1068.
49. Rambo, B. M.; Kim, S. K.; Kim, J. S.; Bielawski, C. W.; Sessler, J. L. *Chem. Sci.* 2010, *1*, 716–722.
50. Zhang, A. Y.; Xiao, C. L.; Hu, Q. H.; Chai, Z. F. *Solv. Extraction Ion Exchange* 2010, *28*, 526–542.
51. Pellet-Rostaing, S.; Chitry, F.; Spitz, J. A.; Sorin, A.; Favre-Reguillon, A.; Lemaire, M. *Tetrahedron* 2003, *59*, 10,313–10,324.
52. Mohapatra, P. K.; Lakshmi, D. S.; Bhattacharyya, A.; Manchanda, V. K. *J. Hazard. Mater.* 2009, *169*, 472–479; Raut, D. R.; Mohapatra, P. K.; Ansari, S. A.; Manchanda, V. K. *Separ. Sci. Technol.* 2007, *44*, 3664–3678.
53. Wang, X. J.; Brusseau, M. L. *Environ. Sci. Technol.* 1993, *27*, 2821–2825.
54. Villaverde, J.; Maqueda, C.; Morillo, E. *J. Agric. Food Chem.* 2006, *54*, 4766–4772.
55. Petitgirard, A.; Djehiche, M.; Persello, J.; Fievet, P.; Fatin-Rouge, N. *Chemosphere* 2009, *75*, 714–718.
56. Cathum, S. J.; Dumouchel, A.; Punt, M.; Brown, C. E. *Soil Sediment Contamin.* 2007, *16*, 15–27.
57. Sawicki, R.; Mercier, L. *Environ. Sci. Technol.* 2006, *40*, 1978–1983.
58. Allabashi, R.; Arkas, M.; Hormann, G.; Tsiourvas, D. *Water Res.* 2007, *41*, 476–486.
59. Baruch-Teblum, E.; Mastai, Y.; Landfester, K. *Eur. Polym. J.* 2010, *46*, 1671–1678.
60. Rölling, P.; Lamers, M.; Staudt, C. *J. Membr. Sci.* 2010, *362*, 154–163.
61. Kida, T.; Nakano, T.; Fujino, Y.; Matsumura, C.; Miyawaki, K.; Kato, E.; Akashi, M. *Anal. Chem.* 2008, *80*, 317–320.
62. Mamba, B. B.; Krause, R. W.; Malefetse, T. J.; Nxumalo, E. N. *Environ. Chem. Lett.* 2007, *5*, 79–84.
63. Salipira, K. L.; Mamba, B. B.; Krause, R. W.; Malefetse, T. J.; Durbach, S. H. *Water* 2008, *34*, 113–118.
64. Shao, D. D.; Sheng, G. D.; Chen, C. L.; Wang, X. K.; Nagatsu, M. *Chemosphere* 2010, *79*, 679–685.
65. Brusseau, M. L.; Wang, X. J.; Wang, W. Z. *Environ. Sci. Technol.* 1997, *31*, 1087–1092.
66. Ehsan, S.; Prasher, S. O.; Marshall, W. D. *Chemosphere* 2007, *68*, 150–158.
67. Yang, C. J.; Zeng, Q. R.; Wang, Y. Z.; Liao, B. H.; Sun, J. A.; Shi, H.; Chen, X. D. *J. Environ. Sci. China* 2010, *22*, 1910–1915.
68. Mahlambi, M. M.; Malefetse, T. J.; Mamba, B. B.; Krause, R. W. *J. Polym. Res.* 2010, *17*, 589–600.

# 8 Chemomechanical Materials*

*Hans-Jörg Schneider*

## CONTENTS

## INTRODUCTION AND OVERVIEW

Chemomechanical polymers are smart materials that respond to external chemical compounds in the surrounding medium through a mechanical action, usually with a change in volume. Figure 8.1 illustrates the appearance and possible applications of such smart materials: expansion as observed with a chitosan hydrogel particle (a), its possible use as an artificial muscle (b), and its use for drug delivery (c and d).

The word chemomechanical and similar expressions are also used in other contexts, which are not in the scope of the present chapter. Mechanoresponsive materials or mechanotransduction play a great role in living systems[1] and, recently, also in synthetic systems in which they allow, for example, control of reactions by external force[2,3] or in polymer brushes as mechanosensitive surfaces.[4] The influence of surface adsorbates on ceramic materials, which can induce changes in surface plasticity and fracture behavior, has also been called a chemomechanical effect.[5] Chemical force microscopy couples intermolecular interaction directly with microscopic mechanical motion.[6,7] With atomic force microscope probe tips, chemomechanical nanografting on surfaces has been greatly advanced.[8,9] Viscoelastic properties of the interfacial layer in microgel particle emulsions depend on pH and temperature.[10] Monodisperse microcapsules allow precise manipulation of loading levels and release of encapsulated substances.[11] In dentistry, so-called chemomechanical methods are used for caries removal with, for example, amino acid–containing gels.[12]

---

* Dedicated to Professor Michael Hanack on the occasion of his 80th birthday.

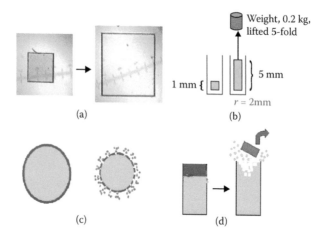

**FIGURE 8.1** Chemomechanical polymers and their possible uses; for explanations, see text. (From Kato, K., Schneider, H.-J., *J. Mater. Chem.*, 19, 569–573, 2009. Reproduced by permission of The Royal Society of Chemistry.)

The motions of droplets or vesicles can be driven by wetting transitions or by ion exchange reactions.[13]

Many polymers have been described which are responsive to electric, thermal, light, magnetic, and solvent/medium stimuli.[14] The present chapter deals with polymers which respond specifically by molecular recognition, and not, for example, by pH changes: supramolecular complex formation between functional groups of the polymer matrix with effector molecules in the external medium will be shown to lead to often dramatic changes of polymer particle size. This principle will be illustrated with examples that also highlight possible practical applications, which can be seen in the increasing number of patents in the field.

In a wider sense, molecular and supramolecular switches can also be seen as chemomechanical entities in which the input of a chemical signal in the form of a specific compound leads to a change of geometry and behavior of a molecule. Most systems have been studied in solution,[15,16] but also recently on surfaces[17,18]; in many cases, the input used is pH, but light, magnetic fields and electric fields can also be considered. Of particular promise is their implementation in and on nanoparticles, specifically with mesoporous silica nanoparticles. These can be made as hollow containers with a stopper, which react on external signals and then release, for example, drugs to the environment.[19,20] A related application is based on mesopores that are equipped with aminoaryl "stalks" for complexation with cyclodextrins. At lower pH, the cyclodextrin loses its affinity to the neutral and hydrophobic stalk, it is decomplexed and the pores of such nanovalves open, thereby releasing molecules from the hollow inside of the nanoparticle (Figure 8.2; related patents: WO2009097439 and WO2009094580).[21]

Two patents (USP 5654006 and 5753261) describe how condensed-phase microparticles, including comb polymer glycoproteins, can be used for compound storage and release; the addition of multivalent counterions such as $Ca^{2+}$ or histamine in the case of a polyanionic polymer matrix will force the particle into a condensed

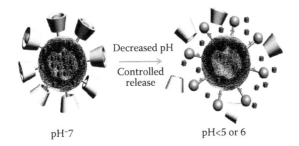

**FIGURE 8.2**  pH-controlled drug release from mesopores; for explanations, see text. (Reproduced with permission from Du, L. et al., *J. Am. Chem. Soc.*, 131, 15136–15142, 2009. Copyright 2009 American Chemical Society.)

phase. Binding of an analyte ligand may then cause lysis of a corresponding membrane, with rapid release of an entrapped reporter compound. With bioelastic polypeptides, one can produce chemomechanical pumps, in which a swollen, drug-laden bioelastic matrix is chemically driven to contraction and expulsion of drug on contact with a preselected physiological condition (EP0449592). Until recently, drug release was typically triggered by pH changes or unspecific salt effects, but a specific host function at the side chain of the peptide may allow for more selective reactions.

The selective response in chemomechanical polymers toward external stimuli has been previously used for optical sensors. In a patented application (PCT/US1998/004566), a crown-ether moiety was incorporated in polymerized crystalline colloidal arrays (PCCA),[22] in which swelling of the gel occurs in the presence of potassium and other cations including lead ions. The volume increase changes the periodicity of the crystalline colloidal array, which results in a shift in the diffracted wavelength. The PCCA is a crystalline colloidal array of spherical polystyrene colloids polymerized within a thin polymer hydrogel film, which secures selective recognition. With the crown-ether derivative in the gel, the response of the PCCA particles toward $Pb^{2+}$ is approximately linear up to 1 mM, with a sensitivity of 90 nm/mM $Pb^{2+}$ (Figure 8.3), which may be present in concentrations as low as 10 µM.[23] Crown-containing microcapsules from poly(N-isopropylacrylamide) with K ions exhibit a decreased lower critical solution temperature and swelling or shrinking.[24,25]

Many biomimetic polymers, mostly in the form of gels, which bind biomolecules, have been described and are often based on molecular imprinting, which ensures selectivity.[26,27] Such recognitive macroporous materials can be used for drug and protein delivery, and possibly for the elimination of undesirable compounds. In most cases, the release of the contents is diffusion-controlled but could also be initiated by changes in temperature or pH.[28–31] Core-multishell nanoparticles can spontaneously encapsulate and transport drugs.[32] Block copolymer hydrogels can exhibit a sol–gel phase transition in response to external stimuli like temperature or pH.[33] Gel–sol transitions can occur in response to temperature, pH, and $Ca^{2+}$ as input stimuli.[34] Nanocomposite membranes composed of layer-by-layer assembled polymers and single-walled carbon nanotubes can be used as transducers to selectively detect trace DNA.[35]

Supramolecular gels, which, in contrast with polymer gels, are based on low molecular mass compounds and exist without cross-linking,[36–38] can be formed or

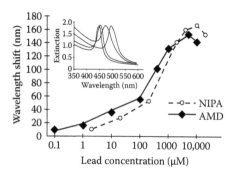

**FIGURE 8.3** Response of 18-crown-6–containing PCCA sensors made with acrylamide (AMD) or *N*-isopropylacrylamide (NIPA) toward $Pb(NO_3)_2$. The shift in the diffracted wavelength is relative to the wavelength diffracted in pure water. Inset shows the extinction spectra of the AMD sensor in water for 0.1, 10, and 100 mM $Pb^{2+}$. (Reproduced with permission from Holtz, J. H. et al., *Anal. Chem.*, 70, 780–791, 1998. Copyright 1998 American Chemical Society.)

broken by external stimuli. Such gels are not chemomechanical materials in the narrow sense, and only a few principles and applications can be mentioned here. Formation or breakage of such gels may be triggered by pH, temperature, light, etc., as the underlying weak noncovalent bonds constituting the gel are sensitive to the corresponding environment. With metallogelators, the addition of metal ions triggers gelation.[39] Gelation may also be triggered by the presence of specific anions.[40] The use of host structures, such as derivatives of cyclodextrins[41] or cavitands, as gelating agents holds particular promise for switching the gel state by specific effectors. Membranes based on poly(*N*-isopropylacrylamide) and β-cyclodextrin can switch from a "closed" to an "open" state by recognition of guest molecules with a hydrophobic side group.[42] One can also equip a gelating structure with an enzyme-susceptible linker, which, in the gel, is protected from the enzyme. When the gel is dissociated by, for example, pH or temperature change, the enzyme can cleave the linker and release a drug that was covalently bound to a gelling scaffold.[43] Many systems have already been described in which the sensitivity of self-assembly or of supramolecular complexation in gels toward external stimuli was exploited.[44]

Particular attention has been paid to the development of hydrogels for oral insulin delivery[45–50]; insulin release from related implants could be regulated by external glucose-sensing units.[49,50] In contrast, chemomechanical polymers could act simultaneously as a sensing and delivery unit within one single particle.

## GLUCOSE-RESPONSIVE CHEMOMECHANICAL POLYMERS

Smart materials, which can measure glucose levels in the blood and at the same time deliver insulin to the body, are obviously of great medicinal and commercial interest. If such a material is biocompatible and can be used as an implant, this would free diabetics from continuous testing and injections. Glucose-responsive materials have been developed on the basis of lectins, specifically of concanavalin,[51] and by using

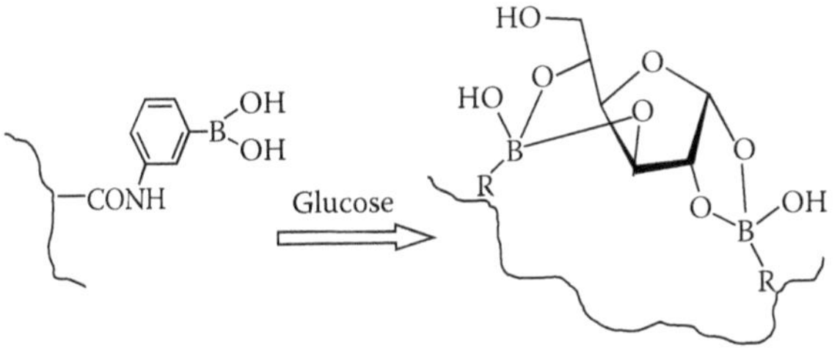

**FIGURE 8.4**  Cross-linking and gel contraction by boronic ester formation with glucose.

enzymes such as glucose oxidase.[52] Glucose and galactose sensors were obtained using biotinylated gels on PCCAs, which bind avidinated glucose oxidase or β-D-galactosidase as recognition elements. Glucose oxidase converts glucose to gluconic acid anions, which is believed to cause glucose sensor swelling. The sensor detects glucose in the 0.1 to 0.5 mM (18–90 ppm) concentration range in the presence of oxygen and detects as little as $10^{-12}$ M of glucose (0.18 ppt) in the absence of oxygen. Typical response times of such sensors are 1 to 5 minutes with a 150-μm-thick gel, decreasing with thinner gels.[22]

Recognition elements such as boronic acid were also used in a photonic crystal hydrogel for optical sugar detection. Glucose reacts rather selectively with boronic acid, more than with galactose, mannose, or fructose, resulting in cross-linking of the hydrogel (Figure 8.4). This shifts the photonic crystal diffraction across the visible spectral region.[53]

Cross-linking of boronic acid–containing polymers by selective reaction with glucose can be used not only for sensing but also for the delivery of drugs such as insulin. After Kitano and coworkers[54] proposed such a system in 1992, approximately 25 related publications have appeared (for an overview, see Schneider et al. 2007,[55] selected patents: USP 7316999, WO/2001/092334, and WO/2006/102762). The efficiency has been demonstrated with tiny tubes containing a dye (for visibility) which has a boronic acid–containing polymer as a stopper. In the presence of glucose, the stopper shrinks and thus opens the container with the release of the interior solution (as in Figure 8.1d). In human blood plasma, millimole concentrations of added glucose lead to contraction within 5 to 10 minutes.[56] Such a chemical corkscrew demonstrates how chemomechanical polymers combine sensor and actuator within one single device.

## NONCOVALENT INTERACTIONS AS THE BASIS OF CHEMOMECHANICAL POLYMERS

In principle, all types of noncovalent interaction mechanisms known from supramolecular chemistry[57] can be used in chemomechanical materials that bear suitable binding elements.[58,59] In hydrogel **I** (Figure 8.5) produced from polymethyl(methyl) acrylate and different amines, the randomly distributed side chain functionalities can exert a manifold of interactions; several materials such as chitosan (**II**) or polyallylamine (**III**) provide for distinct interactions without further modification.

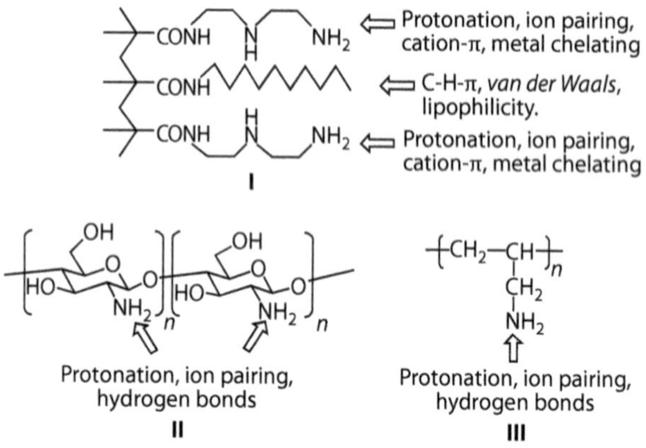

**FIGURE 8.5** Noncovalent interactions with some chemomechanical polymers (**I**, poly-methacrylic acid with supramolecular binding functions; **II**, chitosan; and **III**, polyallyl-amine). (From Schneider, H.-J. et al., *Sensors*, 7, 1578–1611, 2007. With permission.)

Hydrogels derived from polyamines such as polyalylamine or polyethyleni-mine show, after protonation, contraction by noncovalent cross-linking between the ammonium ion side chains and the necessarily always present anions inside the gels. If smaller anions, such as, chloride, are replaced by larger ions, one observes an expansion with hydrogel **I**, which increases with the size of the anion[60] (pat-ent WO002005003179A2/A3); moreover, even isomeric carboxylates can be distin-guished (Figure 8.6). The size changes depend on the pH value, due to protonation

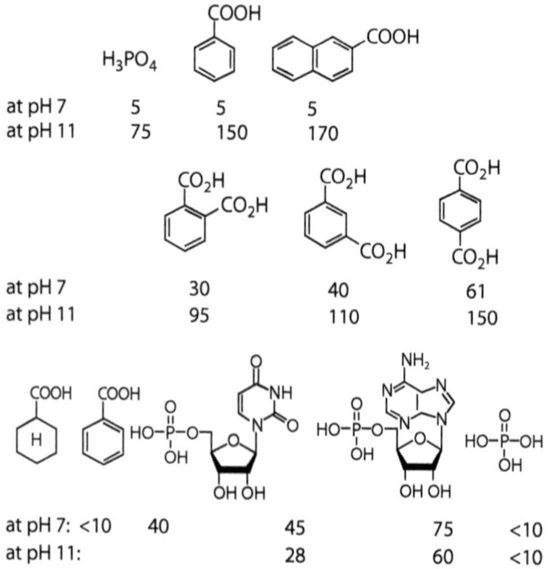

**FIGURE 8.6** Expansions of hydrogel **I** with different organic anions (as a percentage of expansion in one dimension).

and deprotonation of host and guest. Most chemomechanical hydrogels bear charges because of the presence of base or acid functions. In consequence, their response is therefore pH-dependent, and neutral guest compounds usually do not lead to specific size changes. For the intricate mechanisms involved in such volume changes, we refer to pertinent publications.[58,59,61] It is essentially the water content which determines the size changes; gravimetric measurements indicate a continuous change of the water content in the hydrogel particles and no sudden changes, which are typical for phase transitions.

Noncovalent cross-linking can be used to trigger contractions of hydrogel particles in many ways. An antigen-responsive gel was obtained based on reversible binding between antigen and antibody as a cross-linking mechanism in a semi-interpenetrating network hydrogel. Competitive binding of the free antigen triggers a change in gel volume because of the breaking of noncovalent cross-links.[62,63] A hydrogel that covalently incorporates a single-stranded DNA as a cross-linker expands or contracts in response to a complementary single-stranded DNA as effector.[64]

Noncovalent interactions between the covalently bound host and the mobile guest molecules are not the only clue to gel particle changes; they can also occur between the guest molecules inside the polymer. With polyallylamine **III** and aromatic acid anions, one observes contraction, which increases from 20% size reduction (with benzoate) to 60% and more with additional substituents at the phenyl rings. Naphthyl instead of phenyl-carboxylic acid leads to contraction by as much as 67%, which must be ascribed to the significant stacking between the aromatic units within the gel network (see Figure 8.7). Nitro substituents lead to similarly large contraction if the nitro function is not ortho to the carboxylate: here, steric hindrance prevents a flat orientation, which secures stacking van der Waals effects between the nitrobenzoates.[65] Nitro groups are known to exhibit almost as significant a stacking effect as do phenyl groups.[57]

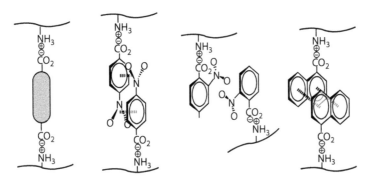

**FIGURE 8.7**  Supramolecular interactions between effector guest molecules lead to large differences in gel particle size, see text. (From Kato, K., Schneider, H.-J., *Eur. J. Org. Chem.*, 1378–1382, 2008. With permission.)

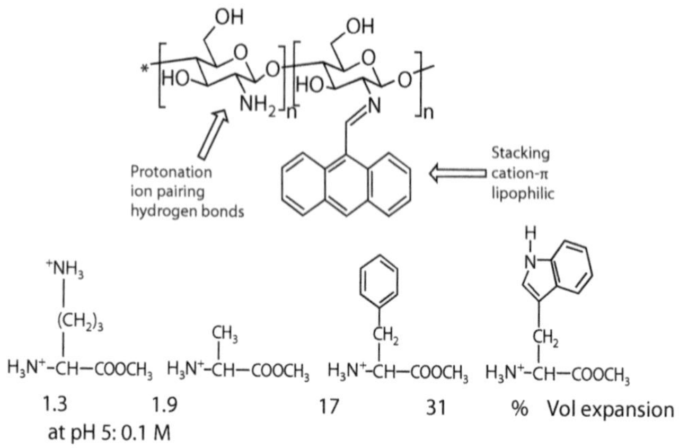

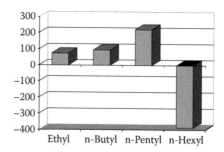

**FIGURE 8.8**   Chemomechanical effects of amino acids. Stacking interactions leading to
volume increase, see text. (From Lomadze, N., Schneider, H.-J., *Tetrahedron*, 61, 8694–8698,
2005. With permission.)

Stacking interactions are also the clue to the distinction of amino acids with
a hydrogel derived from chitosan, when this is equipped with anthracene units.[66]
Larger and aromatic side chains in the amino acids lead to increased volumes
(Figure 8.8).

Lipophilic interactions with the long alkyl substituents in the polymer **I** lead
to an abrupt change from expansion to contraction of the gel by different tetraalkylammonium effectors (Figure 8.9). A normal expansion, which increases
with the size of the ammonium effector, as observed with the examples in
Figure 8.6, changes to contraction with the tetrahexyl compound. The stronger
association between the longer effector molecule chains with the polymer alkyl

**FIGURE 8.9**   Expansion can change to contraction: volume changes (%) of the polymer gel
**I** by tetraalkylammonium hydroxides as effectors. (From Schneider, H.-J. et al., *Eur. J. Org.
Chem.*, 677–692, 2006. With permission.)

chain leads to dominating lipophilic contributions and collapse of the polymer network.[61]

## COOPERATIVE EFFECTS IN CHEMOMECHANICAL POLYMERS/LOGICAL GATE FUNCTIONS

The dependence of a response on the simultaneous presence of two or more compounds is a more common observation in chemomechanical polymers than in solution chemistry, in which this effect is usually confined to allosteric systems. With chemomechanical polymers, cooperativity in the sense of a logical AND gate is much more frequent. In the first place, the influence of pH on the effect of guest compounds can be observed (see Figure 8.6). Beyond this, the presence of inorganic ions on the response of nucleotides also shows mutual dependence,[61,67] as does the action of, for example, naphthoate on polyethylenimine gel, which shows a strong dependence on the simultaneous presence of amino acids as effectors.[68] With chitosan gels, basic amino acids such as His, Arg, or Lys lead to sizable volume changes only on the simultaneous presence of benzoic acid, but not with inorganic acids (Figure 8.10).

In terms of practical applications, the cooperativity between metal ions and guest molecules, such peptides are of particular interest. Gels containing ethylenediamine functions show, with various metal ions, large size changes, with polymer **I** culminating, for example, in a 390% volume expansion with ($Pb^{2+}$) ions.[69] Ions such as $Cu^{2+}$ or $Zn^{2+}$ bound into hydrogel **I** provide a secondary binding site, which can then interact with, for example, peptides. This allows a chemomechanical effect with guest molecules which otherwise are completely silent.[70] The alkyl chains in the polymer **I** lead to discrimination according to the lipophilicity of the amino acid side chain, affording volume changes between 125% and 245% (Figure 8.11).

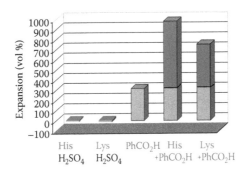

**FIGURE 8.10** Basic amino acids such as His, Arg, or Lys lead to sizable volume changes of chitosan gels only in the presence of benzoic acid. (From Kato, K., Schneider, H.-J., *Eur. J. Org. Chem.*, 1042–1047, 2009. With permission.)

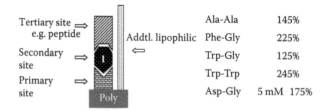

| | | Ala-Ala | 145% |
| Tertiary site ⟹ e.g. peptide | Addtl. lipophilic ⟸ | Phe-Gly | 225% |
| Secondary site ⟹ | | Trp-Gly | 125% |
| Primary site ⟹ | | Trp-Trp | 245% |
| | Poly | Asp-Gly | 5 mM  175% |

**FIGURE 8.11**  Ternary complexes with cooperativity between metal ions and peptides, see text; examples for expansion (in one dimension) triggered by 0.25 mM peptide; volume changes (%) of polymer **I** loaded with $Cu^{2+}$. (From Lomadze, N., Schneider, H. M., *Tetrahedron Lett.*, 46, 751–754, 2005. With permission.)

## CHIRAL DISCRIMINATION

Discrimination between enantiomers has been, until now, essentially restricted to solution or the solid state, as exemplified by chromatographic separations or crystallization techniques. The use of optically active hydrogels makes it possible to translate the molecular recognition of enantiomers directly into macroscopic motions. The aminoglucose units in the optically active natural polymer chitosan allow such a transmission; the protonated ammonium centers can bind tartaric acids by ion pairing with an effector carboxylate. At the same time, the $NH_3^+$ group exhibits a cation–π interaction with phenyl groups of the $O$-dibenzoyl derivative of the tartaric acid (Figure 8.12). This is evidenced by the absence of any chemo-mechanical effect if the substrate carries no phenyl substituent and is visible in the nuclear magnetic resonance spectrum of the supramolecular complex in the hydrogel.[71] With 1 mM concentrations of the $O$-dibenzoyl tartrates, the D-isomer shows a contraction of the chitosan gel by 90%, whereas the L-enantiomer is almost inactive.

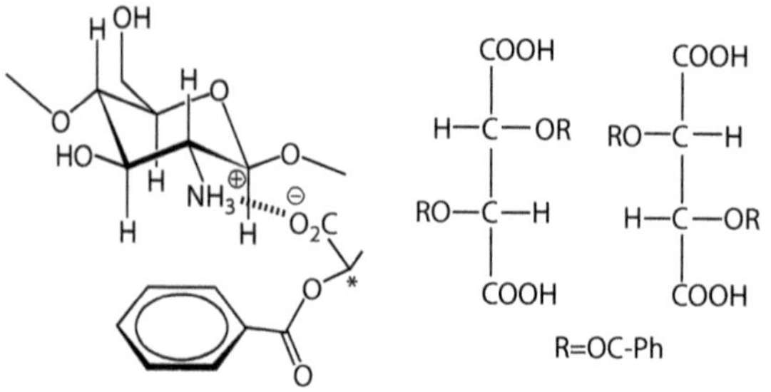

**FIGURE 8.12**  Complexation of enantiomeric $O$-dibenzoyl tartrates with the aminoglucose units in chitosan, see text. (From Schneider, H.-J., Kato, K., *Angew. Chem.-Int. Ed.*, 46, 2694–2696, 2007. With permission.)

## KINETICS AND SENSITIVITY

The speed and sensitivity of response are critical issues for the practical use of che-momechanical materials. The kinetics are essentially controlled by the diffusion rates of the substrates in the gel, after a short initial phase of loading the surface of the gel particles (for details, see Schneider et al. 2009,[58] Schneider and Kato 2007,[59] and Schneider et al.[61]). Figure 8.13 demonstrates how the response rates increase with the surface to volume ratio of gel particles.[72]

The sensitivity of chemomechanical materials is analogous to that of sensor par-ticles[73]: on the one hand, it is a function of the particle size; on the other hand, it depends on the binding force between polymer and effector molecule. Only if the affinity is high enough can one expect that all binding sites within one particle be occupied by the effector molecules (Figure 8.14).[74]

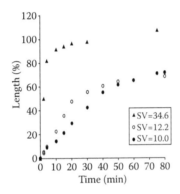

**FIGURE 8.13** Rate dependence of chitosan gel particle elongation induced by 50 mM L-histidine on the surface to volume ratio (S/V). Approximate half-life ($t_{1/2}$) for 50% of the maximum expansion: $t_{1/2} = 42$, 32, and 3 minutes for S/V = 10.0, 12.2, and 34.6, respectively. (Kato, K., Schneider, H.-J., *Eur. J. Org. Chem.*, 2009, 1042–1047. Copyright Wiley-VCH Verlag GmbH & Co. KGaA. Reproduced with permission.)

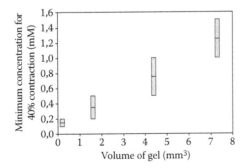

**FIGURE 8.14** Sensitivity increase with decreasing particle size. Sensitivity of 40% contrac-tion (in one dimension) induced by D-*O*-dibenzoyl tartrate as function of the gel particle size. (Schneider, H.-J., Kato, K., *Angew. Chem.-Int. Ed.*, 2007, 46, 2694–2696. Copyright Wiley-VCH Verlag GmbH & Co. KGaA. Reproduced with permission.)

## CONCLUSIONS

Smart materials can be made responsive to an effector molecule in the surrounding medium, ranging from inorganic anion and cations to proteins and nucleic acids. Flexible polymers containing functions for molecular recognition allow one to engineer actuator devices, which can drive a micromachine or, for example, deliver a drug in an entirely self-controlled manner. The speed of response can be enhanced significantly by enlarging the surface to volume ratio of the chemomechanical hydrogel particles, whereas the sensitivity is increased dramatically by downsizing the polymer particles. The response of such materials will nevertheless be slower than that of conventional optical detectors. Although this may pose a restriction for sensing, the major advantage of chemomechanical polymers is that they can act within tiny particles at the same time as actuators, without the need for any transducer or power supply. Such polymers can be made biocompatible or biodegradable,[75] and have a promising future, in particular, for drug delivery and targeting.

## REFERENCES

1. Orr, A. W., B. P. Helmke, B. R. Blackman, M. A. Schwartz. 2006. *Dev. Cell*, 10: 11–20.
2. Davis, D. A., A. Hamilton, J. L. Yang, L. D. Cremar, D. Van Gough, S. L. Potisek, M. T. Ong, P. V. Braun, T. J. Martinez, S. R. White, J. S. Moore, N. R. Sottos. 2009. *Nature*, 459: 68–72.
3. Ribas-Arino, J., M. Shiga, D. Marx. 2010. *J. Am. Chem. Soc.*, 132: 10609–10614.
4. Bünsow J., T. S. Kelby, W. T. S. Huck. 2010. *Acc. Chem. Res.*, 43: 466–474.
5. See also for glass surface modulation Belde, K. J., S. J. Bull. 2006. *Thin Solid Films*, 515: 859–865.
6. Huang, Z., R. Boulatov. 2010. *Pure Appl. Chem.*, 82: 931–951.
7. Noy, A. 2006. *Surface Interface Anal.*, 38: 1429–1441.
8. Lee, M. V., M. T. Hoffman, K. Barnett, J. M. Geiss, V. S. Smentkowski, M. R. Linford, R. C. Davis. 2006. *J. Nanoscience Nanotechnology*, 6: 1639–1643.
9. Ruiz, S. A., C. S. Chen. 2007. *Soft Matter*, 3: 168–177.
10. Brugger, B., J. Vermant, W. Richtering. 2010. *Phys. Chem. Chem. Phys.* 12: 14573–14578.
11. Chu, L. Y., A. S. Utada, R. K. Shah, J. W. Kim, D. A. Weitz. 2007. *Angew. Chem. Int. Ed. Engl.*, 46: 8970–8974.
12. Peric, T., D. Markovic, B. Petrovic. 2009. *Acta Odont. Scand.*, 67: 277–283.
13. Shioi, A., T. Ban, Y. Morimune. 2010. *Entropy*, 12: 2308–2332.
14. Meng, H., J. L. Hu. 2010. *J. Intell. Material Systems Struct.*, 21: 859–885.
15. Balzani, V., A. Credi, M. Venturi. 2008. *Molecular devices and machines*, Wiley-VCH, Weinheim. *Molecular Switches*, ed. B. L. Feringa, Wiley-VCH, Weinheim, 2001.
16. Balzani, V., A. Credi, M. Venturi. 2009. *Chem. Soc. Rev.*, 38: 1542–1550.
17. Browne, W. R., B. L. Feringa. 2009. *Annu. Rev. Phys. Chem.*, 60: 407.
18. Coronado, E., P. Gavina, S. Tatay. 2009. *Chem. Soc. Rev.*, 38: 1674.
19. Klajn, R., J. F. Stoddart, B. A. Grzybowski. 2010. *Chem. Soc. Rev.*, 39: 2203–2237.
20. Coti, K. K., M. E. Belowich, M. Liong, M. W. Ambrogio, Y. A. Lau, H. A. Khatib, J. I. Zink, N. M. Khashab, J. F. Stoddart. 2009. *Nanoscale*, 1: 16–39.
21. Du, L., S. J. Liao, H. A. Khatib, J. F. Stoddart, J. I. Zink. 2009. *J. Am. Chem. Soc.*, 131: 15136–15142.
22. Holtz, J. H., J. S. W. Holtz, C. H. Munro, S. A. Asher. 1998. *Anal. Chem.*, 70: 780–791.
23. Reese, C. E., M. E. Baltusavich, J. P. Keim, S. A. Asher. 2001. *Ibid.*, 73: 5038–5042.

24. Pi, S. W., X. J. Ju, H. G. Wu, R. Xie, L. Y. Chu. 2010. *J. Colloid Interface Sci.*, 349: 512–518 (and references cited therein).
25. Mi, P., X. J. Ju, R. Xie, H. G. Wu, J. Ma, L. Y. Chu. 2009. *Polymer*, 51: 1648–1653 (and references cited therein).
26. Xia, F., L. Jiang. 2008. *Advanced Materials*, 20: 2842–2858.
27. Yoon, H. J., W. D. Jang. 2010. *J. Materials Chem.*, 20: 211–222.
28. Byrne, M. E., E. Oral, J. Z. Hilt, N. A. Peppas. 2002. *Polymers Adv. Technol.*, 13: 798–816.
29. Herrero, E. P., E. M. M. Del Valle, N. A. Peppas. 2010. *Ind. Engin. Chem. Res.*, 49: 9811–9814.
30. Gehrke S. H., L. H. Uhden, J. F. McBride. 1998. *J. Control. Release*, 55: 21–33.
31. Foster, J. A., J. W. Steed. 2010. *Angew. Chem. Int. Ed. Engl.*, 49: 6718–6724.
32. Quadir, M. A., M. R. Radowski, F. Kratz, K. Licha, P. Hauff, R. Haag. 2008. *J. Control. Release*, 132: 289–294.
33. He, C., S. W. Kim, D. S. Lee. 2008. *J. Control. Release*, 127: 189–207.
34. Komatsu, H., S. Matsumoto, S. Tamaru, K. Kaneko, M. Ikeda, I. Hamachi. 2009. *J. Am. Chem. Soc.*, 131: 5580–5585.
35. Kang, T. J., D. K. Lim, J. M. Nam, Y. H. Kim. 2010. *Sensors Actuators B*, 147: 691–696.
36. Foster, J. A., J. W. Steed. 2010. *Angew. Chem. Int. Ed. Engl.*, 49: 6718–6724.
37. Banerjee, S., R. K. Das, U. Maitra. 2009. *J. Mater. Chem.*, 19: 6649–6687.
38. Hirst, A. R., B. Escuder, J. F. Miravet, D. K. Smith. 2008. *Angew. Chem. Int. Ed. Engl.*, 47: 8002–8018.
39. Piepenbrock, M. O. M., G. O. Lloyd, N. Clarke, J. W. Steed. 2010. *Chem. Rev.*, 110: 1960–2004.
40. Becker, T., C. Y. Goh, F. Jones, M. J. McIldowie, M. Mocerino, M. I. Ogden. 2008. *Chem. Comm.*, 3900–3902.
41. Zhou, J. W., H. Ritter. 2010. *Polymer Chem.*, 1: 1552–1559.
42. Yang, M., R. Xie, J. Y. Wang, X. J. Ju, L. H. Yang, L. Y. Chu. 2010. *J. Membrane Sci.*, 355: 142–150.
43. van Bommel, K. J. C., M. C. A. Stuart, B. L. Feringa, J. van Esch. 2005. *Org. Biomol. Chem.*, 3: 2917–2920.
44. Osada, K., R. J. Christie, K. Kataoka. 2009. *J. Royal Soc. Interface*, 6: S325–S3390.
45. Siegel, R. A., Y. D. Gu, M. Lei, A. Baldi, E. E. Nuxoll, B. Ziaie. 2010. *J. Control. Release*, 141: 303–313.
46. Finotelli, P. V., D. Da Silva, M. Sola-Penna, A. M. Rossi, M. Farina, L. R. Andrade, A. Y. Takeuchi, M. H. Rocha-Leao. 2010. *Colloids Surfaces B-Biointerfaces*, 81: 206–211.
47. Sajeesh, S., K. Bouchemal, V. Marsaud, C. Vauthier, C. P. Sharma. 2010. *J. Control. Release*, 147: 377–384.
48. Sharma, G., K. Wilson, C. F. van der Walle, N. Sattar, J. R. Petrie, M. Kumar. 2010. *Eur. J. Pharmaceut. Biopharm.*, 76, 159–169.
49. Tuesca, A., K. Nakamura, M. Morishita, J. Joseph, N. Peppas, A. Lowman. 2008. *J. Pharm. Sciences*, 97: 2607–2618.
50. Richter, A., C. Klenke, K. F. Arndt. 2004. *Macromol. Sympos.*, 210: 377–384.
51. Miyata, T., A. Jikihara, K. Nakamae, A. S. Hoffman. 2004. *J. Biomat. Sci.-Polymer Ed.*, 15, 1085–1098 (and references cited therein).
52. Suzuki, H., A. Kumagai, K. Ogawa, E. Kokufuta. 2004. *Biomacromolecules*, 5: 486–491.
53. Alexeev, V. L., S. Das, D. N. Finegold, S. A. Asher. 2004. *Clin. Chem.*, 50: 2353.
54. Kitano, Y., K. Koyama, O. Kataoka, T. Kazunori, Y. Okano, Y. Sakurai. 1992. *J. Control. Release*, 19: 161–170.
55. Schneider, H.-J., K. Kato, R. M. Strongin. 2007. *Sensors*, 7: 1578–1611.
56. Samoei, G. K., W. H. Wang, J. O. Escobedo, X. Y. Xu, H.-J. Schneider, R. L. Cook, R. M. Strongin. 2006. *Angew. Chem. Int. Ed. Engl.*, 45: 5319–5322.

57. Schneider, H.-J. 2009. *Angew. Chem. Int. Ed. Engl.*, 48: 3924–3977.
58. Schneider, H.-J., R. L. Cook, R. M. Strongin. 2009. *Acc. Chem. Res.*, 42: 1489–1500.
59. Schneider, H.-J., K. Kato. 2007. In *Intelligent Materials*, edited by H.-J. Schneider, M. Shahinpoor, 100–120. Cambridge, UK: Royal Society of Chemistry.
60. Schneider, H.-J., T. J. Liu, N. Lomadze. 2003. *Angew. Chem. Int. Ed. Engl.* 42: 3544–3546.
61. Schneider, H.-J., T. J. Liu, N. Lomadze. 2006. *Eur. J. Org. Chem.*, 677–692.
62. Miyata, T., N. Asami, T. Uragami. 1999. *Nature*, 399: 766–769.
63. Miyata, T. 2010. *Polymer J.*, 42: 277–289.
64. Murakami, X., M. Maeda. 2005. *Biomacromolecules*, 6: 2927–2929.
65. Kato, K., H.-J. Schneider. 2008. *Eur. J. Org. Chem.*, 1378–1382.
66. Lomadze, N., H.-J. Schneider. 2005. *Tetrahedron*, 61: 8694–8698.
67. Schneider, H.-J., T. J. Liu, N. Lomadze, B. Palm. 2004. *Advanced Materials*, 16: 613–615.
68. Kato, K., H.-J. Schneider. 2007. *Langmuir*, 23: 10741–10745.
69. Schneider, H.-J., T. J. Liu. 2004. *Chem. Comm.*, 100–101.
70. Lomadze, N., H. M. Schneider. 2005. *Tetrahedron Lett.*, 46: 751–754.
71. Schneider, H.-J., K. Kato. 2007. *Angew. Chem.-Int. Ed.*, 46: 2694–2696.
72. Kato, K., H.-J. Schneider. 2009. *Eur. J. Org. Chem*, 1042–1047.
73. Clark, H. A., R. Kopelman, R. Tjalkens, M. A. Philbert. 1999. *Anal. Chem*, 71: 4837–4843.
74. Schneider, H.-J., L. Tianjun, N. Lomadze. 2004. *Chem. Commun.*, 2436–2437.
75. Ju, X-L., R. Xie, L. Yang, L.-Y. Chu. 2009. *Expert Opin. Therap. Patents*, 19: 683–696.

# 9 Supramolecular Structures in Organic Electronics

*Dirk Beckmann and Klaus Müllen*

## CONTENTS

## MOTIVATION

Organics may exhibit isolating, electrically conducting, or semiconducting characteristics which feature their application in electronic devices such as field-effect transistors (FETs), light-emitting diodes, and solar cells without the need for rare and expensive materials like copper or gold. This opens a new variety of information technology in which various functional components can be combined within a single platform. Many future applications have already been envisioned, which mainly include mobile electronic equipment enabled by the mechanical flexibility, (semi)transparency, and the low weight of organic materials (Figure 9.1).

Realizing those visions requires the development of high-performance organic films, which represents the most difficult as well as the most interesting parts of ongoing research. It is essential to generate knowledge about the impact of the molecular structure on the formation of supramolecular structures during the

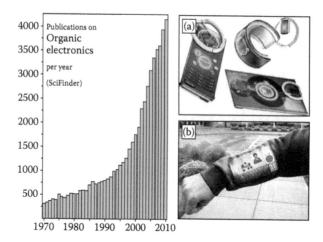

**FIGURE 9.1** Growing research and envisioned applications of organic electronics. (Left) Increasing number of publications on "organic electronics" since 1970. (From SciFinder 01-2011) (Right) (a) Two examples of the envisioned mobile, low-weight, flexible, and semitransparent electronics (right): Nokia Morph. (From Nokia GmBH, Copyright 2011 Nokia; http://research.nokia.com/morph.) (b) Illustration by Derek Wilson of the Center for Advanced Engineering Environments at Old Dominion University. (From *Mechanical Engineering* magazine. With permission.)

device-processing steps and about the effect of all these structures on intramolecular and intermolecular charge transport and thus on the electronic device's properties. In general, organic layers have to provide a charge transport between two electrodes. In FETs, this occurs parallel to the transistor's substrate; in light-emitting diodes and photovoltaic cells, this is perpendicular to the substrate. Liquid crystalline displays (LCDs) and electrochromic devices are based on the transmission of light which is controlled by electrical field–induced supramolecular assemblies. In organic lasers, fuel cells, or batteries, additional supramolecular arrangements are used. Because of the necessity for device-specific tailoring of supramolecularity, it is reasonable to elucidate supramolecular structures in organic electronics separately for each device species and intended function.

In the following subchapters, present-day organic π-conjugated molecules, as they are usually used in organic electronic devices, their supramolecular assemblies, and the influence of processing on supramolecular structures, are presented. Then, supramolecularity in FETs, LCDs, light-emitting diodes, photovoltaic cells, and additional electronic devices (such as electrochromic devices, lasers, sensors, and fuel cells) are discussed.

## Π-CONJUGATED MOLECULES

π-Conjugated molecules are the most promising materials for cost-efficient and flexible electronic devices which, in the case of organic light-emitting diodes (OLEDs), have already been partially commercialized. The properties of organic-based devices are dictated by the chemical structure, purity, and supramolecular organization of the conjugated material. The chemical structure consists of alternating single and double or triple bonds. Herein, overlap of atomic $\pi$-orbitals leads to delocalized electrons which enable current flow within electrical fields. A few classic representatives of conjugated molecules are depicted in Figure 9.2. Polyacetylene **1** was first produced in 1974 by the polymerization of acetylene. Despite the metallic appearance of that black powder, the polymer was not conductive. This changed, however, after oxidation in halogen gas (in a procedure created by Shirakawa, MacDiarmid, and Heeger in 1977). The conductivity of the thus doped polyacetylene was increased by $10^9$ times up to $10^5$ Sm$^{-1}$. By comparison, Teflon exhibits a conductivity of $10^{-16}$ Sm$^{-1}$, whereas copper has a conductivity of $10^8$ Sm$^{-1}$. Today, conductive polymer films and layers made of low-molecular weight, organic substances are used in corrosion protection, antistatic layers, electromagnetic shielding, and (semi)transparent electrodes.[1] Poly-*p*-phenylene **4** is another example of conjugated polymers. In a formal

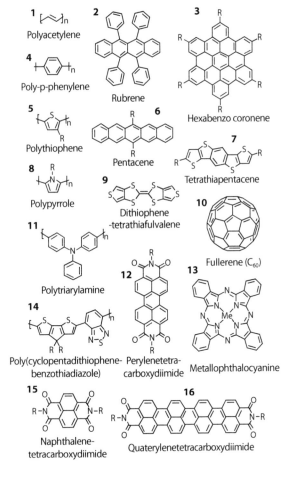

**FIGURE 9.2** Chemical structures of conjugated molecules.

sense, when copolymerized with polyacetylene, it yields poly(p-phenylene vinylene), an electroluminescent polymer applied in light-emitting diodes and organic lasers because of its high conductivity and luminance.[2,3] A second generation of electrically active, organic molecules came up in 1990. These molecules exhibited semiconducting properties without additional dopants. The conductivity of semiconductors usually ranks between $10^{-6}$ Sm$^{-1}$ and $10^2$ Sm$^{-1}$ and strongly depends on the temperature as well as on the applied potentials. These energy inputs lead to the excitation of electrons from the valence band (VB), through the bandgap, to the conduction band (CB). Both electrons in the CB and residual holes in the VB allow for electrical conductivity. In this way, the amount of the applied external energy permits control of the electrical current without the need for mechanically movable parts such as in relays or dimmers. By comparison, the bandgap of insulators is too high, thus no electrons can be transferred from the VB to the CB and no current can be achieved. In conductors, either there is no bandgap, so VB and CB are overlapping, or the CB is (independent of the external energy input) already occupied by electrons. Electrical current is enabled in both cases. Because only semiconducting molecules facilitate (externally controllable) isolating as well as conductive properties, they are the most important, active components in electronic devices, for example, in FETs, light-emitting diodes, photovoltaic cells, and so on. Enormous ongoing research efforts will surely continue within the next few decades. A good case can be made for pentacene **6** and rubrene **2**. These benzene-based, low-molecular weight semiconductors yield high charge carrier mobilities[4,5] necessary for fast switching in electronic devices. Because the basic chemical structure of pentacene (R = H) is not soluble in common organic solvents, different strategies for increasing the solubility while preserving the electronic benefits had to be developed to enable the cost-efficient fabrication of these so-called "printed electronics."

The most prominent pentacene derivative (R = triisopropylsilylethynyl), which provides good solubility and excellent high-charge transport, was developed in 2001.[6,7] Analogue substitutions of conjugated molecules are, in the meantime, common practice because they do not only allow for control of the solubility and thus diverse solution-processing techniques, but also enable precise manipulation of the supramolecular organization or morphology and thus of the (opto)electronic properties ("Supramolecular Structures" section). Extensive research has likewise been dedicated to the structure–property relationships of the polycyclic aromatic hydrocarbon hexabenzocoronene **3** (HBC).[8,9] Some alkyl chain–substituted HBCs exhibit a thermotropic liquid crystalline phase. In this regime, the material is neither solid nor liquid but combines the optical and electrical characteristics of crystals with the flow behavior of fluids and allows further processing/orientation steps (e.g., zone-melting, "Processing-Aided Assembly" section) or application in LCDs ("Liquid Crystalline Displays" section). Note that the impact of the substituents on supramolecular organization, whether the alkylation might be short or long, linear or branched, always strongly depends on the molecule's core structure. Next to the benzene-based compounds, thiophene-containing semiconductors have been proven to provide excellent properties in organic electronic devices as well. This is mainly because of their electron-rich character. Dithiophene-tetrathiafulvalene **9** is an example of a very sulfur-rich semiconductor that benefits, on the one hand, from a high electron density, and on the other hand, from a high crystallinity and large overlap of the diffuse electron orbitals of the sulfur atoms, which lead to a high degree of electron delocalization and thus excellent charge transport characteristics.[10] Sexithiophene **5** ($n = 6$, R = H) and poly-3-hexylthiophene **5** ($n > 30$, R = $C_6H_{13}$) (P$_3$HT) are also distinguished representatives of the thiophene family. A challenge in synthesizing semiconducting polymers like polyalkylthiophenes is the achievement of a high degree of regioregularity. Head-to-tail regioregular polymers exhibit enhanced microstructure and crystallinity in the solid state and thus improved (opto)electronic properties.[11] Furthermore, regioregularity leads to a lower bandgap between the highest occupied molecular orbital (HOMO) and the lowest unoccupied molecular orbital (LUMO), thus enabling easier excitation of electrons. Nevertheless, electron-rich oligo- and polythiophenes exhibit a high HOMO energy, which yields a stronger oxidation tendency and thus a stability problem. The substitution of thiophene by phenyl rings was found to lower the HOMO and yield improved stability of the devices under test. An interesting approach has been developed with the combination of the molecular shape of pentacene and thiophene end groups. This strategy led to semiconductors such as anthradithiophene (or "dithiapentacene") and tetrathiapentacene **7**. Next to enhanced stability, thiophene end groups enable the attachment of solubilizing substituents within the molecule's long axis. For example, tetrathiapentacene **7** bearing two hexyl-chains (R = $C_6H_{13}$) is soluble in common organic solvents and forms highly crystalline films with increased charge-carrier mobility. According to the outstanding properties of thiophene-based semiconductors, selenophene- and tellurophene-containing compounds, which exhibit even larger electron orbital overlap, have also been successfully investigated.[12] Furthermore, nitrogen has been used as a heteroatom in conjugated molecules as exemplified by polypyrrole **8** and polytriarylamine **11**. Herein, the free electron pair at the nitrogen yields an increase of the electron density, thus an amplification of the electron-donating tendency of the molecule (the so-called donor). Polypyrrole **8** is a conductive polymer, which has mainly been used as a microwave-absorbing, radar-invisible coating and as an active layer in sensors.[1] Polytriarylamine **11** shows semiconducting characteristics and has been applied as an active layer in FETs, light-emitting diodes, and photovoltaic cells.[13] Another part of current research focuses on the semiconducting properties of metallophthalocyanines **13**, the most popular representative of which is copper-phthalocyanine. Next to its use as an active layer in FETs,[14] its most spectacular application, and a great example of the capability of organic electronics, was presented in 2005. The authors reported the manufacture of a pressure- and temperature-sensitive artificial skin based on a combination of copper-phthalocyanine and pentacene **6** or perylenetetracarboxydiimide **12**.[15] Perylenetetracarboxydiimide, naphthalenetetracarboxydiimide **15**, and derivatives are nitrogen-containing semiconductors which, in contrast with the above-mentioned compounds, exhibit an electron-poor character (the so-called acceptors).[16] Fullerene ($C_{60}$) **10** is an acceptor as well. Although holes are the majority charge carriers in donor molecules, electron transport is preferred in acceptors. Both electron and hole transport layers are necessary for the functionality of light-emitting diodes, photovoltaic cells, and complementary electronic circuits. An extension of the perylenetetracarboxydiimide's rylene core structure yields quaterylenetetracarboxydiimide **16** (QDI), which is also an electron acceptor. Dependent on the processing-related film morphology, ambipolar transport (electrons and holes at the same time) have been achieved for QDI.[17] In general, ambipolar transport materials are considered to be highly economical because layers of this single component enable complex electronics which, in the case of unipolar transport materials, require at least two components as well as multiple processing steps. Nevertheless, in case of a different processing technique and film structure, QDI yielded electron transport only, thus emphasizing the impact of morphology on the electronic properties of the semiconductor film. Cyclopentadithiophene-benzothiadiazole copolymer **14** consists of alternating donor and acceptor components. Because of the presence of the electron-rich and electron-poor units, a high degree of intermolecular order is established, which leads to excellent charge-carrier mobility. The degree of the supramolecular order and the corresponding transport characteristics were found to be dependent on the length of the macromolecules. Herein, increase of the molecular weight yielded enhanced supramolecularity and thus improved charge-carrier mobility in FETs.[18] A large number of such copolymers as well as semiconductor blends have been developed and successfully tested within the last few years.[19] However, there is still a myriad of more promising, combinatorial possibilities for well-directed syntheses of novel semiconductors, which provide desired properties but require a profound knowledge of the (molecular and supramolecular) structure–property relationship. The organization principles of molecules and the generated supramolecular structures are therefore discussed in detail in the "Supramolecular Structures" section.

## SUPRAMOLECULAR STRUCTURES

An ideal charge transporting semiconductor film demands, on the one hand, a pronounced conjugation within the molecules (intramolecular) and, on the other hand, a high degree of intermolecular $\pi$-overlap. A detailed description of intermolecular, noncovalent interactions would exceed the volume of this section, which is why the reader is referred to the literature.[20–22] Additional and equally important factors are given by the spatial arrangement of the $\pi$-stacking direction, so the electronic coupling's anisotropy and by large-scale supramolecular structures as they are generated during film formation (Figure 9.3). In the first level of Figure 9.3, the most common $\pi$–$\pi$ interaction-based molecule–molecule geometries are illustrated. The typical packing mode of low-molecular weight semiconductors usually varies between cofacial (face-to-face) $\pi$-stacking (Figure 9.3, 1a) with different degrees of offset and a brick-like geometry (Figure 9.3, 1c). Cofacial packing with molecular offsets close to zero is assumed to exhibit the most effective charge transport because of its large $\pi$-overlap and short intermolecular distance, which benefits charge carrier hopping from one molecule to another. The major drawback of the zero offset cofacial arrangement (Figure 9.3, 1a) is the $\pi$-stacking disruption within the generated columns, which act as dead ends for charge transfer. In addition, the intercolumnar conjugation is generally very weak and creates a large electronic barrier for intercolumnar charge transfer, which hinders the circumvention of columnar defects. In comparison with zero offset cofacial assemblies, offset or face-to-edge structures exhibit enhanced intercolumnar charge transport.

**FIGURE 9.3** Supramolecular structures in π-conjugated molecules. 1, Molecule–molecule geometries: (a) face-to-face π-stacking, (b) herringbone structure as result of edge-to-face and offset geometry, and (c) brickwork structure as result of the offset-arrangement. 2, Molecule-interface geometries: (a) edge-on, (b) face-on, and (c) arrangement of the conjugated long-axes of macromolecules (here, parallel to the substrate). 3, Large-scale film structures: (a) grain boundaries, (b) regions of dewetting, and (c) film gaps.

The tendency of molecules to pack in face-to-edge mode oftentimes yields the so-called herringbone structure, whose popular representatives are pentacene and rubrene (Figure 9.2, R = H), both high-performance semiconductors. For brickwork structures, such as triisopropylsilylethynyl-anthradithiophene (**22a** in Figure 9.11), and slightly offset arrangements, such as triisopropylsilylethynyl-pentacene (**19a** in Figure 9.11) and bis-silylethynylanthradithiophene (**22d** in Figure 9.11), excellent charge transport characteristics have also been measured. These examples highlight the important impact of intercolumnar charge transport on the performance of organic electronic devices.

Single crystals of organic semiconductors usually exhibit anisotropic charge-carrier mobility with the highest values in the direction of the maximum π-overlap. This behavior requires knowledge of the arrangement of the molecules/columns with respect to the device plane (level 2 of Figure 9.3). Dependent on the device species (if the charge transport is intended to occur parallel or perpendicular to the substrate) either the molecular edge-on (Figure 9.3, 2a) or the face-on (Figure 9.3, 2b) orientation is preferred. In case of macromolecular semiconductors, in which the charge transport is preferred along the polymer-chain, the orientation of the molecular long-axis has a similar effect on the device's performance (Figure 9.3, 2c). Control of the direction of best charge transport can be achieved by clever manipulation of semiconductor–substrate interactions, that is, surface modification of the substrate, as well as by the application of specific semiconductor processing techniques. How the preferred charge transport direction has to be adjusted to provide optimal functionality to the different device configurations is discussed in detail in the respective device's section.

Because of the economic constraints of miniaturization, semiconductor films have to span areas in the micrometer regime. For both low-molecular weight and polymer semiconductors, the manufacture of micrometer-large, highly oriented monodomains turned out to be an intractable challenge. The mean size of crystal domains is of course dependent on the chemical structure but mostly about just a few hundred nanometers. This inevitably leads to the formation of grain/domain boundaries (Figure 9.3, 3a), which represent charge transport barriers based on scarce π–π interactions between the domains. In general, domain boundaries are classified as film defects, to which regions of dewetting (Figure 9.3, 3b) and film gaps (Figure 9.3, 3c) also belong. These rather macroscopic structures can be influenced by surface chemistry and thus by semiconductor–substrate interactions and by the type of the semiconductor's film formation technique.

## PROCESSING-AIDED ASSEMBLY

Supramolecular structures of π-conjugated molecules are not only dictated by molecular interactions, but also by the processing technique used for the active layer formation in organic electronic devices. In principle, one can differentiate between six types of processing such as that illustrated in Figure 9.4.

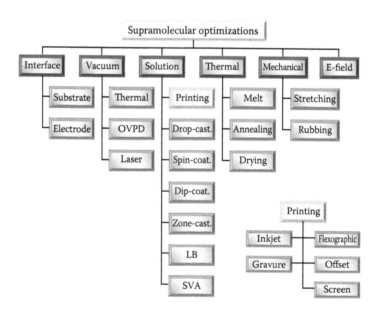

**FIGURE 9.4**  Film optimization techniques which enable manipulation of the supramolecularity within the active layer of organic electronic devices (OVPD, organic vapor phase deposition; Cast., casting; Coat., coating; LB, Langmuir–Blodgett film formation; SVA, solvent vapor annealing).

Manipulation of the semiconductor layer via modifying the semiconductor–substrate/electrode interface generally relies on the adjustment of the substrate/electrode surface roughness and chemical nature of the semiconductor and substrate/electrode materials. Rough surfaces induce, on the one hand, an extended contact area; on the other hand, they yield a decrease of the molecular diffusion length during deposition and a reduction of the nucleation energy. Both factors result in a multicrystalline semiconductor film morphology and thus in a multitude of grain boundaries which lead to a strongly hindered charge transport.[23] Hence, it is aimed for surfaces/interfaces which are as planar as possible. In contrast, preoriented substrates (specified roughness) have been used to structurally modify the semiconductor layer by surface-guided assembly of the molecules. This technique is mainly used for the fabrication of the active layer in LCDs (as light guidance; see "Liquid Crystalline Displays" section) but has been applied for the generation of preferred charge transfer orientations as well.[24–27]

Chemical modifications of the substrate/electrode surfaces are mainly executed for the optimization of charge transport along/through the interfaces.[28–30] The interactions between the chemical surface groups of the substrate and the semiconductor molecules have also been found to yield variations in the supramolecular structure of the semiconductor film. This principle has become a common tool of present-day interface engineering[31–33] and is exemplified in Figure 9.5.

Semiconductor films are mainly formed by vacuum or solution-based processing techniques. Vacuum processing, for example, thermal evaporation/sublimation, organic vapor phase deposition, and laser-aided deposition, enables control of the film supramolecularity by adjusting the deposition rate and substrate temperature. Herein, the semiconductor grain size proceeds inversely proportional with the deposition rate. Evaporation under ultra-high vacuum conditions is sometimes called organic molecular beam deposition or organic molecular beam epitaxy. At very low deposition rates, this technique, in combination with *in situ* structure analysis, even allows for research into film growth mechanisms with monolayer resolution. In contrast to vacuum deposition methods, solution processing does not require expensive equipment and long pumping times for an appropriate vacuum level, thus it offers a low-cost and fast film formation. Furthermore, control of morphology can be gained by variation of solvent, solution concentration, rheological additives, and precise regulation of the solidification zone. From an economical point of view, especially the continuous mass printing techniques, such as flexographic, gravure, inkjet, offset, and screen printing, are the focus of interest,[34,35] whereas film formation via drop-casting, spin-coating,[36–38] dip-coating,[38–40] zone-casting,[41] or the Langmuir–Blodgett technique[42,43] is mainly applied in research laboratories. Drop-casting and spin-coating are most commonly used. Film formation by drop-casting requires the evaporation of solvent and thus represents a relatively slow process in comparison to spin-coating, in which the film is rapidly formed under high centrifugal force. Because of the difference in film formation, drop-casting generally yields more crystalline domains within the semiconductor layer, whereas less crystalline but macroscopically more homogeneous films, especially from polymer solutions, are achieved by spin-coating. Dip-coating, zone-casting (or solution shearing[44]), and Langmuir–Blodgett deposition allow

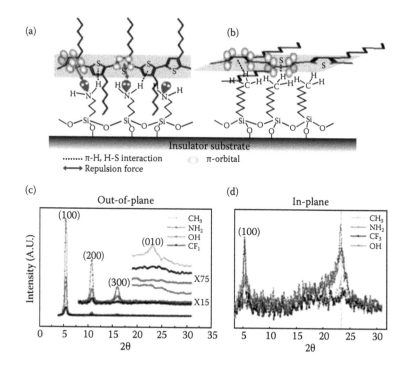

**FIGURE 9.5** Schematics of P$_3$HT (5 with R = C$_6$H$_{13}$, Figure 9.2) chain conformations, (a) edge-on and (b) face-on, according to interfacial characteristics. (c) Out-of-plane and (d) in-plane grazing incidence angle x-ray diffraction intensities as a function of the scattering angle 2θ for regioregular P$_3$HT thin films crystallized on insulator substrates modified with different chemical surface groups. (Kim, D. H. et al.: *Adv. Func. Mater.* 2005. 15. 77. Copyright Wiley-VCH Verlag GmbH & Co. KgaA. Reproduced with permission; Reprinted from *Mater. Today*, 10, Park, Y. D. et al., 46, Copyright 2007, with permission from Elsevier.)

for precise control of supramolecular orientations with high anisotropy and huge crystalline domains because of the regulation of the solidification zone. Herein, the liquid–solid phase transition occurs at the meniscus between solution and substrate such as that demonstrated by the illustration of the zone-casting technique in Figure 9.6.

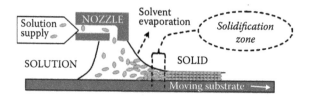

**FIGURE 9.6** Schematic presentation of the zone-casting technique. (Reprinted from Tracz, A. et al.: *Mater. Sci. (Poland)*, 22, 415, 2004. With permission from Oficyna Wydawnicza Politechniki Wrocławskiej.)

Active layers can be structurally modified even after the film formation has already been finished. Solvent vapor annealing[45,46] is one such postprocessing method whereby amorphous to crystalline reorganization or polymorphous transformations are initiated. Therefore, solvents are commonly used in which the film material is just weakly soluble. Similar changes of the supramolecular film structure have been obtained after thermal annealing as well. Herein, the annealing process has to be clearly distinguished from the low-temperature drying process (which is normally applied for removing residual solvent molecules) and the high-temperature melt-processing. Solvent vapor annealing and thermal annealing are stationary processes which do not allow the generation of a preferred orientation within the semiconductor film. This is enabled by the zone-crystallization technique,[47] in which highly anisotropic films can be achieved out of the material's molten or liquid crystalline state. In general, processing of a material within its liquid crystalline, thermotropic phase can be described as a kind of self-healing mechanism,[48] in which local disorders or defects are repaired because of thermally activated molecular rearrangements.

Semiconductor films have also been mechanically oriented by the application of tension or shear forces, such as, stretching[49,50] or rubbing techniques.[51] Both treatments have been successfully applied to polythiophenes in FETs and yielded high gear anisotropy of the charge-carrier mobility with up to 800% enhanced values in the direction of the applied force. The most popular use of mechanical orientation is rubbing used within the fabrication of LCDs. Herein, the liquid crystals are oriented with respect to the preoriented substrate, which yields an improved display quality and allows for lower voltage input. The electrical field applied during the use of LCDs also effects the crystal orientation and thus the light guidance through the films, which then yields the voltage-controlled display information. Electrical fields can be used for supramolecular structuring during solution-based film formation as well. Herein, the applied field induces a dipole within the molecules, which then become aligned in the direction of the highest field gradient and retains this orientation, whereas the solvent evaporates and the film becomes solid. This method has been successfully applied for the orientation of $P_3HT$ (5 in Figure 9.2; $n > 30$, $R = C_6H_{13}$).[52]

## SUPRAMOLECULAR STRUCTURES IN ORGANIC ELECTRONIC DEVICES

In the following pages, an overview of current organic electronic devices in which designed supramolecular interactions are used in the manufacture of the active layer is presented. The devices are subdivided according to application, which focus on supramolecular materials for FETs, LCDs, light-emitting diodes, photovoltaic cells, and additional electronic devices such as electrochromic layers, lasers, sensors, fuel cells, and batteries.

### FIELD-EFFECT TRANSISTORS

FETs are basic elements of electronic circuits and one of the most important devices in the electronics industry. The conventional application of Si/GaAs as a semiconductor

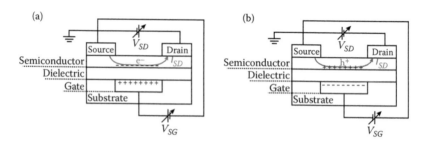

**FIGURE 9.7**  OFET function: (a) $V_{SD}$, $V_{SG} > 0$ V results in electron transport "n-type"; and (b) $V_{SD}$, $V_{SG} < 0$ V results in hole transport "p-type."

layer in FETs requires high-temperature fabrication steps and yields mechanically steep and opaque devices as they are integrated in present-day computer boards. FETs based on organic semiconductors can be fabricated at ambient temperature, which is why on the one hand the manufacture of organic FETs (OFETs) is more cost-efficient, and on the other hand, a multitude of materials like flexible polymer foils can be used as substrates. Therewith, OFETs widen the practical application of transistors by large-area, flexible, and (semi)transparent devices.

An OFET basically consists of gate electrode, dielectric, and semiconductor layers in a sandwich-like structure. Furthermore, source and drain electrodes are directly connected to the semiconductor film, as shown schematically in Figure 9.7. OFETs are usually operated in accumulation mode, with the source always grounded. With an applied source-drain potential, $V_{SD}$, but no gate voltage, $V_{SG}$, the intrinsic conductivity of most organic semiconductors is low and the device is in the off-state. When a gate voltage, $V_{SG}$, is applied, a potential gradient between gate and source is built, thus yielding the accumulation of charges at the dielectric–semiconductor interface. Those charges are mostly mobile and lead to the formation of conducting channels between source and drain. As a consequence, these charges move in response to the applied $V_{SD}$ thus generating the transistor current $I_{SD}$. The transistor is in the on-state.[53] When an OFET is active, on the application of positive $V_{SG}$ and $V_{SD}$, the organic material is said to be n-type because electrons are the majority charge carriers (Figure 9.7a). On the other hand, when a (negative) SD current is observed on the application of negative $V_{SG}$ and $V_{SD}$, the semiconductor is p-type because the holes are mobile (Figure 9.7b). In a few cases, OFETs operate for both $V_{SG}$ and $V_{SD}$ polarities and the semiconductor is said to be ambipolar. It is important to stress that the categorization of an organic semiconductor as n-type or p-type has no absolute meaning but is strongly related to the FET device structure/material combination on which the transport characteristics are measured.[54] To characterize the transistor behavior, a constant $V_{SG}$ can be chosen and $I_{SD}$ measured as a function of $V_{SD}$. The current in the saturation region ($I_{SD,sat}$), that is, where $I_{SD}$ no longer increases linearly with $V_{SD}$, can be described by

$$I_{SD,sat} = \frac{W}{2L} \mu_{sat} C_i \left( V_{SG} - V_T \right)^2,$$                    (9.1)

with channel width $W$, channel length $L$, capacitance per unit area of the gate insulator $C_i$, the saturated charge-carrier field-effect mobility $\mu_{sat}$, and the threshold voltage $V_T$. $V_T$ is a parameter that describes the nonideal behavior of OFETs, that is, for cases in which channel conductance is still considerably large for $V_{SG} = 0$ V or in which channel conductance is still low for $|V_{SG}| \gg 0$. $V_T$ can be positive or negative, and includes effects related to unwanted doping, charge carrier trapping, mismatches of energy levels at the organic/contact interfaces, or (supra)molecular connectivity. The ratio between the transistor current $I_{SD,sat}$ at maximum $V_{SD}$ and $V_{SG}$ (so-called on-current $I_{ON}$) and the transistor current at $V_{SG} = 0$ V (so-called off-current $I_{OFF}$), that is, $I_{ON/OFF}$, presents the third major evaluation factor of FETs. To achieve acceptable performance, that is, $\mu_{sat} > 0.1$ cm$^2$ V$^{-1}$ s$^{-1}$, $V_T$ close to zero and $I_{ON/OFF} > 10^6$, organic semiconductors must satisfy general criteria regarding both the injection and current-carrying characteristics, in particular[54]:

i. The molecules should exhibit continuous conjugation for efficient intramolecular charge transport and the whole material itself should be extremely pure because impurities can act as trapping sites and hinder charge carrier transport.

ii. The crystal structure of the material must provide sufficient overlap of frontier orbitals to allow efficient intermolecular charge transfer.

iii. The HOMO/LUMO energies of the individual molecules must be at levels at which holes/electrons can be induced at accessible applied electric fields.

iv. Low-molecular weight semiconductors should be preferentially oriented with their $\pi$–$\pi$ stacking direction, polymeric semiconductors with their long axis, parallel to the FET substrate and in the direction of the transport channel because the most effective charge transport occurs along the $\pi$–$\pi$ stacking axis in low-molecular weight semiconductors and along the polymeric backbone in macromolecular semiconductors.

v. The crystalline domains of the semiconductor must cover the area between the source and drain electrodes uniformly; hence, the film should possess a single domain-like morphology to avoid domain boundary-founded charge transport barriers.

Perhaps the most significant advances in the understanding of the intrinsic electronic properties of organic materials have arisen with the development of efficient methods to grow large, high-quality single crystals of rubrene **2** (Figure 9.2), allowing the fabrication of semiconductor devices directly on their surfaces. Fabrication of transistors on flexible stamps facilitates the removal and replacement of rubrene crystals without degradation of electronic properties. Simple rotation of the crystal on the device enabled the measurement of the hole mobility across a large sampling of molecular orientations. Mobility was highest along the molecular $\pi$-stacking axis and lowest perpendicular to this axis (where interactions are predominantly edge-to-face; Figure 9.8).[55]

The understanding generated by such fundamental research led to the development of a large number of low-molecular weight and polymeric semiconductors

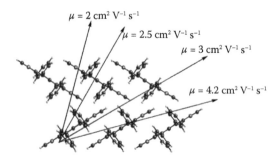

$\mu = 2 \text{ cm}^2 \text{ V}^{-1} \text{ s}^{-1}$

$\mu = 2.5 \text{ cm}^2 \text{ V}^{-1} \text{ s}^{-1}$

$\mu = 3 \text{ cm}^2 \text{ V}^{-1} \text{ s}^{-1}$

$\mu = 4.2 \text{ cm}^2 \text{ V}^{-1} \text{ s}^{-1}$

**FIGURE 9.8**  Mobility in single crystals of rubrene as a function of source–drain electrode orientation. (Anthony, J. E.: *Angew. Chem. Int. Ed.* 2008. 47. 452. Copyright Wiley-VCH Verlag GmbH & Co. KGaA. Reproduced with permission.)

since the beginning of the 21st century (see examples in Figure 9.2). Many of these new materials yielded excellent charge carrier mobilities. Values up to 35 cm² V⁻¹ s⁻¹ were reported for highly pure pentacene **6** (Figure 9.2, R = H) single crystal OFETs and impressive values higher than 1 cm² V⁻¹ s⁻¹ have been measured for vacuum-deposited pentacene transistors. As the performance of these organic semiconductors already surpasses that of amorphous silicon, it is not surprising that current research in this field has focused considerably on the engineering of acene derivatives toward more stable and easier-to-process materials. For the production of low-cost, flexible electronic devices for mass applications, simple solution casting techniques are more attractive than vacuum deposition technology. In recent years, efforts have been made to use soluble precursors of acenes, in particular of pentacene, enabling the deposition of pentacene from solution. The preparation of such precursors is, however, synthetically challenging, and a technically demanding annealing process is required to realize high-order thin films. Alternatively, semiconducting molecules may be structurally modified to achieve better solution-processability, and thus, a preferred supramolecular order in the solid state via the enabled application of techniques for processing-aided assembly (see related section). Remarkably, substituted acene derivatives bearing two trialkylsilylethynyl groups at the central ring from the smallest member, the anthracene **17** up to the heptacene **21** have been developed (Figure 9.9).[55,56] The molecular structures differ only slightly from each other, but allow to derive important structure–property relationships because the new bis-silylethynylated acenes not only exhibit improved solubility and solution-processibility but also show considerable differences in the supramolecular order and thus in the electronic behavior.

For example, the parent pentacene **6** (Figure 9.2, R = H) falls into the class of herringbone structures owing to the dominance of π - σ attraction (see "Supramolecular Structures" section). In contrast, for many silylethynylated pentacenes and dithienoacenes, one-dimensional offset stack (e.g., dithienoanthracene **22a**) or two-dimensional brickwork arrangements (e.g., pentacene **19a** and dithienoanthracene **22d**) have been observed. For solution-processed bis-silylethynylated pentacene **19a**, charge carrier mobilities of up to 0.2 cm² V⁻¹ s⁻¹ could be measured. FETs

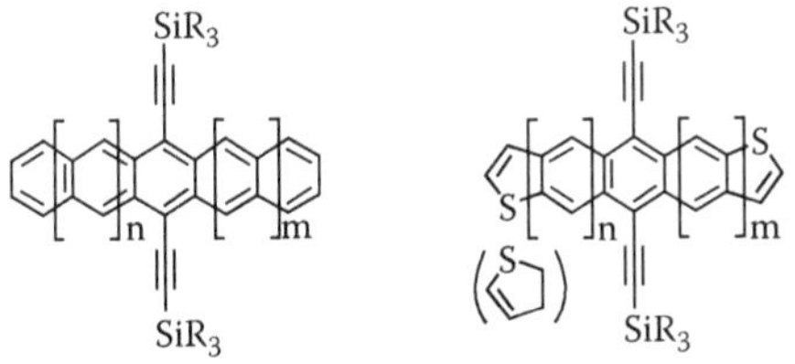

**17**: n, m = 0
**18**: n = 0, m = 1
**19**: n, m = 1
**20**: n = 1, m = 2
**21**: n, m = 2

**a**: R = i–Pr
**b**: R = t–Bu
**c**: R = $(Me)_3Si$
**d**: R = Et
**e**: R = Me

**22**: n, m = 1
**23**: n = 1, m = 2
**24**: n, m = 2

**FIGURE 9.9** Novel bis-silylethynylated acene and dithienoacene derivatives. (From Würthner, F., Schmidt, R., *ChemPhysChem*, 7, 793, 2006. With permission.)

based on solution-processed dithienoanthracene **22d** yielded mobilities higher than 1 $cm^2 V^{-1} s^{-1}$, on/off current ratios of $10^7$ and threshold voltages of approximately 5 V,[7] improved oxidative stability, and favorable HOMO energy for the injection/extraction of holes. Although the acquired charge carrier mobilities of these structurally modified acenes undermatch the values of vacuum-processed pentacene **6** (Figure 9.2, R = H), they are still sufficiently high for application in commercial products. Thus, dithienoanthracene **22d** can be considered as a prime example of a rationally designed organic electronic material. The influence of the supramolecular ordering in the solid state is further demonstrated by dithienoanthracene **22e**, which does not pack properly. For transistors based on **22e** no charge carrier mobility was achieved. This demonstrates that finding the best substituents to direct crystal packing is still a matter of trial and error.

With benzobisbenzothiophene **25** and dithienobenzodithiophene **7**, the consequent continuation of the acene studies has been represented (Figure 9.10). Both molecules allow for the evaluation of the impact of the substituent position on the supramolecular order because, in contrast with the above described acene derivatives, substitution within the molecule's long axis could be synthetically achieved. The packing modes for both molecules, that is, herringbone structure[58] for **25** and offset motif[40] for **7**, were found to remain stable, independent of the substitution size.

The best charge carrier mobilities were measured for solution-processed **7c** with 4.2 $cm^2 V^{-1} s^{-1}$ and 0.3 $cm^2 V^{-1} s^{-1}$ for **25b**, both with on/off current ratios of $10^6$. Substituting each of the core structures from either molecule with longer alkyl

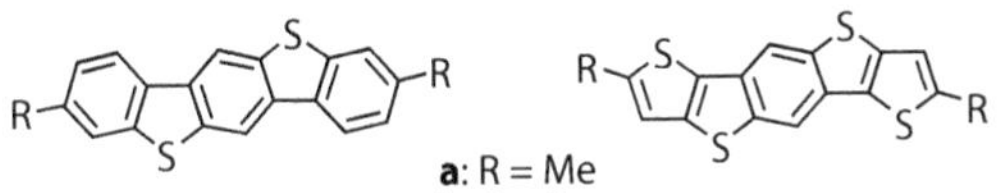

**25**

**a**: R = Me
**b**: R = Bu
**c**: R = Hex
**d**: R = Non

**7**

**FIGURE 9.10** Rod-shaped benzothiophenes.

chains, on the one hand, leads to enhanced solubility and solution-processability and, on the other hand, increases the nonconductive material fraction within the active transistor layer. Shorter alkylation also yields downgraded transport characteristics because of inhomogeneous film formation due to low solubility and thus hindered processing. The extracted very high-mobility of **7c** is attended by a very high threshold voltage of $V_T = -75$ V, which points toward severe charge carrier trapping problems (the origin of the underlying mechanism is still not fully clarified), and surely avoids application in low-power handheld devices.

Supramolecular order and electronic properties crucially depend on the solution-based processing technique, especially in highly crystalline, low-molecular weight semiconductors. Different film formation types may result in charge transport variations of a few decades. In Figure 9.11, charge carrier mobilities as a function of crystal quality (full-width at half-maximum of the 100-Bragg reflex) in **7c** films are illustrated.

Dip-coating solutions of **7c** yields several square millimeters of broad crystal domains and allows for uninterrupted π–π connection of the source and drain electrode. Because of the absence of grain boundaries, almost ideal charge transport is enabled, which is why one of the highest OFET mobilities have been recently achieved (4.2 cm$^2$ V$^{-1}$ s$^{-1}$).

A few years ago, similar processing-aided assembly had already been developed for HBC **3** by drop-casting HBC solution onto preoriented polytetrafluoroethylene or by application of the zone-casting technique. Parallel to oriented HBC columns (face-to-face motif with zero offset) and along the π-stacking direction, charge carrier mobilities of 10$^{-3}$ cm$^2$ V$^{-1}$ s$^{-1}$ have been achieved, whereas only 10$^{-5}$ cm$^2$ V$^{-1}$ s$^{-1}$ were measured in unaligned films.[27,59] In comparison to molecules **6**, **7**, and **17–25**, the relatively low transport characteristics of HBCs follow from intracolumnar defects as depicted in Figure 9.12. In general, intracolumnar transport barriers have to be overcome by redirecting charge carriers into neighboring columns, which most often vanish because of the isolating properties of the solubilizing substituents.

A supramolecular approach afforded highly efficient n-channel organic semiconductor materials has been recently presented for thermally evaporated,

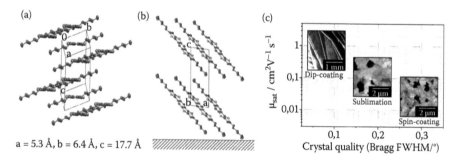

a = 5.3 Å, b = 6.4 Å, c = 17.7 Å

**FIGURE 9.11**   7c-based films in FETs: (a) crystal structure, (b) orientation of the crystals with respect to the device plane, and (c) charge-carrier mobility as a function of the crystal quality, which has been defined as full-width at half-maximum of the (100) Bragg reflexes. The pictures in *c* represent the surface topography of the active layer.

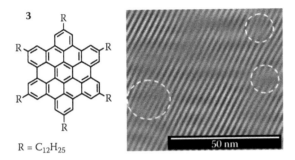

3

R = C₁₂H₂₅

50 nm

**FIGURE 9.12** Chemical structure of dodecyl-substituted HBC (HBC-C$_{12}$) with an inverse FFT-filtered HRTEM image of a zone-cast HBC-C$_{12}$ film. (Pisula, W. et al.: *Adv. Mater.* 2005. 17. 684. Copyright Wiley-VCH Verlag GmbH & Co. KGaA. Reproduced with permission.)

chlorine-modified perylenetetracarboxydiimides.[60] Herein, the combination of hydrogen bonding and contortion of the π core directs two-dimensional π–π stacked percolation paths yielding $\mu_{sat} = 0.9$ cm$^2$ V$^{-1}$ s$^{-1}$, $V_T = 28$ V, and $I_{ON/OFF} = 10^7$.

In another study, ambipolar devices were realized by simply drop-casting a solution of a quaterrylenetetracarboxdiimide.[17] In ambipolar transistors, both electrons and holes are transported. It is commonly believed that organic semiconductors are generally conductors of both charge carrier types, and that their mostly unipolar behavior is caused by interface trapping of one of the carrier types or by a too-high charge injection barrier resulting from a drastic mismatch of the HOMO and LUMO energy levels with the work function of the source and drain electrodes. Reducing the HOMO–LUMO bandgap and thus the injection barrier in case of metal electrodes with the Fermi level lying inside this bandgap is one way toward the realization of ambipolar devices. The swallow-tailed quaterrylenetetracarboxdiimide **16** molecules (SWQDI) used as an active layer in thin-film transistors exhibited quite a low bandgap of approximately 2 eV and clean ambipolar device performance with an electron and hole mobility of 10$^{-3}$ cm$^2$ V$^{-1}$ s$^{-1}$. However, after thermally annealing these samples, hole transport was lost, giving rise to a solely n-type performance of 10$^{-4}$ cm$^2$ V$^{-1}$ s$^{-1}$. Although no changes of the HOMO level and no increased hole trapping at the SWQDI/dielectric interface was detected, the observed absence of p-type behavior was assigned to the thermally induced change in film morphology. No macroscopic order in as-cast films, but improved supramolecular order in thermally annealed films, could be detected. It was assumed that heating the devices triggered the formation of ordered columnar structures, leading to local well-organized domains within the **16** layer, as schematically illustrated in Figure 9.13. The emergence of such a polycrystalline-like structure went hand in hand with the formation of local macroscopic domain boundaries. Thus, the dramatic decrease of hole mobility was attributed to charge-carrier trapping at those boundaries. This lowering of hole transport has been further reinforced by the appearance of one-dimensional columns, which are considered to be particularly sensitive to structural defects, ultimately causing the loss of p-type behavior. These limiting factors are expected to equally affect electron current. However, because **16** has a strong affinity

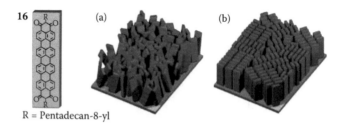

R = Pentadecan-8-yl

**FIGURE 9.13** Chemical structure of SWQDI and schematic illustration of the supramolecular organization before (a) and after (b) annealing. (Reprinted from Tsao, H. N. et al.: *Adv. Mater.* 2008. 20. 2715. Copyright Wiley-VCH Verlag GmbH & Co. KGaA. Reproduced with permission.)

for electrons, the latter was able to diffuse through the macroscopic and microscopic defects within the columns, resulting in the observed 10 times decrease of electron mobility but not in the entire absence of n-type performance.

A comprehensive study has been performed on OFETs based on cyclopentadithiophenebenzothiadiazole copolymer **14**. First, the authors focused on proper processing techniques that could be used to achieve enhanced long-range molecular order such as that previously described for low-molecular weight semiconductors. As depicted in Figure 9.14a, the as-spun polymers form a film, with spot-like features of average size approximately 50 nm diameter. Interestingly, these small domains consist of rings, as indicated by the arrows in Figure 9.14a, suggesting that the polymer chains self-assemble into circular structures when spin-coated (see illustration Figure 9.14a). Such donut-shaped nanostructures lie next to each other, establishing a polymer network that seems to provide efficient charge-carrier transport with a close intermolecular π-stacking distance of 0.37 nm and a high mobility of 0.67 cm² V⁻¹ s⁻¹. Conversely, for the dip-coated polymers, a film containing a fibrous morphology (Figure 9.14b) was present instead of the circular assembly. In fact, a significant

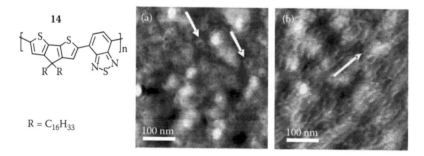

**FIGURE 9.14** Chemical structure of cyclopentadithiophenebenzothiadiazole copolymer **14** and the obtained film topology (AFM) of the polymer processed by (a) spin-coating and (b) dip-coating. The arrows in (a) highlight the donut-like structures of spin-coated **14** films. The arrow in (b) indicates the dip-coating direction. (Reprinted from Tsao, H. N. et al.: *Adv. Mater.* 2009. 21. 2715. Copyright Wiley-VCH Verlag GmbH & Co. KGaA. Reproduced with permission.)

number of the fibers have been found to be aligned parallel to the dip-coating direction, indicated by the arrow in Figure 9.14b, suggesting that the plane of conjugation of the polymers lies parallel to this direction as well. Hole mobilities in the range of 1.0 to 1.4 cm$^2$ V$^{-1}$ s$^{-1}$ were measured along the dip-coating direction, reflecting charge transport through the plane of conjugation and representing the highest mobilities reported thus far for polymer-based OFETs. Measurements perpendicular to the dip-coating direction, in this way along the π-stacking of the polymer backbones, revealed a lower mobility of 0.5 to 0.9 cm$^2$ V$^{-1}$ s$^{-1}$.[18]

More research on **14** focused on the influence of the molecular weight on the film crystallinity, morphology, and corresponding transistor performance. The charge-carrier mobility was found to scale proportionally with the molecular weight, yielding ultra-high record mobilities of up to 3.3 cm$^2$ V$^{-1}$ s$^{-1}$ ($V_T$ = −20 to −30 V; $I_{ON/OFF}$ = 10$^5$–10$^6$).[61] This enhancement has been addressed to increasing π-stacking interactions of the planar extended conjugated backbones; thus, increasing film crystallinity with the molecular weight. Herein, the π-stacking distance was observed to be independent of the molecular weight and minor changes in alkylation. Furthermore, it has been stressed that the influence of the film morphology on the transistor performance is not necessarily proportional with the effect of the film crystallinity even in the case of equally processed polymers which only differ in molecular weight ($M_n$). This phenomenon can be attributed to differences in the crystalline domain size, and thus the influence of domain boundaries. Films of the high-performance naphthalenetetracarboxydiimide–dithiophene copolymer **26** (Figure 9.15) exhibit similar fiber-like morphologies, as demonstrated for **14** in Figure 9.14b, but only negligible crystallinity.[62] On the one hand, this leads to the absence of domain boundaries and thus enhanced charge carrier transport; on the other hand, it is hard to understand (and widely discussed at the moment) how noncrystalline polymer films with hindered intermolecular charge transport enable mobilities close to 1 cm$^2$ V$^{-1}$ s$^{-1}$ ($V_T$ = 5–10 V; $I_{ON/OFF}$ = 10$^6$–10$^7$), when keeping in mind that a single molecule is definitely too short to interconnect the source and drain electrode.

However, from most examples described in this section, there is one great outcome. It is apparent that for the fabrication of high-performance OFETs, control of organization of the organic compound plays a crucial role. This control can be achieved by clever modification/combination of the semiconductor's chemical

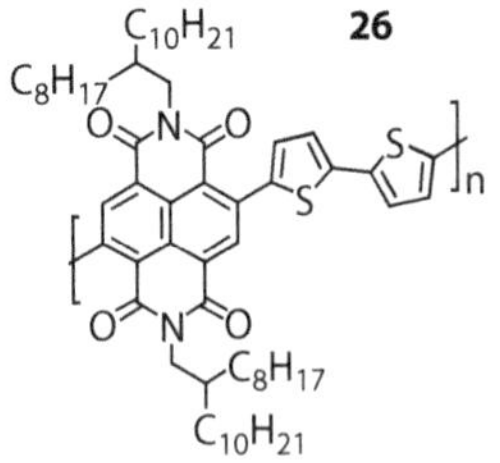

**FIGURE 9.15**   Chemical structure of naphthalenetetracarboxydiimide-dithiophene copolymer **26**.

structure and processing technique and by the impact of the substrate's properties on the semiconductor's supramolecular arrangement, which has to enable efficient charge transport between the source and drain electrode.

## LIQUID CRYSTALLINE DISPLAYS

LCDs have been developed since the mid-1960s and are already well established on the market today. The basic functionality of the liquid crystal cells (the control of light passing through it) has been described in numerous reviews, of which we would like to recommend *The History of Liquid-Crystalline Displays*.[63] There are two ways to produce an image with liquid crystal cells: the segment driving method and the matrix driving method. The segment driving method displays characters and pictures with cells defined by patterned electrodes and is used for simple displays, such as those in calculators. The matrix driving method displays characters and pictures in sets of dots and is used for high-resolution displays, such as those in portable computers and monitors. The dot-matrix LCDs are subdivided into passive-matrix and active-matrix driven monitors. In passive-matrix LCDs (PM-LCDs) there are no switching devices, which results in a slow response time, blurred images, and a reduction of the maximum contrast ratio. In active-matrix LCDs (AM-LCDs), thin-film transistors as switching devices (see "Field-Effect Transistors" section), and storage capacitors are integrated at each cross-point of the electrodes, which yields fast responses and high-contrast imaging. Present-day color LCD TVs and LCD monitors consist of similar sandwich-like structures (illustrated in Figure 9.16). Each pixel in a color LCD is subdivided into three subpixels, in which one set of red (R), green (G), and blue (B) filtered subpixels is equal to one pixel. Because the subpixels are too small to distinguish independently, the RGB elements appear to the human eye as a mixture of the three colors. Any color, with some qualifications, can be produced by mixing these three primary colors. In a TFT-LCD's subpixel, the liquid crystal layer on the pixel electrode forms a capacitor whose counterelectrode is the common electrode on the color-filter substrate. A storage capacitor ($C_S$) and the liquid crystal capacitor ($C_{LC}$) are connected as a load on the TFT (Figure 9.16b). Applying a positive pulse to the

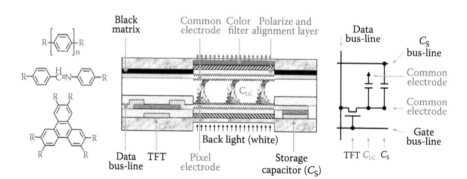

**FIGURE 9.16** TFT-based AMLCD subpixel setup and circuit.

transistor gate electrode through a gate bus-line turns the TFT on. $C_{LC}$ and $C_S$ are charged and the voltage level on the pixel electrode rises to the signal voltage level applied to the data bus-line. The liquid crystals are oriented to the electrical field, the backlight cannot pass through the cell anymore and the subpixel appears black. The main function of $C_S$ is to maintain the voltage on the pixel electrode until the next signal voltage is applied.

Because the material properties of liquid crystals are the result of their molecular properties amplified by the long-range molecular interactions, the correct and application-specific design of liquid crystalline molecules is very important. Three examples of liquid crystalline core structures are presented in Figure 9.16. It has been shown that rod-like liquid crystals for AM-LCDs normally consist of two to four 1,4-disubstituted cyclohexane or benzene rings, linking units and terminal alkyl (alkenyl) and halogenated chains.[64] The degree of overlap of two molecules when forming a dimer depends on the extension of the permanent dipole moment (related to the amount of conjugation), and on the creation of induced dipole moments in the most polarizable parts of the molecule. It has been shown that LCDs require liquid crystals with low viscosity (for reducing the response times), high dielectric anisotropy, suitable elastic constants (for decreasing threshold voltage of the twist-effect and consequent driver voltage), and adjustable optical properties (for the operation in the transmission minimum for optimum display viewing angle). In the LCDs, at least ten different liquid crystalline material parameters have to be optimized to achieve optimal display performance. Because the single liquid crystal shows at best one or two distinguished properties, mixtures consisting of up to twenty and more components have to be developed for a given application.[64]

In contrast to rod-like (calamitic) liquid crystals, disc-like (columnar) liquid crystals cannot be used as switching units in LCDs.[65,66] This is mainly because these disk-shaped molecules do not have a central dipole moment, typically have a much higher viscosity than rod-shaped molecules of similar molecular weight, and are therefore difficult to orient by an electric field. Even if switching is possible, that is, reorientation of the bulk sample, switching times are too slow as compared with the corresponding nematic or smectic compounds. However, this does not mean that columnar mesogens are completely useless for display applications. One serious limitation of current LCDs is the problem of the viewing angle: the brightness, contrast, and sharpness of focus are only optimal when the display is viewed at a certain angle. Furthermore, an image inversion is observed because of the positive birefringence of the liquid crystal layer. These effects can be suppressed by the use of compensation films, which should ideally have negative birefringence. The most promising materials for negative birefringence films indeed are columnar liquid crystals.[9] Although modern AM-LCDs have an excellent range of acceptable viewing angles, this is not the case for many of the simpler LCDs that are being considered for flexible products. Thus, the viewing angle effect may already be dramatic. By far the most difficult challenge for the LCD arises from cell gap variations. The visual quality of a LCD relies heavily on a constant cell gap, as this parameter figures directly in the equations of light management. Flexing the panel creates forces that cause the liquid crystal to flow, resulting in cell gap variations across the panel. This produces visual distortions and artifacts, which can easily become irreversible. It is likely that cell gap problems will prevent LCDs from serving truly dynamically flexible markets, such as realistic electronic paper for which displays based on OLEDs are definitely the more promising candidates. However, LCDs face fewer restrictions in less challenging applications such as curved auto dashboard and, not to be forgotten, LCDs have already enabled most handheld devices (like cell phones or PDA) that we see on the market today.[67]

## ORGANIC LIGHT-EMITTING DIODES

OLED technology offers many advantages over traditional LCDs. OLED displays are self-luminescent, eliminating the requirement for backlighting and allowing them to be thinner, lighter, and more efficient than LCDs. Light is emitted only from the required pixels rather than the entire panel, reducing the overall power consumption by 20% to 80% of that of LCDs.[68] Because of the solid-state construction, OLEDs are extremely robust, exhibiting no cell gap problem such as in LCDs,[67] and may be deposited on most substrates, rigid or flexible, introducing the possibility of many new applications. Finally, OLED displays are aesthetically superior to LCDs, providing true colors, higher contrast, and wider viewing angles.[69] In general, OLEDs are evaluated by their external quantum efficiency $\Phi_{ext}$ of the electroluminescence that is defined as the number of photons emitted per number of injected charge carriers and is expressed as

$$\Phi_{ext} = \frac{\pi L \int \frac{F'(\lambda)\lambda}{hc} d\lambda}{\int F'(\lambda)K_m y(\lambda)d\lambda} \div \frac{J}{e}, \tag{9.2}$$

where $L$ is the luminance (cd m$^{-2}$), and $J$ is the current density (A m$^{-2}$) needed to obtain the luminance, $\lambda$ is the wavelength, $e$ is the elementary charge (C), $h$ is the Planck constant (J s), $c$ is the velocity of light (m s$^{-1}$), $F'(\lambda)$ is the electroluminescence spectrum, $K_m$ is the maximum luminous efficacy, and $y(\lambda)$ is the normalized photopic spectral response function. In 1990, a single-layer OLED using a thin film of poly($p$-phenylene vinylene) with a photoluminescence quantum yield of approximately 8% was reported.[2] Today, maximum external quantum efficiencies of approximately 25% have been achieved for the best OLEDs.[70] Poly($p$-phenylene vinylene) **29** is next to anthracene **27**, 8-hydroxyquinoline aluminum (**28**), and polyfluorene **30**, one of the commonly applied emitter materials (Figure 9.17).

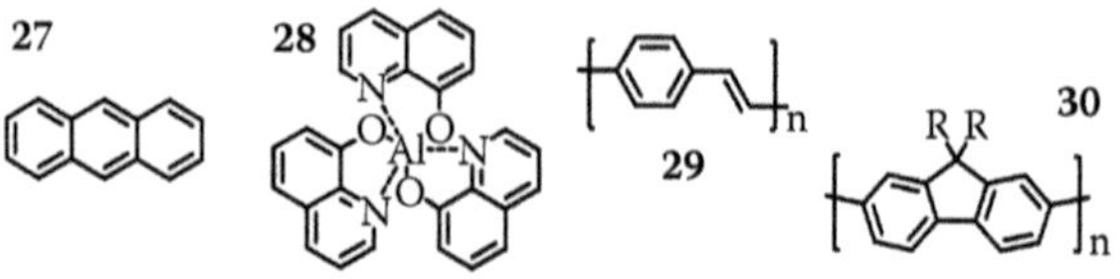

**FIGURE 9.17** Examples of light-emitting conjugated molecules [anthracene (**27**), 8-hydroxyquinoline aluminum (**28**), poly($p$-phenylene vinylene) (**29**), and polyfluorene (**30**)].

OLEDs are double-charge injection devices, requiring the simultaneous supply of both electrons and holes to the electroluminescent material sandwiched between two electrodes (Figure 9.18).

To achieve an efficient OLED with the single-layer configuration shown in Figure 9.18a, the organic electroluminescent material would ideally have a high luminescence quantum yield and be able to facilitate injection and transport of electrons and holes. This demand of multifunctional capabilities from a single organic material is a very difficult one to meet by nearly all current materials. Most highly fluorescent or phosphorescent organic materials of interest in OLEDs tend to have either p-type (hole transport) or n-type (electron transport) charge transport characteristics. A consequence of this is that the simplest OLED configuration shown in Figure 9.18a, in which an organic emitter layer is sandwiched between a transparent anode and a metallic cathode, gives very poor efficiency and brightness. The use of two or more different materials to perform the required functions of efficient light emission and good electron and hole injection and transport properties in an OLED has resulted in improvements of several orders of magnitude in device performance, albeit with the attendant more complex OLED architectures shown in Figure 9.18b through d.[71] When an electric field is applied to an OLED, electrons are injected from the cathode into the LUMO of the adjacent molecules, concomitant with holes being injected from the anode to the HOMO. The two types of carriers migrate toward each other, some of which recombine to form excitons, a fraction of which decay radiatively to the ground-state emitting light.

It has been shown that nonradiative processes that limit the efficiency of radiative decay can be controlled by the design of the emissive layer. One of the key challenges of novel high-performance OLEDs is the design and synthesis of readily processible and thermally robust emissive and charge transport materials with improved multifunctional properties. For the application of OLEDs, the clarity of the emissive film is important and a total light transmission of more than 85% coupled with a haze of less than 0.7% are typically required.[67] Crystalline domains and particles often cause light-scattering and thus diminished device efficiency, which is why mainly amorphous layers are used in OLEDs. Nevertheless, there are a few cases in which supramolecular order has been used for enhanced device performance. In this section, such rare exceptions of the structureless mainstream are demonstrated. Supramolecular interactions can potentially come into play in active layers by enhancing electron and hole transport, as well as assuring stability and even the colors obtained. Structuring active layers are described in the "Processing-Aided Assembly" section; additionally, a range of growth strategies (e.g., oblique incidence

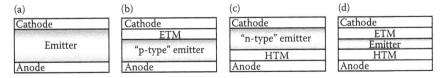

**FIGURE 9.18**   Common OLED architectures, (a) single-layer configuration, (b–d) multi-material OLEDs of higher efficiency with additional hole transport (HTM) and an electron transport material (ETM).

molecular beam deposition, hot-wall deposition, and electrophoretic deposition) for the growth of molecularly ordered thin films were reported.[72–74] Equally, the strong tendency of organic and organometallic nanoparticles/molecular clusters to rapidly self-assemble into highly aligned, stable superlattices at room temperature when solution-cast from dispersions or spray-coated directly onto various substrates was harnessed in OLEDs.[75] Organic–metallic hybrid polymers were formed by the complexation of metal ions with organic ligands or polymers bearing coordination sites. Figure 9.19 illustrates one of the first of such materials. The luminescent polymers based on **33** to **35** were formed by self-assembly of the emissive units through zinc ion complexation between the terpyridine ligands at each end of the chromophore. The photoluminescence maxima have been observed to rank from 450 nm for **33** to 567 nm for **35**. Polymers **33** and **34** have been used to make blue ($\lambda_{max} = 450$ nm) and green ($\lambda_{max} = 572$ nm) emitting LEDs, respectively.[76]

Hybrid polymers consisting of bis(terpyridine)s **31** and **32** and metal ions such as Fe(II) or Ru(II) instead of Zn(II) yield specific colors based on metal-to-ligand charge-transfer absorption and emission. Solid-state devices based on these polymer films display electrochromic properties (see "Electrochromic Devices" section).

Next to the complexation of metal ions with organic ligands or polymers, hydrogen bonded complexes have been applied in light-emitting diodes. The energies of H-bonds are significantly weaker than those of ionic bonds but stronger than van der Waals interactions. Thus, mesomorphic and photophysical properties of H-bonded polymer networks and trimers can be easily adjusted by tuning the H-bonded donors **36** and acceptor emitters **37** and **38** in the H-bonded complexes (Figure 9.20). In addition, the emission properties of bis(pyridyl) acceptor emitters can be manipulated by their surrounding nonphotoluminescent proton donors. Compared with pure bis(pyridyl) acceptor emitters, red shifts of emission wavelengths occurred in most of the H-bonded complexes, showing even widely distributed emission from blue to red colors (from 396 to 642 nm). These systems were further elaborated on incorporating electron-transporting dendritic H-donors in addition to the light-emitting H-acceptors.[77]

Light-emitting materials are gradually increasing in number as well as in quality of the photoluminescence and stability. An overview of present-day organic

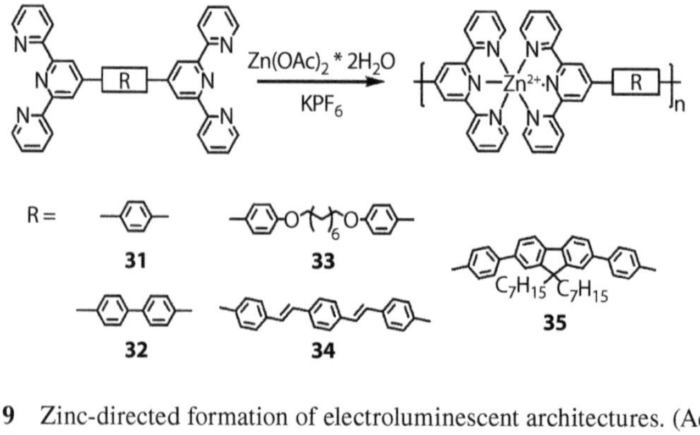

**FIGURE 9.19**   Zinc-directed formation of electroluminescent architectures. (Adapted from Yu, S. C. et al., *Adv. Mater.*, 15, 1643, 2003.)

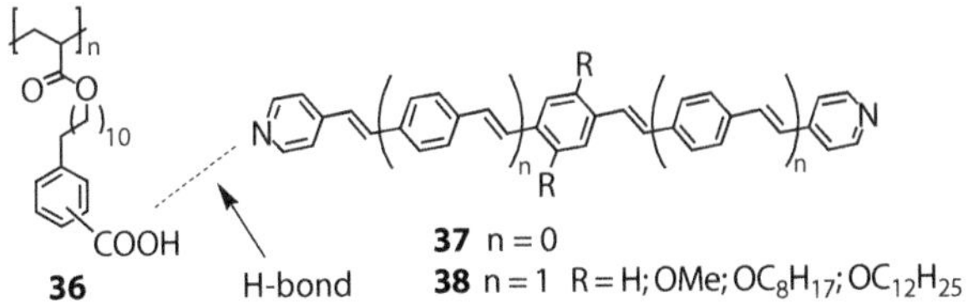

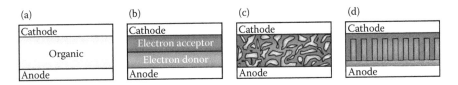

**FIGURE 9.20** Ladder-like structures formed via H-bond interactions. (Adapted from Lin, H. C. et al., *Macromolecules*, 39, 557, 2006.)

emitters is given elsewhere.[79,80] In terms of energy consumption, the photoluminescent organic materials, especially polycyclic aromatic compounds, are very promising. Despite the favorable attributes in comparison with LCDs, significant chemical and physicochemical hurdles must be overcome to realize the full potential of OLEDs. These challenges include (1) effective mediation of charge carrier injection and, where necessary, charge carrier blocking, (2) thermal stability, and (3) operational lifetime. Modifications have been applied to both cathode-organic and anode-organic interfaces, all influence charge injection and afford varying degrees of enhanced device response (e.g., in turn-on voltage, maximum luminance, and quantum efficiency). Notably, the exact interfacial structures and mechanistic roles remain incompletely defined,[81] which can be addressed by applying additional chemical assembly methodologies and well-defined supramolecular structures.

## ORGANIC PHOTOVOLTAIC CELLS

Solar cells can be seen as the opposite to light-emitting diodes. They are classified into photoelectrochemical and photovoltaic cells. Dye-sensitized organic photoelectrochemical solar cells using nanocrystalline, porous $TiO_2$, on which an organic dye is adsorbed, and $I^{3-}/I^-$ redox species in solution or gels.[82,83] Organic solid-state photovoltaic devices (OPVs) consist of thin films of organic photoactive materials sandwiched between two metal electrodes. Both Schottky-type and pn-heterojunction cells have been studied (Figure 9.21). In Schottky-type cells, a single-layer organic photoactive material is sandwiched between two dissimilar electrodes to form a Schottky barrier in the organic layer at the interface with the metal electrode.

**FIGURE 9.21** Organic solid-state photovoltaic devices, (a) Schottky-type cell, (b–d) pn-heterojunction cells with (b) a bilayer heterojunction, (c) a bulk heterojunction, and (d) an ordered heterojunction.

Pn-heterojunction cells are typically based on double layers of organic thin films, in which the organic–organic interface plays an important role in the performance. The electrodes simply provide ideally ohmic contacts with the organic layers.

Generally, higher quantum yields for the photogeneration of charge carriers have been attained for pn-heterojunction cells compared with Schottky-type devices because of electron donor–acceptor interactions between the two kinds of organic semiconductors, that is, electron-donating and -accepting organic semiconductors. The performance of OPVs is evaluated by power conversion efficiency ($\eta$) defined as

$$\eta = \frac{(VJ)_{max}}{I},$$

(9.3)

where $(VJ)_{max}$ is the maximum product of the cell's voltage and current density and $I$ is the incident light power. The basic operation processes of pn-heterojunction OPVs are as follows: (1) light absorption by organic semiconductors to form excitons, (2) diffusion of excitons, (3) charge carrier generation and separation at the organic–organic interface, (4) and charge transport through the organic layers.[84] From these processes, general design principles for the realization of efficient OPV cells have been derived[85–87] and are collected in the following paragraphs.

In process 1, on photon absorption within one of the organic layers, an electron is excited from the HOMO to the LUMO, thus creating an electro-hole pair. This electron–hole pair then relaxes and is known as an exciton. In general, the absorption coefficients of organic semiconductors are high and almost 100% of the incoming light can be absorbed in films a few hundred nanometers in thickness. However, because of the highly anisotropic electronic properties of many organic semiconductors, the orientation of the relevant transition dipoles with respect to the incoming light has to be considered, in particular, in highly ordered organic thin films.[88]

In process 2, the excitons must migrate to an interface where there is a sufficient chemical potential energy drop to redissociate into an electron–hole pair. This process is mostly efficient directly at an organic–organic heterojunction where the electron–hole pair spans the interface across the donor and acceptor. The desired diffusion of excitons toward the organic–organic interface poses stringent requirements on thin-film morphology. Exciton diffusion lengths in polymeric and disordered molecular solids are typically in the range of a few nanometers and increase to a few hundred nanometers for highly ordered crystalline samples.[89] Consequently, the average spacing of organic heterojunctions should be on the length scale of these exciton diffusion lengths. This led to the development of bilayer, bulk, and ordered heterojunctions. Commonly, bulk heterostructures can be fabricated, for example, via spin-coating where phase separation occurs on the required length scale. However, there are only a few or no continuous percolation paths for charge transport for either electrons or holes within one single phase. Supramolecular organization provides a direct method for assembling large numbers of molecules into structures that can bridge length scales from nanometers to macroscopic dimensions for efficient long-distance charge transport.[90,91] However, only a few examples of solid-state all-organic photovoltaic devices based on supramolecular interactions have been reported thus far. Promising candidates include discotic liquid crystalline materials based on extended π-aromatic structures, which can lead to the formation of one-dimensional columnar superstructures that allow efficient exciton transport.[92,93] Power conversion efficiencies of up to 2% were reported for bulk heterojunctions containing the vertically segregated nonmesomorphic electron-accepting perylene-tetracarboxydiimide **12** (Figure 9.2) and the columnar-phase forming hole-accepting HBC **3** (Figure 9.2).[94] Regarding polymer-based OPV cells, an approach consisting of laminating a cyano derivative of poly(*p*-phenylene vinylene) **29** (Figure 9.17) as an electron acceptor, and a derivative of polythiophene **5** (Figure 9.2) as an electron donor layer allowed a power conversion efficiency of 1.9%.[95] Subsequent annealing of such structures led to laterally phase-separated areas (~300 nm) much larger than the exciton diffusion length, which decreased overall efficiency. However, postproduction annealing can have a beneficial influence on the material pair comprising the donor polymer poly(3-hexylthiophene) **5** (R = hexyl, Figure 9.2) and the molecular fullerene acceptor [6,6]-phenyl-C61 butyric acid methyl ester ($C_{61}$-PCBM) (similar to **10**, Figure 9.2). The major advantage of these polymer/fullerene systems is a favorable phase separation of the semiconducting polymer and the fullerenes on an ideal length scale of approximately 20 nm. This nanophase separation could be indeed further optimized by thermal annealing processes. Increased crystallinity of the polymer (resulting in better charge-carrier mobility; see "Field-Effect Transistors" section), and lower electrical resistance have been achieved, thus resulting in $\eta$ = 5%.[96] However, the formation of well-defined and long-term stable interpenetrating networks for such blend materials remains a challenge. Hydrogen-bonding (H-bond) interactions offer a convenient approach toward precisely controlling the geometry of supramolecular assemblies, and numerous conjugated molecules bearing groups for H-bonds have been reported.[97,98] For example, triaminotriazine-substituted oligo(phenylene-vinylenes) (similar to **29**, Figure 9.17) will bind perylenetetracarboxydiimides **12** (Figure 9.2) to form H-bond trimers, which then self-assemble into stacks conducive for charge separation.[90,99] Besides perylenetetracarboxydiimides, other electron acceptors such as fullerene derivatives **10** (Figure 9.2) incorporating H-bond units have been described.[100] To enhance the control of heterojunction spacings, different approaches have shown promise for controlled nano-patterning in solid-state OPV cells leading to ordered heterojunctions (Figures 9.21d and 9.22). Bottom-up approaches are often based on the self-assembly of block copolymers or polymer brushes such as those reported for polypyrrole[101] or poly(triphenylamine acrylate).[102] An additional bottom-up method was presented by the rarely applied vacuum deposition of material onto a substrate rotating around its surface normal, called glancing angle deposition.[103–105] More control of nanostructure dimensions and molecular packing can be gained by top-down methods, for example, hard template production and a transfer of this structure into the organic material.[106] The top-down approach has been applied, for example, to polytriarylamines **11** (Figure 9.2),[107] perylenetetracarboxydiimides **12** (Figure 9.2),[108] and HBC **3** (Figure 9.2).[109] Investigations on heterojunction solar cells based on these nanostructured films

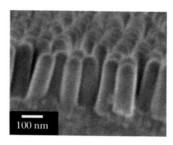

**FIGURE 9.22**  Scanning electron microscope image of a structured polytriarylamine **11** (Figure 9.2) layer for ordered heterojunctions. (Reprinted with permission from Haberkorn, N. et al., 2009, 1415. Copyright 2009 American Chemical Society.)

filled with a second organic layer have not been reported thus far or are currently underway.

Besides the necessity of phase separation, a second issue needs to be considered on the supramolecular level, and this is the self-assembly of the same molecular building blocks. Well-ordered aggregates seem promising, in particular, if the molecular packing favored the desired intermolecular couplings for exciton transport to the heterojunction interface and charge transport to the electrodes. Both excitons and charge carriers were found to migrate via a hopping mechanism[110,111] that requires intermolecular order and crystallinity of the organic phases such as that described for charge carrier transport only in the "Field-Effect Transistors" section.[112] The requirement of high charge-carrier mobility is of great importance for the following two basic operation processes of pn-heterojunction OPVs.

At first, process 3 in OPV cells requires that exciton dissociation across the organic–organic interface into free charge carriers is stable, and that recombination of carriers is inhibited. Therefore, charge carrier mobilities should be higher than $10^{-4}$ cm$^2$ V$^{-1}$ s$^{-1}$. In optimized donor–acceptor material pairs (e.g., polythiophene **5** and C$_{60}$ **10** in Figure 9.2), charge transfer across the interface happens on a picosecond timescale,[113] thus giving rise to the often-achieved high efficiency ranking of between 4.8% and 5.1% in OPV cells.[96,114] However, the most important parameter governing stable charge transfer is the energy level offset at the organic–organic heterojunction, which should exceed several tenths of an electron volt for the valence bands (HOMOs) and conduction bands (LUMOs), respectively. Another important issue regarding charge separation involves the possible formation of exciplexes. These metastable complexes of donors and acceptors across the interface provide an unwanted loss channel. It is suggested that this effect could be suppressed by increased intermolecular distances.[115]

Finally, process 4 calls for high charge-carrier mobility, to facilitate fast transport of separated charges away from the organic–organic interface and thus minimize exciton regeneration. Charge-carrier mobility is strongly linked to intermolecular order and crystallinity of the organic phases (see "Field-Effect Transistors" section). However, the formation of optimal domains and percolation paths for charge carriers to the electrodes over macroscopic dimensions (length scale of 100 nm), is probably the level that is most difficult to address by rational means. Thus, even for optimally mastered functional properties of the molecules and most desirable supramolecular arrangements, the proper intermixing and macroscopic orientation of the two components is a big challenge, for which barely rational design concepts exist and accordingly time-consuming optimization by trial-and-error is warranted.[112] Variations of the solvent, the deposition technique and, in particular, thermal post-treatment (annealing) of the cast photoactive layers are the most commonly applied strategies for successfully approaching this issue. In addition, it has been found that the presence of excess surfactant has a dramatic effect on film morphology in poly(3-hexylthiophene):$C_{61}$-PCBM blends.[116] Measurements of steady-state and transient photoconductivities, thin-film transistor mobility, optical transmission, and x-ray diffraction demonstrated that the incorporation of short-chain alkylthiols in the casting solution markedly enhances the photoconductive response of the resulting films because of the increased structural order in the polymer phase. The efficiency of blends based on a cyclopentadithiophene-benzothiadiazole copolymer derivative **14** (Figure 9.2) and a $C_{71}$-PCBM (similar to **10**, Figure 9.2) has been enhanced by gaining control of the morphology.[117] This control was gained by mixing alkanedithiols into the solution, and yields an efficiency of 5.5% in bulk heterojunction solar cells. In contrast with the previous results of poly(3-hexylthiophene):$C_{61}$-PCBM blends, no increase in hole mobility and no indication of crystallization in the **14**:$C_{71}$-PCBM films either with or without thiol processing were achieved. Thus, the increased efficiencies were addressed to changes in the heterojunction morphology and light absorption, implying possible optimizations of every basic operation process of pn-heterojunction OPVs and no clear classification. In general, structure–property relationships are not easily accessible on this level because it is rather difficult to get insight into the organization of the constituent molecules in the interior of soft matter.

Nevertheless, the performance of organic photovoltaic cells is continuously increasing because of the ongoing enormous research effort of scientists all over the world. Accordingly, the psychological barrier of $\eta = 8\%$ has recently been broken with an efficiency of 8.13%, which will surely not be the end of further success.

## ADDITIONAL ELECTRONIC DEVICES

In this section, additional organic electronic devices (aside from FETs, LCDs, OLEDs, and OPVs) in which designed supramolecular interactions are used in the manufacture of the active layer are presented. These devices are subdivided according to the application that focuses on supramolecular materials for electrochromic elements, lasers, sensors, and fuel cells.

### Electrochromic Devices

Dynamic tintable or the so-called smart windows can change properties such as the solar factor and the transmission of radiation in the solar spectrum in response to an electric current or to changing environmental conditions. The application of such windows may lead to a drastic reduction of the energy consumption of highly glazed buildings by reducing cooling loads, heating loads, and the demand for electric lighting. At present, three different technologies with external triggering signals are commonly known for this purpose and have been available on the market: chromic materials, liquid crystals, and electrophoretic or suspended-particle devices. Here, the chromic devices may be divided into four categories, that is, electrochromic, gasochromic, photochromic, and thermochromic devices.

The electrochromic device mostly consists of a transparent conducting film, on which one (or multiple) cathodic electroactive layer(s) is (are) affixed, followed by layers of an ion conductor, an ion-storage film or one (or multiple) complimentary anodic electroactive layer(s), and another transparent conducting film. By combining different types of electrochromics, ion-storage films, and ion conductors, different properties can be obtained for the device. That is why electrochromism may be seen as a device characteristic instead of a material property.[118]

Many different polymers have been incorporated in prototype electrochromic devices, for example, the bis(terpyridine) polymers **31** and **32** (see Figure 9.19), polyaniline **39**, poly(3-methylthiophene) **40**, and poly(X-dioxypyrrole) (PXDOP) **41–43** or poly(X-dioxythiophene) (PXDOT) **44–45** based electrochromics (with X being 3,4-ethylene or 3,4-propylene), as well as related carbazole **46** and pyridopyrazine **47** polymers (Figure 9.23). Recently, the main focus has gone toward

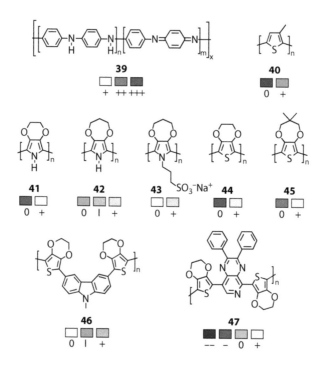

**FIGURE 9.23** Electrochromic polymers. Color swatches are representations of thin films based on measured CIE 1931 Yxy color coordinates. Key: 0 = neutral; I = intermediate; +, ++, and +++ = oxidized; – and –– = reduced. (Adapted from Thompson, B. C. et al., *Chem. Mater.*, 12, 1563, 2000; Argun, A. A. et al., *Chem. Mater.*, 16, 4401, 2004.)

polyaniline (PANI) **39** and poly(3,4-ethylene-dioxythiophene) (PEDOT) **44**. Polyaniline is a conducting polymer which may undergo color changes from a transparent state to violet by both a redox process and proton doping. PANI is one of the most extensively researched electrochromic materials thus far, owing to its good electrochemical cycling stability in nonaqueous electrolytes above $10^6$ cycles and its low cost and ease of processing by electrodeposition or liquid casting techniques. Recently, a new class of hybrid polymers has been developed based on aniline, with a 40% enhancement in electrochromic contrast at their $\lambda_{max}$ compared with PANI because of more accessible doping sites and with increased electrochemical stability. The polymer polyaniline-tethered polyhedral oligomeric silsesquioxane was found to form a loosely packed structure because of covalent binding of the PANI chains to the polyhedral oligomeric silsesquioxane nano-cages, creating a nano-porous structure and allowing easy ion movement during redox switching. In addition to the enhanced contrast of up to a total value of 44%, the bleaching times have been found slightly shorter.[119,120] Electrochromic applications based on PEDOT and its derivates have also drawn a lot of attention because of their ease of coloring, high electrochromic contrast, and fast response times. PEDOT switches from blue in the neutral state to transparent in the oxidized state. The quality of PEDOT films electropolymerized onto ITO can be greatly enhanced (in homogeneity) by the presence of a predeposited conductive layer of PEDOT/PSS.[121] Devices based on this approach exhibited higher contrasts than devices constructed on bare ITO, which was attributed to the difference in film morphology. Furthermore, in the presence of dodecyl sulfate and lithium triflate, a nanostructured network of nearly spherical grains of PEDOT (50–150 nm in size) and pores facilitating the ionic diffusion was achieved, which yielded up to 62% contrast.[122]

Switchable devices based on liquid crystals offer another approach besides electrochromic devices. The mechanism of optical switching is a change in the orientation of liquid crystal molecules between two conductive electrodes by applying an electric field, resulting in a change of their transmittance.[125,126] A remarkable transmittance of 80–1% (at 1.0–1.6 V $\mu m^{-1}$) has been achieved by using the nematic liquid crystalline monomer **48**, which is known for its ability to allow and keep a good homeotropic (perpendicular to the surface) alignment of liquid crystals on ITO-covered substrates after polymerization (Figure 9.24).[127]

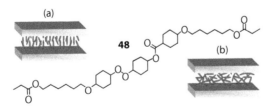

**FIGURE 9.24** Schematic operation of a 48-based switchable device: (a) optically clear state induced by dielectric orientation of the liquid crystal and (b) highly scattering texture because of an applied electrical field.

Nevertheless, liquid crystal windows need an electrical field to be maintained as long as the transparent mode of the glass is required, resulting in a higher energy consumption compared with electrochromic devices which normally only require an electrical field during switching. Based on this, electrochromic windows seem to be the most promising state-of-the-art technology for daylight and solar energy purposes.[118]

## Organic Lasers

Organic semiconductors combine novel optoelectronic properties with simple fabrication and the scope for tuning the chemical structure to give desired features, making them attractive candidates as active materials in lasers.[128] There are four types of organic semiconductors that have been studied as laser materials (1) small molecules, for example, anthracene **27** (Figure 9.17), perylenetetracarboxydiimide **12** (Figure 9.2), and aluminum tris(quinolate) **28** (Figure 9.17); (2) conjugated polymers, for example, poly(phenylene-vinylene)s **29** and polyfluorenes **30** (Figure 9.17); (3) conjugated dendrimers **49**; and (4) spiro-compounds **50** (Figure 9.25).[128–130]

At high concentrations or in solid state, conjugated organic molecules can interact with their neighbors, leading to the formation of dimers, aggregates, or excimers, which can quench light emission. Hence, many laser dyes, which are extremely fluorescent materials in dilute solution, are almost nonemissive in the solid state. The approaches to avoiding quenching generally involve increasing the spacing of the chromophores (light-emitting units). For small molecules, this is frequently done by blending with a host material. In conjugated polymers, ground-state absorption losses can be reduced by using polymer blends.

Another approach is based on controlling interchain interactions, as both charge generation and singlet state annihilation—two major loss mechanisms—are mediated by interchain coupling. Control of molecular interactions has been applied by the use of bulky side groups to increase interchain distances. Furthermore, light-emitting dendrimers **49** have been designed with the chromophore at the core $R_1$ and the dendrons $R_{2/3}$ acting as spacers. In spiro-compounds **50**, the spiro-linkage imposes a geometry that makes dense packing of the chromophores difficult, thereby

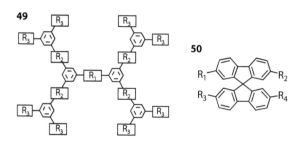

**FIGURE 9.25**  Basic structures of conjugated dendrimers **49** and spiro-compounds **50**.

controlling intermolecular interactions (Figure 9.25). In addition, one focused on molecules that form a J-type excitonic coupling and thus a strong displacement of the chromophores in the solid state.[131] However, the use of organic semiconductor lasers for commercial applications requires further improvements in efficiency by reducing losses and significantly improving the organic material's lifetime.[128,130]

## Organic Sensors

Supramolecular self-assembly has a general impact on the sensing device's output (as described in previous sections) and, thus, to its response to analytes. For example, in organic thin-film transistor sensors, the relative response has been found to be amplified as the grain size became smaller.[132–134] This was attributed to enhanced grain boundary trapping[135] mediated by analyte molecules and the consequent reduction in drain current because of reduced charge-carrier mobility and shifted threshold voltage (see the "Field-Effect Transistors" section for transistor function).

Organic semiconductors are excellent when used as transducers because they can be tailored to be both sensitive and highly selective. Selectivity is achieved by exploiting the mechanical or chemical nature of the organic material itself, or through chemical or biological functionalization, which is made easy because of its organic nature. Sensitivity can be achieved by exploiting either the nature of thresholds or efficiencies by measuring intensity changes or the emission by measuring wavelength changes. Small changes in the environment, for example, the oxygen or humidity, may switch off the device function, thus giving a large response to a small stimulus. Over the past three decades, extensive efforts have been devoted to understanding the interactions of gaseous, chemical, or biological analytes with semiconductors as well as to exploiting these effects in developing a variety of sensors,[136–138] for example, immunosensing, glucose monitoring, and genetic analysis such as the detection of single-nucleotide polymorphisms and deoxyribonucleic acid sequencing. An overview of supramolecular interactions between organic analytes and sensing elements is given in the chapter "Organic and Biological Analytes."

As a technical example, a FET-based glucose sensor is presented here. Phenylboronic acid has been chemically introduced onto the FET gate surface in the form of a thin copolymer gel layer as a totally synthetic and stable glucose-sensing moiety (Figure 9.26).[138] Excellent transistor characteristics were confirmed after the surface modification. Glucose-induced changes in the FET's electric characteristics were obtained in quantitative as well as reversible manners. It was also demonstrated that the prepared FET is able to continuously perceive the change in the glucose concentration in the milieu. The detected signals were attributed to the faction change of the gate-introduced phenylborate anions, also presumably involving other parameter changes such as permittivity and conductivity.

It is noteworthy that the phenylborate moiety is also known to interact with the natural biological membranes of a variety of cells, viruses, bacteria, and fungi through the membrane constituting carbohydrate moieties. With the potential ability to perceive those biological membrane interactions, and to do so in a real-time manner, the phenylborate-modified transistor may also become applicable to noninvasive types of cytology.[138]

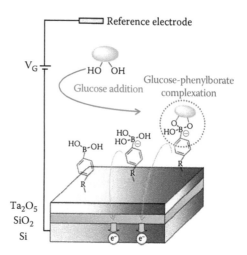

**FIGURE 9.26** Conceptual scheme for glucose detection by the gate surface-modified FET with phenylboronic acid moieties. (With kind permission from Springer Science+Business Media: *J. Solid State Electr.* 13, 2009, 165, Matsumoto, A. et al., Figure 1.)

## Organic Fuel Cells

Organic fuel cells as well as organic-based batteries are further examples of energy-related devices which enable the generation, storage, and conservation of energy.

Fuel cells are an interesting alternative to existing combustion engine/energy conversion systems, and some types are already in the market. Generally, fuel cells combine high efficiency to transfer chemical energy into electric energy with fuels that have the potential to be renewable. Various fuel cell types, differing, for example, in their operation temperature or proton-conducting material, are known.[139,140] One type of organic fuel cell under active development is the polymeric-electrolyte membrane fuel cell. The "golden standard" of a polymeric-electrolyte membrane, Nafion (**51**), was developed by DuPont and introduced in 1966 (Figure 9.27).[141] The fact that Nafion, although quite old, can still be regarded as the benchmark for new polymeric membranes is because of its favorable nanostructure.[142] The combination of a perfluorinated backbone and sulfonic acid side groups is responsible for a microphase separation within Nafion membranes on hydration. These interconnected water channels are finally responsible for the high proton conductivity.[143] The proton transport is governed by the vehicle mechanism in which the proton does not migrate as $H^+$ but as $OH_3^+$, $NH_4^+$, etc., bonded to a "vehicle" such as $H_2O$, $NH_3$, etc. The "unladen" vehicles move in the opposite direction and show a diffusion coefficient corresponding to the proton conduction.[144] The drawback of Nafion (and of most other membranes based on sulfonated aromatics) is its low operating temperature because of the loss of the required water molecules at temperatures higher than 100°C. Alternatively, heterocycles, such as imidazole and benzimidazole, have been used as vehicles leading to proton conductivities of up to 250°C. Herein, the volatility of the heterocycle as the proton solvent required the immobilization of the solvent in the polymer membrane while keeping the protons mobile. The proton mobility

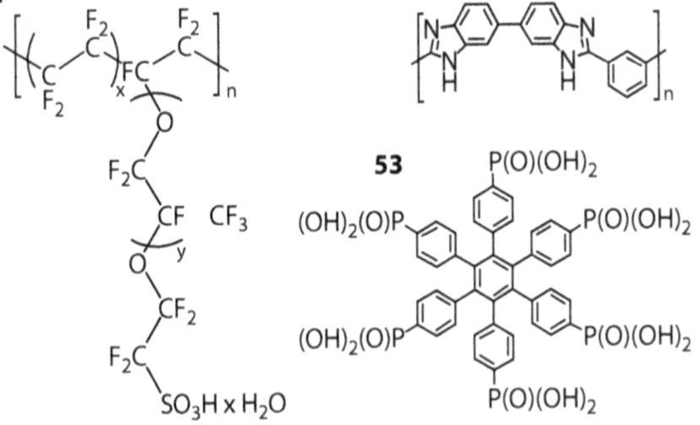

**FIGURE 9.27** Examples of organic membrane materials for fuel cells: Nafion (**51**), poly(benzimidazole) **52**, and phosphonic acid functionalized hexaphenylbenzene **53**.

in such immobilized heterocycles mainly relies on structure diffusion, called the Grotthuss-type mechanism.[143]

Further development of membranes focused on poly(benzimidazole) **52** and phosphoric acid. Analogue to Nafion-type membranes, the achievement of defined biphasic nanostructure of a scaffold (PBI) and proton carrier ($H_3PO_4$) is crucial. Recently, mesoporous PBI networks with 10-nm pores were synthesized, predefining the latter membrane morphology. The synthetic strategy was based on hard-templating of silica nanoparticles. After dissolving the silica spheres, the resulting pores could be filled with phosphoric acid to yield a highly proton-conducting material that could be easily operated up to high temperatures of 200°C.[145] A major problem dealing with the diluting and washing out of $H_3PO_4$ by the water formed during the operation of the fuel cell was solved by the development of phosphonic acid–functionalized polymers. Also, for such functionalized polymers, it has been shown that the microstructure plays a significant role, as aggregation of phosphonic acid groups is necessary to obtain good proton transport properties in this dry state.[146] In another approach, the self-assembly of phosphonic acid functionalized hexaphenylbenzenes **53** have been applied. Herein, high and temperature-independent proton conductivity has been achieved in a crystalline state, thus satisfying one of the prerequisites for novel fuel cell membranes.[147]

## CONCLUSION

Electronic and optoelectronic devices using organic materials have gained huge attention because of coupled characteristics like light weight, (semi)transparency, flexibility, and pronounced low cost and large-area fabrication. As fundamental parts of future electronic equipment, OFETs, light-emitting diodes and lasers, LCDs, and electrochromic devices, photovoltaic cells, and fuel cells have to enable basic functions such as power control and energy conversion from electricity to light or to molecular motion, or the other way around, from light to electricity as well as from fuel to electricity. Herein, charge or light transport represents the main operation process that has been elucidated to be directly linked to the molecular and supramolecular structuring of the organic layers. Although significant progress has already been made, further investigations of structure–property relationships still remain necessary to improve important characteristics, for example, stability (and toxicity in sensing elements), and finally realize the envisioned applications.

## REFERENCES

1. Heeger, A. J., A. G. MacDiarmid, H. Shirakawa. 2000. Conductive polymers. *The Nobel Prize in Chemistry*.
2. Burroughes, J. H., D. D. C. Bradley, A. R. Brown, R. N. Marks, K. Mackay, R. H. Friend, P. L. Burns, A. B. Holmes. 1990. Light-emitting-diodes based on conjugated polymers. *Nature*, 347(6293), 539–541.
3. Friend, R. H., R. W. Gymer, A. B. Holmes, J. H. Burroughes, R. N. Marks, C. Taliani, D. D. C. Bradley, D. A. Dos Santos, J. L. Bredas, M. Logdlund, W. R. Salaneck. 1999. Electroluminescene in conjugated polymers. *Nature*, 397(6715), 121–128.
4. Jurchescu, O. D., J. Baas, T. T. M. Palstra. 2004. Effect of impurities on the mobility of single crystal pentacene. *Applied Physics Letters*, 84(16), 3061–3063.
5. Sundar, V. C., J. Zaumseil, V. Podzorov, E. Menard, R. L. Willett, T. Someya, M. E. Gershenson, J. A. Rogers. 2004. Elastomeric transistor stamps: Reversible probing of charge transport in organic crystals. *Science*, 303(5664), 1644–1646.
6. Anthony, J. E., J. S. Brooks, D. L. Eaton, S. R. Parkin. 2001. Functionalized pentacene: Improved electronic properties from control of solid-state order. *Journal of the American Chemical Society*, 123(38), 9482–9483.
7. Payne, M. M., S. R. Parkin, J. E. Anthony, C. C. Kuo, T. N. Jackson. 2005. Organic field-effect transistor from solution-deposited functionalized acenes with mobilities as high as 1 cm(2)/V-s. *Journal of the American Chemical Society*, 127(14), 4986–4987.
8. Simpson, C. D., J. S. Wu, M. D. Watson, K. Müllen. 2004. From graphite molecules to columnar superstructures—an exercise in nanoscience. *Journal of Materials Chemistry*, 14(4), 494–504.
9. Laschat, S., A. Baro, N. Steinke, F. Giesselmann, C. Hägele, G. Scalia, R. Judele, E. Kapatsina, S. Sauer, A. Schreivogel, M. Tosoni. 2007. Discotic liquid crystals: From tailor-made synthesis to plastic electronics. *Angewandte Chemie (International ed. in English)*, 46(26), 4832–4887.
10. Mas-Torrent, M., M. Durkut, P. Hadley, X. Ribas, C. Rovira. 2004. High mobility of dithiophene-tetrathiafulvalene single-crystal organic field effect transistors. *Journal of the American Chemical Society*, 126(4), 984–985.
11. Sirringhaus, H., P. J. Brown, R. H. Friend, M. M. Nielsen, K. Bechgaard, B. M. W. Langeveld-Voss, A. J. H. Spiering, R. A. J. Janssen, E. W. Meijer, P. Herwig, D. M. de Leeuw. 1999. Two-dimensional charge transport in self-organized, high-mobility conjugated polymers. *Nature*, 401(6754), 685–688.
12. Takimiya, K., Y. Kunugi, Y. Konda, N. Niihara, T. Otsubo. 2004. 2,6-diphenylbenzo[1,2-b: 4,5-b']dichalcogenophenes: A new class of high-performance semiconductors for organic field-effect transistors. *Journal of the American Chemical Society*, 126(16), 5084–5085.
13. Thelakkat, M., J. Hagen, D. Haarer, H. W. Schmidt. 1999. Poly(triarylamine)s-synthesis and application in electroluminescent devices and photovoltaics. *Synthetic Metals*, 102(1–3), 1125–1128.
14. Tang, Q. X., H. X. Li, M. He, W. P. Hu, C. M. Liu, K. Q. Chen, C. Wang, Y. Q. Liu, D. B. Zhu. 2006. Low threshold voltage transistors based on individual single-crystalline submicrometer-sized ribbons of copper phthalocyanine. *Advanced Materials*, 18(1), 65–68.

15. Someya, T., Y. Kato, T. Sekitani, S. Iba, Y. Noguchi, Y. Murase, H. Kawaguchi, T. Sakurai. 2005. Conformable, flexible, large-area networks of pressure and thermal sensors with organic transistor active matrixes. *Proceedings of the National Academy of Sciences of the United States of America*, 102(35), 12321–12325.

16. Gao, X. K., C. A. Di, Y. B. Hu, X. D. Yang, H. Y. Fan, F. Zhang, Y. Q. Liu, H. X. Li, D. B. Zhu. 2010. Core-expanded naphthalene diimides fused with 2-(1,3-dithiol-2-ylidene) malonitrile groups for high-performance, ambient-stable, solution-processed n-channel organic thin film transistors. *Journal of the American Chemical Society*, 132(11), 3697–3699.

17. Tsao, H. N., W. Pisula, Z. H. Liu, W. Osikowicz, W. R. Salaneck, K. Müllen. 2008. From ambi- to unipolar behavior in discotic dye field-effect transistors. *Advanced Materials*, 20(14), 2715–2719.

18. Tsao, H. N., D. Cho, J. W. Andreasen, A. Rouhanipour, D. W. Breiby, W. Pisula, K. Müllen. 2009. The influence of morphology on high-performance polymer field-effect transistors. *Advanced Materials*, 21(2), 209–212.

19. Hamilton, R., J. Smith, S. Ogier, M. Heeney, J. E. Anthony, I. McCulloch, J. Veres, D. D. C. Bradley, T. D. Anthopoulos. 2009. High-performance polymer-small molecule blend organic transistors. *Advanced Materials*, 21(10–11), 1166–1171.

20. Hunter, C. A., J. K. M. Sanders. 1990. The nature of pi-pi interactions. *Journal of the American Chemical Society*, 112(14), 5525–5534.

21. Meyer, E. A., R. K. Castellano, F. Diederich. 2003. Interactions with aromatic rings in chemical and biological recognition. *Angewandte Chemie (International ed. in English)*, 42(11), 1210–1250.

22. Schneider, H. J. 2009. Binding mechanisms in supramolecular complexes. *Angewandte Chemie (International ed. in English)*, 48(22), 3924–3977.

23. Steudel, S., S. De Vusser, S. De Jonge, D. Janssen, S. Verlaak, J. Genoe, P. Heremans. 2004. Influence of the dielectric roughness on the performance of pentacene transistors. *Applied Physics Letters*, 85(19), 4400–4402.

24. Wittmann, J. C., B. Lotz. 1990. Epitaxial Crystallization of polymers on organic and polymeric substrates. *Progress in Polymer Science*, 15, 909–948.

25. Chen, X. L., A. J. Lovinger, Z. N. Bao, J. Sapjeta. 2001. Morphological and transistor studies of organic molecular semiconductors with anisotropic electrical characteristics. *Chemistry of Materials*, 13(4), 1341–1348.

26. Amundson, K. R., B. J. Sapjeta, A. J. Lovinger, Z. N. Bao. 2002. An in-plane anisotropic organic semiconductor based upon poly(3-hexyl thiophene). *Thin Solid Films*, 414(1), 143–149.

27. van de Craats, A. M., N. Stutzmann, O. Bunk, M. M. Nielsen, M. Watson, K. Müllen, H. D. Chanzy, H. Sirringhaus, R. H. Friend. 2003. Meso-epitaxial solution-growth of self-organizing discotic liquid-crystalline semiconductors. *Advanced Materials*, 15(16), 495–499.

28. Yoon, M. H., C. Kim, A. Facchetti, T. J. Marks. 2006. Gate dielectric chemical structure-organic field-effect transistor performance correlations for electron, hole, and ambipolar organic semiconductors. *Journal of the American Chemical Society*, 128(39), 12851–12869.

29. Facchetti, A., M. H. Yoon, T. J. Marks. 2005. Gate dielectrics for organic field-effect transistors: New opportunities for organic electronics. *Advanced Materials*, 17(14), 1705–1725.

30. de Boer, B., A. Hadipour, M. M. Mandoc, T. van Woudenbergh, P. W. M. Blom. 2005. Tuning of metal work functions with self-assembled monolayers. *Advanced Materials*, 17(5), 621–625.

31. Kim, D. H., Y. D. Park, Y. S. Jang, H. C. Yang, Y. H. Kim, J. I. Han, D. G. Moon, S. J. Park, T. Y. Chang, C. W. Chang, M. K. Joo, C. Y. Ryu, K. W. Cho. 2005. Enhancement of field-effect mobility due to surface-mediated molecular ordering in regioregular polythiophene thin film transistors. *Advanced Functional Materials*, 15(1), 77–82.

32. Park, Y. D., J. A. Lim, H. S. Lee, K. Cho. 2007. Interface engineering in organic transistors. *Materials Today*, 10(3), 46–54.

33. Gundlach, D. J., J. E. Royer, S. K. Park, S. Subramanian, O. D. Jurchescu, B. H. Hamadani, A. J. Moad, R. J. Kline, L. C. Teague, O. Kirillov, C. A. Richter, J. G. Kushmerick, L. J. Richter, S. R. Parkin, T. N. Jackson, J. E. Anthony. 2008. Contact induced crystallinity for high-performance soluble acene-based transistors and circuits. *Nature Materials*, 7(3), 216–221.

34. Berggren, M., D. Nilsson, N. D. Robinson. 2007. Organic materials for printed electronics. *Nature Materials*, 6(1), 3–5.

35. Forrest, S. R. 2004. The path to ubiquitous and low-cost organic electronic appliances on plastic. *Nature*, 428(6986), 911–918.

36. Bornside, D. E., C. W. Macosko, L. E. Scriven. 1989. Spin coating—One-dimensional model. *Journal of Applied Physics*, 66(11), 5185–5193.

37. Schubert, D. W., T. Dunkel. 2003. Spin coating from a molecular point of view: Its concentration regimes, influence of molar mass and distribution. *Materials Research Innovations*, 7(5), 314–321.

38. Scriven, L. E., C. J. Brinker, D. E. Clark, III. 1988. Physics and applications of dip coating and spin coating. *Materials Research Society Symposium on Better Ceramics through Chemistry*, 121, 717–729.

39. Brinker, C. J., G. C. Frye, A. J. Hurd, C. S. Ashley. 1991. Fundamentals of sol-gel dip coating. *Thin Solid Films*, 201(1), 97–108.

40. Gao, P., D. Beckmann, H. N. Tsao, X. L. Feng, V. Enkelmann, M. Baumgarten, W. Pisula, K. Müllen. 2009. Dithieno[2,3-d;2′,3′-d′]benzo[1,2-b;4,5-b′]dithiophene (DTBDT) as Semiconductor for high-performance, solution-processed organic field-effect transistors. *Advanced Materials*, 21(2), 213–216.

41. Tracz, A., T. Pakula, J. K. Jeszka. 2004. Zone casting—a universal method of preparing oriented anisotropic layers of organic materials. *Materials Science (Poland)*, 22(4), 415–421.

42. Langmuir, I. 1932. Surface chemistry. *The Nobel Prize in Chemistry*.

43. Blodgett, K. B. 1935. Films built by depositing successive monomolecular layers on a solid surface. *Journal of the American Chemical Society*, 57(1), 1007–1022.

44. Becerril, H. A., M. E. Roberts, Z. H. Liu, J. Locklin, Z. N. Bao. 2008. High-performance organic thin-film transistors through solution-sheared deposition of small-molecule organic semiconductors. *Advanced Materials*, 20(13), 2588–2594.

45. Mascaro, D. J., M. E. Thompson, H. I. Smith, V. Bulovic. 2005. Forming oriented organic crystals from amorphous thin films on patterned substrates via solvent-vapor annealing. *Organic Electronics*, 6(5–6), 211–220.

46. Dickey, K. C., J. E. Anthony, Y. L. Loo. 2006. Improving organic thin-film transistor performance through solvent-vapor annealing of solution-processable triethylsilylethynyl anthradithiophene. *Advanced Materials*, 18(13), 1721.

47. Pfann, W. G. T. 1955. Temperature gradient zone melting. *American Institute of Mining and Metallurgical Engineers*, 203(9), 961–964.

48. Sergeyev, S., W. Pisula, Y. H. Geerts. 2007. Discotic liquid crystals: A new generation of organic semiconductors. *Chemical Society Reviews*, 36(12), 1902–1929.

49. Dyreklev, P., G. Gustafsson, O. Inganas, H. Stubb. 1992. Aligned polymer-chain field-effect transistors. *Solid State Communications*, 82(5), 317–320.

50. Dyreklev, P., G. Gustafsson, O. Inganas, H. Stubb. 1993. Polymeric field-effect transistors using oriented polymers. *Synthetic Metals*, 57(1), 4093–4098.

51. Heil, H., T. Finnberg, N. von Malm, R. Schmechel, H. von Seggern. 2003. The influence of mechanical rubbing on the field-effect mobility in polyhexylthiophene. *Journal of Applied Physics*, 93(3), 1636–1641.

52. Mas-Torrent, M., D. den Boer, M. Durkut, P. Hadley, A. P. H. J. Schenning. 2004. Field effect transistors based on poly(3-hexylthiophene) at different length scales. *Nanotechnology*, 15(4), S265–S269.
53. Katz, H. E., J. Huang. 2009. Thin-film organic electronic devices. *Annual Review of Materials Research*, 39, 71–92.
54. Facchetti, A. 2007. Semiconductors for organic transistors. *Materials Today*, 10(3), 28–37.
55. Anthony, J. E. 2008. The larger acenes: Versatile organic semiconductors. *Angewandte Chemie (International ed. in English)*, 47(3), 452–483.
56. Anthony, J. E. 2006. Functionalized acenes and heteroacenes for organic electronics. *Chemical Reviews*, 106(12), 5028–5048.
57. Würthner, F., R. Schmidt. 2006. Electronic and crystal engineering of acenes for solution-processible self-assembling organic semiconductors. *ChemPhysChem*, 7(4), 793–797.
58. Gao, P., D. Beckmann, H. N. Tsao, X. L. Feng, V. Enkelmann, W. Pisula, K. Müllen. 2008. Benzo[1,2-b:4,5-b']bis[b]benzothiophene as solution processible organic semiconductor for field-effect transistors. *Chemical Communications*, (13), 1548–1550.
59. Pisula, W., A. Menon, M. Stepputat, I. Lieberwirth, U. Kolb, A. Tracz, H. Sirringhaus, T. Pakula, K. Müllen. 2005. A zone-casting technique for device fabrication of field-effect transistors based on discotic hexa-peri-hexabenzocoronene. *Advanced Materials*, 17(6), 684–689.
60. Gsänger, M., J. H. Oh, M. Könemann, H. W. Höffken, A.-M. Krause, Z. Bao, F. Würthner. 2010. A crystal-engineered hydrogen-bonded octachloroperylene diimide with a twisted core: An n-channel organic semiconductor. *Angewandte Chemie (International ed. in English)*, 122(4), 752–755.
61. Tsao, H. N., D. M. Cho, I. Park, M. R. Hansen, A. Mavrinskiy, D. Y. Yoon, R. Graf, W. Pisula, H. W. Spiess, K. Müllen. 2011. Ultrahigh mobility in polymer field-effect transistors by design. *Journal of the American Chemical Society*, 133, 2605–2612.
62. Yan, H., Z. H. Chen, Y. Zheng, C. Newman, J. R. Quinn, F. Dotz, M. Kastler, A. Facchetti. 2009. A high-mobility electron-transporting polymer for printed transistors. *Nature*, 457(7230), 679–686.
63. Kawamoto, H. 2002. The history of liquid-crystal displays. *Proceedings IEEE*, 90(4), 460–500.
64. Petrov, V. F. 1995. Liquid-crystals for AMLCD and TFT-PDLCD applications. *Liquid Crystals*, 19(6), 729–741.
65. Boden, N., R. J. Bushby, J. Clements, B. Movaghar. 1999. Device applications of charge transport in discotic liquid crystals. *Journal of Materials Chemistry*, 9(9), 2081–2086.
66. Borchard-Tuch, C. 2004. Flüssigkristall-Anzeigen. *Chemie in Unserer Zeit*, 38(1), 58–89.
67. Allen, K. 2005. Thin film transistors. In *Flexible Flat Panel Displays*, edited by G. P. Crawford, 507. Wiley-SID Series in Display Technology. Chichester: John Wiley & Sons, Ltd.
68. Borchardt, J. K. 2004. Developments in organic displays. *Materials Today*, 7(9), 42–46.
69. Evans, R. C., P. Douglas, C. J. Winscom. 2006. Coordination complexes exhibiting room-temperature phosphorescence: Evaluation of their suitability as triplet emitters in organic light emitting diodes. *Coordination Chemistry Reviews*, 250(15–16), 2093–2126.
70. Su, S.-J., E. Gonmori, H. Sasabe, J. Kido. 2008. Highly efficient organic blue- and white-light-emitting devices having a carrier- and exciton-confining structure for reduced efficiency roll-off. *Advanced Materials*, 20(21), 4189–4194.
71. Kulkarni, A. P., C. J. Tonzola, A. Babel, S. A. Jenekhe. 2004. Electron transport materials for organic light-emitting diodes. *Chemistry of Materials*, 16(23), 4556–4573.

72. Khan, R. U. A., O. P. Kwon, A. Tapponnier, A. N. Rashid, P. Günter. 2006. Supramolecular ordered organic thin films for nonlinear optical and optoelectronic applications. *Advanced Functional Materials*, 16(2), 180–188.

73. Tada, K., M. Onoda. 2003. Preparation and application of nanostructured conjugated polymer film by electrophoretic deposition. *Thin Solid Films*, 438, 365–368.

74. Tada, K., M. Onoda. 2006. Preparation of nanostructured conjugated polymer films from suspension-based technique and their applications. *Thin Solid Films*, 499(1–2), 19–22.

75. Jagannathan, R., G. Irvin, T. Blanton, S. Jagannathan. 2006. Organic nanoparticles: Preparation, self-assembly, and properties. *Advanced Functional Materials*, 16(6), 747–753.

76. Yu, S. C., C. C. Kwok, W. K. Chan, C. M. Che. 2003. Self-assembled electroluminescent polymers derived from terpyridine-based moieties. *Advanced Materials*, 15(19), 1643–1647.

77. Yang, P. J., C. W. Wu, D. Sahu, H. C. Lin. 2008. Study of supramolecular side-chain copolymers containing light-emitting H-acceptors and electron-transporting dendritic H-donors. *Macromolecules*, 41(24), 9692–9703.

78. Lin, H. C., C. M. Tsai, G. H. Huang, Y. T. Tao. 2006. Synthesis and characterization of light-emitting H-bonded complexes and polymers containing bis(pyridyl) emitting acceptors. *Macromolecules*, 39(2), 557–568.

79. Türker, L., A. Tapan, S. Gümüs. 2009. Electroluminescent properties of certain polyaromatic compounds: Part 1-Characteristics of OLED devices based on fluorescent polyaromatic dopants. *Polycyclic Aromatic Compounds*, 29(3), 123–138.

80. Türker, L., A. Tapan, S. Gümüs. 2009. Electroluminescent properties of certain polyaromatic compounds: Part 2-Organic emitters. *Polycyclic Aromatic Compounds*, 29(3), 139–159.

81. Veinot, J. G. C., T. J. Marks. 2005. Toward the ideal organic light-emitting diode. The versatility and utility of interfacial tailoring by cross-linked siloxane interlayers. *Accounts of Chemical Research*, 38(8), 632–643.

82. Oregan, B., M. Grätzel. 1991. A Low-cost, high-efficiency solar-cell based on dye-sensitized colloidal TIO2 films. *Nature*, 353(6346), 737–740.

83. Nazeeruddin, M. K., A. Kay, I. Rodicio, R. Humphrybaker, E. Müller, P. Liska, N. Vlachopoulos, M. Grätzel. 1993. Conversion of light to electricity by cis-X2bis(2,2'-bipyridyl-4,4'-dicarboxylate)ruthenium(II) charge-transfer sensitizers (X = Cl−, Br−, I−, CN−, and SCN−) on nanocrystalline $TiO_2$ electrodes. *Journal of the American Chemical Society*, 115(14), 6382–6390.

84. Shirota, Y., H. Kageyama. 2007. Charge carrier transporting molecular materials and their applications in devices. *Chemical Reviews*, 107(4), 953–1010.

85. Koch, N. 2007. Organic electronic devices and their functional interfaces. *ChemPhysChem*, 8(10), 1438–1455.

86. Mayer, A. C., S. R. Scully, B. E. Hardin, M. W. Rowell, M. D. McGehee. 2007. Polymerbased solar cells. *Materials Today*, 10(11), 28–33.

87. Howard, I. A., F. Laquai. 2010. Optical probes of charge generation and recombination in bulk heterojunction organic solar cells. *Macromolecular Chemistry and Physics*, 211(19), 2063–2070.

88. Zojer, E., N. Koch, P. Puschnig, F. Meghdadi, A. Niko, R. Resel, C. Ambrosch-Draxl, M. Knupfer, J. Fink, J. L. Bredas, G. Leising. 2000. Structure, morphology, and optical properties of highly ordered films of para-sexiphenyl. *Physical Review B*, 61(24), 16538–16549.

89. Bulovic, V., S. R. Forrest. 1995. Excitons in crystalline thin-films of 3,4,9,10-perylenetetracarboxylic dianhydride studied by photocurrent response. *Chemical Physics Letters*, 238(1–3), 88–92.

90. Beckers, E. H. A., Z. J. Chen, S. C. J. Meskers, P. Jonkheijm, A. Schenning, X. Q. Li, P. Osswald, F. Würthner, R. A. J. Janssen. 2006. The importance of nanoscopic ordering on the kinetics of photoinduced charge transfer in aggregated pi-conjugated hydrogen-bonded donor-acceptor systems. *Journal of Physical Chemistry B*, 110(34), 16967–16978.

91. Bullock, J. E., R. Carmieli, S. M. Mickley, J. Vura-Weis, M. R. Wasielewski. 2009. Photoinitiated charge transport through pi-stacked electron conduits in supramolecular ordered assemblies of donor-acceptor triads. *Journal of the American Chemical Society*, 131(33), 11919–11929.

92. Wu, J. S., W. Pisula, K. Müllen. 2007. Graphenes as potential material for electronics. *Chemical Reviews*, 107(3), 718–747.

93. Li, J., M. Kastler, W. Pisula, J. W. F. Robertson, D. Wasserfallen, A. C. Grimsdale, J. Wu, K. Müllen. 2007. Organic bulk-heterojunction photovoltaics based on alkyl substituted discotics. *Advanced Functional Materials*, 17(14), 2528–2533.

94. Schmidt-Mende, L., A. Fechtenkötter, K. Müllen, E. Moons, R. H. Friend, J. D. MacKenzie. 2001. Self-organized discotic liquid crystals for high-efficiency organic photovoltaics. *Science*, 293(5532), 1119–1122.

95. Granstrom, M., K. Petritsch, A. C. Arias, A. Lux, M. R. Andersson, R. H. Friend. 1998. Laminated fabrication of polymeric photovoltaic diodes. *Nature*, 395(6699), 257–260.

96. Ma, W. L., C. Y. Yang, X. Gong, K. Lee, A. J. Heeger. 2005. Thermally stable, efficient polymer solar cells with nanoscale control of the interpenetrating network morphology. *Advanced Functional Materials*, 15(10), 1617–1622.

97. Hoeben, F. J. M., P. Jonkheijm, E. W. Meijer, A. P. H. J. Schenning. 2005. About supramolecular assemblies of pi-conjugated systems. *Chemical Reviews*, 105(4), 1491–1546.

98. Mishra, A., C. Q. Ma, P. Bäuerle. 2009. Functional oligothiophenes: Molecular design for multidimensional nanoarchitectures and their applications. *Chemical Reviews*, 109(3), 1141–1276.

99. Würthner, F., Z. J. Chen, F. J. M. Hoeben, P. Osswald, C. C. You, P. Jonkheijm, J. von Herrikhuyzen, A. Schenning, P. van der Schoot, E. W. Meijer, E. H. A. Beckers, S. C. J. Meskers, R. A. J. Janssen. 2004. Supramolecular p-n-heterojunctions by co-self-organization of oligo(p-phenylene vinylene) and perylene bisimide dyes. *Journal of the American Chemical Society*, 126(34), 10611–10618.

100. Sanchez, L., N. Martin, D. M. Guldi. 2005. Hydrogen-bonding motifs in fullerene chemistry. *Angewandte Chemie (International ed. in English)*, 44(34), 5374–5382.

101. Lee, J. I., S. H. Cho, S. M. Park, J. K. Kim, J. K. Kim, J. W. Yu, Y. C. Kim, T. P. Russell. 2008. Highly aligned ultrahigh density arrays of conducting polymer nanorods using block copolymer templates. *Nano Letters*, 8(8), 2315–2320.

102. Snaith, H. J., G. L. Whiting, B. Q. Sun, N. C. Greenham, W. T. S. Huck, R. H. Friend. 2005. Self-organization of nanocrystals in polymer brushes. Application in heterojunction photovoltaic diodes. *Nano Letters*, 5(9), 1653–1657.

103. Robbie, K., M. J. Brett, A. Lakhtakia. 1996. Chiral sculptured thin films. *Nature*, 384(6610), 616–616.

104. Hrudey, P. C. P., K. L. Westra, M. J. Brett. 2006. Highly ordered organic Alq3 chiral luminescent thin films fabricated by glancing-angle deposition. *Advanced Materials*, 18(2), 224–228.

105. Zhang, J., I. Salzmann, S. Rogaschewski, J. P. Rabe, N. Koch, F. J. Zhang, Z. Xu. 2007. Arrays of crystalline C-60 and pentacene nanocolumns. *Applied Physics Letters*, 90(19), 193117.

106. Steinhart, M. 2008. Supramolecular organization of polymeric materials in nanoporous hard templates. *Advances in Polymer Science*, 220, 123–187.

107. Haberkorn, N., J. S. Gutmann, P. Theato. 2009. Template-assisted fabrication of freestanding nanorod arrays of a hole-conducting cross-linked triphenylamine derivative: Toward ordered bulk-heterojunction solar cells. *ACS Nano*, 3(6), 1415–1422.

108. Bai, R., M. Ouyang, R. J. Zhou, M. M. Shi, M. Wang, H. Z. Chen. 2008. Well-defined nanoarrays from an n-type organic perylene-diimide derivative for photoconductive devices. *Nanotechnology*, 19(5), 055604.

109. Hesse, H. C., D. Lembke, L. Dössel, X. Feng, K. Müllen, L, Schmidt-Mende. 2011. Nanostructuring discotic molecules on ITO support. *Nanotechnology*, 22(5), 055303.
110. Herz, L. M., C. Daniel, C. Silva, F. J. M. Hoeben, A. P. H. J. Schenning, E. W. Meijer, R. H. Friend, R. T. Phillips. 2003. Fast exciton diffusion in chiral stacks of conjugated p-phenylene vinylene oligomers. *Physical Review B*, 68(4), 045203.
111. Bayer, A., J. Hübner, J. Kopitzke, M. Oestreich, W. Rühle, J. H. Wendorff. 2001. Timeresolved fluorescence in 3-dimensional ordered columnar discotic materials. *Journal of Physical Chemistry B*, 105(20), 4596–4602.
112. Würthner, F., K. Meerholz. 2010. Systems chemistry approach in organic photovoltaics. *Chemistry—A European Journal*, 16(31), 9366–9373.
113. Sariciftci, N. S., L. Smilowitz, A. J. Heeger, F. Wudl. 1992. Photoinduced electron-transfer from a conducting polymer to buckminsterfullerene. *Science*, 258(5087), 1474–1476.
114. Hoppe, H., N. S. Sariciftci. 2006. Morphology of polymer/fullerene bulk heterojunction solar cells. *Journal of Materials Chemistry*, 16(1), 45–61.
115. Morteani, A. C., P. Sreearunothai, L. M. Herz, R. H. Friend, C. Silva. 2004. Exciton regeneration at polymeric semiconductor heterojunctions. *Physical Review Letters*, 92(24), 247402.
116. Peet, J., C. Soci, R. C. Coffin, T. Q. Nguyen, A. Mikhailovsky, D. Moses, G. C. Bazan. 2006. Method for increasing the photoconductive response in conjugated polymer/fullerene composites. *Applied Physics Letters*, 89(25), 252105.
117. Peet, J., J. Y. Kim, N. E. Coates, W. L. Ma, D. Moses, A. J. Heeger, G. C. Bazan. 2007. Efficiency enhancement in low-bandgap polymer solar cells by processing with alkane dithiols. *Nature Materials*, 6(7), 497–500.
118. Baetens, R., B. P. Jelle, A. Gustavsen. 2010. Properties, requirements and possibilities of smart windows for dynamic daylight and solar energy control in buildings: A state-ofthe-art review. *Solar Energy Materials and Solar Cells*, 94(2), 87–105.
119. Xiong, S. X., P. T. Jia, K. Y. Mya, J. Ma, F. Boey, X. H. Lu. 2008. Star-like polyaniline prepared from octa(aminophenyl) silsesquioxane: Enhanced electrochromic contrast and electrochemical stability. *Electrochimica Acta*, 53(9), 3523–3530.
120. Zhang, L. Y., S. X. Xiong, J. Ma, X. H. Lu. 2009. A complementary electrochromic device based on polyaniline-tethered polyhedral oligomeric silsesquioxane and tungsten oxide. *Solar Energy Materials and Solar Cells*, 93(5), 625–629.
121. Wang, X. J., K. Y. Wong. 2006. Effects of a base coating used for electropolymerization of poly(3,4-ethylenedioxythiophene) on indium tin oxide electrode. *Thin Solid Films*, 515(4), 1573–1578.
122. Deepa, M., S. Bhandari, M. Arora, R. Kant. 2008. Electrochromic response of nano-structured poly(3,4-ethylenedioxythiophene) films grown in an aqueous micellar solution. *Macromolecular Chemistry and Physics*, 209(2), 137–149.
123. Thompson, B. C., P. Schottland, K. W. Zong, J. R. Reynolds. 2000. In situ colorimetric analysis of electrochromic polymers and devices. *Chemistry of Materials*, 12(6), 1563–1571.
124. Argun, A. A., P. H. Aubert, B. C. Thompson, I. Schwendeman, C. L. Gaupp, J. Hwang, N. J. Pinto, D. B. Tanner, A. G. MacDiarmid, J. R. Reynolds. 2004. Multicolored electro-chromism polymers: Structures and devices. *Chemistry of Materials*, 16(23), 4401–4412.
125. Doane, J. W., G. Chidichimo, N. A. Vaz. 1987. Light modulating material comprising a liquid dispersion in a plastic matrix. U.S. Patent 4688900.
126. Fergason, J. L. 1984. Encapsulated liquid crystal and method. U.S. Patent 4435047.
127. Cupelli, D., F. P. Nicoletta, S. Manfredi, G. De Filpo, G. Chidichimo. 2009. Electrically switchable chromogenic materials for external glazing. *Solar Energy Materials and Solar Cells*, 93(3), 329–333.
128. Samuel, I. D. W., G. A. Turnbull. 2007. Organic semiconductor lasers. *Chemical Reviews*, 107(4), 1272–1295.

129. Calzado, E. M., P. G. Boj, M. A. Diaz-Garcia. 2010. Amplified spontaneous emission properties of semiconducting organic materials. *International Journal of Molecular Sciences*, 11(6), 2546–2565.
130. Clark, J., G. Lanzani. 2010. Organic photonics for communications. *Nature Photonics*, 4(7), 438–446.
131. Würthner, F., T. E. Kaiser, C. R. Saha-Möller. 2011. J-aggregates: From serendipitous discovery to supramolecular engineering of functional dye materials. *Angewandte Chemie (International ed. in English)*, 50(15), 3376–3410.
132. Torsi, L., A. J. Lovinger, B. Crone, T. Someya, A. Dodabalapur, H. E. Katz, A. Gelperin. 2002. Correlation between oligothiophene thin film transistor morphology and vapor responses. *Journal of Physical Chemistry B*, 106(48), 12563–12568.
133. Someya, T., H. E. Katz, A. Gelperin, A. J. Lovinger, A. Dodabalapur. 2002. Vapor sensing with alpha,omega-dihexylquarterthiophene field-effect transistors: The role of grain boundaries. *Applied Physics Letters*, 81(16), 3079–3081.
134. Wang, L., D. Fine, T. H. Jung, D. Basu, H. von Seggern, A. Dodabalapur. 2004. Pentacene field-effect transistors with sub-10-nm channel lengths. *Applied Physics Letters*, 85(10), 1772–1774.
135. Verlaak, S., V. Arkhipov, P. Heremans. 2003. Modeling of transport in polycrystalline organic semiconductor films. *Applied Physics Letters*, 82(5), 745–747.
136. Ampuero, S., J. O. Bosset. 2003. The electronic nose applied to dairy products: A review. *Sensors and Actuators B: Chemistry*, 94(1), 1–12.
137. Goepel, W., J. Hesse, J. N. Zemel. 1991. *Sensors, a comprehensive survey*. New York: VCH Weinheim.
138. Matsumoto, A., N. Sato, T. Sakata, K. Kataoka, Y. Miyahara. 2009. Glucose-sensitive field effect transistor using totally synthetic compounds. *Journal of Solid State Electrochemistry*, 13(1), 165–170.
139. Steele, B. C. H., A. Heinzel. 2001. Materials for fuel-cell technologies. *Nature*, 414(6861), 345–352.
140. Kreuer, K. D., S. J. Paddison, E. Spohr, M. Schuster. 2004. Transport in proton conductors for fuel-cell applications: Simulations, elementary reactions, and phenomenology. *Chemical Reviews*, 104(10), 4637–4678.
141. Connolly, D. J., W. F. Gresham. 1966. Fluorocarbon vinyl ether polymers. U.S. Patent 3282875.
142. Thomas, A., P. Kuhn, J. Weber, M. M. Titirici, M. Antonietti. 2009. Porous polymers: enabling solutions for energy applications. *Macromolecular Rapid Communications*, 30(4–5), 221–236.
143. Kreuer, K. D. 2001. On the development of proton conducting polymer membranes for hydrogen and methanol fuel cells. *Journal of Membrane Science*, 185(1), 29–39.
144. Kreuer, K. D., A. Rabenau, W. Weppner. 1982. Vehicle mechanism, a new model for the interpretation of the conductivity of fast proton conductors. *Angewandte Chemie (International ed. in English)*, 21(3), 208–209.
145. Weber, J., K. D. Kreuer, J. Maier, A. Thomas. 2008. Proton conductivity enhancement by nanostructural control of poly(benzimidazole)-phosphoric acid adducts. *Advanced Materials*, 20(13), 2595–2598.
146. Steininger, H., M. Schuster, K. D. Kreuer, A. Kaltbeitzel, B. Bingol, W. H. Meyer, S. Schauff, G. Brunklaus, J. Maier, H. W. Spiess. 2007. Intermediate temperature proton conductors for PEM fuel cells based on phosphonic acid as protogenic group: A progress report. *Physical Chemistry Chemical Physics*, 9(15), 1764–1773.
147. Jiménez-García, L., A. Kaltbeitzel, W. Pisula, J. S. Gutmann, M. Klapper, K. Müllen. 2009. Phosphonated hexaphenylbenzene: A crystalline proton conductor. *Angewandte Chemie (International ed. in English)*, 48(52), 9951–9953.

# 10 Molecular Tectonics
## *An Approach to Crystal Engineering*

### Mir Wais Hosseini

## CONTENTS

## INTRODUCTION

Molecular crystals are solid and periodic entities. They are defined by the nature of their molecular components and interactions between them in the solid state. Although crystals are described by the translation of the unit cell into all three directions of space (periodicity in three space directions); nevertheless, by considering crystals as supramolecular entities,[1,2] one may describe crystals in terms of molecular lattices (a particular case of networks) by analyzing intermolecular interactions (specific recognition patterns) and their geometrical features. Because the term molecular network has been widely used, we will continue to use it in replacement of molecular lattice, which would be a more appropriate term. In marked contrast with discrete molecules, molecular networks display translational symmetry. Thus, they may be regarded as infinite molecular assemblies for which specific interaction patterns are repeated through space. In other words, the recognition motifs or supramolecular synthons[3] taking place between molecular components comprising the crystal become the assembling nodes of the network. The dimensionality of molecular network depends on the number of translations operating at the level of the assembling nodes.[4,5] Thus, one-dimensional (1-D) or α-networks are formed when a single translation takes place. Similarly, β-networks (two-dimensional; 2-D) and γ-networks (three-dimensional; 3-D) are defined when two or three translations, respectively, of the same or different assembling nodes are present. Although in principle molecular networks may be obtained in any type of condensed media such as solution, gel, or solid state, the crystalline phase has been extensively used because of the possibility of accurate structural studies by X-ray diffraction methods.

This way of analyzing molecular crystals in terms of networks is called molecular tectonics.[6-9] This approach leads to the definition of tectons[10] or construction units, which are structurally and energetically defined active molecular building blocks bearing within their backbone an assembling program based on molecular recognition processes. The molecular programming leading to the formation of molecular networks with predefined dimensionality and connectivity is controlled by the nature and localization of recognition sites within the structure of tectons.

The strength of molecular tectonics is related to the fact that this approach not only permits the description of a given crystal in terms of networks but also, and more interestingly, allows predefined molecular networks to be conceived through the specific design of tectons. At that stage, one may draw a parallel between molecular synthesis, which deals with the connection of atoms by covalent bonds to generate molecules and supramolecular synthesis,[2] which in this context concerns the formation of networks through reversible intermolecular interactions. Tectons are prepared through covalent synthesis, allowing the construction of the backbone (geometrical information) to which recognition sites (interaction information) are connected. In principle, tectons are configurationally robust entities and thus survive the assembling step into networks. The latter step is governed by principles guiding supramolecular synthesis, in particular reversible interactions between recognition sites located on complementary tectons.

It is worth noting that because of our limited knowledge of all intermolecular interactions governing the formation of the crystalline phases, the complete understanding of the packing of molecular entities is currently impossible.[11] However, with our current level of understanding of intermolecular interactions and our ability to master some of them, one may conceive molecular networks with predicted structures in terms of connectivity with an acceptable degree of confidence.

## FROM TECTONS TO NETWORKS

Molecular networks may be generated in the crystalline phase through self-assembly[2,12,13] processes engaging either a self-complementary or several complementary tectons. Although a mono component system composed of a single self-complementary tecton would be the most attractive situation in terms of atoms and synthetic economy, in practice, often, such a system produces insoluble powders difficult to characterize structurally. The use of a poly component system composed of several complementary tectons is a more viable strategy. Indeed, for poly component systems, both thermodynamic and kinetic parameters leading to the formation of the crystalline material may be monitored. Consequently, one may use X-ray diffraction methods for structural investigations. However, it is worth noting that increasing the number of components would lead to a larger possibility space, and in terms of predictability, the formation of molecular networks with projected connectivity and dimensionality would require rather fine tecton design.

For illustration purposes and simplicity of schematic representation, we shall focus on 1-D and 2-D networks. The formation of 1-D networks requires tectons bearing at least two interaction sites arranged in a divergent fashion. The divergent orientation is essential for avoiding the formation of discrete species. A 1-D network may be generated using a self-complementary tecton (Figure 10.1a). Such a network may also be designed using a two-component system on the basis of a combination of two complementary tectons (Figure 10.1b and c). For this design, one may use two centric tectons leading thus to a single recognition pattern (Figure 10.1b) or two noncentric tectons generating two different assembling nodes (Figure 10.1c). One may increase the number of tectons. For example, one may use a combination of three different tectons leading to two different recognition patterns (Figure 10.1d) or a tetra-component system on the basis of four different tectons generating two different assembling nodes (Figure 10.1e and f). The analysis given here is not exhaustive. One may indeed imagine a large variety of cases. The only limitation is the imagination and the synthetic ability of chemists to prepare the tectons.

As stated earlier, for 1-D networks, the information coded within the structure of tectons concerns only the recognition event generating the assembling nodes and the iteration of the process leading to its translation into one direction of space. In principle, the arrangements in the other two directions of space leading thus to the crystal are not a priori controlled. In other words, the system is free to find its own way to pack the 1-D networks in the other two directions.

For the design of 2-D networks, tectons bearing at least three divergently oriented recognition sites are required. Again bidimensional networks may either be

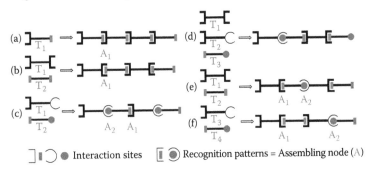

FIGURE 10.1   A nonexhaustive schematic representation of a variety of 1-D molecular networks based on self-assembly of either (a) self-complementary or (b–f) complementary tectons. $T_i$ and $A_i$ represent tectons and assembling nodes.

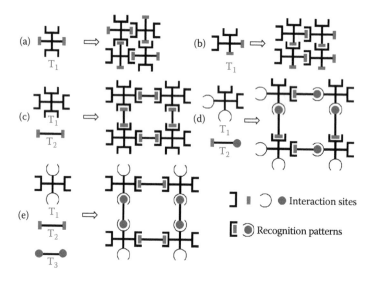

**FIGURE 10.2**  A nonexhaustive schematic representation of a variety of 2-D molecular networks based on self-assembly of either (a, b) self-complementary or (c–e) complementary tectons. $T_i$ and $A_i$ represent tectons and assembling nodes.

generated using a single component system based on a unique self-complementary tecton (Figure 10.2a and b) or poly component systems based on combinations of complementary tectons (Figure 10.2c through e). For example, depending on the number of recognition sites, a two-component system may be based on two complementary tectons leading to one assembling node (Figure 10.2c) or generating two different recognition patterns (Figure 10.2d). As in the case of 1-D networks mentioned earlier, one may also increase the number of components. For example, one may design a combination based on three tectons generating two different assembling nodes (Figure 10.2e). The formal analysis presented here is not exhaustive, and only some selected examples are given here. Again as mentioned for 1-D networks, the increase in the number of tectons might generate diversity (several connectivity) if the design of different tectons is not accurately carried out.

As in the case of 1-D networks, 2-D networks possess lower dimensionality than the crystal. The information coded within the structure of tectons concerns in that case the formation of bidimensional architecture (control of two space dimensions). Consequently, the arrangement in the third dimension, that is, the packing of 2-D sheets, in principle is not programmed and thus cannot be predicted.

Only in the case of 3-D networks the connectivity between tectons in all three dimensions of space is programmed. Because of this lack of freedom, the generation of 3-D networks requires rather precise tecton design in terms of both geometry and metrics.

## SELF-ASSEMBLY AND SELF-HEALING

The difference between the size of tectons (around a nanometer) and the molecular networks (micrometer to millimeter) is a factor of $10^3$ to $10^6$. In other words, to generate a molecular network, thousands to millions of tectons are needed. This clearly demonstrates that a step-by-step type strategy cannot be viable to assemble tectons into networks. The only strategy that one may use is self-assembly.[2,12,13] As stated earlier, the formation of molecular networks in the crystalline phase requires a large number of assembling steps. To avoid the generation of diversity, that is, a large collection of structures, one must use reversible interactions allowing self-healing processes. Indeed, reversible interactions would allow the system to self-repair mistakes generated during the assembling process. In other words, the system composed of tectons would find its own way to the most stable situation under given conditions (temperature, pressure, concentration, solvent, etc.).

The use of nonreversible interactions such as covalent bond formation, because of large activation barriers between different states, would generate a series of energy minima and thus a variety of different structures. The use of reversible processes, because of low energy barriers between local minima corresponding to different intermediate states, would allow the system to reach a final state with the lowest energy. The number of final states depends on the quality of the design of tectons. A perfect design must lead to a single final state. For a less accurate design, one may obtain two or several final states, which may be convertible structures if the energy barrier between them is low or nonconvertible states if the barrier between them is high. Using this simplistic approach, one may also describe the formation of polymorphs and supramolecular structural isomers.

## MOLECULAR RECOGNITION AND REVERSIBLE INTERACTIONS

The recognition event between complementary sites may take place through a variety of reversible attractive intermolecular interactions (Figure 10.3). The toolbox available to supramolecular chemists is composed of van der Waals interactions, charge-assisted electrostatic interactions, π–π interactions, charge transfer interactions, metal–metal interactions, hydrogen (H) bonding, and coordination bonding. In terms of energy, van der Waals interactions are among the weakest (approximately 0.5–2 kcal/mol) and for that reason are rather difficult to master. H bonding, one of the most attractive interactions, ranges from weak to rather strong (approximately 2–10 kcal/mol). Metal–metal interactions such as aurophilic interactions may be as strong as H bonding. Electrostatic charge–charge interactions may be substantially strong (up to 50 kcal/mol). Finally, coordination bonding, which is another

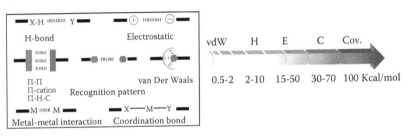

**FIGURE 10.3** Schematic representation of toolbox of reversible intermolecular interactions (left) and their energy (right).

commonly used interaction, ranges from weak to rather strong (30–70 kcal/mol). The terminology used thus far defines molecular networks in roughly three categories: inclusion networks mainly based on van der Waals interactions, H-bonded networks and charge-assisted H-bonded networks (combination of H bonding and charge–charge electrostatic interactions), and finally coordination networks or polymers mainly based on the use of coordination bonds.

As stated earlier, although in principle any type of reversible intermolecular interactions may be used for the design of molecular networks, most cases reported to date are based on either H bonding[3–5,7,8,10,13–25] or coordination bonds.[26–33] A few examples of inclusion networks mainly based on van der Waals interactions in the solid state between concave and convex tectons have been also reported.[34,35]

## Molecular Tectonics and Crystal Engineering

We have shown that molecular crystals may be described as molecular networks by considering some specific recognition patterns (assembling nodes) and their translations. At that stage, one must reflect on the recurrent and important question dealing with the predictability of crystal structures. This matter has always been debated in the community of "crystal engineers."[3,15] As discussed previously by Gavezzotti[11] and Dunitz,[36] with our current level of understanding, by no means, one may predict crystal structures from the knowledge of the molecular components. In particular, we are unable to predict the adopted conformation of flexible tectons, the presence of solvent molecules, their arrangement, interpenetration, polymorphism, and so forth. However, as stated by Dunitz,[1] by considering a crystal as a "supermolecule *par excellence*," by using concepts and principles developed in supramolecular chemistry,[2] one may, through the control of molecular recognition events, predict some of the connectivity patterns between tectons and thus the formation of molecular networks. This is precisely the aim of molecular tectonics. On the basis of continuous investigations by several research groups, today, one may design tectons and predict their interconnections into molecular networks in the crystalline phase with an acceptable degree of precision. The reliability of the prediction depends on the quality of the design of tectons, in other words, the nature of the energy associated with interactions between tectons, the structural rigidity of tectons (restricted conformational space), the number of tectons involved, and so forth. A further step, which must be taken, is related to the precise energetic analysis of molecular networks formation, in other words, the quantification of contributions of different inter-tecton interactions in the solid phase.[35] Today, molecular tectonics is in its infancy, and much more systematic investigations are needed before reaching maturity and before claiming to fully understand the formation of molecular networks in the crystalline phase.

Here, we present some selected examples of strategies and networks following an increasing order of energy of interaction.

### Inclusion Networks: An Interplay between Concave and Convex Tectons

Inclusion processes between concave and convex unites lead to molecular complexes in solution and/or in the solid state. Many receptors possessing a cavity have been designed to bind substrates through inclusion processes. On the other hand, the formation of clathrates in the solid state has been also extensively explored. We have extended the concept of inclusion in solution and clathrates formation in the solid state to the design of inclusion networks called koilates.[38] The design of koilates is based on the use of koilands (from the Greek word *koilos* meaning hollow),[39] which are multicavity tectons possessing at least two cavities arranged in a divergent fashion, and connectors capable of being included within the cavities of the koilands and thus connecting consecutive units by inclusion processes. van der Waals interactions are the driving forces for the formation of koilates. It is interesting to note that recognizing a network in a molecular crystal is a matter of translation of the recognition pattern. For inclusion networks exclusively based on the van der Waals interaction, spotting a network is not trivial because the packing of all components is ensured by the same type of interactions. However, by considering the inclusion between the koiland and the connector as a specific recognition pattern, this difficulty may be overcome.

For the design of koilates, we have explored a variety of possibilities in terms of geometry and dimensionality.[31] For the purpose of demonstration, here we shall restrict ourselves to 1-D koilates on the basis of two-component systems, that is, a combination of koilands possessing two divergent cavities and linear connectors. For 1-D koilates, as for any 1-D networks, two possibilities (directional or nondirectional) may be envisaged (Figure 10.4). A combination of centrosymmetric koilands and connectors will lead to nondirectional koilates because of the translation of centers of symmetry (Figure 10.4a). To impose the directionality, one may envisage a combination of nonsymmetric koilands and or connectors (Figure 10.5b through f). When combining centrosymmetric koilands and non-centrosymmetric connectors, either directional (Figure 10.4b) or nondirectional (Figure 10.4c) koilates may be generated. Similarly, when using non-centrosymmetric koilands and symmetric connectors, directional (Figure 10.4d) or nondirectional (Figure 10.4e) koilates may be obtained. Finally, the most promising strategy, based on specific recognition between koilands and connectors, is a combination of non-centrosymmetric koilands and connectors (Figure 10.4f).

## Design of Koilands

Calix[4]arene derivatives[40] are appropriate candidates for the design of koilands because these compounds accommodate in the solid state a variety of neutral guest

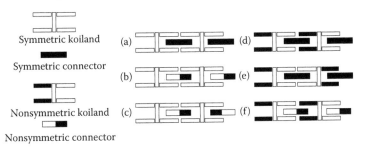

**FIGURE 10.4** Schematic representations of centric (a, c, e) and acentric (b, d, f) 1-D inclusion networks based on interconnection of koilands and connector molecules.

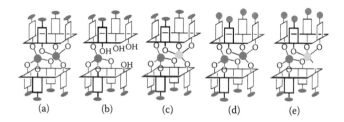

(a)          (b)          (c)          (d)          (e)

**FIGURE 10.5** Schematic representation of centrosymmetric (a) and non-centrosymmetric (b–e) koilands.

molecules in their cavity (Scheme 10.1). Indeed, these compounds offer a preorganized and tunable hydrophobic pocket as well as four hydroxy groups for further functionalization. However, the calix[4]arene backbone presents conformational flexibility. Among the four possible conformations, only the cone conformation is of interest for the design of koilands. Furthermore, these compounds offer the possibility of controlling both the entrance and the depth of the cavity by the nature of the substituents X. Although a large number of calix[4]arene derivatives have been used for the synthesis of koilands, here we shall focus only on three of them (**1–3**).

Taking into account that a calix[4]arene moiety offers four OH groups, two of such a molecule in cone conformation and in a face-to-face type arrangement were fused by two silicon atoms.[39] Four examples of koilands are given in Scheme 10.1.

*Centric and Noncentric Koilands*

Centrosymmetric koilands (Figure 10.5a) may be generated on double fusion of two identical calix[4]arene derivatives by two identical fusing elements such as silicon atom (see koilands **4** and **5** in Scheme 10.3). The design of nonsymmetric koilands may be based on electronic, geometric, or both electronic and geometric differentiation of the two cavities (Figure 10.5b through e). Electronic differentiation may be achieved using a single silicon atom to connect two calix units (Figure 10.5b). Indeed,

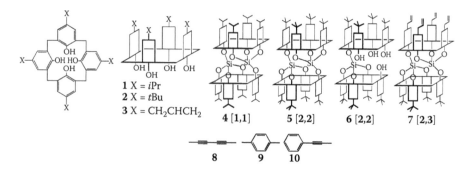

**SCHEME 10.1** Chemical structures of calix[4]arene derivatives **1–3** in cone conformation and connectors **8–10** as well as schematic representations of koilands **4–7**.

in that case, one of the two calix units is triply coordinated to a Si atom, whereas the other one is coordinated by a single bond to the Si atom. Another possibility may be based on two different fusing atoms (same oxidation state IV) such as Si and Ti or Ge (Figure 10.5c). Although all three cases have been investigated, here we only present the case of koiland **6**.[41] Geometric differentiation may be accomplished while keeping the same fusing element such as Si by connecting two different calix units. Here, we only describe the behavior of the geometrically differentiated koiland **7**.[42] The third possibility consisting of fusing two different calix derivatives by two different atoms has not been yet exploited.

### Nondirectional and Directional Koilates

Relevant examples of koilates using koilands **4–7** and connectors **8–10** are given in Figure 10.6. The formation of nondirectional koilates in the crystalline phase was demonstrated by combining the centric koilands **4** (Figure 10.6a)[43] or **5** (Figure 10.6b)[44] with **8**, a rod-type molecule possessing both the requested linear geometry and the distance between its two terminal $CH_3$ groups. As expected, linear koilates are formed through the interconnection of consecutive koilands by the connector molecules. Each connector bridges two consecutive koilands by penetrating their cavities through its terminal $CH_3$ groups. The methyl groups of **8** are deeply inserted into the cavity of the koiland, establishing between 8 and 12 van der Waals contacts.

Because **8** is the most appropriate connector for koiland **5**, the same connector was used to generate directional koilate with **6** (Figure 10.6c).[45] The connectivity between **6** and the **8** is roughly the same as the one described earlier. Because of the non-centrosymmetric nature of **6**, the network generated is directional. The packing of koilates within the crystal is parallel. However, because of the opposite orientation of linear arrays, a nondirectional arrangement is obtained.

Dealing with the geometrically differentiated koiland **7**, the formation of directional koilates was studied using **9** as a centrosymmetric and **10** as a non-centrosymmetric connector.[46] For a combination of **7** and **9** (Figure 10.6d), a directional koilate formed by the bridging of consecutive koilands **7** by **9** was obtained. Interestingly, because the **9** is centric, the noncentric nature of the network is due to nonsymmetrical

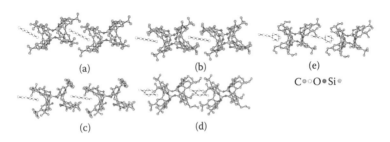

(a)          (b)          (e)

C ⊙ ○ O ● Si ⊚

(c)          (d)

**FIGURE 10.6** Portions of the x-ray structures of koilates formed between koilands (a) **4**, (b) **5**, and (c) **6** and the connector **8**, koiland **7** with connectors (d) **9** and (e) **10**. For the sake of clarity, the carbon atoms of the connector are colored in yellow, and H atoms are not represented.

arrangement of consecutive koilands **7**. Unfortunately, again, the directional linear koilates are packed in antiparallel fashion.

The final example deals with the combination of the noncentric koiland **7** and the noncentric connector **10** (Figure 10.6e). As in the previous case discussed earlier, again a directional koilate is observed. In addition, the 1-D inclusion networks are positioned in a parallel fashion but with opposite orientation.

## H-BONDED NETWORKS

H bonding is one of the most used interactions for the design of molecular networks.[3,14–24,47] The H bond (DH⋯A), formed between two electronegative atoms D (donor) and A (acceptor), is defined by the nature of D and A, the distance between them, and by the (DHA) angle. For design of H-bonded networks, one may use different modes of H bonding. The monohapto mode, although synthetically easier to achieve because of the large angular (DHA) distribution,[48] is not always a viable mode because often it does not allow a fine control of geometrical features of the recognition pattern and thus the prediction of the connectivity and the overall topology of the network. However, by restricting the number of possible geometrical arrangements using di- or tri-hapto mode of H bonding, one may control, to a certain extent, and thus predict the connectivity pattern between tectons.

### Neutral 1-D Helical H-Bonded Networks

For single-stranded 1-D networks, one may obtain linear, stair type, or helical geometries. The design of helical H-bonded networks is a subject of current investigations. Although few cases of enantiomerically pure H-bonded helical networks have been published,[49] most examples reported deals with achiral or a racemic mixture of tectons, which leads to the formation of racemates.

Molecular networks may be generated using either a self-complementary tecton (single component system) or two or more complementary tectons (multi component system). Let us focus here on the first possibility. Neutral tectons **11** and **12** (Figure 10.7) are based on the optically active isomannide backbone bearing both H-bond

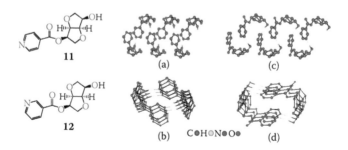

**FIGURE 10.7** Portions of the structure of enantiomerically pure 1-D helical H-bonded networks formed on the self-bridging of tectons (a, b) **11** or (c, d) **12**. H atoms, except those involved in H bonding, are omitted for clarity.

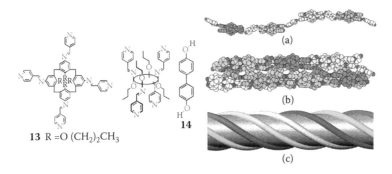

**FIGURE 10.8** A portion of the structure of the single-stranded helical H-bonded network formed between (a) tectons **13** and **14**, (b) the braid generated by five helical strands, and its schematic representation (c). H atoms are omitted for clarity.

donor (OH) and acceptor (pyridine) sites. Because of the stereochemistry of the iso-mannide unit, in both cases the pyridine and the OH groups are divergently oriented toward the concave face of the bicyclic unit. The difference between enantiomeri-cally pure tectons **11** and **12** results from the orientation of the nitrogen atom of the pyridine ring.[50]

Because of their self-complementarily and enantiomerically pure nature, both tectons **11** (Figure 10.7a and b) or **12** (Figure 10.7c and d) lead to the formation of single-stranded H-bonded helical networks. The mutual interconnection of consecu-tive tectons indeed takes place through H bonding between the OH and the pyridine groups. For both helical networks, the P helicity is imposed by the predefined stereo-chemistry of isomannide. For both cases, the helical strands are packed in parallel mode. The difference between the two helical structures results from the location of the nitrogen atom on the pyridine ring inducing thus different pitches.

The second possibility deals with the formation of helical arrangements using a two-component system based on of two complementary tectons. The tecton **13**, possessing four H-bond acceptor sites of the pyridine type occupying the apices of a pseudo-tetrahedron, was combined with the linear tecton **14** possessing two H-bond donor site of the phenolic OH type (Figure 10.8).[51]

The H-bond acceptor **13** and donor **14** form by mutual bridging through H-bonds a single-stranded helical network (Figure 10.8a). Rather interestingly and unexpect-edly, probably for best compaction reasons, five helical strands of the same handed-ness are assembled into a quintuple helical braid (Figure 10.8b and c). Finally, the quintuple helices are packed in a parallel mode. Three levels may be used to describe this rather complex system. The formation of a single-stranded H-bonded helical network, the formation of the quintuple helical braided network, and the lateral asso-ciation of braided networks leading to the crystal.

## CHARGE-ASSISTED H-BONDED NETWORKS

Most reported H-bonded molecular networks are based on nonionic H bonds.[4,16–22] To increase the robustness of the recognition pattern between H-bond donors and

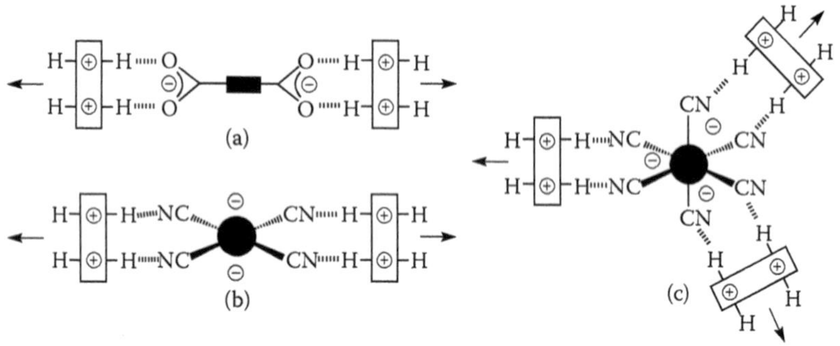

**SCHEME 10.2**  Schematic representations of recognition of dicarboxylate, tetracyano-metallate and hexacyanometallate anions by dicationic tectons through a dihapto mode of H-bonding.

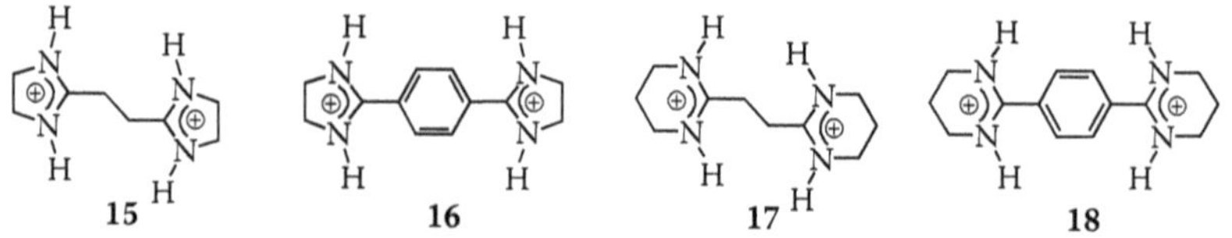

**SCHEME 10.3**  Chemical structures of bisamidinium dications **15–18**.

acceptors, one may combine directional H bonds with strong but less directional ionic interactions (Scheme 10.2).[25,52]

The *bis*-amidinium dications such as **15–18** (Scheme 10.3) are interesting building blocks for the design of charge-assisted H-bonded networks. Indeed, these dications, due to the presence of four acidic N-H protons oriented in a divergent fashion, may act as tetra H-bond donors in the presence of H-bond acceptors such as carboxylates, tetracyanometallates, and hexacyanometallates. By tuning the distance between the acidic H atoms on the same side of these units, one may induce the recognition of the above-mentioned anions by a dihapto mode of interaction (Scheme 10.2).

## H-Bonded Networks Based on Recognition of Carboxylates

The formation of 1-D H-bonded networks based on dicarboxylates was demonstrated using tectons such as **17** (Figure 10.9).[25,53–58] Because of the proper spacing of the two cyclic amidinium moieties (($CH_2$)$_2$), compound **17** recognizes on its each face the carboxylate moiety by a dihapto mode of H bonding (Scheme 10.4). This mode of recognition was also established between **17** and monocarboxylate for the formation of discrete exo-binuclear complexes.

As expected, the formation of a variety of 1-D H-bonded networks between **17** and different dicarboxylates such as **19–23** was demonstrated (Figure 10.9). In all cases, the common features are as follows: the neutral 1-D H-bonded networks are

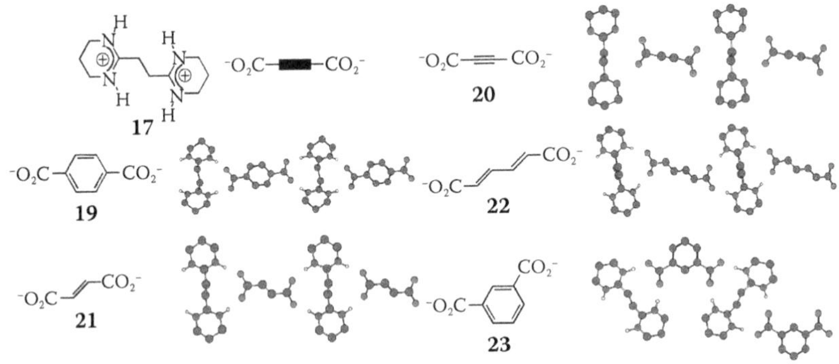

**FIGURE 10.9** Portions of structures of the 1-D networks formed between **17** and dianions **19–23**. For the sake of clarity, H atoms, except those involved in H bonding, are not represented.

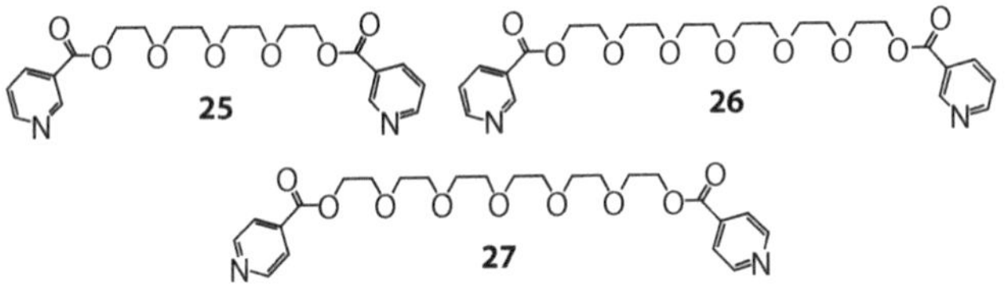

**SCHEME 10.4** Chemical structures of bismonodentate tectons **25–27**.

formed through the mutual interconnection of dicationic and dianionic tectons; the dicationic tecton **17** adopts a centrosymmetric conformation; all four acidic protons are localized on **17**; and the recognition by the dication **17** of the carboxylate moieties of the dianionic tecton takes place on each side of the tecton.

Using other combinations of cationic and anionic tectons, other 1-D as well as 2-D and 3-D H-bonded networks have been obtained.[56]

## H-Bonded Networks Based on Recognition of Tetracyanometallate

The *bis*-amidinium tecton **18** has been designed to generate H-bonded networks with $[M(CN)_4]^{2-}$.[59] Indeed, the distance between the two nitrogen atoms located on the same side of **18** is optimum for the binding of square planar $[M(CN)_4]^{2-}$ anions by a dihapto mode of H bonding. This mode of recognition may be regarded as the organization of the second coordination sphere around the transition metal (Scheme 10.4).[60,61]

The combination of **18** with $M(CN)_4^{2-}$, (M = Pd, Pt, Ni) leads, as predicted, to almost identical 1-D H-bonded networks (Figure 10.10).[62,63] These 1-D neutral networks are generated by H bonds in the dihapto mode between consecutive cationic and anionic tectons and are packed in a parallel fashion in the crystalline phase. Interestingly, all three networks obtained are isostructural.

The same type of approach was used by combining $[MCl_4]^{2-}$ anions with protonated *bis*-pyridine derivatives.[62]

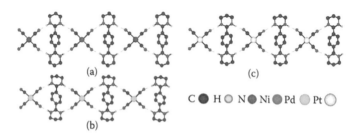

**FIGURE 10.10** Portions of structures of 1-D networks formed between Ni(CN)$_4^{2-}$ (a), Pd(CN)$_4^{2-}$ (b), and Pt(CN)$_4^{2-}$ (c) dianions and **18**. Only H atoms involved in H-bonds are represented for clarity.

## H-Bonded Networks Based on Recognition of Hexacyanometallate

Because the distance between CN groups in syn-localization in [M(CN)$_6$]$^{3-}$ anions is almost the same as the one in [M(CN)$_4$]$^{2-}$, a similar mode of recognition may be expected with dicationic tectons such as **18** (Scheme 10.4). However, for charge neutrality reasons, an **18**/[M(CN)$_6$]$^{3-}$ ratio of 3/2 is expected. Because of the octahedral geometry around the metal center, the interconnection of [M(CN)$_6$]$^{3-}$ complexes by **18** leads to a neutral 2-D network (Figure 10.11).

A combination of the tecton **18** with [Fe(CN)$_6$]$^{3-}$ (Figure 10.11a) or [Co(CN)$_6$]$^{3-}$ (Figure 10.11b) leads indeed to the formation of the expected 2-D networks.[60] The 2-D networks are formed by the interconnection of dicationic and dianionic units through dihapto mode of H bonding. The geometrical features of both the organic tecton **18** and the anionic [M(CN)$_6$]$^{3-}$ complex are almost identical. In both cases, the packing of the neutral 2-D networks leads to channels that are filled with seven water molecules. The latter, in both cases, form H-bonded 1-D networks composed of hexagons interconnected by tetragons.

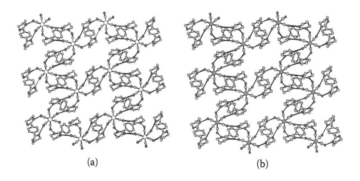

**FIGURE 10.11** Portions of x-ray structures of 2-D H-bonded networks formed between **18** and Fe(CN)$_6^{3-}$ (a) and Co(CN)$_6^{3-}$ (b) anions. H atoms are omitted for clarity.

It is worth noting that because of the dihapto or chelate mode of H bonding between **18** and $[M(CN)_6]^{3-}$, supramolecular chirality taking place within the second coordination sphere around the metal is generated. By analogy, with $\Delta$ and $\Lambda$-type chirality defined for octahedral complexes surrounded by two or three chelating ligands, the supramolecular chirality of the type $\Delta'$ and $\Lambda'$ occurring within the second coordination sphere is defined as resulting from noncovalent reversible interactions through a chelate mode of H bonding. Because **18** is achiral, both $\Delta'$ and $\Lambda'$ enantiomers are present in the 2-D network, thus leading to achiral crystals. Again, as in the case of $[M(CN)_4]^{2-}$, both networks are isostructural.

## FROM H-BONDED NETWORKS TO COMPOSITE CRYSTALS

Because crystals obtained on combining **18** with $[Fe(CN)_6]^{3-}$ (crystal **A** yellowish) and with $[Co(CN)_6]^{3-}$ (crystal **B** colorless) are isostructural, the formation of composite crystals (3-D epitaxial growth) was demonstrated using these two combinations.[63] Because of the difference in color between crystals **A** and **B**, the growth process was monitored by changes in color at the interface (Figure 10.12). Starting with **A**, the growth of **B** on **A** generates colorless surfaces (**A–B**), whereas starting with **B**, after epitaxial growth of **A** on **B**, yellow crystals (**B–A**) are obtained (**B–A**).

X-ray diffraction on the composite (**A–B**) crystal revealed that the two crystalline systems are oriented in the same space directions. However, some twinned reflections resulting from the slight difference in cell parameters between the crystalline systems of **A** and **B** were observed. The obtained composite crystals present the following features: (a) the components of (**A–B**) or (**B–A**) are molecules, (b) the material generated by the 3-D epitaxial growth being crystalline possesses periodicity and thus short- and long-range orders, and (c) (**A–B**) or (**B–A**) possesses the dicationic tecton **18** as the common component, and the two crystalline systems differ only by the nature of the metal center $Fe(CN)_6^{3-}$ for **A** and $Co(CN)_6^{3-}$ for **B**. A similar strategy has been used previously to generate other composite crystals.[64,65]

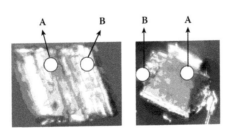

**FIGURE 10.12**  Pictures of an epitaxially grown composite (**A** and **B**) crystal (see text for definition of **A** and **B** and growth strategy).

## COORDINATION NETWORKS: AN INTERPLAY BETWEEN ORGANIC AND METALLIC TECTONS

Coordination networks are infinite metallo-organic architectures resulting from mutual bridging between organic and metallic tectons. The design of this type of network with predefined dimension (1-D, 2-D, or 3-D) is based on the interplay between both electronic and geometric requirements of the organic and the metallic tectons. Because the combination of organic and metallic tectons offers endless structural possibilities and furthermore because metal centers may present a variety of physical properties (redox, optical, and magnetic), the formation of coordination networks or polymers is a subject of intensive investigations.[66–70] Although we have reported many examples of 1-D, 2-D,[71–73] and 3-D[74,75] coordination networks, here we shall only describe design principles dealing with the formation of 1-D networks.

### TUBULAR COORDINATION NETWORKS

Infinite tubular architectures are 1-D networks. This type of assemblies may be obtained under self-assembly conditions using reversible coordination bonding (Figure 10.13).[76] The tecton **24** was designed because it adopts the 1,3-alternate conformation presenting thus four nitrile groups located below and above the main plane of the macrocycle in an alternate fashion (Figure 10.13a). As predicted, the combination of **24** with $Ag^+$ leads to the formation of a cationic tubular structure through the bridging of consecutive tectons **24** by $Ag^+$ cations adopting a linear coordination geometry (Figure 10.13b and c). The cationic networks are packed in a parallel mode. The tubular networks are interconnected in a dihapto mode by strong metal–$\pi$ interactions between $Ag^+$ cation and the aromatic moiety belonging to the next tubular strand. The tubular network is not empty but occupied by solvent molecules.

### HELICAL COORDINATION NETWORKS

The formation of helical networks has been a subject of intensive investigations.[77] Tectons **25–27** are *bis*-monodentate ligands based on two coordinating pyridine units interconnected by the oligoethyleneglycol moieties (Scheme 10.4).

The combination of **25–27** with $Ag^+$ leads to the formation of double-stranded intertwined infinite linear networks (Figure 10.14).[78,79] The bridging of consecutive

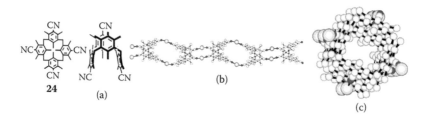

**FIGURE 10.13** Tecton **24** adopting the (a) alternate conformation leads in the presence of silver cation to (b, c) tubular coordination network.

(a)                    (b)                    (c)

**FIGURE 10.14** Portions of x-ray structures of double-stranded helical coordination networks formed between silver cation and tectons (a) **25**, (b) **26**, and (c) **27**. H atoms are omitted for clarity.

organic tectons by silver cations leads to the formation of an infinite helical polycationic network. Interestingly but unpredicted, adjacent linear networks are interwoven. The driving force for the formation of these assemblies is related to the loop-type disposition (pseudo-crown ether) of the oligoethyleneglycol units, allowing interactions between silver cations belonging to one strand and ether oxygen atoms of the other strand. It was demonstrated that the pitch of the helical arrangement could be controlled by the length of the oligoethyleneglycol spacer and by the orientation of the nitrogen atom of the pyridine ring.

The three examples of helical assemblies mentioned earlier deal with achiral tectons. Consequently, in all three cases, racemates have been obtained. This is indeed the case for most helical coordination networks reported. To generate optically pure helical strands, we used enantiomerically pure tectons such as **28** and **29** (Scheme 10.5).[80]

Both tectons are based on isomannide backbone and bear two pyridine units as neutral coordination sites. The difference between the two tectons results from the orientation of the N atom of the pyridine ring. A combination of **28** with $HgCl_2$ leads to the formation of a single-stranded helical coordination network with P helicity. The latter is imposed by the stereochemistry of isomannide.[81] The infinite helical arrangement is obtained by the bridging of consecutive tectons **28** by mercury cations (Figure 10.15a). It is interesting to note that the helical networks are packed in the syn-parallel fashion, thus leading to a polar crystal (Figure 10.15b).

A combination of **29** and $HgCl_2$ leads again to the formation of an enantiomerically pure helical strand with P helicity. Again, the network is generated by the bridging of consecutive tectons **29** by mercury cations (Figure 10.16). Again, interestingly but unexpectedly, because of interstrand interactions, a triple-stranded helical arrangement is observed. The latter is not cylindrical but rather flat. The triple-stranded helices are arranged in a parallel fashion.[80]

The formation of helical strand based on another combination using the exo-bischelate tectons **30–33** (Scheme 10.6) and again sliver cation was also demonstrated.[82] These macrocyclic ligands, because of the spacer groups connecting the two 2,2′-bipyridine units at positions 4 and 4′, adopt a roof-type conformation.

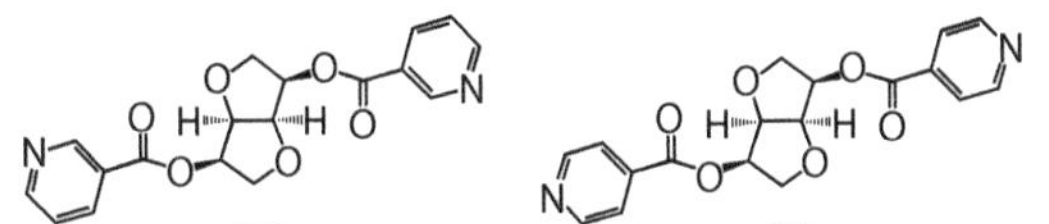

**28**                    **29**

**SCHEME 10.5** Chemical structures of chiral and optically pure bismonodentate tectons **28** and **29**.

(a)          C ● N ● O ● Hg ○ Cl ○          (b)

**FIGURE 10.15**  Portions of the structure of the single-stranded helical coordination networks (perpendicular (a) and parallel (b) views) formed between HgCl$_2$ and tecton **28** (a) and the syn-parallel packing of 1-D networks affording a polar crystal.

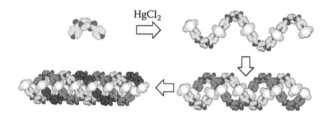

**FIGURE 10.16**  Generation of the triple-stranded helical coordination networks formed between HgCl$_2$ and tecton **29**.

For that reason, when the tectons **31** are interconnected by cations such as silver adopting the tetrahedral coordination geometry, a single-stranded helical network is formed (Figure 10.17). Because of the achiral nature of the organic tecton, both P and M helices are present in the crystal affording thus a racemate.

## DIRECTIONAL 1-D COORDINATION NETWORKS

The design of directional networks is an important issue for the exploitation of directional physical properties. Here, we shall focus only on 1-D networks. The strategy investigated was to combine dissymmetric organic tectons bearing a monodentate and a tridentate coordination poles such as **34** and **35** (Scheme 10.7) with a transition metal complex (CoCl$_2$), adopting an octahedral coordination geometry and offering four coordination sites in square arrangement.[83]

Indeed for such a strategy, interestingly, the mutual binding between the organic and the metallic tectons leads to a neutral directional 1-D network (Figure 10.18). For directional 1-D networks, two packing arrangements (syn-parallel and antiparallel) may be envisaged. In the case of the achiral tecton **34**, as expected, the centrosymmetric packing is observed, thus leading to an apolar crystal.

To investigate the possibility of avoiding the antiparallel packing, the introduction of chirality within the backbone of the organic tecton was explored. Thus, the neutral acentric and $c_2$-chiral tecton **35** was designed. The latter is based on two coordination poles composed of a pyridine unit connected at the four-position to a pyridine bearing at the two- and six-positions—two optically active oxazoline moieties.[84]

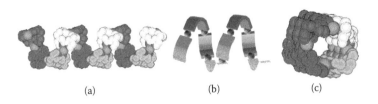

**SCHEME 10.6**   Chemical structures of bipyridine based bischelate tectons **30–33**.

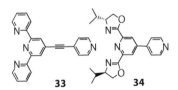

(a)    (b)    (c)

**FIGURE 10.17**   Portions of the x-ray structure of the single-stranded helical coordination networks formed between (a) silver cation and tecton **31**, (b) its schematic representation, and (c) a lateral view. H atoms and phenyl groups are not represented for clarity.

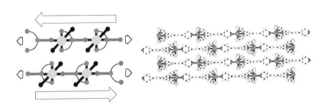

**SCHEME 10.7**   Chemical structures of ditopic (mono-tridentate) tectons **34** and **35**.

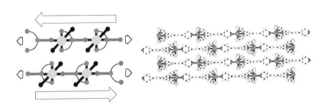

**FIGURE 10.18**   Schematic representation of two directional 1-D coordination networks packed in an antiparallel fashion (left) and a portion of the x-ray structure of the neutral network formed between $CoCl_2$ and tecton **34** (right). Carbon atoms belonging to consecutive strands are differentiated by color, and H atoms are not represented for clarity.

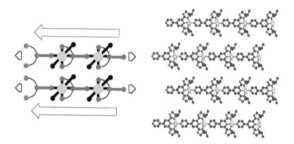

**FIGURE 10.19**   Schematic representation of two directional 1-D coordination networks packed in a syn-parallel fashion (left) and a portion of the x-ray structure of the neutral network formed between CoCl$_2$ and tecton **35** (right) showing the polar arrangement. H atoms are not represented for clarity.

By combining the chiral tecton **31** with CoCl$_2$, a chiral and directional 1-D coordination network is obtained. As stated earlier, the directionality of the network is due to the acentric nature of the organic tecton **31**. In marked contrast with the case mentioned earlier, here, a syn-parallel arrangement of the chiral 1-D networks is observed. This packing mode leads to a polar solid (Figure 10.19).

## SUMMARY AND OUTLOOK

In this short report, we have attempted to demonstrate that molecular tectonics is a viable approach for the design and generation of molecular networks in the crystalline phase. Indeed, by designing programmed and active molecular tectons capable of mutual interactions (molecular recognition) and by allowing the translation of the recognition event (iteration), one may predict, in some cases, the connectivity pattern between tectons and thus the geometry, topology, and directionality of molecular networks. An important issue here is the reliability of the prediction. The latter is the consequence of the quality of the design of the molecular building blocks used. This obviously depends on the nature and thus the energy associated with interactions used, the number of components, and the structural flexibility of tectons. Our experience showed that, as one may expect, using strong interactions such as a combination of charge–charge electrostatic interactions and dihapto mode of H bonding or coordination bonds based on polyhapto mode of coordination, the reliability in the prediction of the network is approximately 60% to 70%. This may be further increased by using tectons with rather restricted conformational space (rigid structures). In our opinion, today this domain is in its infancy and thus not fully explored and mastered. More systematic investigations are needed before reaching maturity. Now about the interesting and recurrent question, strongly under discussion is the relationship between this approach and crystal engineering.[3] In other words, are crystal structures predictable?[85,86] To our opinion, the response is *no*. With our present level of knowledge and understanding, we are not capable of predicting the crystal structure of a set of tectons, in particular the adopted conformation of flexible tectons, the presence of solvent molecules, polymorphs, and so forth. However, as stated by Dunitz,[1] by considering a crystal as a "supermolecule *par excellence*," using concepts and principles developed in supramolecular chemistry,[2] one may, through the control of molecular recognition events, predict some of the connectivity patterns between tectons and thus the formation of molecular networks. After many years of investigations by several research groups, today one may design molecular networks in the crystalline phase with an acceptable degree of precision. So far, the description of molecular networks is based on geometrical features. We have analyzed networks formation using extensive energy analysis.[33] This type of analysis must be pursued and shall give a more precise description. A further important issue is obviously to use this knowledge for the design of molecular networks with predicted properties. In other terms, the molecular tectonics approach must extend structural networks to functional networks. This aspect is of prime importance in terms of applications and remains a challenge. However, although restricted to metal-organic frameworks (coordination networks), several applications such as gas storage,[87-89] enantioselective catalysis,[90,91] and polymerization in confined space[92,93] have been reported over the last few years.

## REFERENCES

1. Dunitz, J. D. 1991. *Pure and Applied Chemistry*, 63, 177.
2. Lehn, J.-M. *Supramolecular Chemistry, Concepts and Perspectives*, VCH, Weinheim, 1995.
3. Desiraju, G. R. 1995. *Angewandte Chemie (International ed. in English)*, 34, 2311.
4. Etter, M. C. 1990. *Accounts of Chemical Research*, 23, 120.
5. Fowler, F. W., and J. W. Lauher. 1993. *Journal of the American Chemical Society*, 115, 5991.
6. Mann, S. 1993. *Nature*, 365, 499.
7. Hosseini, M. W. 2004. *Accounts of Chemical Research*, 38, 313.
8. Brand, G., M. W. Hosseini, O. Félix, P. Schaeffer, and R. Ruppert. 1995. In *Magnetism: A Supramolecular Function*, edited by O. Kahn, 129–142. Dordrecht: Kluwer.
9. Delaigue, X., E. Graf, F. Hajek, M. W. Hosseini, and J.-M. Planeix. In *Crystallography of Supramolecular Compounds*, edited by G. Tsoucaris, J. L. Atwood, and J. Lipkowski, Vol. 480, 159–180. Dordrecht: Kluwer, 1996.
10. Simard, M., D. Su, and J. D. Wuest. 1991. *Journal of the American Chemical Society*, 113, 4696.
11. Gavezzotti, A. 1994. *Accounts of Chemical Research*, 27, 309.
12. Lindsey, J. S. 1991. *New Journal of Chemistry*, 15, 153.
13. Whitesides, G. M., J. P. Mathias, and T. Seto. 1991. *Science*, 254, 1312.
14. Ermer, O. 1988. *Journal of the American Chemical Society*, 110, 3747.
15. Desiraju, G. D. *Crystal Engineering: The Design of Organic Solids*, Elsevier, New York, 1989.
16. Aakeröy, C. B., and K. R. Seddon. 1993. *Chemical Society Reviews*, 22, 397.
17. Subramanian, S., and M. J. Zaworotko. 1994. *Coordination Chemistry Reviews*, 137, 357.
18. Lawrence, D. S., T. Jiang, and M. Levett. 1995. *Chemical Reviews*, 95, 2229.
19. Stoddart, J. F., and D. Philip. 1996. *Angewandte Chemie (International ed. in English)*, 35, 1155.
20. Fredericks, J. R., and A. D. Hamilton. 1996. In *Comprehensive Supramolecular Chemistry*, edited by J. P. Sauvage and M. W. Hosseini, Vol. 9, 565–594. Oxford: Pergamon Press.
21. Aakeroy, C. B., and A. M. Beatty, 2004. In *Comprehensive Coordination Chemistry II*, Eds. J. A. McCleverty, T. J. Meyer, 1, 679.
22. Braga, D., and F. Grepioni. 2000. *Accounts of Chemical Research*, 33, 601.
23. Holman, K. T., A. M. Pivovar, J. A. Swift, and M. D. Ward. 2001. *Accounts of Chemical Research*, 34, 107.

24. Brammer, L. 2003. In *Perspectives in Supramolecular Chemistry*, edited by G. Desiraju, Vol. 7, 1. Wiley, West Sussex, UK.

25. Hosseini, M. W. 2003. *Coordination Chemistry Reviews*, 240, 157.

26. Batten, S. R., and R. Robson. 1998. *Angewandte Chemie (International ed. in English)*, 37, 1461.

27. Yaghi, O. M., H. Li, C. Davis, D. Richardson, and T. L. Groy. 1998. *Accounts of Chemical Research*, 31, 474.

28. Biradha, K., and M. Fujita. 2003. In *Perspectives in Supramolecular Chemistry*, Ed. G. Desiraju, Wiley, West Sussex, UK, 7, 211.

29. Blake, A. J., N. R. Champness, P. Hubberstey, W.-S. Li, M. A. Withersby, and M. Schröder. 1999. *Coordination Chemistry Reviews*, 183, 117.

30. Carlucci, L., G. Ciani, and D. M. Proserpio. 2003. *Coordination Chemistry Reviews*, 246, 247.

31. Swegers, G. F., and T. J. Malefetse. 2000. *Chemical Reviews*, 100, 3483.

32. Hosseini, M. W. 1999. In *NATO ASI Series*, edited by D. Braga, F. Grepiono, G. Orpen, Serie C, Vol. 538, 181. Dordrecht, Netherlands: Kluwer.

33. Moulton, B., and M. J. Zaworotko. 2001. *Chemical Reviews*, 101, 162.

34. Hosseini, M. W., and A. De Cian. 1998. *Chemical Communications*, 727.

35. Martz, J., E. Graf, A. De Cian, and M. W. Hosseini. 2003. In *Perspectives in Supramolecular Chemistry*, edited by G. Desiraju, Vol. 7, 177. Wiley.

36. Dunitz, J. D. 2003. *Chemical Communications*, 545.

37. Henry, M., and M. W. Hosseini. 2004. *New Journal of Chemistry*, 28, 897.

38. Hosseini, M. W., and A. De Cian. 1998. *Chemical Communications*, 727.

39. Delaigue, X., M. W. Hosseini, A. De Cian, J. Fischer, E. Leize, S. Kieffer, and A. Van Dorsselaer. 1993. *Tetrahedron Letters*, 34, 3285.

40. Gutsche, C. D. 1989. *Calixarenes*. Monographs in Supramolecular Chemistry, edited by J. F. Stoddart. London: R.S.C.

41. Delaigue, X., M. W. Hosseini, E. Leize, S. Kieffer, and A. van Dorsselaer. 1993. *Tetrahedron Letters*, 34, 7561.

42. Hajek, F., E. Graf, M. W. Hosseini, A. De Cian, and J. Fischer. 1997. *Tetrahedron Letters*, 38, 4555.

43. Martz, J., E. Graf, M. W. Hosseini, A. De Cian, and J. Fischer. 1998. *Journal of Materials Chemistry*, 8, 2331.

44. Hajek, F., E. Graf, M. W. Hosseini, X. Delaigue, A. De Cian, and J. Fischer. 1996. *Tetrahedron Letters*, 37, 1401.

45. Martz, J., E. Graf, M. W. Hosseini, and A. De Cian. Unpublished results.

46. Martz, J., E. Graf, M. W. Hosseini, A. De Cian, and J. Fischer. 2000. *Journal of the Chemical Society Dalton Transactions*, 3791.

47. Hosseini, M. W. 2003. *Coordination Chemistry Reviews*, 240, 157.

48. Taylor, R., and O. Kennard. 1984. *Accounts of Chemical Research*, 17, 320.

49. Saladino, R., and S. Hanessian. 2003. In *Crystal Design and Function, Perspectives in Supramolecular Chemistry*, edited by G. R. Desiraju, Vol. 7, 77, Wiley, West Sussex, UK.

50. Grosshans, P., A. Jouaiti, V. Bulach, J.-M. Planeix, M. W. Hosseini, and J.-F. Nicoud. 2003. *CrystEngComm*, 5, 414.

51. Jaunky, W., M. W. Hosseini, J.-M. Planeix, A. De Cian, N. Kyritsakas, and J. Fischer. 1999. *Chemical Communications*, 2313.

52. Holman, K. T., A. M. Pivovar, J. A. Swift, and M. D. Ward. 2001. *Accounts of Chemical Research*, 34, 107.

53. Hosseini, M. W., R. Ruppert, P. Schaeffer, A. De Cian, N. Kyritsakas, and J. Fischer. 1994. *Chemical Communications*, 2135.

54. Félix, O., M. W. Hosseini, A. De Cian, and J. Fischer. 1997. *Tetrahedron Letters*, 38, 1933.

55. Félix, O., M. W. Hosseini, A. De Cian, and J. Fischer. 1997. *Tetrahedron Letters*, 38, 1755.

56. Hosseini, M. W., G. Brand, P. Schaeffer, R. Ruppert, A. De Cian, and J. Fischer. 1996. *Tetrahedron Letters*, 37, 1405.
57. Félix, O., M. W. Hosseini, A. De Cian, and J. Fischer. 1997. *Angewandte Chemie (International ed. in English)*, 36, 102.
58. Félix, O., M. W. Hosseini, A. De Cian, and J. Fischer. 2000. *Chemical Communications*, 281.
59. Félix, O., M. W. Hosseini, A. De Cian, and J. Fischer. 1997. *New Journal of Chemistry*, 21, 285.
60. Ferlay, S., O. Félix, M. W. Hosseini, J.-M. Planeix, and N. Kyritsakas. 2002. *Chemical Communications*, 702.
61. Ferlay, S., V. Bulach, O. Félix, M. W. Hosseini, J.-M. Planeix, and N. Kyritsakas. 2002. *CrystEngComm*, 4, 447.
62. Lewis, G. R., and A. G. Orpen. 1998. *Chemical Communications*, 1873.
63. Ferlay, S., and M. W. Hosseini. 2004. *Chemical Communications*, 787.
64. MacDonald, J. C., P. C. Dorrestein, M. M. Pilley, J. M. Foote, L. Lundburg, R. W. Henning, A. J. Schultz, and J. L. Manson. 2000. *Journal of the American Chemical Society*, 122, 11692.
65. Noveron, J. C., M. S. Lah, R. E. Del Sesto, A. M. Arif, and J. S. Miller. 2002. *Journal of the American Chemical Society*, 124, 6613.
66. Batten, S. R., and R. Robson. 1998. Interpenetrating nets: ordered periodic entanglement. *Angewandte Chemie (International ed.)*, 37, 1460.
67. Blake, A. J., N. R. Champness, P. Hubberstey, W.-S. Li, M. A. Withersby, and M. Schröder. 1999. *Coordination Chemistry Reviews*, 183, 117.
68. Hosseini, M. W. 1999. In *Crystal Engineering: From Molecules and Crystals to Materials, NATO ASI Series*, edited by D. Braga, F. Grepioni, and G. Orpen, Vol. 538, 181. Serie c, Kluwer, Dordrecht, Netherlands.
69. Eddaoudi, M., D. B. Moler, H. Li, B. Chen, T. M. Reineke, M. O'Keeffe, and O. M. Yaghi. 2001. *Accounts Chemistry Research*, 34, 319.
70. Moulton, B. and M. J. Zaworotko. 2001. *Chemical Reviews*, 101, 1629.
71. Akdas, H., E. Graf, M. W. Hosseini, A. De Cian, and J. McB. Harrowfield. 2000. *Chemical Communications*, 2219.
72. Zimmer, B., V. Bulach, M. W. Hosseini, A. De Cian, and N. Kyritsakas. 2002. *European Journal Inorganic Chemistry*, 3079.
73. Grosshans, P., A. Jouaiti, M. W. Hosseini, and N. Kyritsakas. 2003. *New Journal of Chemistry*, 27, 793.
74. Klein, C., E. Graf, M. W. Hosseini, A. De Cian, and J. Fischer. 2001. *New Journal of Chemistry*, 25, 207.
75. Ferlay, S., S. Koenig, M. W. Hosseini, J. Pansanel, A. De Cian, and N. Kyritsakas. 2002. *Chemical Communications*, 218.
76. Klein, C., E. Graf, M. W. Hosseini, A. De Cian, and J. Fischer. 2000. *Chemical Communications*, 239.
77. Munakata, M., L. P. Wu, and T. Kuroda-Sowa. 1999. *Advances in Inorganic Chemistry*, 46, 173.
78. Schmaltz, B., A. Jouaiti, M. W. Hosseini, and A. De Cian. 2001. *Chemical Communications*, 1242.
79. Jouaiti, A., M. W. Hosseini, and N. Kyritsakas. 2003. *Chemical Communications*, 473.
80. Grosshans, P., A. Jouaiti, V. Bulach, J.-M. Planeix, M. W. Hosseini, and J.-F. Nicoud. 2003. *Chemical Communications*, 1336.
81. Grosshans, P., A. Jouaiti, V. Bulach, J.-M. Planeix, M. W. Hosseini, and J.-F. Nicoud. 2004. *Comptes Rendus Chimie*, 7, 189.
82. Kaes, C., M. W. Hosseini, C. E. F. Rickard, B. W. Skelton, and A. H. White. 1998. *Angewandte Chemie (International ed. in English)*, 37, 920.

83. Jouaiti, A., M. W. Hosseini, and A. De Cian. 2000. *Chemical Communications*, 1863.
84. Jouaiti, A., M. W. Hosseini, and N. Kyritsakas. 2002. *Chemical Communications*, 1898.
85. Gavezzotti, A. 1994. *Accounts of Chemical Research*, 27, 309–314.
86. Dunitz, J. D. 2003. *Chemical Communications*, 545–548.
87. Chen, B., D. S. Contreras, N. W. Ockwig, and O. M. Yaghi, 2005. *Angewandte Chemie (International ed.)*, 44, 4745.
88. Millward, A. R., and O. M. Yaghi. 2005. *Journal of the American Chemical Society*, 127, 17998.
89. Férey, G. 2008. *Chemical Society Reviews*, 37, 191.
90. Seo, J.-S., D. Whang, H. Lee, S.-I. Jun, J. Oh, Y.-J. Jeon, and K. Kim. 2000. *Nature*, 404, 982.
91. Ma, L., C. Abney, and W. Lin. 2009. *Chemical Society Reviews*, 38, 1248.
92. Uemura, T., R. Kitaura, Y. Ohta, M. Nagaoka, and S. Kitagawa. 2006. *Angewandte Chemie (International ed.)*, 45, 4112.
93. Uemura, T., D. Hiramatsu, Y. Kubota, M. Takata, and S. Kitagawa. 2007. *Angewandte Chemie (International ed.)*, 46, 4987.

# 11 Supramolecular Complex Design and Function for Photodynamic Therapy and Solar Energy Conversion via Hydrogen Production

## Common Requirements for Molecular Architectures for Varied Light-Activated Processes

Jing Wang, Shamindri Arachchige, and Karen J. Brewer

**CONTENTS**

# INTRODUCTION

Supramolecular complexes are ideally suited for the development of novel materials to perform complex functions. The application of supramolecular complexes in this forum has been limited by the synthetic ability to construct complex assemblies and a detailed understanding of the perturbations of the individual component's properties upon incorporation into the supramolecular assembly. The synthetic methodology to prepare supramolecular complexes of significant structural diversity has been expanding rapidly in the last decade. This has led to fundamental studies of the perturbation of subunit properties upon incorporation into supramolecular assemblies and is providing a database of properties that has allowed preliminary systems to be designed to perform complicated functions. The structural arrangement of subunits within the assemblies as well as the electronic coupling of subunits is critical to understand and ultimately predict and control for the successful optimization of supramolecular complex function. Many fundamental studies have provided the background for the work described herein and the contributions of these researchers is paramount to the current state of the art in supramolecular design, construction, and application.

## SCOPE AND LIMITATIONS

The properties of supramolecular complexes provide for the ability to use molecular design to produce complex molecular systems that perform a wide array of functions. In the context of this chapter, supramolecular complexes are large molecular assemblies that are composed of individual components coupled through coordinate covalent bonds where the individual subunits bring to the supramolecular assembly a particular act that when combined leads to the performance of a complex function by the supramolecule. Supramolecular complexes can be designed to be activated to function by absorption of light providing a class of complexes known as photochemical molecular devices.[1] The activation of supramolecules through light absorption requires the coupling of a light absorber (LA) subunit into the molecular construct. A variety of light-activated functions are of considerable current interest, including the conversion of solar energy into chemical fuels from abundant feedstocks and the treatment of disease via light-activated destruction of biomolecules. These areas of solar energy conversion via hydrogen production from water and photodynamic therapy (PDT) have related requirements for supramolecule molecular architecture, both requiring the coupling of LAs to reactive centers. This chapter will focus on the molecular architectures that are of similar constructs that have been applied in both forums: solar energy conversion and PDT. The modulation of subunit identity and the mechanism of functioning will be highlighted, focusing on the overall similarity of the structures and varied mechanism of action for these seemingly diverse applications. The focus will be on metal complexes with similar molecular architectures that have been applied in both solar energy conversion and PDT. A wide array of interesting supramolecular complexes have been used independently in each of these forums, but an analysis of similar systems used in both forums will provide for considerable insight into the role of subunit identity and order of assembly on the light-activated function of supramolecular complexes.

## SUPRAMOLECULAR DEVICES AND THEIR COMPONENTS

Supramolecular complexes are large molecular systems that are linked via coordinate covalent bonds. The nature of the subunits, the order of their assembly, and the acts performed using each subunit provide for the complex function performed by the supramolecular assembly. The systems described herein have their function initiated by the absorption of light and are often termed photochemical molecular devices. The absorption of light is accomplished by a LA subunit. Light absorption provides for a unique mechanism for initiation of function.

The absorption of light leads to the production of an electronic excited state (ES) from the ground state (GS) molecule. Electronic ESs are different from their GSs possessing a low energy electron hole and a high energy electron. This provides for electronic ESs that are both more powerful oxidizing and reducing agents than the GS, capable of undergoing oxidative or reductive quenching.

Supramolecular complexes that are used in solar energy conversion typically possess several subunits or building blocks. These may include LA, bridging ligand (BL), electron donor (ED), electron acceptor (EA), electron collector (EC), and catalytically active components (CATs). As described earlier, the LA provides for the absorption of light and the activation of the device. ED components are engaged to donate electrons to another part of the supramolecule, typically providing for spatial separation of the promoted electron and electron hole. EA similarly accept electrons providing for additional spatial separation. A site that can accept more than one electron is coined an EC. ECs are often used in solar energy conversion to allow for multielectron reactions such as the reduction of water to produce hydrogen. CATs are metal centers that are involved in a catalytic reaction and are engaged in the transfer of electrons to substrate as well as facilitating bond breaking and formation needed to produce the fuel from the chemical feedstock.

Supramolecular complexes that are used in PDT possess many of the same subunits as those described earlier for solar energy conversion. Both light-activated processes require the coupling of a LA unit. Supramolecules for solar energy conversion often possess multiple LAs, whereas systems applied as PDT agents often use a single LA unit. PDT applications often use ED or EA to move charge within a molecule but do not make use of EC units as multielectron reactions are not typically required. Bioactive sites (BAS) are used in PDT to provide for the site of biological activity, including the metal engaged in biomolecule binding or damage. These can be the same metals used as CAT in supramolecules for solar energy conversion. The order of subunit attachment is critical to the functioning of the supramolecules.

## LIGHT ABSORPTION AND ELECTRONIC EXCITED STATES

The absorption of the photon of light results in the generation of electronic ES from their respective GS molecules. In supramolecular systems, this ES is typically more localized on one part of the larger molecular assembly, termed the LA unit. ES molecules can relax to the GS in a unimolecular manner via radiative, $k_r$, or nonradiative processes, $k_{nr}$, where radiative processes involve the emission of a photon of light and nonradiative processes release heat.

In the presence of another molecule or quencher, $Q$, bimolecular deactivation pathways are possible whereby the excited LA transfers energy to the $Q$ via energy transfer, $k_{en}$. Biomolecular reactions can also occur via electron transfer, $k_{et}$, as the ES is both a more powerful oxidizing and reducing agent than the electronic GS.

The thermodynamic driving force ($E_{redox}$) of an ES electron-transfer process via oxidative quenching is dictated by the ES reduction potential of the LA ($E(*LA^{n+}/LA^{(n-1)+})$) and the GS oxidation potential of the ED ($E(ED^{0/+})$) for ES reduction. Similarly for oxidative quenching, the driving force depends on the ES oxidation potential of the LA ($E(*LA^{n+}/LA^{(n+1)+})$) and the GS reduction potential of the EA ($E(EA^{0/-})$) for ES oxidation. The ES redox potentials, $E(*LA^{n+}/LA^{(n-1)+})$ and $E(*LA^{n+}/$

$LA^{(n+1)+}$), differ significantly from the GS potentials with electronic ESs being better oxidizing and reducing agents. The energy gap between the ground vibronic state of the electronic ground and the ESs is termed $E^{0-0}$.

Supramolecular complexes display complex electronic absorption spectroscopy with transitions observed for each subunit introduced into the supramolecular assembly with the added caveat that the perturbation of subunit properties upon incorporation into the supramolecular assembly is typically observed. If the BL provides a significant electronic coupling of coordinated metal centers, larger perturbations are observed. The binding of metals to polyazine ligands, both terminal ligand (TL) and BL, provide for significant stabilization of the ligand $\pi^*$ orbitals through electron donation to the electropositive metal centers. This provides for perturbations of subunit properties upon assembly into supramolecules. Metal complexes of polyazine ligands display several electronic transitions and ESs including internal ligand or $\pi \rightarrow \pi^*$, metal-to-ligand charge transfer (MLCT), ligand-to-metal charge transfer, and ligand field or $d \rightarrow d$ transitions. Internal ligand transitions occur in the ultraviolet (UV) and high energy visible with BL-centered transitions typically lower in energy than TL bands. CT bands are typically in the visible region of the spectrum with high intensity that correlates to the number of CT LA subunits. The MLCT transitions are often significantly red-shifted upon the bridging of two metals upon supramolecular assembly. Ligand field transitions are weak transitions that typically lie under the more intense CT bands in polyazine complexes. These ESs can still play a role in the ES reactivity of some polyazine metal complexes.

The electronic states of molecules are often represented using Jablonski or state diagrams where electronic states are represented by lines. Radiative processes that provide for conversion between states are represented by straight arrows, and nonradiative processes for conversion between states are shown as wavy arrows. The nonradiative conversion between states of the same spin multiplicity is termed internal conversion ($k_{ic}$) and that between states of differing spin multiplicity is termed intersystem crossing ($k_{isc}$). The radiative relaxation without spin change gives rise to fluorescence ($k_f$) and with spin change provides phosphorescence ($k_p$). This is illustrated below with the state diagram of the prototypical Ru polyazine charge transfer LA, $[Ru(bpy)_3]^{2+}$ (bpy = 2,2′-bipyridine), which possesses a lowest lying MLCT ES that is long lived and emissive in the visible region of the spectrum (Figure 11.1).[2]

A variation of the ligand bound to the Ru center is used to tune the properties of this class of LA with polyazine BLs being incorporated to allow for coupling to additional metal centers (Figure 11.2).

The quantification of light-activated processes is typically described in terms of quantum yields, $\Phi$, with timescale provided by ES lifetimes, $\tau$. These quantities can be related to the rates of reactions and conversions between states. The quantum yield is a measure of the efficiency of a process expressed as a ratio with the maximum efficiency being 1. The $\Phi$ for emission is defined as $k_r$ divided by the sum of the rate constants of all pathways depopulating that state, $\sum k$ ($\sum k = k_r + k_{nr} + k_{rxn}$, where the subscripts r, nr, and rxn are the radiative decay, nonradiative decay, and photochemical reaction, respectively) times the quantum yield for population of that

2+
Ru
N
1MLCT
3MLCT
E
hν
kisc
krxn
kr
knr
1GS

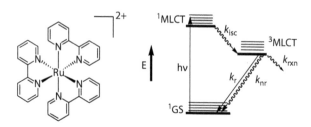

**FIGURE 11.1** Representation of the structure and state diagram for the prototypical Ru polyazine LA $[Ru(bpy)_3]^{2+}$ (bpy = 2,2′-bipyridine).

state. For emission by $[Ru(bpy)_3]^{2+}$ from the $^3$MLCT state, $\Phi$ is given by Equation 11.1 as:

$$\Phi = \Phi_{^3\mathrm{MLCT}} \frac{k_r}{\sum k} \tag{11.1}$$

Ruthenium polyazine LAs typically display $\Phi_{^3\mathrm{MLCT}}$ of unity, so this term is often omitted from Equation 11.1. The ES lifetime of any state, $\tau$, is defined as the inverse of the sum the rate constants of all pathways depopulating that state, $\sum k^{-1}$. For the $^3$MLCT state of $[Ru(bpy)_3]^{2+}$, the lifetime is provided by Equation 11.2,

$$\tau = \frac{1}{k_r + k_{nr} + k_{rxn}} \tag{11.2}$$

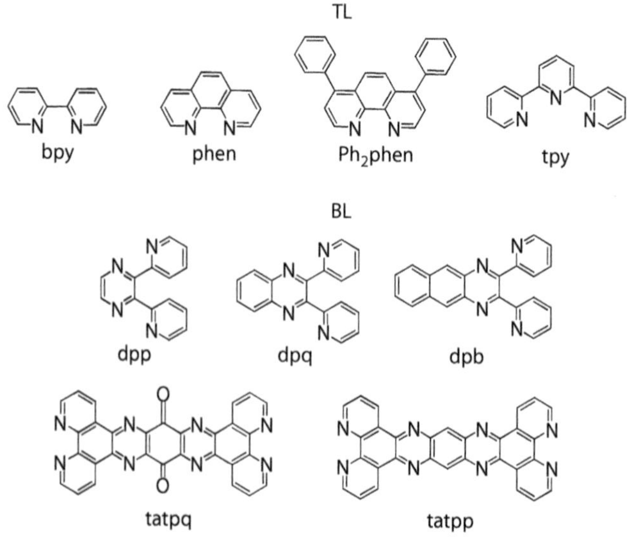

TL
bpy
phen
Ph2phen
tpy
BL
dpp
dpq
dpb
tatpq
tatpp

**FIGURE 11.2** Representative TLs and BLs used in the construction of supramolecular systems for solar energy conversion and PDT.

The Stern–Volmer kinetic relationship is used to study the kinetics and efficiency of bimolecular processes between excited LA and other components of a system. $\Phi^o$ and $\Phi$ are the quantum yields of the observable (often emission) in the absence and presence of $Q$, respectively, $k_q$ is the overall quenching rate constant, and $\tau$ is the ES lifetime of the $^3$MLCT state. The Stern–Volmer relationship, which plots $\Phi^o/\Phi$ as a function of the quencher concentration, provides a linear relationship with a slope of $\tau k_q$, coined $K_{sv}$. The value of $k_q$ affords the rate of quenching by $Q$.

$$\Phi^o/\Phi = 1 + \tau\, k_q [Q] \tag{11.3}$$

## REDOX CHEMISTRY OF SUPRAMOLECULAR METAL COMPLEXES

The redox properties of supramolecular complexes of transition metals with coordinated redox active polyazine ligands are very complex. Typically, metal complexes of polyazine ligands display reversible metal-based oxidations and reversible ligand-based reductions for each polyazine ligand ordered by the energy of the $\pi^*$ acceptor orbital of the ligands. Polyazine ligands that are bridging between two electropositive metal centers typically display $BL^{0/-}$ and $BL^{-/2-}$ couples before TL reduction.[3] Metal centers can also display reductive processes that are typically irreversible such as the $Rh^{III/II/I}$ couples seen in supramolecular Rh complexes with $Rh^{III}(NN)_2X_2$ subunits.[3] The redox properties of supramolecular metal complexes are typically studied by cyclic voltammetry (CV) in deoxygenated solution in a three electrode configuration. The choice of solvent, electrolyte, and electrode material is important to the observation of electrochemistry of supramolecular complexes. Solvent and electrolyte are important to provide for a sufficient electrochemical window to observe the important redox processes of the supramolecule. The supramolecule must also be soluble in the solvent and not react with either the solvent or the electrolyte not only in the synthesized redox state but also those generated upon electrochemical analysis. The pulse sequence of CV and a typical reversible redox couple is illustrated in Figure 11.3.

The CV method is a variation of linear sweep voltammetry whereby one potential scan from the initial potential to the switching potential is followed by a switching in the direction of potential scan moving to the final potential. Current is measured, and the CV displays $E$ versus $i$. For a reversible redox couple, the $\Delta E_p$ ($E_p^a - E_p^c$) at room temperature (RT) is 59 mV/n (where $n$ is the number of electrons in the redox process). The peak current for the anodic peak, $i_p^a$, is equal to the cathodic peak current, $i_p^c$, for a reversible redox process. For a reversible couple, the $E_{1/2}$, ($E_p^a + E_p^c$)/2, is reflective of the thermodynamic driving force for this redox process.

Supramolecular complexes display several redox processes in the typical solvent window. Figure 11.4 displays the electrochemistry of a Ru monometallic and the corresponding bimetallic complex.

The CV for the monometallic shows one reversible $Ru^{II/III}$ oxidation. The bimetallic displays two Ru couples indicative of electronic coupling of the two Ru centers through the BL. Both complexes display Ru-based oxidations, indicative of a Ru-centered highest occupied molecular orbital (HOMO) in this structural motif.

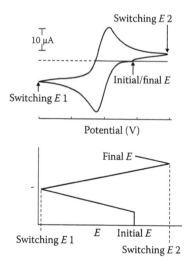

**FIGURE 11.3** Pulse sequence for a typical CV and the resultant CV, *E* versus *i*, response for an analyte with a single reversible redox couple.

Reductively, both complexes display a series of reversible reductions with the BL (dpp$^{0/-}$) couple occurring first at the least negative potential. In the monometallic complex, the dpp$^{0/-}$ couple is followed by two TL$^{0/-}$ couples indicative of the electronic coupling of the ligands bound to the same metal center. For the bimetallic complex, the dpp$^{0/-}$ couple is followed by the dpp$^{-/2-}$ couple, indicative of the BL being coordinated to two electropositive Ru metal centers. The BL couples are followed by couples for the simultaneous reduction of one TL on each Ru center, 2 TL$^{0/-}$ couples. This is indicative of the electronic isolation of the TLs on separate Ru centers. The BL reduction before the TL couples illustrates the lower energy $\pi^*$ acceptor orbitals on the BL providing for a BL-based lowest unoccupied molecular orbital (LUMO) in this motif.

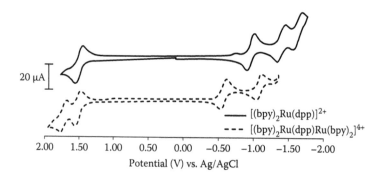

**FIGURE 11.4** CV of the complexes [(bpy)$_2$Ru(dpp)](PF$_6$)$_2$ and [(bpy)$_2$Ru(dpp)Ru(bpy)$_2$](PF$_6$)$_4$ in RT CH$_3$CN with 0.1 M of Bu$_4$NPF$_6$ recorded with a glassy carbon working electrode, a Pt wire auxiliary electrode, and a Ag/AgCl reference electrode.

## PDT BY SUPRAMOLECULAR COMPLEXES

Supramolecular complexes are ideally suited for applications in PDT as multiple subunits can be incorporated into one molecule. This allows the targeting of specific biomolecules, cell types, or diseases. Most work to date focuses on cancer treatment via DNA targeting. Typical supramolecular complexes couple some number of LA units through BL to BAS. Highlighted herein are supramolecular complexes displaying PDT actions that have structure similarity to solar $H_2$ photocatalysts. Interestingly, the major classes of supramolecular motifs that are used in PDT with multifunctional actions display close structural analogs to systems used in solar energy conversion. In addition, the LA subunits in many solar $H_2$ photocatalysts are structurally similar to active PDT agents that function via sensitization of $^1O_2$ generation.

### DNA as a Target for PDT

DNA contains the inherited information of cells.[4] It is a polymer chain of nucleotides, which consist of a nucleobase (guanine, cytosine, adenine, and thymine), a sugar (deoxyribose), and a phosphate. DNA plays a central role in the cellular life, controlling cell structure and functions, especially the replication and transcription processes. DNA becomes a major target in PDT processes.[5]

### Development of PDT Drugs

There are requirements for drugs to be used in PDT applications. They should be nontoxic and stable in the dark. They should absorb light within the "phototherapeutic window" (600–850 nm, where light can efficiently penetrate body tissues) and should generate reactive, toxic species after light excitation.[6]

In the past 30 years, work has been reported developing organic photosensitizers including porphyrins, chlorins, and phthalocyanines. The first photosensitizer family discovered was the hematoporphyrin and its derivative, HpD. Its commercial product, Photofrin (Figure 11.5), is composed of different porphyrins including monomers and dimers.[7] Photofrin was approved for use in PDT in 1995. Tetraphenylchlorin (Foscan as shown in Figure 11.5) belongs to the chlorin family and can effectively control tumors.[8] Phthalocyanines and their derivatives are clinical photosensitizers.[9]

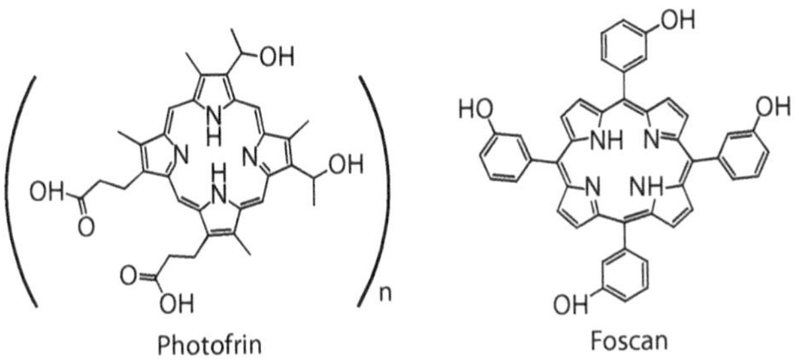

Photofrin          Foscan

**FIGURE 11.5** Molecular structure for Photofrin and Foscan.

## Ru and Os Polyazine Complexes as Light Absorbers for PDT

Metal-based chromophores such as the prototypical $[Ru(bpy)_3]^{2+}$ are another class of photosensitizer of interest to develop PDT drugs. They are efficient LAs with high extinction coefficients in the UV and visible regions of the spectrum. The chromophores are photostable, and their redox and spectroscopic properties are tunable by using different ligands. The ruthenium polyazine complex, $[Ru(bpy)_3]^{2+}$, is well studied and shown to display interesting photophysical and photochemical properties.[2,10] $[Ru(bpy)_3]^{2+}$ displays intense absorption bands in the UV region assigned to bpy-based $\pi \rightarrow \pi^*$ transitions and MLCT transitions in the visible region, $Ru(d\pi) \rightarrow bpy(\pi^*)$ CT in nature. When irradiated in the UV or visible region, $[Ru(bpy)_3]^{2+}$ is excited to its lowest-lying singlet $Ru(d\pi) \rightarrow bpy(\pi^*)$ CT ES ($^1$MLCT) and undergoes intersystem crossing to the triplet ES ($^3$MLCT) with unit efficiency. The $^3$MLCT ES can undergo a nonradiative decay, a radiative decay (emitting phosphorescence), or a photochemical reaction to the GS ($^1$GS). The $^3$MLCT emission of $[Ru(bpy)_3]^{2+}$ is observed at 605 nm ($\Phi_{RT}^{em} = 0.10$) with a long ES lifetime ($\tau = 860$ ns) in room temperature $CH_3CN$.[2] Because of the long-lived $^3$MLCT state, $[Ru(bpy)_3]^{2+}$ can photocleave pBR322 DNA (DNA created by Bolivar and Rodriguez) with visible light ($\lambda > 450$ nm) radiation through type II mechanism (singlet oxygen generation) in an oxygen-dependent environment.[11–13]

## Supramolecular Complexes Coupling Ru(II) Polyazine Light Absorbers to Rh(III) Metal Centers for PDT

Ru(II),Rh(III) polyazine complexes consist of coupling one or two ruthenium polyazine LAs to a *cis*-Rh$^{III}$Cl$_2$-reactive metal center through dpp (2,3-bis(2-pyridyl)pyrazine) BLs.[3,14–19] These Ru(II),Rh(III) supramolecular complexes are efficient LAs, have complex redox properties because of energetically close frontier orbitals, and possess an emissive $^3$MLCT ES. Because most aggressive cancer cells exist in low-oxygen environments, it is important to develop oxygen-independent mechanisms of action.[20] Ru(II),Rh(III) supramolecular complexes display $O_2$-independent reactivity in contrast to most Ru-based systems and hold promise as future PDT agents via a novel mechanism of action.[3]

### Trimetallic Complex [{(bpy)$_2$Ru(dpp)}$_2$RhCl$_2$]$^{5+}$

The [{(bpy)$_2$Ru(dpp)}$_2$RhCl$_2$]$^{5+}$ trimetallic complex shown in Figure 11.6 has two ruthenium polyazine LAs attached to a *cis*-Rh$^{III}$Cl$_2$ center through two dpp BL.[15]

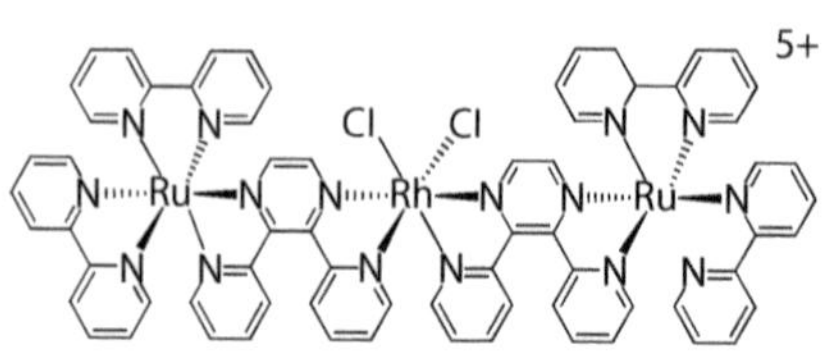

**FIGURE 11.6**   The mixed-metal trimetallic complex [{(bpy)$_2$Ru(dpp)}$_2$RhCl$_2$]$^{5+}$.

The electrochemical, spectroscopic, photophysical, and photochemical properties are reported by the Brewer group.[3,17,19]

*Electrochemical Properties of [{(bpy)₂Ru(dpp)}₂RhCl₂]⁵⁺*

The electrochemical properties were investigated using a three-electrode system in 0.1 M $Bu_4NPF_6$ acetonitrile solution.[17] The oxidative electrochemistry of $[\{(bpy)_2Ru(dpp)\}_2RhCl_2]^{5+}$ shows the reversible overlapping $Ru^{II/III}$ couples at 1.61 V versus Ag/AgCl due to the nearly simultaneous oxidation of the two equivalent Ru centers. Reductively, an irreversible two-electron reduction occurs at −0.35 V versus Ag/AgCl corresponding to $Rh^{III/II/I}$ couple, followed by two reversible reductions at −0.76 and −1.01 V versus Ag/AgCl attributed to two $dpp^{0/-}$ couples (Table 11.1).[3,17] The electrochemistry data show that the HOMO is $Ru(d\pi)$ based whereas the LUMO is $Rh(d\sigma^*)$ based. The fact that the rhodium center can be reduced more easily than the dpp BL indicates the presence of a low-lying Ru→Rh metal-to-metal charge transfer (MMCT) ES.[3]

*Spectroscopic Properties of [{(bpy)₂Ru(dpp)}₂RhCl₂]⁵⁺*

The electronic absorption spectrum of $[\{(bpy)_2Ru(dpp)\}_2RhCl_2]^{5+}$ displays the characteristic absorptions from the ruthenium polyazine LAs in the UV and visible region of the spectrum as shown in Figure 11.7. The spectroscopy in the UV region is dominated by one major transition ($\lambda_{max}^{abs}$ = 284 nm) assigned to bpy-based π→π* transition, and the shoulder ($\lambda_{max}^{abs}$ = 338 nm) corresponds to dpp-based π→π* transitions also in the UV region. One of two MLCT bands in the visible region is centered at 416 nm attributed to the Ru→bpy CT transition, whereas the other lowest energy band at 518 nm is assigned as a Ru→dpp CT transition (Table 11.1).[17]

*Photophysical Properties of [{(bpy)₂Ru(dpp)}₂RhCl₂]⁵⁺*

Emission spectroscopy and ES lifetime measurements are used to investigate the ES dynamics of $[\{(bpy)_2Ru(dpp)\}_2RhCl_2]^{5+}$. The emission spectra of $[\{(bpy)_2Ru(dpp)\}_2RhCl_2]^{5+}$ were observed in room temperature acetonitrile solution and 4:1 ethanol/methanol glass at 77 K.[18] At room temperature, a weak ($\Phi^{em}$ = 2.6 × $10^{-4}$) emission is observed ($\lambda_{max}^{em}$ = 776 nm, $\tau$ = 38 ns) assigned to the ³MLCT emission. At 77 K, the emission displays a longer lifetime ($\tau$ = 1.86 μs) and blue shifts to 730 nm also from the ³MLCT ES. The different observations at room temperature and at 77 K are attributed to RT electron transfer quenching of the ³MLCT through dpp to Rh intramolecular electron transfer to populate the Ru→Rh ³MMCT ES (Figure 11.8). At 77 K in a rigid media, the intramolecular electron transfer is hindered providing for dramatically enhanced lifetimes of the emissive ³MLCT ESs.[21]

*Photochemical Properties of [{(bpy)₂Ru(dpp)}₂RhCl₂]⁵⁺*

The $[\{(bpy)_2Ru(dpp)\}_2RhCl_2]^{5+}$ is the first complex that cleaves DNA via a ³MMCT ES, and this cleavage occurs via a novel oxygen-independent mechanism.[3,17]

The photocleavage of DNA is investigated by the agarose gel electrophoresis method. The agarose gel electrophoresis is a common method used to study the DNA modification. It can separate biological macromolecules including DNA, RNA, and proteins according to their molecular size, shape, mass, and charges by

**TABLE 11.1**

**Electrochemical, Spectroscopic, Photophysical, and Photochemical Properties of Metal-Based Complexes**

| Complex[a] | Electrochemical Properties (V) | $\lambda_{max}^{abs}$ (nm) | $\lambda_{max}^{em}$ (nm)[b] | $\tau$ (ns)[b] | PDT Activity, $\lambda_{irr}$ (nm) | H$_2$ Photocatalyst, $\lambda_{irr}$ (nm) | H$_2$ TON mol H$_2$/mol catalyst | $\Phi_{H_2}$ | Reference |
|---|---|---|---|---|---|---|---|---|---|
| [Ru(bpy)$_3$]$^{2+\,c}$ | Ru$^{II/III}$ +1.27<br>bpy$^{0/-}$ −1.24<br>bpy$^{0/-}$ −1.43<br>bpy$^{0/-}$ −1.67 | 452 | Ru(dπ)→bpy(π*) CT | 605 | 850 | 400 Photocleavage with O$_2$ | 400. Active in multicomponent systems | | | 2<br>10 |
| [{(bpy)$_2$Ru(dpp)}$_2$RhCl$_2$]$^{5+\,c}$ | 2Ru$^{II/III}$ +1.61<br>Rh$^{III/II}$ −0.35<br>dpp$^{0/-}$ −0.76<br>dpp$^{0/-}$ −1.01 | 284<br>338 (sh)<br>416<br>518 | bpy π→π*<br>dpp π→π*<br>Ru(dπ)→bpy(π*) CT<br>Ru(dπ)→dpp(π*) CT | 776 | 38 | >475 Photocleavage PDT with Vero cell | 470 | 22 | 0.01 | 3<br>17<br>19<br>81 |
| [{(bpy)$_2$Os(dpp)}$_2$RhCl$_2$]$^{5+\,c}$ | 2Os$^{II/III}$ +1.21<br>Rh$^{III/II}$ −0.39<br>dpp$^{0/-}$ −0.76<br>dpp$^{-/2-}$ −1.00 | 284<br>336<br>412<br>534<br>798 | bpy π→π*<br>dpp π→π*<br>Os(dπ)→bpy(π*) CT<br>Os(dπ)→dpp(π*) CT<br>Os(dπ)→dpp(π*) CT | | | >475 Photocleavage | 470 | 0.6 | | 3<br>79 |
| [{(bpy)$_2$Ru(dpb)}$_2$IrCl$_2$]$^{5+\,c}$ | 2Ru$^{II/III}$ +1.56<br>dpb$^{0/-}$ 0.03<br>dpb$^{0/-}$ −0.13 | 666 | Ru(dπ)→dpb(π*) CT | | | Not active | Not active | | | 3<br>79 |
| [{(bpy)$_2$Ru(dpp)}$_2$RhBr$_2$]$^{5+\,c}$ | 2Ru$^{II/III}$ +1.61<br>Rh$^{III/II}$ −0.32<br>dpp$^{0/-}$ −0.71<br>dpp$^{0/-}$ −1.01 | 284<br>344<br>520 | bpy π→π*<br>dpp π→π*<br>Ru(dπ)→dpp(π*) CT | 776 | 34 | d | 470 | 27 | 0.01 | 78<br>79<br>81 |
| [{(phen)$_2$Ru(dpp)}$_2$RhCl$_2$]$^{5+\,c}$ | 2Ru$^{II/III}$ +1.61<br>Rh$^{III/II}$ −0.35<br>dpp$^{0/-}$ −0.75<br>dpp$^{0/-}$ −1.02 | 520 | Ru(dπ)→dpp(π*) CT | 760 | 35 | d | 470 | 20 | 0.01 | 79<br>81 |

| Complex | $E_{1/2}$ (V) | Assignment | $\lambda_{abs}$ (nm) | Assignment | | | Photochemistry | | | | Ref |
|---|---|---|---|---|---|---|---|---|---|---|---|
| [{(phen)₂Ru(dpp)}₂RhBr₂]⁵⁺ c | +1.62, −0.32, −0.71, −1.01 | $2Ru^{II/III}$, $Rh^{III/II,II/I}$, $dpp^{0/-}$, $dpp^{0/-}$ | 262, 346, 418, 520 | phen π→π*, dpp π→π*, Ru(dπ)→phen(π*) CT, Ru(dπ)→dpp(π*) CT | 760 | 30 | d | 470 | 24 | 0.01 | 81 |
| [{(tpy)OsCl(dpp)}₂RhCl₂]³⁺ c | +0.85, −0.51, −0.86, −1.20 | $2Os^{II/III}$, $Rh^{III/II,II/I}$, $dpp^{0/-}$, $dpp^{-/2-}$ | 272, 317, 426, 535, 856 | tpy π→π*, dpp π→π*, Os(dπ)→tpy(π*) CT, Os(dπ)→dpp(π*) CT, Os(dπ)→dpp(π*) CT | | | 455 Photocleavage | Not active (No photoreduction) | | | 79 |
| [(bpy)₂Ru(dpp)RhCl₂(phen)]³⁺ c | +1.60, −0.39, −0.74, −0.98 | $Ru^{II/III}$, $Rh^{III/II}$, $Rh^{II/I}$, $dpp^{0/-}$ | 279, 338 (sh), 418, 509 | phen π→π*, bpy π→π*, dpp π→π*, Ru(dπ)→bpy(π*) CT, Ru(dπ)→dpp(π*) CT | 786 | 32 | >450 Photobinding, photocleavage | c | | | 18 |
| [(phen)₂Ru(dpp)RhCl₂(bpy)]³⁺ c | +1.62, −0.44, −0.79, −1.03 | $Ru^{II/III}$, $Rh^{III/II}$, $Rh^{II/I}$, $dpp^{0/-}$ | 262, 356, 413, 505 | phen π→π*, bpy π→π*, dpp π→π*, Ru(dπ)→phen(π*) CT, Ru(dπ)→dpp(π*) CT | 766 | 42 | d | 470 Not active | 0 | 0 | 25 |
| [Rh₂(O₂CCH₃)₄(H₂O)₂] f | +1.02 | $Rh^{II/III}$ | 443, 585 | Rh–Rh(π*)→Rh–O(σ*), Rh–Rh(π*)→Rh–Rh(σ*) | | | >395 Photocleavage with py+ | c | | | 35, 36 |
| [Rh₂⁰,⁰(dpfma)₃(PPh₃)CO] | | | 338 | | | | d | ≥338 | 27 | 0.01 | 87 |
| [(bpy)₂Ru(dpp)PtCl₂]²⁺ f | +1.57, +1.47, −0.54, −1.11, −1.49 | $Ru^{II/III}$, $Pt^{II/III}$, $dpp^{0/-}$, $dpp^{0/-}$, $bpy^{0/-}$ | 422, 509 | Ru(dπ)→bpy(π*) CT, Ru(dπ)→dpp(π*) CT | 786 | 450 | d | c | | | 59 |
| [Ru(bpy)₂[m-bpy-(CONH-(CH₂)₃-NH₂)₂]PtCl₂]²⁺ | | | 470 | Ru(dπ)→bpy(π*) CT | 670 | 518 | >470 Photocleavage with O₂ | UV Not active | 0 | 0 | 55 |

(continued)

## TABLE 11.1 (Continued)
### Electrochemical, Spectroscopic, Photophysical, and Photochemical Properties of Metal-Based Complexes

| Complex[a] | Electrochemical Properties (V) | | $\lambda_{max}^{abs}$ (nm) | $\lambda_{max}^{em}$ (nm)[b] | $\tau$ (ns)[b] | PDT Activity, $\lambda_{irr}$ (nm) | $H_2$ Photocatalyst, $\lambda_{irr}$ (nm) | $H_2$, TON mol $H_2$/mol catalyst | $\Phi_{H_2}$ | Reference |
|---|---|---|---|---|---|---|---|---|---|---|
| [(bpy)$_2$Ru(phenNHCO(COOHbpy))PtCl$_2$]$^{2+ g}$ | Ru$^{II/III}$<br>COOHbpy$^{0/-}$ | +0.89<br>−1.20 | 450 | 610 | | d | >350 | 5 | 0.01 | 95 |
| [(tpy)RuCl(dpp)PtCl$_2$]$^{+ c}$ | Ru$^{II/III}$<br>dpp$^{0/-}$<br>dpp$^{-/2-}$<br>tpy$^{0/-}$ | +1.10<br>−0.55<br>−1.15<br>−1.43 | 236   tpy π→π*<br>317   tpy π→π*<br>354 (sh)   dpp π→π*<br>462   Ru(dπ)→tpy(π*) CT<br>544   Ru(dπ)→dpp(π*) CT | | | >450<br>Photocleavage with O$_2$<br>Inhibition of E. coli replication in dark | c | | | 56 |
| [(Ph$_2$phen)$_2$Ru(dpp)PtCl$_2$]$^{2+ f}$ | Ru$^{II/III}$<br>Pt$^{II/IV}$<br>dpp$^{0/-}$<br>dpp$^{0/-}$<br>Ph$_2$phen$^{0/-}$<br>Ph$_2$phen$^{0/-}$ | +1.57<br>+1.47<br>−0.50<br>−1.06<br>−1.37<br>−1.56 | 279   Ph$_2$phen π→π*<br>310 (sh)   dpp π→π*<br>424   Ru(dπ)→Ph$_2$phen(π*) CT<br>520   Ru(dπ)→dpp(π*) CT | 740 | 40 | 455<br>Photocleavage with O$_2$ | c | | | 63 |
| [(Ph$_2$phen)$_2$Ru(dpq)PtCl$_2$]$^{2+ f}$ | Ru$^{II/III}$<br>Pt$^{II/IV}$<br>dpq$^{0/-}$<br>dpq$^{0/-}$<br>Ph$_2$phen$^{0/-}$<br>Ph$_2$phen$^{0/-}$ | +1.58<br>+1.47<br>−0.23<br>−0.99<br>−1.37<br>−1.56 | 279   Ph$_2$phen π→π*<br>320 (sh)   dpp π→π*<br>424   Ru(dπ)→Ph$_2$phen(π*) CT<br>600   Ru(dπ)→dpq(π*) CT | | | 455<br>Photocleavage with O$_2$ | c | | | 63 |

| Complex | Potentials (V) | Assignment | λabs (nm) | Assignment | | | | | | | Ref |
|---|---|---|---|---|---|---|---|---|---|---|---|
| [(Mephtpy)RuCl(dpp)PtCl$_2$]$^{1+}$ [c] | +1.10<br>−0.55<br>−1.15<br>−1.44 | Ru$^{II/III}$<br>dpp$^{0/-}$<br>dpp$^{-/2-}$<br>MePhtpy$^{0/-}$ | 227<br>318<br>354 (sh)<br>464<br>548 | Mephtpy π→π*<br>Mephtpy π→π*<br>dpp π→π*<br>Ru(dπ)→Mephtpy(π*) CT<br>CT<br>Ru(dπ)→dpp(π*) CT | | | c | >450<br>Photocleavage with O$_2$ | | | 56 |
| [{(bpy)$_2$Ru(dpp)}$_2$Ru(dpp)PtCl$_2$]$^{6+}$ [c] | +1.58<br>−0.40<br>−0.60<br>−0.70 | 2Ru$^{II/III}$<br>dpp$^{0/-}$<br>dpp$^{0/-}$<br>dpp$^{0/-}$ | 290<br>320 (sh)<br>416<br>520–540 | bpy π→π*<br>dpp π→π*<br>Ru(dπ)→bpy(π*) CT<br>Ru(dπ)→dpp(π*) CT | 750 | 100 | c | >450<br>Photocleavage with O$_2$ | | | 58 |
| [{(bpy)$_2$Ru(dpp)}$_2$Ru(dpq)PtCl$_2$]$^{6+}$ [c] | +1.59<br>−0.08<br>−0.45<br>−0.62 | 2Ru$^{II/III}$<br>dpq$^{0/-}$<br>dpp$^{0/-}$<br>dpp$^{0/-}$ | 420<br>540<br>540 | Ru(dπ)→bpy(π*) CT<br>Ru(dπ)→dpp(π*) CT<br>Ru(dπ)→dpq(π*) CT | 745 | 92 | 530 | d | 40 | 0.01 | 101 |
| [{(phen)$_2$Ru(dpp)}$_2$Ru(dpq)PtCl$_2$]$^{6+}$ [c] | +1.58<br>−0.05<br>−0.42<br>−0.59 | 2Ru$^{II/III}$<br>dpq$^{0/-}$<br>dpp$^{0/-}$<br>dpp$^{0/-}$ | 262<br>350 (sh)<br>350 (sh)<br>415<br>541<br>541 | phen π→π*<br>dpp π→π*<br>dpq π→π*<br>Ru(dπ)→bpy(π*) CT<br>Ru(dπ)→dpp(π*) CT<br>Ru(dπ)→dpq(π*) CT | 752 | 120 | 470 | d | 115 | 0.01 | 100 |

[a] bpy = 2,2'-bipyridine; dpp = 2,3-bis(2-pyridyl)pyrazine; dpb = 2,3-bis(2-pyridyl)benzoquinoxaline; dpq = 2,3-bis(2-pyridyl)quinoxaline; Mephtpy = (4'-(4-methylphenyl)-2,2':6',2"-terpyridine; phen = 1,10-phenanthroline; Ph$_3$phen = 4,7-diphenyl-1,10-phenanthroline; PPh$_3$ = triphenylphosphine; py+ = 3-cyano-1-methylpyridinium; tpy = 2,2':6',2"-terpyridine.

[b] Recorded at room temperature.

[c] In CH$_3$CN at 100 mV/s scan rate and reported as V versus Ag/AgCl (3 M NaCl solution). Standard reference versus Ag/AgCl: NHE = −0.209 V, SCE = 0.035 V, FeCp$_2^{0/+}$ = 0.461 V.

[d] DNA photocleavage experiments not reported.

[e] H$_2$ photocatalyst experiments not reported.

[f] Potentials versus SCE.

[g] Potentials versus FeCp$_2^{0/+}$.

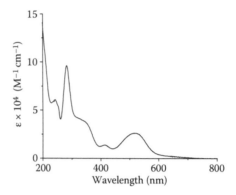

**FIGURE 11.7**   The electronic absorption spectrum of $[\{(bpy)_2Ru(dpp)\}_2RhCl_2]^{5+}$.

applying a potential to an agarose gel matrix. When a potential is applied, the negatively charged DNA moves toward the positive electrode. The rate of DNA migration
depends on the molecular size and shape. The compact, supercoiled DNA moves
faster than open-circular relaxed form, illustrated in Figure 11.9.[22] The gel pictures
(Figure 11.9) show a molecular weight standard (lane λ), a DNA control (lane C), a
complex DNA solution at 5:1 base pair to metal complex (BP/MC) ratio in the dark
(lane MC), and a complex DNA solution at 5:1 BP/MC ratio irradiated for 20 minutes
under argon (lane hν MC). The $[\{(bpy)_2Ru(dpp)\}_2RhCl_2]^{5+}$ complex photocleavage of
pUC18 plasmid and pBluescript DNA was observed in an oxygen-independent environment irradiated by light at λ ≥ 475 nm at 5:1 BP/MC ratio for 20 minutes (lane
hν MC). This is confirmed by converting compact, supercoiled pUC18 DNA (lane
C) to its nicked open-circular relaxed form.[23] The open-circular relaxed form has
a large size and migrates more slowly than the native supercoiled DNA. The function of $[\{(bpy)_2Ru(dpp)\}_2RhCl_2]^{5+}$ complex that photocleaves DNA with or without
molecular oxygen via visible light excitation is unusual. This complex is excited by
a low-energy visible light and then populates a low-lying ³MMCT ES. The ³MMCT
ES has a formally $Ru^{III}$ and $Rh^{II}$ center, both reactive centers. DNA cleavage likely
occurs at the Rh(II) site.

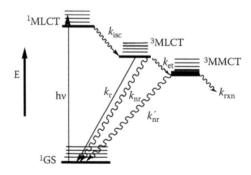

**FIGURE 11.8**   State diagram of $[\{(bpy)_2Ru(dpp)\}_2RhCl_2]^{5+}$.

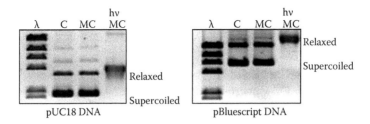

**FIGURE 11.9** Agarose gel showing the photocleavage of pUC18 plasmid and pBluescript DNA by [{(bpy)$_2$Ru(dpp)}$_2$RhCl$_2$]$^{5+}$.

The PDT functions of [{(bpy)$_2$Ru(dpp)}$_2$RhCl$_2$]$^{5+}$ were further observed in African green monkey kidney epithelial (Vero) cell cultures. Vero cells replicate rapidly doubling the cell population in 24 hours. The trimetallic complex [{(bpy)$_2$Ru(dpp)}$_2$RhCl$_2$]$^{5+}$ can inhibit Vero cell growth *in vitro* after exposure to visible light ($\lambda > 460$ nm).[14] Vero cells were treated by 3 to 120 μM of [{(bpy)$_2$Ru(dpp)}$_2$RhCl$_2$]$^{5+}$ with growth media solutions for 48 hours, media were removed, cells were treated with light for 4 minutes (light samples) or were not treated with light (dark controls), and then the cells were incubated with fresh media for another 48 hours. The dark control (cells treated with metal complex but no light) and light control (cells treated with no metal complex but with light) showed normal growth, whereas Vero cells treated by different concentrations of [{(bpy)$_2$Ru(dpp)}$_2$RhCl$_2$]$^{5+}$, then exposed to light, show increased inhibition of cell replication as the complex concentration increased. Cell death was observed as shown in Figure 11.10, when the complex concentration was

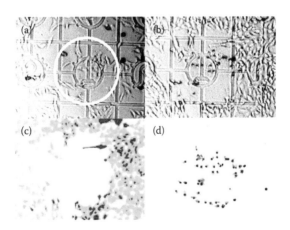

**FIGURE 11.10** Micrographs of Vero cells after treatment with 122 μM of [{(bpy)$_2$Ru(dpp)}$_2$RhCl$_2$]Cl$_5$, rinsed with a clean medium, irradiated by focused light ($\lambda > 460$ nm) for 4 minutes: (a) immediately after photolysis (the circle shows the border of the irradiation spot); (b) after a 48-hour growth period in the dark; (c) after a 48-hour growth period with live cells (green), visualization with calcein AM fluorescence; and (d) after a 48-hour growth period with dead cells (red), visualization with ethidium homodimer-1 fluorescence.

higher than 10 μM. The [{(bpy)$_2$Ru(dpp)}$_2$RhCl$_2$]$^{5+}$ displays high phototoxicity and low dark cytotoxicity to Vero cells, further indication of an effective PDT agent.

*Other Trimetallic Complexes*

Other trimetallic complexes including [{(bpy)$_2$Os(dpp)}$_2$RhCl$_2$]$^{5+}$ and [{(tpy)ClRu(dpp)}$_2$ RhCl$_2$]$^{3+}$ photocleave DNA under visible irradiation through an oxygen-independent mechanism.[3] However, the trimetallic complexes such as [{(bpy)$_2$Ru(bpm)}$_2$RhCl$_2$]$^{5+}$ and [{(bpy)$_2$Ru(dpp)}$_2$IrCl$_2$]$^{5+}$ cannot photocleave DNA because they cannot populate $^3$MMCT ESs after visible irradiation.[3]

## Bimetallic Complex [(bpy)$_2$Ru(dpp)RhCl$_2$(phen)]$^{3+}$

The bimetallic complex [(bpy)$_2$Ru(dpp)RhCl$_2$(phen)]$^{3+}$ couples only one ruthenium polyazine LA to a *cis*-Rh$^{III}$Cl$_2$ center through a dpp BL, as shown in Figure 11.11. The bimetallic complex has a smaller size and a lower cationic charge than the Ru,Rh,Ru trimetallic complexes. This Ru,Rh bimetallic displays many similar electrochemical, photophysical, and photochemical properties compared with Ru,Rh,Ru trimetallic systems.[18]

*Electrochemical Properties of [(bpy)$_2$Ru(dpp)RhCl$_2$(phen)]$^{3+}$*

The electrochemical properties of [(bpy)$_2$Ru(dpp)RhCl$_2$(phen)]$^{3+}$ are somewhat different from that of [{(bpy)$_2$Ru(dpp)}$_2$RhCl$_2$]$^{5+}$.[18] The oxidative electrochemistry of [(bpy)$_2$Ru(dpp)RhCl$_2$(phen)]$^{3+}$ shows a reversible one electron couple at 1.60 V versus Ag/AgCl assigned to the Ru$^{II/III}$ couple. The first two quasi-reversible reductive waves with one electron transfer occur at −0.39, −0.74 V corresponding to rhodium-based reductions before the reversible dpp$^{0/-}$ couple at −0.98 V versus Ag/AgCl (Table 11.1). Both the bimetallic and the trimetallic complexes display a Ru(dπ)-based HOMO and a Rh(dσ*)-based LUMO, indicating the Ru(II),Rh(III) bimetallic possess a low-lying Ru→Rh MMCT ES.[18]

*Spectroscopic Properties of [(bpy)$_2$Ru(dpp)RhCl$_2$(phen)]$^{3+}$*

The [(bpy)$_2$Ru(dpp)RhCl$_2$(phen)]$^{3+}$ complex has one ruthenium polyazine LA displaying approximately half the extinction coefficient values relative to the Ru,Rh,Ru trimetallics both in the UV and visible regions of the electronic absorption spectrum.[18] The bimetallic complex strongly absorbs at 279 nm (bpy or phen π→π*) with a shoulder at 338 nm (dpp π→π*) in the UV region. The transitions in the visible region are assigned to lower energy MLCT transitions, Ru→bpy ligand CT at 418 nm, and Ru→dpp CT transition attributed to 509 nm (Table 11.1).

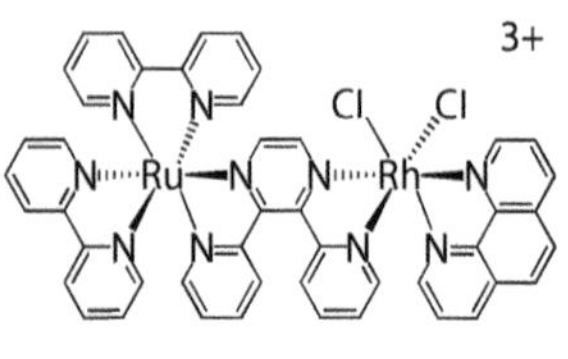

**FIGURE 11.11**   The mixed-metal bimetallic complex [(bpy)$_2$Ru(dpp)RhCl$_2$(phen)]$^{3+}$.

*Photophysical and Photochemical Properties of [(bpy)$_2$Ru(dpp)RhCl$_2$(phen)]$^{3+}$*

The emission spectra and ES lifetime of [(bpy)$_2$Ru(dpp)RhCl$_2$(phen)]$^{3+}$ are similar to that of [{(bpy)$_2$Ru(dpp)}$_2$RhCl$_2$]$^{5+}$ trimetallic complex at RT and at 77 K.[18] At RT, a weak, short-lived, emission band is observed at 786 nm ($\tau$ = 32 ns) assigned to Ru(d$\pi$)→dpp($\pi$*) CT ES ($^3$MLCT). The quantum yield of [(bpy)$_2$Ru(dpp)RhCl$_2$(phen)]$^{3+}$ emission ($\Phi_{RT}^{em} = 2.3 \times 10^{-4}$) and the quantum yield of [{(bpy)$_2$Ru(dpp)}$_2$RhCl$_2$]$^{5+}$ emission ($\Phi_{RT}^{em} = 2.6 \times 10^{-4}$) at RT are much smaller than that of [(bpy)$_2$Ru(dpp)Ru(bpy)$_2$]$^{4+}$ ($\Phi_{RT}^{em} = 1.4 \times 10^{-3}$). This is attributed to a low-lying Rh(d$\sigma$*) acceptor orbital allowing the $^3$MLCT ESs to populate $^3$MMCT ES through intramolecular electron transfer at room temperature. At 77 K, the emission of [(bpy)$_2$Ru(dpp)RhCl$_2$(phen)]$^{3+}$ blue shifts to 716 nm. The ES lifetime of [(bpy)$_2$Ru(dpp)RhCl$_2$(phen)]$^{3+}$ ($\tau$ = 1.80 µs) is close to that of [(bpy)$_2$Ru(dpp)Ru(bpy)$_2$]$^{4+}$ at 77 K ($\tau$ = 2.45 µs), indicating that intramolecular electron transfer is impeded at 77 K in a rigid media.

The photochemical properties of the [(bpy)$_2$Ru(dpp)RhCl$_2$(phen)]$^{3+}$ bimetallic complex are somewhat different from the trimetallic [{(bpy)$_2$Ru(dpp)}$_2$RhCl$_2$]$^{5+}$.[24] [(bpy)$_2$Ru(dpp)RhCl$_2$(phen)]$^{3+}$ not only photocleaves pUC18 plasmid DNA similar to [{(bpy)$_2$Ru(dpp)}$_2$RhCl$_2$]$^{5+}$ but also photobinds to DNA under visible radiation. The pUC18 plasmid DNA photocleavage was observed by using a gel shift assay after the [(bpy)$_2$Ru(dpp)RhCl$_2$(phen)]$^{3+}$ DNA solution at a 5:1 BP/MC ratio was irradiated by $\lambda$ > 450 nm light under argon flow.[24] [(bpy)$_2$Ru(dpp)RhCl$_2$(phen)]$^{3+}$ photobinding was investigated using calf thymus DNA via a selective precipitation experiment, at a 5:1 BP/MC ratio. Binding is observed reaching saturation at 60% of metal complex bound to DNA.[24] DNA photobinding occurs via the Rh site, presumably by halide loss and covalent bond to DNA base pairs. Halide loss from the $^3$MMCT state is possible as it populates the Rh(d$\sigma$*) orbital decreasing Rh–Cl bond order. The trimetallic complex is sterically hindered around the Rh center with two large (bpy)$_2$Ru$^{II}$(dpp) subunits bond to each side. The bimetallic complex is not sterically impeded from Rh binding to DNA as only one large (bpy)$_2$Ru$^{II}$(dpp) is bonded to the Rh center. The smaller phen ligand is bonded to the other side of Rh in the Ru,Rh bimetallic, allowing Rh binding to DNA.

Other bimetallic complexes such as [(phen)$_2$Ru(dpp)RhCl$_2$(bpy)]$^{3+}$ also photobind to and photocleave DNA under visible irradiation through an oxygen-independent mechanism.[25,26] Although the bimetallic complex [(bpy)$_2$Ru(bpm)RhCl$_2$(phen)]$^{3+}$ photobinds to DNA, it cannot photocleave DNA under visible irradiation because of the inability to form the $^3$MMCT exited state after visible irradiation.[24]

## DIRHODIUM (II,II) COMPLEXES

A large variety of dirhodium (II,II) complexes with a direct Rh–Rh bond have been synthesized and characterized since 1960.[27–31] The metal–metal bond strongly influences the shape and the properties of the complexes. Typically, the BLs between the two Rh centers can be classified into three different groups: complexes with carboxylato-BLs,[28,30] complexes with other BLs such as sulfato- and phosphate ligands,[32] and complexes without BLs.[33,34] Among the dirhodium (II,II) complexes, tetra-µ-carboxylato-dirhodium(II) complexes [Rh$_2$(O$_2$CR)$_4$(L)$_2$] (L=H$_2$O, CH$_3$OH) have attracted more attention due to their antitumor activity as PDT agents.[35]

## Electrochemical, Spectroscopic, and Photochemical Properties of [Rh₂(O₂CCH₃)₄(H₂O)₂]

The electrochemical properties of $[Rh_2(O_2CCH_3)_4]$ were investigated using a three-electrode system, including a platinum disc electrode as the working electrode, a calomel reference electrode, and a platinum wire auxiliary electrode.[36] The oxidative electrochemistry of $[Rh_2(O_2CCH_3)_4]$ displays a reversible one electron couple at 1.02 V versus SCE due to a $Rh^{II/III}$ oxidation. The electrochemical reduction of $[Rh_2(O_2CCH_3)_4]$ shows an irreversible one electron wave attributed to a $Rh^{II/I}$ reduction. The authors suggest a solvent effect influencing the potential of electrochemical reactions due to different solvent ligands coordinating to the Rh atoms in the axial position. Substituent effect was investigated by using a series of $[Rh_2(O_2CR)_4]$ complexes (R=$(CH_3)_3C$, $C_5H_9$, $C_3H_7$, $C_2H_5$, $CH_3$, $C_6H_5CH_2$, $CH_3OCH_3$, $C_6H_5OCH_2$, $CH_3CHCl$ and $CF_3$).[36] The results show that the electron-donating substituent groups make the complex more easily oxidized and more difficult to be reduced, whereas the electron-withdrawing substituent groups work the opposite way.

The electronic absorption spectrum of $[Rh_2(O_2CCH_3)_4(H_2O)_2]$ in water has been reported.[36] In the UV region, there is an intense absorption band at 220 nm with a shoulder at 250 nm attributed to the transitions from the axial ligand. However, the electronic transitions in the visible region are weak with a transition at 443 nm due to Rh–Rh $\pi^* \rightarrow$ Rh–O $\sigma^*$ transition followed by a transition at 585 nm assigned as an Rh–Rh $\pi^* \rightarrow$ Rh–Rh $\sigma^*$ transition, which is affected by different axial ligands coordinated to the Rh atoms. Martin and co-workers also performed a $[Rh_2(O_2CCH_3)_4(H_2O)_2]$ single-crystal polarized transition experiment with consistent results.[37]

Fu et al.[38] reported that the dirhodium complex, $[Rh_2(O_2CCH_3)_4(H_2O)_2]$, populates a long-lived triplet ES and photocleaves pUC18 plasmid DNA after irradiation with visible light ($\lambda \geq 395$ nm) and in the presence of EAs such as 3-cyano-1-methylpyridinium (py+). The authors proposed that DNA cleavage was caused by an oxidized $[Rh_2(O_2CCH_3)_4(H_2O)_2]^+$ product resulting from the photolysis of the $[Rh_2(O_2CCH_3)_4(H_2O)_2]$ and EA solutions. The dirhodium complex $[Rh_2(O_2CCH_3)_4(H_2O)_2]$ weakly binds to double-stranded DNA through axial coordination ($K_b = 4.6 \times 10^2$ M$^{-1}$).[39] The $[Rh_2(O_2CCH_3)_4(H_2O)_2]$ complex also shows inhibition of HeLa cell culture growth proposed to be a result of the inactivation of DNA polymerase-$\alpha$ causing reduced DNA replication.[35,40]

## Other Dirhodium Complexes

On the basis of the primary complex, $[Rh_2(O_2CCH_3)_4(H_2O)_2]$, Bradley et al.[41] and Friedman et al.[42] synthesized and characterized a series of dirhodium complexes by modifying the BLs and axial ligands to optimize PDT function. For example, the dirhodium complex $[Rh_2(\mu-O_2CCH_3)_2(\eta^1-O_2CCH_3)(CH_3OH)(dppz)]^+$ replaces an acetate BL with an intercalating dppz ligand (dppz = dipyrido[3,2-a;2',3'-c] phenazine) and a labile $CH_3OH$ ligand compared with $[Rh_2(O_2CCH_3)_4(H_2O)_2]$. The purpose of introducing an intercalating dppz ligand into a dirhodium system is to improve the dirhodium complex association with DNA as well as providing a ligand with low-lying $\pi^*$ acceptor orbitals. The complex $[Rh_2(\mu-O_2CCH_3)_2(\eta^1-O_2CCH_3)(CH_3OH)(dppz)]^+$ can intercalate with DNA and photocleave DNA under irradiation ($\lambda > 395$ nm) without EAs and in an oxygen-independent environment. The complexes cis-[Rh$_2$($\mu$-O$_2$CCH$_3$)$_2$(bpy)(dppz)]$^{2+}$ and cis-[Rh$_2$($\mu$-O$_2$CCH$_3$)$_2$(dppz)$_2$]$^{2+}$ display low dark-cytotoxicity and high photocytotoxicity to Hs-27 human skin fibro-blasts.[43,44] Other dirhodium complexes such as cis-[Rh$_2$($\mu$-O$_2$CCH$_3$)$_2$(dppn)(L)]$^{2+}$ (dppn = benzo[i]dipyrido[3,2-a:2′, 3′-h]quinoxaline and L = bpy, phen, dpq, dppz) bind to DNA, photocleave DNA in air under irradiation ($\lambda \geq 375$ nm), and show low dark-cytotoxicity and high photocytotoxicity to Hs-27 human skin fibroblasts.[45] Cis-[Rh$_2$($\mu$-O$_2$CCH$_3$)$_2$(dppn)(dppz)]$^{2+}$ can effectively inhibit the growth of the cancer cell such as HeLa and COLO-316.[46]

## SUPRAMOLECULAR COMPLEXES COUPLING RU(II) POLYAZINE LIGHT ABSORBERS TO A PT(II) METAL CENTER

Cisplatin, cis-[PtCl$_2$(NH$_3$)$_2$], a commonly used anticancer agent, acts via covalent binding to DNA resulting in cell death or apoptosis.[47–51] Cisplatin is currently used clinically, but many cancers have become resistant to cisplatin treatment and sys-temic toxicity occurs with cisplatin treatment.[52,53] In an attempt to develop PDT drugs that have a cellular target, work during the past 10 years has been focused on coupling Ru-based LAs to a cis-PtCl$_2$ BAS for light-activated selectivity and DNA-binding abilities. Several Ru(II),Pt(II) supramolecular architectures have been reported with Ru light-absorbing and Pt BAS moieties coupled by dpp, dpq, dpb, or other spatially separated polyazine BLs.[54–58] These Ru(II),Pt(II) complexes display redox and spectroscopic properties that are tunable by ligand variation and systems have been prepared with effective PDT action.

### Supramolecular Complex [(bpy)$_2$Ru(dpp)PtCl$_2$]$^{2+}$

The bimetallic complex [(bpy)$_2$Ru(dpp)PtCl$_2$]$^{2+}$, as shown in Figure 11.12, has one ruthenium polyazine LA attached to a Pt BAS through a dpp BL.[59] Yam et al.[54] reported the crystal structure of [(bpy)$_2$Ru(dpp)PtCl$_2$]$^{2+}$, which shows a distorted octahedral geometry of a ruthenium center and a distorted square-planar geome-try of a platinum center connected by dpp noncoplanar rings. The electrochemical, spectroscopic, and photophysical properties are also reported.

*Electrochemical Properties of [(bpy)$_2$Ru(dpp)PtCl$_2$]$^{2+}$*

The oxidative electrochemistry of [(bpy)$_2$Ru(dpp)PtCl$_2$]$^{2+}$ displays a reversible couple at 1.57 V versus SCE assigned to Ru$^{II/III}$ oxidation and an irreversible wave at 1.47 V

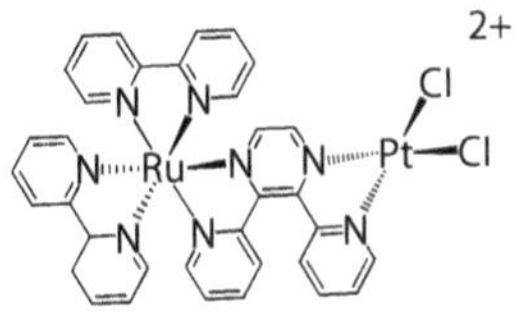

**FIGURE 11.12** The mixed-metal bimetallic complex [(bpy)$_2$Ru(dpp)PtCl$_2$]$^{2+}$.

versus SCE attributed to Pt center-based oxidation.[59] The electrochemical reduction of the complex shows three reversible one electron transfer couple at −0.54, −1.11, and −1.49 V versus SCE due to $dpp^{0/-}$, $dpp^{-/2-}$, and $bpy^{0/-}$ reduction couple, respectively. The electrochemical data imply a Ru(dπ)-based HOMO and a dpp(π*)-based LUMO in [(bpy)$_2$Ru(dpp)PtCl$_2$]$^{2+}$ system.

*Spectroscopic and Photophysical Properties of [(bpy)$_2$Ru(dpp)PtCl$_2$]$^{2+}$*

The electronic absorption spectrum of [(bpy)$_2$Ru(dpp)PtCl$_2$]$^{2+}$ shows the typical absorptions from the ruthenium polyazine LAs in the UV and visible light region of the spectrum.[59] The electronic transitions in the UV region were assigned to bpy- or dpp-based π→π* transitions. One of the MLCT transitions in the visible region at 422 nm is attributed to Ru(dπ)→bpy(π*) CT transition, and another band at 509 nm is due to the Ru(dπ)→dpp(π*) CT transition.

Emission spectroscopy and ES lifetime measurements are used to probe the ES dynamics of [(bpy)$_2$Ru(dpp)PtCl$_2$]$^{2+}$. A weak emission ($\lambda_{max}$ = 786 nm, τ = 0.45 μs) is observed from RT acetonitrile complex solution assigned to Ru(dπ)→dpp(π*) $^3$MLCT emission. At 77 K, the emission has a longer lifetime (τ = 2.20 μs) and blue shifts to 752 nm.

*PDT Actions of Bimetallic Complexes [(bpy)$_2$Ru(BL)PtCl$_2$]$^{2+}$ (BL = dpp, dpq, dpb)*

The biological activity of [(bpy)$_2$Ru(dpp)PtCl$_2$]$^{2+}$ was not reported. Milkevitch et al.[60,61] reported related dpq and dpb complexes prepared by similar methods. These complexes display similar redox and spectroscopic properties modulated by the lower energy π* acceptor orbitals of dpq and dpb versus dpp. These complexes [(bpy)$_2$Ru(dpq)PtCl$_2$]$^{2+}$ and [(bpy)$_2$Ru(dpb)PtCl$_2$]$^{2+}$, where dpq = 2,3-bis(2-pyridyl) quinoxaline) and dpb = (2,3-bis(2-pyridyl)-benzoquinoxaline), bind to DNA through the *cis*-Pt$^{II}$Cl$_2$ moiety (BAS) but do not display light-activated cleavage of DNA likely due to the lowest lying $^3$MLCT being BL-based and strongly coupled to the Pt site.

## Supramolecular Complex [Ru(bpy)$_2${m-bpy-(CONH-(CH$_2$)$_3$-NH$_2$)$_2$}PtCl$_2$]$^{2+}$

The bimetallic complex [Ru(bpy)$_2${m-bpy-(CONH-(CH$_2$)$_3$-NH$_2$)$_2$}PtCl$_2$]$^{2+}$, as shown in Figure 11.13, has one ruthenium LA attached to a *cis*-PtCl$_2$ BAS through a spatially separated amino linker, m-bpy-(CONH-(CH$_2$)$_3$-NH$_2$)$_2$.[55] The complex displays interesting photophysical properties and photocleaves pBR322 DNA under visible light irradiation in air-saturated conditions.

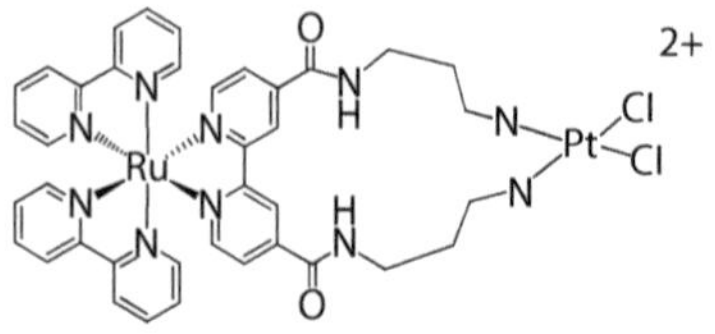

**FIGURE 11.13** The mixed-metal bimetallic complex [Ru(bpy)$_2${m-bpy-(CONH-(CH$_2$)$_3$-NH$_2$)$_2$}PtCl$_2$]$^{2+}$.

*Spectroscopic and Photophysical Properties of*
*[Ru(bpy)₂{m-bpy-(CONH-(CH₂)₃-NH₂)₂}PtCl₂]²⁺*

The spectroscopic photophysical properties of $[Ru(bpy)_2\{m\text{-}bpy\text{-}(CONH\text{-}(CH_2)_3\text{-}NH_2)_2\}PtCl_2]^{2+}$ were investigated by Sakai et al.[55] The electronic absorption spectrum of $[Ru(bpy)_2\{m\text{-}bpy\text{-}(CONH\text{-}(CH_2)_3\text{-}NH_2)_2\}PtCl_2]^{2+}$ is similar to its monometallic synthon displaying ligand-based $\pi \rightarrow \pi^*$ transitions in the UV region and MLCT transitions in the visible region.

The emission of $[Ru(bpy)_2\{m\text{-}bpy\text{-}(CONH\text{-}(CH_2)_3\text{-}NH_2)_2\}PtCl_2]^{2+}$ in water at RT is more intense and longer lived than the emission for its monometallic synthon. The ES lifetime difference between the bimetallic complex ($\tau = 518$ ns) and the monometallic synthon ($\tau = 244$ ns) is attributed to nonradiative decay of the $^3$MLCT ES, which is enhanced by the flexible amine linker unit in the monometallic complex and reduced by the rigid metallocycle in the bimetallic complex.

*Photochemical Properties of [Ru(bpy)₂{m-bpy-(CONH-(CH₂)₃-NH₂)₂}PtCl₂]²⁺*

The bimetallic complex $[Ru(bpy)_2\{m\text{-}bpy\text{-}(CONH\text{-}(CH_2)_3\text{-}NH_2)_2\}PtCl_2]^{2+}$ photocleaves pBR322 DNA under the visible light ($\lambda = 470 \pm 10$ nm) irradiation in air-saturated conditions.[55] No DNA cleavage was observed in dark control solutions at 5:1, 10:1, 20:1, and 50:1 BP/MC ratios, indicating the requirement of light activation.

## Supramolecular Complex [(tpy)RuCl(dpp)PtCl₂]⁺

The bimetallic complex $[(tpy)RuCl(dpp)PtCl_2]^+$ (tpy = 2,2′:6′,2″-terpyridine) couples one tpy-containing ruthenium polyazine LA to a *cis*-PtCl₂ BAS through a dpp BL (Figure 11.14). The tridentate tpy ligand provides control of the stereochemistry of Ru,Pt bimetallics, avoiding the $\Delta$ and $\Lambda$ isomeric mixture, which is unavoidable in the bidentate TL systems. The complex $[(tpy)RuCl(dpp)PtCl_2]^+$ displays interesting electrochemical, spectroscopic properties while maintaining the ability to bind to and photocleave DNA under visible light irradiation in air-saturated conditions. The complex also inhibits *Escherichia coli* JM109 bacterial cell growth, representing the first Ru,Pt bimetallic shown to display antibacterial activity.

*Electrochemical Properties of [(tpy)RuCl(dpp)PtCl₂]⁺*

Electrochemistry was used to investigate relative energies of the frontier orbitals in the $[(tpy)RuCl(dpp)PtCl_2]^+$ system.[56] The electrochemical oxidation of [(tpy) RuCl(dpp)PtCl₂]⁺ occurs at 1.10 V versus Ag/AgCl assigned to Ru$^{II/III}$ couple. The electrochemical reduction of $[(tpy)RuCl(dpp)PtCl_2]^+$ shows two reversible reductions at −0.55 and −1.15 V versus Ag/AgCl attributed to the dpp$^{0/-}$ and dpp$^{-/2-}$ couple, followed by a tpy$^{0/-}$-based reduction at −1.43 V versus Ag/AgCl. The electrochemical

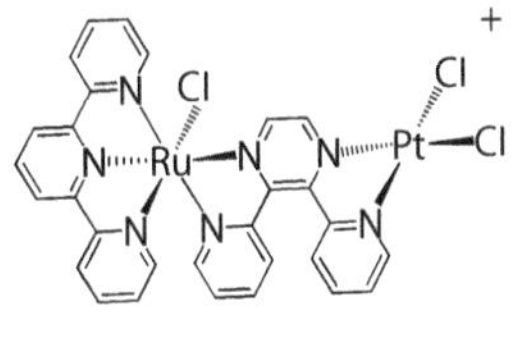

**FIGURE 11.14** The mixed-metal bimetallic complex $[(tpy)RuCl(dpp)PtCl_2]^+$.

data indicate the Ru(dπ)-based HOMO and dpp(π*)-based LUMO in [(tpy)RuCl(dpp) PtCl₂]⁺.

## Spectroscopic Properties of [(tpy)RuCl(dpp)PtCl₂]⁺

The electronic absorption spectrum of [(tpy)RuCl(dpp)PtCl₂]⁺ displays ligand-based π→π* transitions in the UV region and MLCT transition in the visible region. Two strong transitions occur at 236 and 317 nm attributed to tpy-based π→π* transitions with a shoulder at 354 nm due to dpp-based π→π* transitions. A weak MLCT band at 462 nm is assigned to Ru(dπ)→tpy(π*) CT transitions, followed by an intense Ru(dπ)→dpp(π*) CT transition band at 544 nm. The use of tpy and Cl in place of the two diimine bpy ligands shifts the lowest-lying MLCT band to significantly lower energy.

## DNA-Binding Properties of [(tpy)RuCl(dpp)PtCl₂]⁺

The agarose gel electrophoresis experiment is used to investigate the DNA-binding ability of complex [(tpy)RuCl(dpp)PtCl₂]⁺ as shown in Figure 11.15. A series of metal complex and linear pUC18 DNA solutions at different BP/MC ratios were prepared and incubated at 37°C for 1 hour.

The gel picture shows a molecular weight control (lane λ), a linear pUC18 DNA control (lane C), a metal complex and linear pUC18 DNA solution at 5:1 BP/MC ratio (lane 5:1), a metal complex and linear pUC18 DNA solution at 10:1 BP/MC ratio (lane 10:1), and a metal complex and linear pUC18 DNA solution at 20:1 BP/ MC ratio (lane 20:1). A decreasing rate of DNA migration was observed as increasing metal complex concentration from 20:1, to 10:1, to 5:1 in cisplatin as well as in the [(tpy)RuCl(dpp)PtCl₂]⁺ system. The result is consistent with the character of cisplatin, which covalently binds to DNA.[50] The result also indicates that the complex [(tpy)RuCl(dpp)PtCl₂]⁺ binds to the linear DNA through the Pt BAS.[56]

## Photochemical Properties of [(tpy)RuCl(dpp)PtCl₂]⁺

The bimetallic complex [(tpy)RuCl(dpp)PtCl₂]⁺ photocleaves pUC18 DNA under visible light irradiation (λ ≥ 450 nm) in air, as shown in Figure 11.16. The gel picture shows a molecular weight standard (lane λ), a pUC18 DNA dark control (lane C), a complex DNA solution at 20:1 BP/MC ratio incubated for 4 hours in the dark at RT (lane RT), a complex DNA solution at 20:1 BP/MC ratio incubated for 4 hours

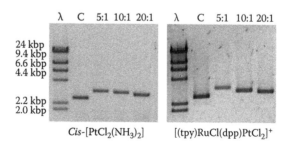

λ   C   5:1  10:1  20:1        λ    C   5:1  10:1 20:1

24 kbp
9.4 kbp
6.6 kbp
4.4 kbp

2.2 kbp
2.0 kbp

Cis-[PtCl₂(NH₃)₂]              [(tpy)RuCl(dpp)PtCl₂]⁺

**FIGURE 11.15** Agarose gel pictures showing cisplatin and the bimetallic complex [(tpy) RuCl(dpp)PtCl₂]⁺ binding to linear pUC18 DNA.

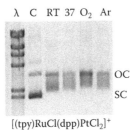

λ   C   RT   37   O₂   Ar

OC

SC

[(tpy)RuCl(dpp)PtCl₂]⁺

**FIGURE 11.16** Agarose gel showing the bimetallic complex [(tpy)RuCl(dpp)PtCl₂]⁺ photo-cleavage pUC18 DNA.

in the dark at 37°C (lane 37), a complex DNA solution at 20:1 BP/MC ratio irradiated for 4 hours in air (lane O₂), and a complex DNA solution at 20:1 BP/MC ratio irradiated for 4 hours under argon (lane Ar). DNA photocleavage was observed in lane O₂, as the supercoiled DNA form I is converted to open-circular DNA form II. Photocleavage was not observed without O₂, indicating the requirement of light and molecular oxygen in this PDT reaction.

*Antibacterial Property of [(tpy)RuCl(dpp)PtCl₂]⁺*

The antibacterial property of [(tpy)RuCl(dpp)PtCl₂]⁺ was investigated by using *E. coli* cell culture.[62] In this study, *E. coli* cells grew in 50 mL of Luria-Bertani media (LB media) at 37°C overnight before the cells were treated by different concentrations of cisplatin, [(tpy)RuCl(dpp)]⁺, and [(tpy)RuCl(dpp)PtCl₂]⁺. The *E. coli* cell growth inhibition was observed in 0.4 mM of [(tpy)RuCl(dpp)PtCl₂]⁺ solution and in 0.2 mM of cisplatin solution. The cell growth curve shows increased inhibition of cell growth as the complex concentration is increased.

## Supramolecular Complexes [(Ph₂phen)₂Ru(BL)PtCl₂]²⁺ (BL = dpp or dpq)

The Ru(II),Pt(II) bimetallic complexes [(Ph₂phen)₂Ru(BL)PtCl₂]²⁺ (Ph₂phen = 4,7-diphenyl-1,10-phenanthroline) (Figure 11.17) couples one Ph₂phen containing ruthenium polyazine LA to a *cis*-PtCl₂ BAS, through a dpp or dpq BL. The

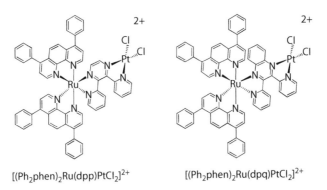

[(Ph₂phen)₂Ru(dpp)PtCl₂]²⁺          [(Ph₂phen)₂Ru(dpq)PtCl₂]²⁺

**FIGURE 11.17** The structures of [(Ph₂phen)₂Ru(dpp)PtCl₂]²⁺ and [(Ph₂phen)₂Ru(dpq)PtCl₂].

complexes $[(Ph_2phen)_2Ru(BL)PtCl_2]^{2+}$ incorporating TL = $Ph_2phen$ into supramolecular systems affords enhanced spectroscopic, photophysical, and photodynamic
properties as compared with the tpy or bpy bimetallic analogs. The supramolecular
bimetallic complexes are efficient LAs with sufficient ES lifetimes to produce singlet
oxygen, a Pt site that binds to linear pUC18 DNA and a Ru LA that facilitates photocleavage of circular pUC18 DNA via an oxygen-dependent mechanism.

## Electrochemical Properties of $[(Ph_2phen)_2Ru(BL)PtCl_2]^{2+}$

The oxidative electrochemistry of $[(Ph_2phen)_2Ru(BL)PtCl_2]^{2+}$ where BL = dpp or
dpq displays a reversible couple at 1.57 V or at 1.58 V versus SCE assigned to $Ru^{II/III}$
oxidation for the dpp or dpq analogs, and an irreversible wave at 1.47 V versus SCE
due to the $Pt^{II/IV}$ oxidation.[63] The reductive electrochemistry of $[(Ph_2phen)_2Ru(dpp)PtCl_2]^{2+}$ consists of four reversible one electron transfer couples at −0.50, −1.06, −1.37,
and −1.56 V versus SCE assigned to $dpp^{0/-}$, $dpp^{-/2-}$, $Ph_2phen^{0/-}$, and $Ph_2phen^{-/2-}$,
respectively. Although the replacement of the dpp BL to the dpq BL in the bimetallic
systems causes BL electrochemical reductions to shift to more positive potentials
at −0.23 V $(dpq^{0/-})$ and −0.99 V $(dpq^{-/2-})$ versus SCE, the TL reduction potentials
remain unperturbed. The electrochemical data indicate a Ru(dπ)-based HOMO and
a BL(π*)-based LUMO in the $[(Ph_2phen)_2Ru(BL)PtCl_2]^{2+}$-based systems.[63]

## Spectroscopic and Photophysical Properties of $[(Ph_2phen)_2Ru(BL)PtCl_2]^{2+}$

The electronic absorption spectra of $[(Ph_2phen)_2Ru(BL)PtCl_2]^{2+}$ display intense
ligand-based π→π* transitions in the UV region and MLCT transitions in the visible
region. An intense transition band observed in the UV region is assigned to $Ph_2phen$
π→π* transitions with a shoulder at 310 nm attributed to dpp-based π→π* transition or at 320 nm is due to dpq-based π→π* transition. The Ru(dπ)→$Ph_2phen$(π*)
CT transitions at 424 nm are observed in both bimetallic complexes. The Ru(dπ)
→dpp(π*) CT transitions were observed at 520 nm from $[(Ph_2phen)_2Ru(dpp)PtCl_2]^{2+}$,
whereas the Ru(dπ)→dpq(π*) CT transitions red shift to 600 nm because of the stabilization of the dpq(π*) orbitals.

Emission ($\Phi^{em}$ = 4.1 × 10⁻⁴, τ = 44 ns) from the ³MLCT state of $[(Ph_2phen)_2Ru(dpp)PtCl_2]^{2+}$ in RT deoxygenated acetonitrile was observed at 740 nm. At 77 K, the emission
from $[(Ph_2phen)_2Ru(dpp)PtCl_2]^{2+}$ blue shifts to 694 nm with a longer lifetime (τ = 2.3 μs).

## DNA Interactions of the Complexes $[(Ph_2phen)_2Ru(BL)PtCl_2]^{2+}$

The bimetallic complexes $[(Ph_2phen)_2Ru(BL)PtCl_2]^{2+}$ bind to linear pUC18 DNA when
the complexes were incubated with linear pUC18 DNA for 1 hour at RT and 37°C. The
evidence from agarose gel electrophoresis shows that rate of DNA migration decreases
as metal complex concentration increases. Similar results were observed for cisplatin.

The bimetallic complexes $[(Ph_2phen)_2Ru(BL)PtCl_2]^{2+}$ where BL = dpp or dpq photocleave circular pUC18 plasmid DNA after the irradiation with a 5-W LED array
(λ = 455 nm) of metal complex–DNA solution at 20:1 BP/MC ratio for 1 hour in the
presence of molecular oxygen.

The gel pictures, as shown in Figure 11.18, show a molecular weight standard (lane
λ), a linear pUC18 DNA control (lane C), a metal complex DNA solution incubated

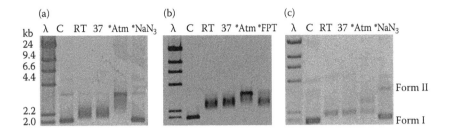

**FIGURE 11.18** Agarose gel showing the bimetallic complexes [(Ph$_2$phen)$_2$Ru(dpp)PtCl$_2$]$^{2+}$ (a and b), and [(Ph$_2$phen)$_2$Ru(dpq)PtCl$_2$]$^{2+}$(c) photocleavage pUC18 DNA.

at RT for 1 hour (lane RT), a metal complex DNA solution incubated at 37°C for 1 hour (lane 37), a metal complex DNA solution irradiated for 1 hour in air-saturated condition (lane *Atm), a metal complex DNA solution irradiated for 1 hour with the singlet oxygen quencher, NaN$_3$ (lane *NaN$_3$), and a metal complex DNA solution irradiated for 1 hour after freeze–pump–thaw (FPT) cycles (lane *FPT).

The complexes binding to circular pUC18 plasmid DNA were observed at RT and 37°C based on the retarded DNA migration through gels in lane RT and lane 37 (Figure 11.18). The complex [(Ph$_2$phen)$_2$Ru(dpp)PtCl$_2$]$^{2+}$ photocleavage of pUC18 DNA in air-saturated conditions was observed by supercoiled DNA (Form I) converting to open circular DNA (Form II) in lane *Atm (Figure 11.18a). The complex [(Ph$_2$phen)$_2$Ru(dpq)PtCl$_2$]$^{2+}$ performs two site scission of circular pUC18 plasmid DNA and forms linear pUC18 plasmid DNA in air-saturated conditions in lane *Atm (Figure 11.18c). The gel shift assays illustrate that both [(Ph$_2$phen)$_2$Ru(BL)PtCl$_2$]$^{2+}$ complexes photocleave DNA through an oxygen-dependent mechanism.

## Other Bimetallic Complexes

Other bimetallic complexes coupling one Ru or Os polyazine LA to a *cis*-PtCl$_2$ BAS through a dpp, dpq, and dpb BL also bind to and photocleave DNA.[56,57,61,64] The complex [(Mephtpy)RuCl(dpp)PtCl$_2$]$^+$ (where Mephtpy = (4′-(4-methylphenyl)-2, 2′:6′, 2″-terpyridine) binds to linear pUC18 DNA and photocleaves circular pUC18 DNA in air-saturated conditions.[56] The complexes [(tpy)RuCl(BL)PtCl]$^+$ (BL = dpq, dpb),[64] [(tpy)Ru(tppz)PtCl]$^{3+}$ (tppz = 2,3,5,6-tetrakis(2-pyridyl)pyrazine),[57] and [(bpy)$_2$Os(dpb)PtCl$_2$]$^{+}$ [61] bind with DNA through a *cis*-PtCl$_2$ or PtCl BAS, but light-activated studies are not reported.

## Supramolecular Complex [{(bpy)$_2$Ru(dpp)}$_2$Ru(dpp)PtCl$_2$]$^{6+}$

The tetrametallic complex [{(bpy)$_2$Ru(dpp)}$_2$Ru(dpp)PtCl$_2$]$^{6+}$, as shown in Figure 11.19, couples a three ruthenium polyazine LA to a reactive *cis*-PtCl$_2$ center BAS through a dpp BL.[58] Moving the Ru that possesses the lowest lying $^3$MLCT state away from the Pt BAS provides for systems with long-lived more photoactive $^3$MLCT states. The tetrametallic complex [{(bpy)$_2$Ru(dpp)}$_2$Ru(dpp)PtCl$_2$]$^{6+}$ covalently binds to DNA and photocleaves DNA through an oxygen-dependent mechanism.

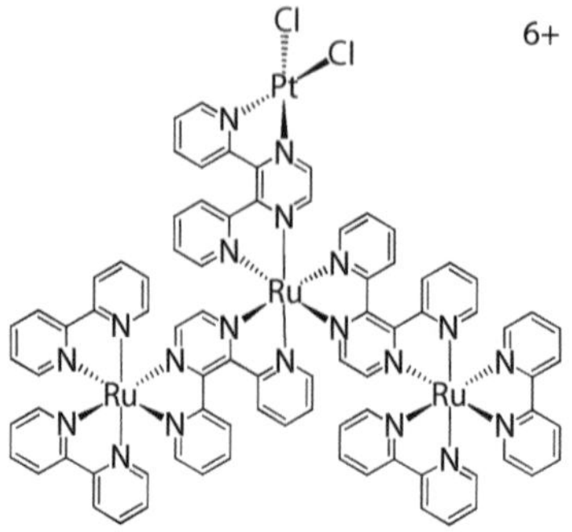

[{(bpy)₂Ru(dpp)}₂Ru(dpp)PtCl₂]⁶⁺

**FIGURE 11.19**  The tetrametallic complex [{(bpy)₂Ru(dpp)}₂Ru(dpp)PtCl₂]⁶⁺.

*Electrochemical Properties of [{(bpy)₂Ru(dpp)}₂Ru(dpp)PtCl₂]⁶⁺*

The electrochemistry of [{(bpy)₂Ru(dpp)}₂Ru(dpp)PtCl₂]⁶⁺ displays metal-based oxidations and ligand-based reductions. The oxidative electrochemistry of [{(bpy)₂Ru(dpp)}₂Ru(dpp)PtCl₂]⁶⁺ shows a reversible oxidative couple at 1.58 V versus Ag/AgCl assigned to $Ru^{II/III}$ oxidation for the two peripheral Ru centers. The reductive electrochemistry of [{(bpy)₂Ru(dpp)}₂Ru(dpp)PtCl₂]⁶⁺ consists of several reversible one electron dpp couples. The first reductive couple at −0.40 V versus Ag/AgCl is assigned to the $dpp^{0/-}$ couple, which connects the *cis*-PtCl₂ moiety to the central Ru, followed by two peripheral dpp BL-based reductions at −0.60 and −0.70 V versus Ag/AgCl, which connect the central Ru to the peripheral Ru.

*Spectroscopic Properties of [{(bpy)₂Ru(dpp)}₂Ru(dpp)PtCl₂]⁶⁺*

The electronic absorption of [{(bpy)₂Ru(dpp)}₂Ru(dpp)PtCl₂]⁶⁺ consists of ligand-based π→π* transitions in the UV region and MLCT transitions in the visible region. A major transition band at 290 nm is assigned to bpy-based π→π* transition, and a shoulder at 320 nm is attributed to dpp-based π→π* transition. The MLCT transition at 416 nm is assigned to Ru(dπ)→bpy(π*) CT transition. The transitions at approximately 520 to 540 nm are attributed to the Ru(dπ)→μ-dpp(π*) CT transitions. The extinction coefficient (35,000 $M^{-1}$ $cm^{-1}$) at 542 nm shows several MLCT transitions overlapping in this region, providing for very efficient light absorption by this complex.

*DNA Interactions of [{(bpy)₂Ru(dpp)}₂Ru(dpp)PtCl₂]⁶⁺*

The DNA-binding and photocleavage ability of the tetrametallic complex [{(bpy)₂Ru(dpp)}₂Ru(dpp)PtCl₂]⁶⁺ was investigated by agarose gel electrophoresis as shown in Figure 11.20.

The gel pictures, as shown in Figure 11.20, show a molecular weight standard (lane λ), a pUC18 DNA control (lane C), a metal complex DNA solution at 5:1 BP/MC ratio incubated in the dark at RT for 1 hour (lane 1), a metal complex DNA solution at 5:1 BP/MC ratio incubated at 37°C for 1 hour (lane 2), a metal complex

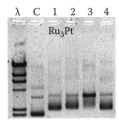

**FIGURE 11.20** Agarose gel showing the tetrametallic complex [{(bpy)$_2$Ru(dpp)}$_2$Ru(dpp) PtCl$_2$]$^{6+}$ photocleavage and binding to pUC18 DNA.

DNA solution at 5:1 BP/MC ratio irradiated for 1 hour in air (lane 3), and a metal complex DNA solution at 5:1 BP/MC ratio irradiated for 1 hour under argon (lane 4). The complex covalently bound to DNA was observed at RT and 37°C on the basis of the retarded DNA migration through gels in lane 1 and lane 2 (Figure 11.20). Photocleavage DNA was observed in lane 3, which shows the supercoiled pUC18 DNA (Form I) converting to open circular DNA (Form II). No DNA photocleavage was observed in lane 4. The tetrametallic complex [{(bpy)$_2$Ru(dpp)$_2$Ru(dpp)PtCl$_2$]$^{6+}$ photocleaves circular pUC18 DNA under visible light (λ > 450 nm) irradiation for 1 hour through an oxygen-dependent mechanism (singlet oxygen generation) and also binds to circular pUC18 DNA through a *cis*-PtCl$_2$ BAS.

## SUPRAMOLECULAR SOLAR HYDROGEN PRODUCTION

This section of the chapter focuses on supramolecular motifs that are structurally similar to supramolecules used in PDT drug development and are applied to the light-activated reduction of H$_2$O to produce H$_2$. Many other systems are emerging as photocatalysts, but our focus herein is on systems with common structures and diverse functions. Interestingly, this focus provides for inclusion of many classes of solar hydrogen photocatalysts including Ru,Pt, Ru,Rh, and Rh,Rh supramolecular architectures. The basic chemical properties of these major classes of supramolecules applied in both forums are highlighted in the description of their PDT action and summarized in Table 11.1 and will not be repeated here.

### THERMODYNAMICS OF SOLAR HYDROGEN PRODUCTION

Solar hydrogen production through water splitting provides an attractive route for hydrogen fuel production. Solar water splitting involves energetically uphill chemical reactions, requiring 1.23 eV.

Water reduction is energetically more favorable via multielectron pathways (1.23 V vs. NHE) than single electron pathways (5 V vs. NHE).[65,66] Most of the solar spectrum, >1.23 eV, can be used for the multielectron splitting of water. Systems known to capture and convert the UV light from the sun are more widely known. Research on designing systems for multielectron photochemistry, which can use the more available visible region of the solar spectrum for photochemical excitation, remains uncommon. Designing complete water splitting systems is complicated; thus, research efforts are geared toward understanding either water oxidation or water reduction by replacing the other half with a sacrificial EA or a sacrificial ED. Supramolecular chemistry provides an ideal forum to the study design and development of solar hydrogen photocatalysts. The individual components within the supramolecule provide the overall photocatalytic activity, each performing a specific act. Ideally, structural components can be selected so that the perturbation of each component upon assembly within the supramolecule is varied in a predictable and controllable fashion with the overall function being maintained, but this knowledge base is still being developed.

## PHOTOINITIATED ELECTRON COLLECTION

Photoinitiated electron collection (PEC) is a process by which at least two electrons are collected on a central site after optical excitation. For PEC, supramolecular complexes can be designed to have an ED-LA-BL-EC-BL-LA-ED structural motif. The presence of the two LAs connected to a single EC allows electron collection with multiple photoexcitation steps and intramolecular electron transfer. Many fuel producing reactions are energetically more favorable via multielectron pathways; thus, PEC is of considerable importance in multielectron photochemistry. Despite the importance of PEC, only few functioning systems exist. System design and development with the proper energetics for multielectron photochemistry remains a challenge. Systems are known, which use light to move a single electron to an EA, establishing the basis of the large number of molecular photovoltaics that are reported. One issue with using these systems as components of PEC devices is that following the movement of one electron to an EA, the movement of the second electron via optical excitation and intramolecular electron transfer is impeded, and systems that function use remote EA sites decoupled from the LA unit.

### Photoinitiated ECs that Collect Electrons on Ligand Orbitals

The first functioning PEC is [{(bpy)$_2$Ru(dpb)}$_2$IrCl$_2$](PF$_6$)$_5$, which collects two electrons on the dpb($\pi^*$) orbitals upon two sequential excitations and electron transfer events (Figure 11.21).[67] Homobimetallic ruthenium systems, [(phen)$_2$Ru(tatpq)Ru(phen)$_2$]$^{4+}$ (phen = 1,10-phenanthroline, tatpq = 9,11,20,22-tetraazatetrapyrido[3,2-a:2'3'-c:3'',2''-1:2''',3'''-n]pentacene-10,21-quinone) and [(phen)$_2$Ru(tatpp)Ru(phen)$_2$]$^{4+}$ (tatpp = 9,11,20,22-tetraazatetrapyrido[3,2-a:2'3'-c:3'',2''-1:2''',3'''-n]pentacene) collect multiple electrons also on the BL($\pi^*$) orbitals (Figure 11.21).[68,69] These systems function, unlike directly coupled systems, via electronic isolation of the Ru→BL LA from the site of collection of the reducing equivalents via the intervening Ir center in the Ru, Ir, Ru system, or remote acceptor orbitals in the Ru,Ru system. These PEC systems are not known to reduce substrates. The monometallic system, [Ru(bpy)$_2$(pbn)]$^{2+}$ (pbn = 2-(2-pyridyl)benzo[b]-1,5-naphthyridine), couples a Ru-based LA to an NAD$^+$ (NAD$^+$ = nicotinamide adenine dinucleotide) model ligand, which acts as a reservoir for two electrons in proton coupled electron transfer (Figure 11.21).[70,71]

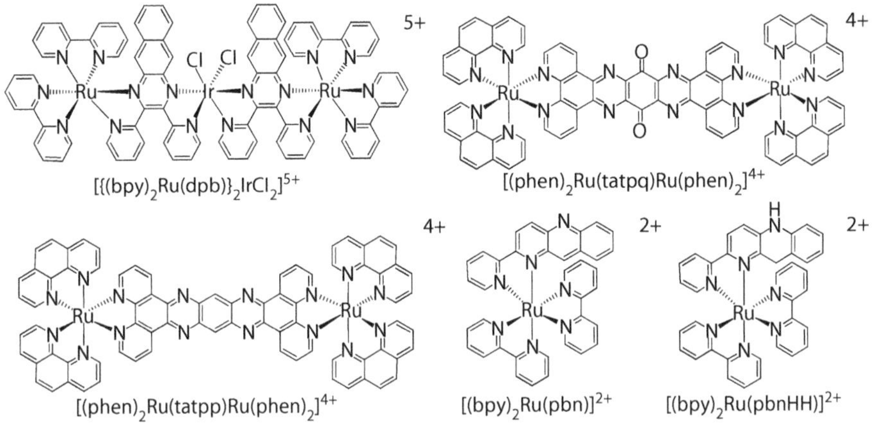

**FIGURE 11.21**   Photoinitiated ECs that collect electrons on ligand orbitals.

## RH-BASED SYSTEMS FOR SOLAR HYDROGEN PRODUCTION

Water splitting to produce hydrogen is complicated and energetically demanding with bond breaking and formation involving multielectron process using multiple photons sequentially. Systems incorporating LAs, typically Ru or Os trischelate systems with ligands such as bpy or phen, that absorb light in the low energy region of the solar spectrum with reactive metal CATs, allow the efficient use of most of the solar spectrum to promote water photocatalysis. It is important to note that Ru and Os MLCT LAs that are used in some PDT drug development are also LA for supramolecular solar energy conversion systems. These systems incorporate Ru or Os-based LAs and reactive metal centers combined with Rh- and Pt-based catalysts. Rh-based systems take advantage of ability of the Rh to cycle through three oxidation states reducing from $Rh^{III}$ to $Rh^{II}$ and $Rh^{I}$, promoting multielectron chemistry. The primary steps of LA excitation and migration of energy or electrons are common to many supramolecular PDT agents and solar hydrogen photocatalysts.

### Supramolecular Rh-Based PDT Agents for Solar Hydrogen Production

Systems that couple a $Rh^{III}$ center to Ru or Os polyazine charge transfer complexes in supramolecular frameworks are reported in solar energy conversion.[72–76] Rh acts as an intramolecular EA after MLCT excitation of the LA. The ligand environment around the Rh center dictates photoreactivity as it is limited in systems in which the Rh centers are bound to only chelating ligands. In PDT applications, these systems typically function after one photon excitation via the $^3$MMCT ES to cleave DNA. In solar hydrogen schemes, multiple excitations and multielectron reduction are needed for these supramolecules to function and enter the catalytic cycle.

### Rhodium-Centered Trimetallic Supramolecular Photocatalysts: Solar Hydrogen Production

The first photoinitiated EC that collects multiple electrons on a metal capable of photo-catalyzing hydrogen production from water, $[\{(bpy)_2Ru(dpp)\}_2RhCl_2]^{5+}$, was reported by Elvington et al.[77] This supramolecular system incorporates two $[(bpy)_2Ru(dpp)]^{2+}$ LAs and a single Rh EC, allowing for multiple electron collection necessary for the multielectron reduction of substrates. The monodentate chloride Rh-bound ligands can be lost after photoreduction, providing a site for substrate binding and photoreactivity to the supramolecular complex. The redox properties of $[\{(bpy)_2Ru(dpp)\}_2RhCl_2]^{5+}$ demonstrate that the Ru LAs are largely electronically isolated, important to PEC and multielectron photochemistry. Component modification by way of changing the TL, LA metal, and Rh-bound halides subsequently afforded a series of additional photoinitiated ECs having a LA-BL-RhX$_2$-BL-LA structural motif (LA = Ru$^{II}$ or Os$^{II}$ polyazine LA, X = Cl$^-$ or Br$^-$, BL = dpp) capable of catalyzing solar hydrogen production.[19,78–81] Visible light irradiation affords PEC, resulting in the photoactivation of the Rh center through halide loss (Figure 11.22). Studies have established $[\{(bpy)_2Ru(dpp)\}_2RhX_2]^{5+}$ and $[\{(phen)_2Ru(dpp)\}_2RhX_2]^{5+}$ (X = Cl$^-$ or Br$^-$) as photochemical molecular devices for PEC hydrogen photocatalysis with a hydrogen yield of $\Phi \approx 0.01$.[79] Changing the LA to Os and/or the TL to tpy and a Cl destabilizes the LA metal–d$\pi$ orbitals to provide systems that absorb lower energy wavelengths.[79] The complexes $[\{(bpy)_2Os(dpp)\}_2RhCl_2]^{5+}$ and $[\{(tpy)RuCl(dpp)\}_2RhCl_2]^{3+}$ are also solar hydrogen photocatalysts but are not as efficient as the $[\{(bpy)_2Ru(dpp)\}_2RhX_2]^{5+}$ and $[\{(phen)_2Ru(dpp)\}_2RhX_2]^{5+}$ (X = Cl$^-$ or Br$^-$) systems because of the lower driving force for the reductive quenching of the photoactive ESs of these systems. The Os-based system containing the tpy and Cl TLs, $[\{(tpy)OsCl(dpp)\}_2RhCl_2]^{3+}$, and the Ir-based system, $[\{(bpy)_2Ru(dpb)\}_2IrCl_2]^{5+}$, do not show hydrogen production in the photocatalytic system. The redox and photophysical properties of these systems assist in evaluating the factors that affect photocatalytic efficiency.

Studies indicate that the visible light irradiation of $[\{(bpy)_2Ru(dpp)\}_2Rh^{III}X_2]^{5+}$ in the presence of the sacrificial ED, $N,N$-dimethylaniline (DMA), provides the two-electron–photoreduced product $[\{(bpy)_2Ru(dpp)\}_2Rh^{I}]^{5+}$ through an intermediate Rh$^{II}$ species.[19] Figure 11.23 outlines the proposed mechanism of photoreduction. The reductive quenching of both the $^3$MLCT and $^3$MMCT states by the ED is feasible. Kinetic investigation using the Stern–Volmer emission quenching as well as kinetic analysis of product formation, focusing on the formation Rh$^{II}$ species for simplicity, show a rate of product formation of $1.9 \times 10^9$ M$^{-1}$ s$^{-1}$, with the $^3$MLCT ES efficiently quenched by DMA close to diffusion limit, $k_q = 2 \times 10^{10}$ M$^{-1}$ s$^{-1}$. The estimated ES reduction potentials of the $^3$MLCT and $^3$MMCT states for $[\{(bpy)_2Ru(dpp)\}_2RhCl_2]^{5+}$ are 1.23 and 0.84 V versus SCE, respectively, with very similar values predicted

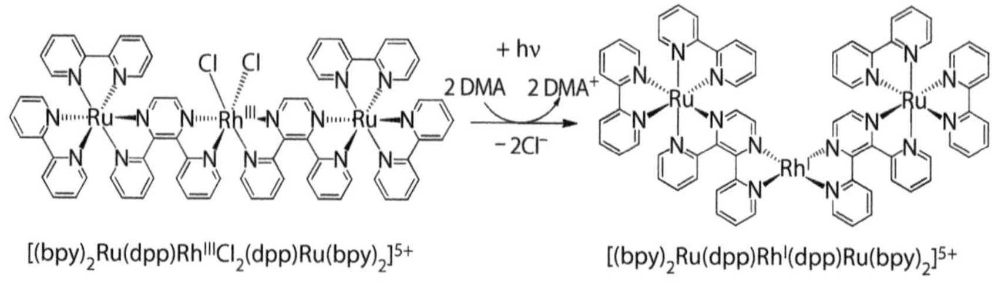

$[(bpy)_2Ru(dpp)Rh^{III}Cl_2(dpp)Ru(bpy)_2]^{5+}$                                                          $[(bpy)_2Ru(dpp)Rh^{I}(dpp)Ru(bpy)_2]^{5+}$

**FIGURE 11.22**   PEC on a metal center.

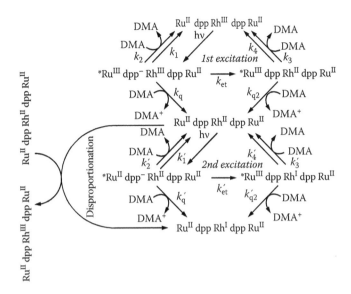

**FIGURE 11.23** Proposed mechanism for photoinitiated electron transfer. bpy, Cl, and charges are omitted for clarity. $Ru^{II}dppRh^{III}dppRu^{II}$ is $[\{(bpy)_2Ru(dpp)_2RhCl_2]^{5+}$, where bpy = 2,2′-bipyridine and dpp = 2,3-bis(2-pyridyl)pyrazine. DMA is dimethylaniline. $k_1$ and $k_4$ are the unimolecular decay pathways from the $^3$MLCT and $^3$MMCT states, respectively, whereas $k_2$ and $k_3$ are the bimolecular deactivation rates from the same states. $k_{et}$ is the rate of electron transfer to generate the $^3$MMCT state. $k_q$ and $k_{q2}$ are the rates of reductive quenching by DMA to generate the $Rh^{II}$ complex from the $^3$MLCT and $^3$MMCT states, respectively.

for $[\{(bpy)_2Ru(dpp)\}_2RhBr_2]^{5+}$ and $[\{(phen)_2Ru(dpp)\}_2RhCl_2]^{5+}$. Given the oxidation potential of DMA ($E_{1/2} \approx 0.81$ V vs. SCE), similar positive driving forces for the reductive quenching of both ESs are expected for all three complexes. The kinetic analysis of the ES properties of a series of Ru,Rh,Ru supramolecules allows the determination of $k_1$, $k_2$, $k_{et}$, and $k_{q2}/k_4$.[82] This analysis shows that rapid electron transfer is seen in all complexes to populate the $^3$MMCT ES. Rapid quenching by DMA is seen for the $^3$MLCT state, and kinetic analysis suggests quenching of the $^3$MLCT state is a dominant pathway in photoreduction and thus likely in hydrogen production from these systems.

The trimetallic complexes $[\{(bpy)_2Ru(dpp)\}_2RhCl_2]^{5+}$, $[\{(bpy)_2Ru(dpp)\}_2RhBr_2]^{5+}$, $[\{(phen)_2Ru(dpp)\}_2RhCl_2]^{5+}$, $[\{(bpy)_2Os(dpp)\}_2RhCl_2]^{5+}$, $[\{(tpy)RuCl(dpp)\}_2RhCl_2]^{3+}$, and $[\{(tpy)OsCl(dpp)\}_2RhCl_2]^{3+}$ and the Ir-based $[\{(bpy)_2Ru(dpb)\}_2IrCl_2]^{5+}$ were evaluated with regard to hydrogen photocatalysts in the presence of water and DMA, $Et_3N$, or TEOA.[79] All complexes demonstrate photocatalytic activity with the exceptions of $[\{(bpy)_2Ru(dpb)\}_2IrCl_2]^{5+}$ and $[\{(tpy)OsCl(dpp)\}_2RhCl_2]^{3+}$. $[\{(bpy)_2Ru(dpb)\}_2IrCl_2]^{5+}$ is also a photoinitiated EC, which collects electrons on BLs. Interestingly, $[\{(bpy)_2Ru(dpb)\}_2IrCl_2]^{5+}$ is not a solar hydrogen photocatalyst, although PEC occurs on the dpb($\pi$*) orbitals and a thermodynamically favorable driving force for the reductive quenching of the $^3$MLCT state by DMA exists. The lack of photocatalysis by $[\{(bpy)_2Ru(dpb)\}_2IrCl_2]^{5+}$ demonstrates the importance of the Rh catalytic center $^3$MLCT and/or $^3$MMCT states with sufficient driving forces for ES reduction by the ED. The photolysis of acetonitrile solutions of $[\{(bpy)_2Ru(dpp)\}_2RhX_2]^{5+}$ or $[\{(phen)_2Ru(dpp)\}_2RhX_2]^{5+}$ (X = Cl or Br) in the presence of DMA and water at $\lambda$ = 470 nm leads to water reduction to produce hydrogen with $\Phi = 0.01$ with 20–27 turnover numbers (TON) in 2 hours (40–54 TON relative to the LA as 2 TON of the LA are needed per catalytic cycle). The rate of halide loss appears to be kinetically important as $[\{(bpy)_2Ru(dpp)\}_2RhBr_2]^{5+}$ shows higher photocatalytic activity relative to the chloride analogue. It is possible that loss of the weaker σ-donor, Br⁻, is more facile relative to Cl⁻, improving catalytic efficiency in the Br system, or the Br ligand may direct the synthetic chemistry to produce certain stereoisomers about the Rh center in this motif. The complexes $[\{(bpy)_2Os(dpp)\}_2RhCl_2]^{5+}$ and $[\{(tpy)RuCl(dpp)\}_2RhCl_2]^{3+}$ yield similar amounts of hydrogen when irradiated at 470 nm in the presence of DMA and water, but with much lower efficiency than $[\{(bpy)_2Ru(dpp)\}_2RhCl_2]^{5+}$. The upper limit on the ES reduction potentials of the $^3$MLCT and $^3$MMCT were predicted as 0.91 and 0.54 V versus SCE, respectively, for $[\{(bpy)_2Os(dpp)\}_2RhCl_2]^{5+}$ and 1.01 and 0.61 V, respectively, for $[\{(tpy)RuCl(dpp)\}_2RhCl_2]^{3+}$. DMA has a favorable thermodynamic driving force to reductively quench only the $^3$MLCT ES. The ES reduction potentials indicate that the reductive quenching of only the $^3$MLCT state by DMA may be sufficient for hydrogen photocatalysis. The lower driving force for the reductive quenching of the $^3$MLCT and the lower $\tau$ relative to the bpy-based systems may account for the lower photocatalytic efficiency of these complexes. $[\{(tpy)OsCl(dpp)\}_2RhCl_2]^{3+}$ does not function as a photocatalyst under the conditions investigated. On the basis of the ES reduction potentials of the $^3$MLCT and $^3$MMCT states, 0.71 and 0.37 V versus SCE, respectively, the reductive quenching of both ESs by DMA are thermodynamically unfavorable consistent with the lack of photocatalytic activity shown by this supramolecular complex. The addition of Hg(l) to the photocatalytic systems does not impair hydrogen production, suggesting that hydrogen production catalyzed by Rh colloid formation is not an active pathway and that the supramolecular architecture remains intact during photocatalysis.[77,79,83,84] The importance of the intact supramolecular architecture in solar $H_2$ production is also established below in the studies of Ru,Rh bimetallic systems for PEC.

The effect of the ED on the most efficient photocatalysts, $[\{(bpy)_2Ru(dpp)\}_2RhCl_2]^{5+}$, $[\{(bpy)_2Ru(dpp)\}_2RhBr_2]^{5+}$, and $[\{(phen)_2Ru(dpp)\}_2RhCl_2]^{5+}$, shows that the efficiency of hydrogen production varies in the order DMA (20–27 TON in 2 hours) > $Et_3N$ (3–6 TON in 2 hours) > TEOA (1 TON in 2 hours).[79] TEOA has a slightly higher driving force for the reductive quenching of the photocatalysts than $Et_3N$ but results in lower hydrogen yields. The effective pH of the photolysis solutions using DMA, $Et_3N$, or TEOA were estimated as approximately 9.1, 14.7, and 11.8, respectively, on the assumption that the $pK_a$ values of their conjugated acids remain unchanged in the photocatalytic solutions relative to aqueous conditions [$pK_a$ = 5.07 (DMAH⁺), 10.75 ($Et_3$NH⁺), and 7.76 (TEOAH⁺)]. The lower catalyst function when TEOA is the ED may be attributed to the higher effective pH of the solution as the reduction potential of water is pH dependent, occurring at lower potentials at lower pH. In addition, the protonation of the Rh center may be critical to photocatalysis, and this equilibrium would be very sensitive to pH. ED–catalyst interactions may also play a key role in hydrogen photocatalysis. The molecular structure of DMA allows donor–LA π-stacking interactions that may be advantageous for a more efficient reductive quenching of the ESs affording the highest hydrogen yields compared with the aliphatic EDs. Recent studies have focused on the scale up of photochemical hydrogen production using the lead photocatalyst [{(bpy)$_2$Ru(dpp)}$_2$RhBr$_2$]$^{5+}$. Studies show that increased DMA concentration, increased headspace and volume, and a dimethylformamide solvent provide enhanced catalyst function.[85]

The bromide salt, [{(bpy)$_2$Ru(dpp)}$_2$RhBr$_2$]Br$_5$, displays PEC in aqueous medium. [{(bpy)$_2$Ru(dpp)}$_2$RhBr$_2$]Br$_5$ also functions as hydrogen photocatalyst in the presence of TEOA with added triflic acid, hydrobromic acid, or phosphoric acid.[80] The photocatalytic efficiency is lower in the aqueous medium, typical of these systems.

## Ruthenium, Rhodium Bimetallic Supramolecules

The bimetallic system [(phen)$_2$Ru(dpp)RhCl$_2$(bpy)](PF$_6$)$_3$, which couples a single LA to a reactive Rh, has been constructed.[26] The study of the redox, spectroscopic, photophysical, and photochemical properties of this complex has been initiated to further understand factors affecting hydrogen photocatalysis using a simpler BL-RhX$_2$-TL structural motif. This Ru,Rh bimetallic motif is synthetically challenging because of the tendency of Rh to produce the Rh$^{III}$(NN)$_2$X$_2$ motif when Rh starting materials are reacted with diimine ligands. [(phen)$_2$Ru(dpp)RhCl$_2$(bpy)](PF$_6$)$_3$ is also a photoinitiated EC, with electron being collected on the Rh producing the Rh$^I$ species, [(phen)$_2$Ru(dpp)Rh$^I$(bpy)]$^{3+}$, upon photolysis with DMA. This represents a new structural motif for PEC, which will allow for the production of a new series of molecules active as PEC molecular devices. Interestingly, the bimetallic system is not a solar hydrogen photocatalyst. This observation is very interesting as it shows the importance for the multiple Ru$^{II}$-polypyridyl light-absorbing subunits for a functioning photocatalyst that establishing the intact trimetallic is required for photocatalysis by the Ru,Rh,Ru motif described earlier. The lack of photocatalysis by the Ru,Rh bimetallics is somewhat surprising given the Ru-based HOMO and Rh-based LUMO as well as the observed photoinitiated electron in the presence of DMA. The reactions of [Rh$^I$(bpy)$_2$]$^+$ have been studied in detail with a Rh,Rh dimer or hydride bridged dimer being formed leading to poor catalytic activity of Rh monometallic systems.[86] The lack of photocatalytic activity by the bimetallic complex may result from Rh–Rh dimer formation, possible in the sterically less demanding bimetallic system upon Rh$^I$ formation and the deactivation of the photosystem resulting in no hydrogen production. Dimer formation is impeded when two sterically demanding LAs are present in the Ru,Rh,Ru trimetallic systems, which do photocatalyze hydrogen production.

## Rhodium, Rhodium Solar Hydrogen Photocatalysts

Interesting metal–metal-bonded systems, including [Rh$_2^{0,0}$(dfpma)$_3$(PPh$_3$)(CO)] (dfpma = MeN(PF$_2$)$_2$ capable of multielectron photochemistry and hydrogen photocatalysis, are known.[32,87] Hydrogen production occurs with $\Phi = 0.01$ with 27 TON/h during the initial 3 hours, in the presence of a halogen trap. The mixed-valence compounds have been constructed to drive multielectron chemistry and are of the form M$^{n+}$···M$^{n+2}$. [Rh$_2$(dfpma)$_3$X$_4$] transforms to a two-electron mixed-valence state, [Rh$_2^{0,II}$(dfpma)$_3$X$_2$(L)] (dfpma = MeN(PF$_2$)$_2$, X = Cl or Br, L = CO, PR$_3$, or CNR), when irradiated at excitation wavelengths between 300 and 400 nm in the presence of excess L and a halogen-atom trap. The dfpma ligand acts as both a $\pi$-acceptor and a $\pi$-donor, providing stability to the mixed-valence system. Further irradiation activates a $Rh^{II}$-X, generating a doubly reduced form, $[Rh_2^{0,0}(dfpma)_3L_2]$. The photolysis of $[Rh_2^{0,0}(dfpma)_3L_2]$ in the presence of HCl results in an intermediate $Rh^{II}$, $Rh^{II}$ dihydride, dihalide, and $[Rh_2^{II,II}(dfpma)_3Cl_2H_2]$, which upon photolysis produces hydrogen with the generation of the $[Rh_2^{I,I}(dfpma)_3Cl_2]$. $[Rh_2^{I,I}(dfpma)_3Cl_2]$ undergoes internal disproportionation affording $[Rh_2^{0,II}(dfpma)_3X_2(L)]$. The overall efficiency for $H_2$ photogeneration is limited by the efficiency of halogen photoelimination by Rh–X bond activation. Recent studies have focused on halogen elimination photochemistry and systems that more efficiently undergo M–X bond cleavage. New systems $[Rh^{I}Au^{I}(tfepma)_2(CN^{t}Bu)_2]^{2+}$, $[Pt^{II}Au^{I}(dppm)_2PhCl]^{+}$ (tfepma = $MeN(P(OCH_2CF_3)_2)_2$ and dppm = $CH_2(PPh_2)_2$), $[Ir^{I}Au^{I}(dcpm)_2(CO)X](PF_6)$ (dcpm = bis(dicyclohexylphosphino)methane, X = Cl, Br), and $[Ir^{I}Au^{I}(dppm)_2(CN^{t}Bu)_2](PF_6)_2$ have been constructed with studies focused on enhancing the M–X bond cleavage.[88–90]

## RUTHENIUM, PLATINUM SUPRAMOLECULES FOR SOLAR HYDROGEN PRODUCTION

Pt-based supramolecules with structural similarities to active PDT agents that are solar hydrogen photocatalysts are known. These systems couple polyazine LAs with bridged Pt CATs.[91,92] The supramolecular systems have been designed to promote light-activated intramolecular electron transfer typically to a polyazine ligand bound to the reactive Pt where catalysis is initiated. The rates of electron transfer and LA properties can be modulated by component modification during the design and construct of the supramolecular architecture. The level of electronic communication is dictated by the choice of the BL used. The structural integrity of supramolecular architectures during photocatalysis has been questioned as the formation of metallic colloids during photoreduction of the supramolecular assembly has been observed in some studies using similar supramolecular assemblies.[93,94] These studies alert researchers that metal colloid formation should be considered. One study shows the photoinduced decomposition of $[(bpy)_2Ru(Mebpy(CH_2)_2C_6H_2(OCH_3)_2(CH_2)_2Mebpy)PdCl_2]^{2+}$ (Figure 11.24) observed by ESI-MS and the formation of Pd colloid observed by TEM and XPS of evaporated reaction mixtures during photochemical hydrogen production.[92] Mercury tests are used to deactivate photoproduced metal colloids

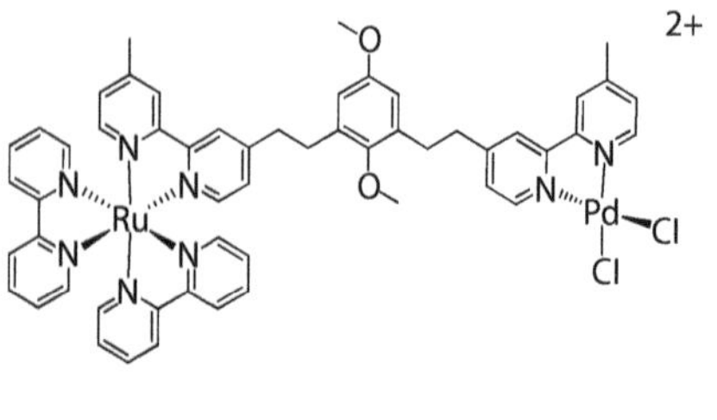

**FIGURE 11.24** The structure of $[(bpy)_2Ru(Mebpy(CH_2)_2C_6H_2(OCH_3)_2(CH_2)_2Mebpy)PdCl_2]^{2+}$ (bpy = 2,2'-bipyridine).

eliminating the effect of metal colloids on photochemical activity.[83,84] Although the effect of metal colloid formation must be seriously considered in systems that use noble metals, its study is quite complicated. The concentration of the medium to study solids through TEM and XPS can also lead to system decomplexation, resulting in the formation of metallic solids. Reduced polyazine complexes including those produced electrochemically are known to adsorb to electrodes or metal surfaces such as Hg(l), and this could deactivate a catalytic system.

Ru,Pt bimetallic systems of the structural motif, $[(bpy)_2Ru(BL)PtCl_2](PF_6)_2$ (BL = bpy(CONH(CH$_2$)$_3$NH$_2$)$_2$ and phenNHCO(COOHbpy)), have been investigated as DNA-binding agents and hydrogen photocatalysts, respectively. The Ru,Pt system $[(bpy)_2Ru(phenNHCO(COOHbpy))PtCl_2]^{2+}$ produces hydrogen from water when illuminated at $\lambda > 350$ nm in aqueous acetate buffer solutions in the presence of EDTA, with a $\Phi = 0.01$ and 5 turnovers in 10 hours.[90,91,95] Intramolecular electron transfer is evident in this system as the emission from the $^3$MLCT state at $\lambda_{max}^{em} = 610$ nm is quenched by 67% relative to the parent $[(bpy)_2Ru(phenNHCO$ $(COOHbpy))]^{2+}$. To address the interference from any photodecomposition products, comparisons are made with analogous multicomponent photosystems. The system consisting of $[Ru(bpy)_3](NO_3)_2$, $N,N'$-dimethyl-4,4'-bipyridinium, and $[PtCl_2(dcbpy)]$ (dcbpy = 4,4'-dicarboxy-2,2'-bipyridine) is an effective hydrogen generating system providing a $\Phi = 0.02$, but hydrogen generation is completely retarded in the absence of $N,N'$-dimethyl-4,4'-bipyridinium. Hydrogen production does not occur when $[Ru(bpy)_2(5-amino-phen)]^{2+}$ and $[PtCl_2(dcbpy)]$ or the parent LA, $[(bpy)_2Ru(phenNHCO(COOHbpy))]^{2+}$, and $K_2PtCl_4$ are the photosystem. The lack of activity by these multicomponent systems suggests that hydrogen production is initiated by intramolecular electron transfer. A variation of orbital energetics through component modification has demonstrated that more electron-withdrawing substituents including COOH on bpy provide higher system efficiency.[96] The hydrogen evolution rate by $[(bpy)_2Ru(phenNHCO(COOHbpy))PtCl_2]^{2+}$ is shown to obey the Michaelis–Menten enzymatic kinetics with regard to the concentration of EDTA.[97] These kinetic studies suggest that the ion-pair formation between $[(bpy)_2Ru(phenNHCO(COOHbpy))PtCl_2]^{2+}$ and EDTA$^{2-}$ is an important step in the photocycle. Bimetallic Ru,Pt systems that incorporate aliphatic linkages within bridges, $[(bpy)_2Ru(bpy(CONH(CH_2)_3NH_2)_2)PtCl_2]^{2+}$ and $[(bpy)_2Ru(bpy(CONH(CH_2)$ py)$_2$)PtCl$_2$]$^{2+}$ (Figure 11.25), do not display photocatalytic activity, implying that the aromaticity of the bridge connecting the LA and the CAT is important for intramolecular electron transfer in $[(bpy)_2Ru(phenNHCO(COOHbpy))PtCl_2]^{2+}$. A more complex tetrametallic system, $[(bpy)_2Ru(phenNHCO(CONH(CH_2)bpy))PtCl_2]_2^{4+}$, shows higher catalytic activity to the original bimetallic system with a $\Phi = 0.014$.[96] Similar Pd-based bimetallic supramolecular system containing a tetrapyridophenazine BL, $[(^tBu_2bpy)_2Ru(tpphz)PdCl_2]^{2+}$ ($^tBu$ = tertiary butyl, tpphz = tetrapyridophenazine), is a solar hydrogen photocatalyst producing hydrogen, with 56 TON in 30 hours, in the presence of the ED, Et$_3$N, in acetonitrile when optically excited at $\lambda = 470$ nm.[98] Recent studies have shown that the hydrogen production efficiency is mediated by irradiation wavelengths, demonstrating the importance of the Franck–Condon point in artificial photosynthesis system design.[99]

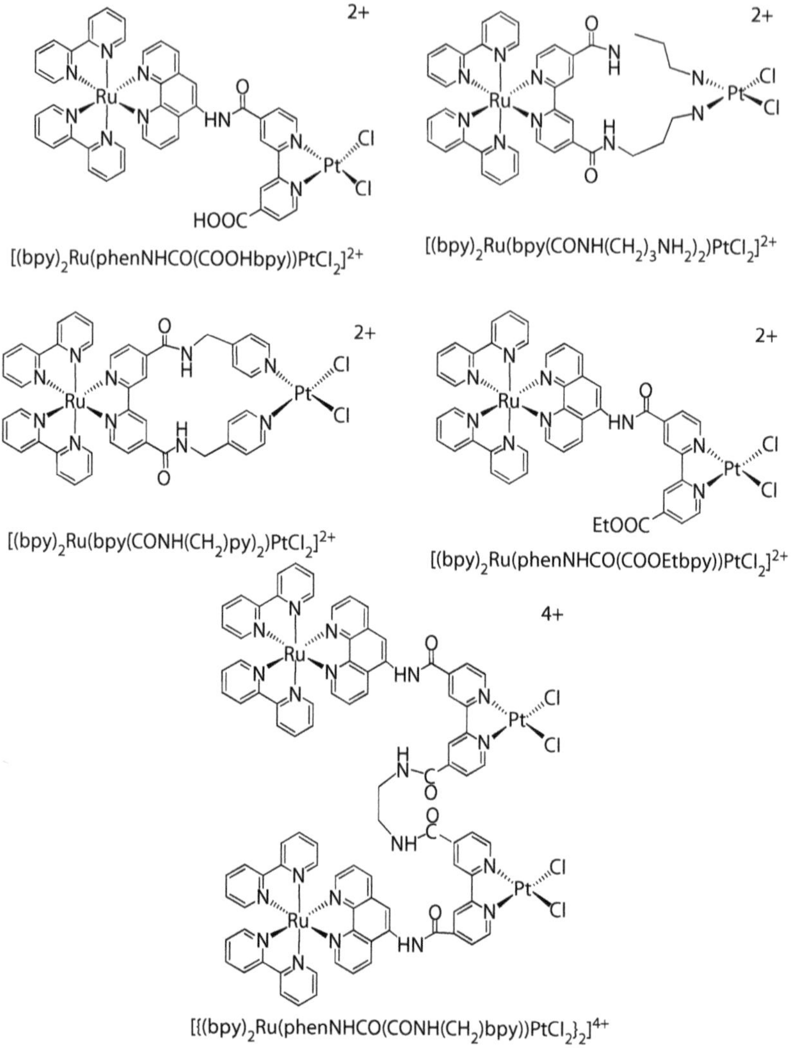

**FIGURE 11.25** Ru,Pt supramolecular systems for solar energy conversion.

Larger four metal systems, [{(phen)$_2$Ru(dpp)}$_2$Ru(dpq)PtCl$_2$]$^{6+}$ and [{(bpy)$_2$Ru(dpp)}$_2$Ru(dpq)PtCl$_2$]$^{6+}$, have been investigated as photoinitiated ECs, which photocatalyze the production of hydrogen from water with $\Phi = 0.01$.[100,101] The analogous [{(bpy)$_2$Ru(dpp)}$_2$Ru(dpp)PtCl$_2$]$^{6+}$ system exhibits both DNA-binding and photocleavage properties.[58] These systems couple two terminal (TL)$_2$Ru$^{II}$(dpp) LA units to a central Ru, which is connected to a reactive Pt through a dpq or dpp bridge (Figure 11.26).

The movement of the active LA units away from the Pt CAT site provides for dramatically enhanced $^3$MLCT ES lifetimes. Both [{(bpy)$_2$Ru(dpp)}$_2$Ru(dpq)PtCl$_2$]$^{6+}$ and [{(phen)$_2$Ru(dpp)}$_2$Ru(dpq)PtCl$_2$]$^{6+}$ are photoinitiated ECs with electrons being

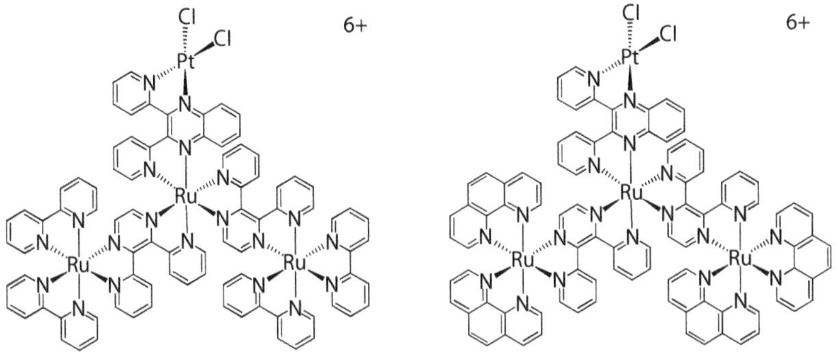

FIGURE 11.26    The structures of $[\{(bpy)_2Ru(dpp)\}_2Ru(dpq)PtCl_2]^{6+}$ and $[\{(phen)_2Ru(dpp)\}_2Ru(dpq)PtCl_2]^{6+}$.

collected on the ligand orbitals surrounding the central Ru. Electrochemical reduction by two electrons and photochemical reduction in the presence of DMA afford species with spectroscopic changes that are identical, indicating that the electrochemically reduced product and the photochemically reduced product are the same. For $[\{(phen)_2Ru(dpp)\}_2Ru(dpq)PtCl_2]^{6+}$, photochemical hydrogen production occurs in the presence of DMA and water when photoexcited at $\lambda = 470$ nm with 115 TON in 5 hours. For $[\{(bpy)_2Ru(dpp)\}_2Ru(dpq)PtCl_2]^{6+}$, photochemical hydrogen production occurs in the presence of DMA and water when photoexcited at $\lambda = 530$ nm with $\Phi = 0.01$ representing 40 turnovers in 3 hours. The amount of hydrogen produced by $[\{(bpy)_2Ru(dpp)\}_2Ru(dpq)PtCl_2]^{6+}$ exceeds that produced by the multicomponent $[\{(bpy)_2Ru(dpp)\}_2Ru(dpq)]^{6+}$ and Pt colloid system, indicating that the supramolecular architecture is important for hydrogen photocatalysis. The addition of excess Hg(1) does not impede catalytic activity, which implies that the supramolecular architecture remains intact during the photocycle and colloidal Pt is not operative. These $[\{(TL)_2Ru(dpp)\}_2Ru(dpq)PtCl_2]^{6+}$ systems display interesting photophysical properties displaying redox properties that demonstrate HOMOs that are based on the terminal Ru LA units and LUMOs on the dpq. This suggests a lowest energy charge separated state in this motif. This state should quench the emission from the terminal Ru→dpp CT state, which is observed in both complexes. The role of this state in photocatalysis in supramolecular complexes is under investigation.

## SUMMARY

Rhodium-centered photoinitiated ECs are known to catalyze water reduction to hydrogen. $[\{(bpy)_2Ru(dpp)\}_2RhCl_2]^{5+}$, $[\{(bpy)_2Ru(dpp)\}_2RhBr_2]^{5+}$, $[\{(phen)_2Ru(dpp)\}_2RhCl_2]^{5+}$, and $[\{(phen)_2Ru(dpp)\}_2RhBr_2]^{5+}$ are the lead photocatalysts in this area, demonstrating higher hydrogen yields and turnover capacities. Photoexcitation followed by intramolecular electron transfer in the presence of an ED leads to PEC and hydrogen photocatalysis. Studies have shown that many factors are important for hydrogen photocatalysis by these Ru,Rh,Ru trimetallic systems, including the presence of two LAs, the Rh center and its coordination environment, the driving force for the reductive quenching of the ESs by the ED, ES lifetime of the photocatalyst, effective pH of the solution, and ED–catalyst interactions.

The work on the reduction of HX by mixed-valence dirhodium complexes in the presence of halogen traps highlights another approach to driving important multi-electron processes. These systems use bidentate ligands that simultaneously stabilize low and high oxidation states of the same metal to provide for facile two-electron processes. Systems that undergo more efficient M–X bond activation are investigated as a means to enhance hydrogen production efficiency.

Recent studies have focused on the coupling of Pt-reactive metal centers to Ru charge transfer LAs. These systems collect multiple electrons at the vicinity of the Pt center where hydrogen photocatalysis occur. The formation of Pt colloid as the CAT species has been observed in some systems alerting researchers to the importance of careful analysis and characterization of photoproducts.

## CONCLUSIONS

A summary of properties and function of supramolecular complexes for PDT and solar energy conversion via $H_2$ production has been highlighted, focusing on systems that have similar structural motifs, providing insight into the similarities and differences of the activation of molecules for different light-activated processes. Supramolecular complexes are composed of a variety of subunits connected in this forum by coordinate covalent bonds. The subunits bring with them into the supramolecular assembly the simple act that can be carried out by that subunit. The combination of a variety of subunits connected in a defined and controlled order or spatial arrangement allows for complex supramolecules that can perform complex functions. PDT or solar energy conversion have as their initiation step absorption of a photon of light and excitation of the supramolecule. This step is then followed by a series of events that lead to the activation of a substrate in both applications that accounts for much of the similarity of structural motifs used in PDT and solar energy conversion.

The supramolecular assemblies described herein for PDT typically are structural motifs that include LA subunits for activation by light coupled to BAS via the use of BLs. The LA units that are used in this review are Ru and Os polyazine complexes with tunable MLCT ESs that are known to undergo ES electron and energy transfer reactions and dirhodium systems often incorporating polyazine ligands. The mechanism of action depends on the structural architecture dictated by the subunits incorporated into the assemblies.

In the PDT systems that use energy transfer, the Ru or Os LA are excited by photon absorption and then undergo energy transfer to dissolved $O_2$ in buffer solutions or organisms to produce the reactive oxygen species singlet oxygen $^1O_2$. These MLCT LAs have properties that are tunable via ligand variation and are often efficient $^1O_2$ generators. The systems that use singlet oxygen generation from Ru and Os LA described herein typically couple Pt BAS to provide for the targeting of the DNA biomolecule. This targeting allows for enhanced reactivity of the photoproduced $^1O_2$ with the DNA target by localizing $^1O_2$ production at the target. This forum reports the development of larger LA assemblies with three Ru centers, in which LA units are spatially separated away from the BAS providing for enhanced $^3$MLCT ES lifetimes and enhanced functioning as DNA cleavage agents. The electronic decoupling of the Ru LA from the Pt BAS enhances PDT action, whereas the electronic coupling for intramolecular electron transfer is needed in the solar $H_2$ application of these supramolecular motifs.

The MLCT LAs are also exploited in this forum for PDT using intramolecular electron transfer reactions. These systems couple EA subunits to the LA typically through BLs. This mode of action is exploited in the Rh containing supramolecular systems that use Ru or Os LA units. These supramolecules undergo excitation with UV or visible light via the intense MLCT transitions. This excitation is followed by electron transfer to the Rh center. The photogenerated MMCT state is then active for reaction with the DNA target, and this produces DNA cleavage. The coupling of a single Ru LA subunit in the Ru,Rh bimetallics results in active PDT agents but not photocatalysts for solar $H_2$ production.

The other major motif used in PDT discussed herein is the Rh bimetallic systems. These complexes use the Rh–Rh bridged systems and their unique orbital energetics to provide for DNA cleavage reactions. In these systems, the Rh core functions as both the LA unit and the BAS. The Rh centers are typically complexed to O donor atoms with acetate BLs that allow for metal–metal binding. The addition of polyazine ligands provides for enhanced functioning and movement of excitation wavelengths into the lower energy visible. Systems have used an organic BAS, the dppz and related ligands that allows targeting to DNA via intercalation into DNA.

The synthetic accessibility of supramolecular architectures is important in the application of these materials in any forum. The commonality of structural motifs in PDT and solar energy conversion is sometimes a result of the synthetic accessibility of a structural motif. Research in one forum can provide supramolecules that are applied in the other forum. Often the synthetic accessibility is a result of the tendency of reactions of the commercially available metal synthons to form certain motifs. This is seen in the Rh chemistry whereby traditional Rh starting materials can be reacted with polyazine BLs to provide for $Rh^{III}(NN)_2X_2$ systems under mild reaction conditions. This provides for the Rh-centered trimetallic systems when coupling Ru or Os LA units with polyazine BLs. In a complementary approach, Rh starting materials react with acetate ligands to produce the commonly used $Rh_2(\mu\text{-}O_2CCH_3)_4$ core. The reaction to couple Pt BAS to Ru or Os LA follows from the synthetic tendency of typical Pt starting materials to form $Pt^{II}(NN)X_2$ motifs. This results in the direct coupling of one Ru or Os LA unit in the systems using Pt BAS. Recent work has expanded these architectures to incorporate three metal LA units, which are synthetically accessible because of the ability to react common Ru starting materials to form $Ru^{II}(NN)_2X_2$ complexes that allow coupling of two Ru or Os LA units to a central Ru core that can then couple an additional BL that is used to bind the Pt BAS.

The supramolecular complexes for solar energy conversion via water reduction to produce hydrogen require molecular architectures that are also activated by the incorporation of an LA. The LAs in this forum typically absorb light and then undergo ES electron transfer reactions to provide for generation of multiple reduced states. This occurs through a series of steps in the typical Ru and Os containing systems via electron transfer to the coupled BL or reactive metal center. In the Rh–Rh bridged systems, new BLs are designed to facilitate the two-electron cycling of the Rh centers to facilitate the multielectron reduction of HX to produce $H_2$. The systems all function using ES reactions, providing for the activation of metal centers to bind to $H_2O$, $H^+$, or HX substrates to promote the initial steps of bond breaking, electron transfer, and ultimately bond formation needed to activate these substrates and produce $H_2$ fuel.

The structural motifs for solar hydrogen production that are directly analogous to systems used in PDT are the Rh-centered trimetallic complexes and the Ru MLCT LA systems that are coupled to Pt sites. The Ru LA with Rh-reactive metal systems for PDT require only the incorporation of one LA unit whereas the systems applied in solar energy conversion require two Ru or Os LA units. These systems, which couple two Ru or Os LA units, are synthetically accessible because of the tendency of Rh to form stable $Rh^{III}(NN)_2X_2$ subunits. The presence of two LA units in the PDT application can have advantages, providing for a symmetric Rh coordination and higher molar absorptivity for the complexes. For the solar energy conversion schemes, two Ru or Os LA units are necessary for function. The supramolecules that couple Pt sites to Ru or Os LA units have many similar systems that function in both forums. Interestingly, the Pt site that binds to DNA in the PDT application is also thought to bind the substrate in solar energy conversion schemes. The Pt containing supramolecules typically incorporate a single Ru or Os LA unit because of the typical $Pt^{II}(NN)Cl_2$ motif that is easily prepared by known synthetic methods. In the PDT action of these supramolecules, the LA units function once to activate the molecules and single photon activation provides for PDT action. The same systems used in solar energy conversion use multiple photon activations in sequence whereby the first excitation moves one electron to the reactive site, and this is followed by subsequent light activation and intramolecular transfer steps. The sequential movement of electrons in these supramolecules is needed for the multielectron activation of substrates to produce fuels. Interestingly, the electronic coupling of these LAs to the Pt sites in these supramolecules must be optimized differently in these two applications. For the best PDT agents, it is necessary to largely electronically uncouple the Ru or Os LA units from the Pt sites. This is evidenced by the enhanced functioning of the tetrametallic complexes in this forum, which provides for longer lived photoactive $^3$MLCT states in this forum. Enhanced $^3$MLCT ES lifetimes in the solar energy application are also valuable, but electronic coupling of the systems is needed to allow for electron transfer to the reactive metal sites to allow for their delivery to substrates. This is illustrated by the use of dpp BLs in the tetrametallic systems for PDT while dpq is used in the solar hydrogen production systems, which allow the promoted electrons to reside on the BL bound to the reactive Pt site.

The structural similarities of supramolecular complexes for a variety of applications provide for the cross fertilization of many fields by research conducted in different forums. The synthetic methods developed in the field of PDT or solar energy conversion can assist in the development of optimized motifs for alternative applications. The fundamental understanding developed about the spectroscopic, redox, photophysical, and photochemical properties of supramolecular complexes can be of significant value in these related forums. The assembly of knowledge on the understanding of the perturbations of subunit properties and function upon incorporation into supramolecular assemblies can provide for the development of a knowledge base applicable in many forums. The continued, long-term study of structurally complicated supramolecular assemblies is needed to provide for the enhanced functioning of molecular devices in these complex forums that require control and manipulation of ground and ES processes of considerable complexity. The interchange of ideas in many interdisciplinary forums is beneficial to the development of the molecular complexity needed for the transition of molecular device functioning into practical applications, approachable with these promising molecular architectures.

## ACKNOWLEDGMENTS

The authors thank all the students and research scientists who have worked in this area in the Brewer Group. Special thanks to Ms. Samantha Higgins, Ms. Jessica Knoll, and Mr. Travis White for help with the manuscript. The authors thank the Chemical Sciences, Geosciences, and Biosciences Divisions, Office of Basic Energy Sciences, Office of Sciences, and the U.S. Department of Energy (DE FG02-05ER15751) for their generous support of our research. They also thank Theralase, Inc., for continued collaboration to investigate the PDT action of supramolecular complexes. Acknowledgment is made to the financial collaboration of Phoenix Canada Oil Company, which holds long-term license rights to commercialize our Rh-based technology.

## REFERENCES

1. Balzani, V., L. Moggi, F. Scandola. 1987. In *Supramolecular Photochemistry*, edited by V. Balzani, L. Moggi, F. Scandola. Dordrecht: Reidel.
2. Lin, C. T., W. Boettcher, M. Chou, C. Creutz, N. Sutin. 1976. *Journal of the American Chemical Society*, 98, 6536.
3. Holder, A. A., S. Swavey, K. J. Brewer. 2004. *Inorganic Chemistry*, 43, 303.
4. Voet, D., J. G. Voet, (Eds.). 2003. In *Biomolecules, Mechanisms of Enzyme Action, and Metabolism*. Hoboken, New Jersey: John Wiley & Sons, Inc.
5. Kirsch-De Mesmaeker, A., J.-P. Lecomte, J. M. Kelly. 1996. *Topics in Current Chemistry*, 177, 25.
6. Szaciłowski, K., W. Macyk, A. Drzewiecka-Matuszek, M. Brindell, G. Stochel. 2005. *Chemical Reviews*, 105, 2647.
7. Stochel, G., A. Wanat, E. Kulis, Z. Stasicka. 1998. *Coordination Chemistry Reviews*, 171, 203.
8. Kennedy, J. C., R. H. Pottier. 1992. *Journal of Photochemistry and Photobiology. B, Biology*, 14, 275.
9. Fan, K. F. M., C. Hopper, P. M. Speight, G. A. Buonaccorsi, S. G. Bown. 1997. *International Journal of Cancer*, 73, 25.
10. Demas, J. N., D. Diemente, E. W. Harris. 1973. *Journal of the American Chemical Society*, 95, 6864.
11. Kelley, J. M., A. B. Tossi, D. J. McConnell, C. Ohuigin. 1985. *Nucleic Acids Research*, 13, 6017.
12. Tossi, A. B., J. M. Kelly. 1989. *Photochemistry and Photobiology*, 49, 545.
13. Dobrucki, J. W. 2001. *Journal of Photochemistry and Photobiology. B, Biology*, 65, 136.

14. Holder, A. A., D. F. Zigler, M. T. Tarrago-Trani, B. Storrie, K. J. Brewer. 2007. *Inorganic Chemistry*, 46, 4760.
15. Molnar, S. M., G. E. Jensen, L. M. Vogler, S. W. Jones, L. Laverman, J. S. Bridgewater, M. M. Richter, K. J. Brewer. 1994. *Journal of Photochemistry and Photobiology. A, Chemistry*, 80, 315.
16. Swavey, S., K. J. Brewer. 2002. *Inorganic Chemistry*, 41, 4044.
17. Swavey, S., K. J. Brewer. 2002. *Inorganic Chemistry*, 41, 6196.
18. Zigler, D. F., J. Wang, K. J. Brewer. 2008. *Inorganic Chemistry*, 47, 11342.
19. Elvington, M., K. J. Brewer. 2006. *Inorganic Chemistry*, 45, 5242.
20. Harris, A. L. 2002. *Nature Reviews, Cancer*, 2, 38.
21. White, T. A., S. M. Arachchige, B. Sedai, K. J. Brewer. 2010. *Materials*, 3, 4328.
22. Vinograd, J., J. Lebowitz. 1966. *Journal of General Physiology*, 49, 103.
23. Armitage, B. 1998. *Chemical Reviews*, 98, 1171.
24. Wang, J., D. F. Zigler, B. S. J. Winkel, K. J. Brewer. 2010. *Journal of Inorganic Biochemistry*. Manuscript in preparation.
25. Wang, J., T. A. White, S. M. Arachchige, K. J. Brewer. 2011. *Chemical Communications*, 47, 4451.
26. Wang, J., T. A. White, S. M. Arachchige, K. J. Brewer. 2011. *Chemical Communications*. 47, 9786.
27. Cotton, F. A., T. R. Felthouse. 1981. *Inorganic Chemistry*, 20, 584.
28. Cotton, F. A., T. R. Felthouse. 1980. *Inorganic Chemistry*, 19, 320.
29. Cotton, F. A., S. A. Koch, M. Millar. 1978. *Inorganic Chemistry*, 17, 2084.
30. Cotton, F. A., B. G. DeBoer, M. D. Laprade, J. R. Pipal, D. A. Ucko. 1970. *Journal of the American Chemical Society*, 92, 2926.
31. Koh, Y. B., G. G. Christoph. 1978. *Inorganic Chemistry*, 17, 2590.
32. Esswein, A. J., A. S. Veige, D. G. Nocera. 2005. *Journal of the American Chemical Society*, 127, 16641.
33. Caulton, K. G., F. A. Cotton. 1969. *Journal of the American Chemical Society*, 91, 6517.
34. Cotton, F. A., K. G. Caulton. 1971. *Journal of the American Chemical Society*, 93, 1914.
35. Rao, P. N., M. L. Smith, S. Pathak, R. A. Howard, J. L. Bear. 1980. *Journal of the National Cancer Institute*, 64, 905.
36. Das, K., K. M. Kadish, J. L. Bear. 1978. *Inorganic Chemistry*, 17, 930.
37. Martin, D. S., T. R. Webb, G. A. Robbins, P. E. Fanwick. 1979. *Inorganic Chemistry*, 18, 475.
38. Fu, P. K. L., P. M. Bradley, C. Turro. 2001. *Inorganic Chemistry*, 40, 2476.
39. Sorasaenee, K., P. K. L. Fu, A. M. Angeles-Boza, K. R. Dunbar, C. Turro. 2003. *Inorganic Chemistry*, 42, 1267.
40. Bear, J. L., H. B. Gray, L. Rainen. 1975. *Cancer Chemotherapy Reports*, 59, 611.
41. Bradley, P. M., A. M. Angeles-Boza, K. R. Dunbar, C. Turro. 2004. *Inorganic Chemistry*, 43, 2450.
42. Friedman, A. E., J. C. Chambron, J. P. Sauvage, N. J. Turro, J. K. Barton. 1990. *Journal of the American Chemical Society*, 112, 4960.
43. Angeles-Boza, A. M., P. M. Bradley, P. K. L. Fu, M. Shatruk, M. G. Hilfiger, K. R. Dunbar, C. Turro. 2005. *Inorganic Chemistry*, 44, 7262.
44. Angeles-Boza, A. M., P. M., Bradley, P. K. L. Fu, S. E. Wicke, J. Bacsa, K. R. Dunbar, C. Turro. 2004. *Inorganic Chemistry*, 43, 8510.
45. Joyce, L. E., J. D. Aguirre, A. M. Angeles-Boza, A. Chouai, P. K. L. Fu, K. R. Dunbar, C. Turro. 2010. *Inorganic Chemistry*, 49, 5371.
46. Aguirre, J. D., A. M. Angeles-Boza, A. Chouai, C. Turro, P. Pellois, K. R. Dunbar. 2009. *Dalton Transactions*, 48, 10806.
47. Rosenberg, B., L. Van Camp, J. E. Trosko, V. H. Mansour. 1969. *Nature*, 222, 385.
48. Cohen, G. L., W. R. Bauer, J. K. Barton, S. J. Lippard. 1979. *Science*, 203, 1014.

49. Cohen, G. L., J. A. Ledner, W. R. Bauer, H. M. Ushay, C. Caravana, S. J. Lippard. 1980. *Journal of the American Chemical Society*, 102, 2487.
50. Sherman, S. E., S. J. Lippard. 1987. *Chemical Reviews*, 87, 1153.
51. Bellon, S. F., J. H. Coleman, S. J. Lippard. 1991. *Biochemistry*, 30, 8026.
52. Jamieson, E. R., S. J. Lippard. 1999. *Chemical Reviews*, 99, 2467.
53. Hambley, T. W. 2001. *Journal of the Chemical Society, Dalton Transactions*, 2711.
54. Yam, V. W. W., V. W. M. Lee, K. K. Cheung. 1994. *Journal of the Chemical Society, Chemical Communications*, 2075.
55. Sakai, K., H. Ozawa, H. Yamada, T. Tsubomura, M. Hara, A. Higuchi, M.-A. Haga. 2006. *Dalton Transactions*, 3300.
56. Jain, A., J. Wang, E. R. Mashack, B. S. J. Winkel, K. J. Brewer. 2009. *Inorganic Chemistry*, 48, 9077.
57. Prussin Ii, A. J., S. Zhao, A. Jain, B. S. J. Winkel, K. J. Brewer. 2009. *Journal of Inorganic Biochemistry*, 103, 427.
58. Miao, R., M. T. Mongelli, D. F. Zigler, B. S. J. Winkel, K. J. Brewer. 2006. *Inorganic Chemistry*, 45, 10413.
59. Yam, V. W.-W., V. W.-M. Lee, K.-K. Cheung. 1997. *Organometallics*, 16, 2833.
60. Milkevitch, M., E. Brauns, K. J. Brewer. 1996. *Inorganic Chemistry*, 35, 1737.
61. Milkevitch, M., H. Storrie, E. Brauns, K. J. Brewer, B. W. Shirley. 1997. *Inorganic Chemistry*, 36, 4534.
62. Jain, A., B. S. J. Winkel, K. J. Brewer. 2007. *Journal of Inorganic Biochemistry*, 101, 1525.
63. Higgins, S. L. H., T. A. White, B. S. J. Winkel, K. J. Brewer. 2011. *Inorganic Chemistry*, 50, 463.
64. Williams, R. L., H. N. Toft, B. Winkel, K. J. Brewer. 2003. *Inorganic Chemistry*, 42, 4394.
65. Bard, A. J., M. A. Fox. 1995. *Accounts of Chemical Research*, 28, 141.
66. Kirch, M., J.-M. Lehn, J.-P. Sauvage. 1979. *Helvetica Chimica Acta*, 62, 1345.
67. Molnar, S. M., G. Nallas, J. S. Bridgewater, K. J. Brewer. 1994. *Journal of the American Chemical Society*, 116, 5206.
68. Konduri, R., H. Ye, F. M. MacDonnell, S. Serroni, S. Campagna, K. Rajeshwar. 2002. *Angewandte Chemie (International ed. in English)*, 41, 3185.
69. Konduri, R., N. R. de Tacconi, K. Rajeshwar, F. M. MacDonnell. 2004. *Journal of the American Chemical Society*, 126, 11621.
70. Polyansky, D. E., D. Cabelli, J. T. Muckerman, T. Fukushima, K. Tanaka, E. Fujita. 2008. *Inorganic Chemistry*, 47, 3958.
71. Fukushima, T., E. Fujita, J. T. Muckerman, D. E. Polyansky, T. Wada, K. Tanaka. 2009. *Inorganic Chemistry*, 48, 11510.
72. Balzani, V., A. Credi, M. Venturi. 1998. *Coordination Chemistry Reviews*, 171, 3.
73. Venturi, M., A. Credi, V. Balzani. 1999. *Coordination Chemistry Reviews*, 185–186, 233.
74. Scandola, F., R. Argazzi, C. A. Bignozzi, M. T. Indelli. 1994. *Journal of Photochemistry and Photobiology. A, Chemistry*, 82, 191.
75. Serroni, S., A. Juris, S. Campagna, M. Venturi, G. Denti, V. Balzani. 1994. *Journal of the American Chemical Society*, 116, 9086.
76. Indelli, M. T., F. Scandola, J.-P. Collin, J.-P. Sauvage, A. Sour. 1996. *Inorganic Chemistry*, 35, 303.
77. Elvington, M., J. Brown, S. M. Arachchige, K. J. Brewer. 2007. *Journal of the American Chemical Society*, 129, 10644.
78. Arachchige, S. M., J. Brown, K. J. Brewer. 2008. *Journal of Photochemistry and Photobiology. A, Chemistry*, 197, 13.
79. Arachchige, S. M., J. R. Brown, E. Chang, A. Jain, D. F. Zigler, K. Rangan, K. J. Brewer. 2009. *Inorganic Chemistry*, 48, 1989.

80. Rangan, K., S. M. Arachchige, J. R. Brown, K. J. Brewer. 2009. *Energy & Environmental Science*, 2, 410.
81. White, T. A., K. Rangan, K. J. Brewer. 2010. *Journal of Photochemistry and Photobiology. A, Chemistry*, 209, 203.
82. White, T. A., J. D. Knoll, S. M. Arachchige, K. J. Brewer. 2011. *Inorganic Chemistry.* Submitted for publication.
83. Anton, D. R., R. H. Crabtree. 1983. *Organometallics*, 2, 855.
84. Baba, R., S. Nakabayashi, A. Fujishima, K. Honda. 1985. *Journal of Physical Chemistry*, 89, 1902.
85. Arachchige, S. M., R. Shaw, T. A. White, V. Shenoy, H. M. Tsui, K. J. Brewer. 2011. *ChemSusChem.* 4, 514–518.
86. Chou, M., C. Creutz, D. Mahajan, N. Sutin, A. P. Zipp. 1982. *Inorganic Chemistry*, 21, 3989.
87. Heyduk, A. F., D. G. Nocera. 2001. *Science*, 293, 1639.
88. Esswein, A. J., J. L. Dempsey, D. G. Nocera. 2007. *Inorganic Chemistry*, 46, 2362.
89. Cook, T. R., A. J. Esswein, D. G. Nocera. 2007. *Journal of the American Chemical Society*, 129, 10094.
90. Teets, T. S., D. A. Lutterman, D. G. Nocera. 2010. *Inorganic Chemistry*, 49, 3035.
91. Sakai, K., H. Ozawa. 2007. *Coordination Chemistry Reviews*, 251, 2753.
92. Ozawa, H., Y. Yokoyama, M.-A. Haga, K. Sakai. 2007. *Dalton Transactions*, 1197.
93. Lei, P., M. Hedlund, R. Lomoth, H. Rensmo, O. Johansson, L. Hammarström. 2007. *Journal of the American Chemical Society*, 130, 26.
94. Du, P., J. Schneider, F. Li, W. Zhao, U. Patel, F. N. Castellano, R. Eisenberg. 2008. *Journal of the American Chemical Society*, 130, 5056.
95. Ozawa, H., M.-A. Haga, K. Sakai. 2006. *Journal of the American Chemical Society*, 128, 4926.
96. Masaoka, S., Y. Mukawa, K. Sakai. 2010. *Dalton Transactions*, 39, 5868.
97. Ozawa, H., M. Kobayashi, B. Balan, S. Masaoka, K. Sakai. 2010. *Chemistry—An Asian Journal*, 5, 1860.
98. Rau, S., B. Schäfer, D. Gleich, E. Anders, M. Rudolph, M. Friedrich, H. Görls, W. Henry, J. G. Vos. 2006. *Angewandte Chemie (International ed. in English)*, 45, 6215.
99. Tschierlei, S., M. Karnahl, M. Presselt, B. Dietzek, J. Guthmuller, L. González, M. Schmitt, S. Rau, J. Popp. 2010. *Angewandte Chemie (International ed. in English)*, 49, 3981.
100. Knoll, J. D., S. M. Arachchige, K. J. Brewer. 2011. *ChemSusChem*, 4, 252–261.
101. Miao, R., D. F. Zigler, K. J. Brewer. 2010. *Inorganic Chemistry.* Submitted for publication.

# 12 Supramolecular Polymers

*Brent E. Dial and Ken D. Shimizu*

## CONTENTS

## INTRODUCTION

Polymers are long chain-like molecules that are composed of simple units called monomers that are connected together in a linear fashion. Polymers are attractive materials in many applications because of their synthetic accessibility and the ability to rationally tailor their physical properties. For example, the strength, flexibility, and stability of the polymers can be controlled by the structure of the monomer units, the number of monomer units in the polymer chain, or the different processing procedures.[1] Recently, an exciting new class of polymers called *supramolecular polymers* has been the subject of considerable research and development. Like conventional polymers, supramolecular polymers are comprised of simple monomer units. However, the monomer units in supramolecular polymers are connected together by weak, reversible noncovalent interactions as opposed to stable covalent bonds. These reversible monomer–monomer interactions include hydrogen bonding, π-stacking, metal coordination, hydrophobic, and electrostatic interactions (Scheme 12.1).[2–6] The monomers and assembled polymer species are in dynamic equilibrium; therefore, they are constantly being formed and disassembled. Although supramolecular polymers generally possess lower mechanical and tensile strengths than conventional polymers, they possess a number of unique properties and applications arising from their dynamic self-assembling nature. For example, the synthesis of supramolecular polymers is an efficient and virtually quantitative process, as the monomers simply assemble themselves into long chains without the aid of catalysts or reagents. Supramolecular polymers also have the ability to automatically repair breaks and fractures because their polymer chains are constantly being broken and reformed. Accordingly, supramolecular polymers have shown utility and have been applied to a wide range of applications. Applications to date include tissue engineering to create biodegradable polymer,[7] drug delivery in which the outside casing of a drug is dissolved on delivery,[8] printing to develop phase changing printer ink,[9–14] adhesive,[15] and personal care products.[16–18]

The goal of this chapter is to describe the key design choices that need to be considered when tailoring a supramolecular polymer for a particular application and to give examples that highlight the utility of supramolecular polymers in specific applications. This chapter is not meant to be a comprehensive review of supramolecular polymers, as there are already many excellent reviews on this topic.[2–6,19,20]

There are many different types of supramolecular polymers. Therefore, to limit the scope of this chapter, the specific focus will be on main-chain supramolecular polymers in which the supramolecular interactions are between adjacent monomer units in the main-chain as shown in Scheme 12.1. Polymers can also be formed via supramolecular interactions between main-chains of adjacent polymer strands or interactions between the main-chain and side-chains. Examples of polymers that form supramolecular main-chain to main-chain interactions include aligned

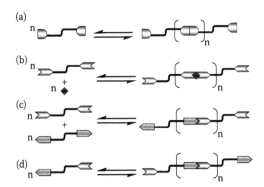

**SCHEME 12.1**   Schematic representation of different types of supramolecular polymerization categorized by the different recognition patterns of the monomer units to form (a) a homo-polymer, (b) a metallo-organic supramolecular polymer, (c) a two-component heteropolymer, and (d) a one-component heteropolymer.

high-density polyethylene and Kevlar, which are used in bulletproof vests,[21] and of course, the double helical structure of DNA, which is made up of two side-by-side polynucleotide polymer chains.[22,23] Examples of polymers that form main-chain to side-chain interactions include the use of a polymer with noncovalent side-chains to form liquid-crystalline and plug-and-play readily tailorable polymers.[24,25] However, these will not be discussed and the readers are directed to recent monographs.[26]

The definition of supramolecular interactions will also be broadly applied to all types of reversible bonding interactions that allow for self-assembly. These include traditional noncovalent interactions such as hydrogen bonds,[27–29] hydrophobic interactions,[30,31] electrostatic,[32] π-stacking, van der Waals, and donor–acceptor interactions.[33] Some examples of the reversible monomer–monomer interactions that have been used in supramolecular polymers are shown in Figure 12.1. The most

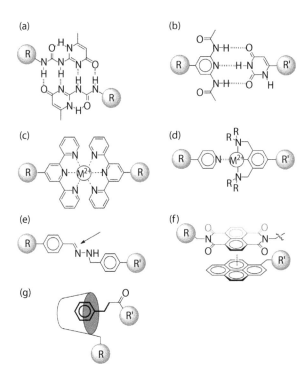

**FIGURE 12.1** Examples of supramolecular motifs (a) hydrogen bonding interactions between 2-ureido-4-[1H]pyrimidone recognition units, (b) uracil-diacylaminopyridine triple hydrogen bond motif, (c) metal–ligand interactions between two terpyridine ligands and metal ion (from Lindsey, J. S., *New Journal of Chemistry*, 15, 153–180, 1991; Beijer, F. H. et al., *Journal of the American Chemical Society*, 120, 6761–6769, 1998; Folmer, B. J. B. et al., *Chemical Communications*, 1847–1848, 1998; Lortie, F. et al., *Macromolecules*, 38, 5283–5287, 2005), (d) single metal–ligand interaction between a pyridine and metallipincer, (e) reversible covalent imine bond, (f) donor–acceptor π-stacking interactions between dimide and pyrene shelves, and (g) inclusion complex formed by hydrophobic interactions.

common classes of supramolecular polymers are those that use hydrogen bonding interactions and metal–ligand interactions. Two different examples are shown in Figure 12.1a and d of each type of these monomer–monomer interactions. Because of the weak nature of the hydrogen bonding interactions, most main-chain supramolecular polymers have used monomers connected together by hydrogen-bonding motifs that form three or more hydrogen bonds. For example, Figure 12.1a is the self-complementary quadruple hydrogen bonding 2-ureido-4-[1H]pyrimidone (UPy) recognition group developed by Sijbesma et al.,[34] and Figure 12.1b is the uracil-diacylaminopyridine triple hydrogen bonding motif developed by Lehn et al.[35,36] The stronger of the noncovalent interactions, the metal–ligand interactions, can also be formed using multiple interactions as shown in Figure 12.1c or by a single interaction as shown in Figure 12.1d. Other types of interactions include reversible covalent bonds such as boronic esters and imines (Figure 12.1e),[37] donor–acceptor $\pi$-stacking interactions (Figure 12.1f),[38,39] and hydrophobic interactions (Figure 12.1g).[40]

Each of the monomer–monomer interactions in Figure 12.1 can be classified as either a self-complementary or heterocomplementary interaction. Figure 12.1a and c are self-complementary, involving two of the same recognition groups; whereas the others are all heterocomplementary. Supramolecular polymers formed using self-complementary interactions require monomers that are homoditopic; this means that the polymer is composed of a single type of monomer (Scheme 12.1a). In contrast, supramolecular polymers formed by heterocomplementary interactions use monomers that are either heteroditopic (Scheme 12.1d) if the monomer has one of each complementary group, or homoditopic (Scheme 12.1c) if composed of two different types of monomers (block copolymer). Of course, more complex supramolecular polymeric architectures can assemble through polymerization of oligotopic monomers. A hyperbranched supramolecular polymer is formed when heterotritopic monomers assemble, in which one of three recognition units must be complementary to the other two recognition units (Scheme 12.2a). Lastly, supramolecular networks are achieved when a homotritopic monomer is mixed in equal portions with a homoditopic monomer with complementary recognition groups (Scheme 12.2b).

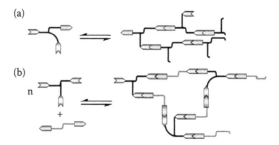

**SCHEME 12.2** Polymerization of (a) a supramolecular hyperbranched polymer and (b) a supramolecular networked polymer.

# PROPERTIES AND POTENTIAL OF SUPRAMOLECULAR POLYMERS

## SIMILARITY TO TRADITIONAL POLYMERS

Although the monomer–monomer interactions in supramolecular polymers have only 10% to 50% of the strength of the covalent monomer–monomer bonds in conventional polymers, supramolecular polymers can form stable films and materials with similar strengths and properties as conventional polymers. For example, monomer **1** (Figure 12.2a), which makes only a single hydrogen bond between the ends of its complementary units, is still able to form stable amorphous films that look and act very similar to those formed using covalent polymers.[41] Another example is shown in Figure 12.3a, which displays a flexible film produced from short poly(ethylene butylene) monomers that are capped with four self-complementary hydrogen bond units similar to that shown in Figure 12.1a. In the absence of the hydrogen bonding recognition units, the short poly(ethylene butylene) monomer is a viscous liquid (Figure 12.3b).[42] The reason that supramolecular polymers are able to produce materials with similar properties as conventional polymers is because the strength of most polymeric materials are determined by the strength of the interstrand polymer–polymer interactions and not by the strength of the main-chain.

## UNIQUE DYNAMIC PROPERTIES OF SUPRAMOLECULAR POLYMERS

In addition to mimicking the properties of conventional polymers, supramolecular polymers possess unique dynamic properties arising from the reversible nature of the supramolecular polymerization process. These include synthetic efficiency, self-repairing ability, and stimuli-responsive properties. Each of these unique properties and their potential utility are described below.

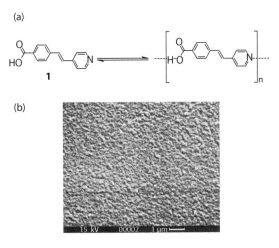

**FIGURE 12.2** (a) Self-assembly of supramolecular monomer **1**, 4-[trans-(pyridin-4-ylvinyl) benzoic acid, via pyridine-carboxylic acid hydrogen bonds. (b) Scanning Electron Microscopy (SEM) image of an amorphous film grown from supramolecular monomer **1** on a glass substrate. (Reproduced from Cai, C. et al., *Adv. Mater.*, 11, 745–749, 1999. With permission.)

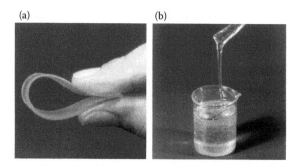

**FIGURE 12.3**  Comparison of (a) supramolecular polymer film formed from a 2-ureido-4-[1H]pyrimidone end-capped poly(ethylene butylene) monomer and (b) a viscous solution of the same poly(ethylene butylene) monomer unit that lacks the supramolecular end-groups. (Reproduced from Folmer, B. J. B. et al., *Adv. Mater.*, 12, 874–878, 2000. With permission.)

## Synthetic Efficiency

Unlike conventional polymer synthesis, the preparation of supramolecular monomers does not require any additional catalysts and reagents. The recognition units in the individual monomers contain all the necessary chemical reactivity and information to form polymeric chains. The small supramolecular monomers are typically readily soluble in a wide range of solvents or are even liquids at elevated temperatures, which can be directly cast into films or poured into molds. This greatly simplifies the synthesis and processing of supramolecular polymers and expands their potential utility. Conventional polymers are at a disadvantage in this area, being viscous at elevated temperatures and requiring extreme temperatures for processing. The ability of monomers to directly and efficiently form stable polymers was nicely illustrated by Berl et al.'s hydrogen-bonding array system (Figure 12.4).[43] Each cyanuric acid recognition group in monomer **2** is able to form six hydrogen bonds with the complementary receptor units in monomer **3**. The strength and fidelity of this interaction was apparent from the formation of the supramolecular heteropolymer spontaneously even in dilute toluene solutions (1 mM).

## Self-Repair

Because of their dynamic nature, supramolecular polymers have the ability to self-repair and heal fractures and breaks. The monomers in supramolecular polymers are continually breaking and reforming monomer–monomer interactions. For example, Tournilhac et al.[44,45] created a supramolecular networked polymer made of fatty carboxylic acid, ethylene diamine, and urea monomers that exhibited self-healing abilities when the fractured polymer ends were brought together, so that no additional stimuli was required to facilitate healing (Figure 12.5). Because the repair process reforms the same monomer–monomer interactions as in the original supramolecular polymer, the repaired material possesses the same strength and properties as the initial material.

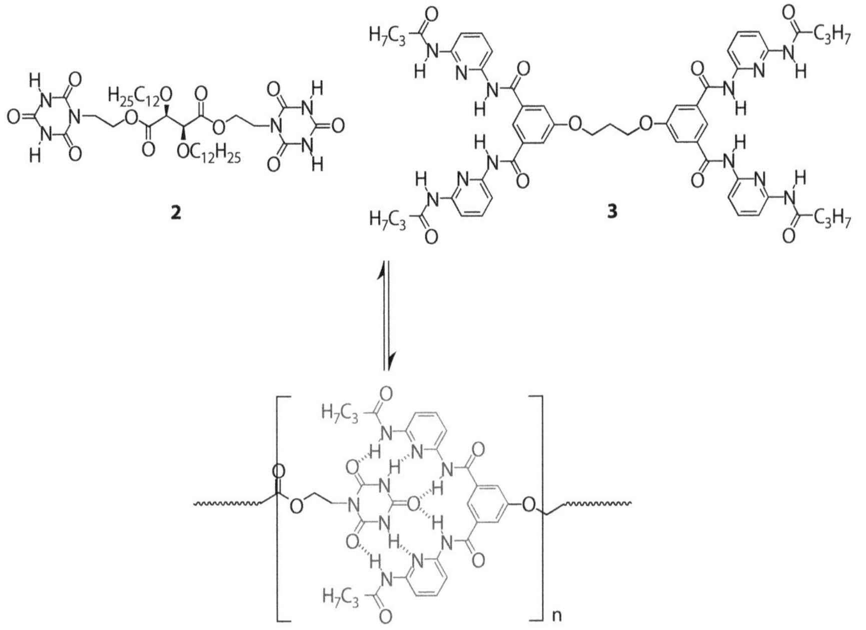

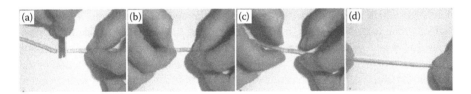

**FIGURE 12.4** Complementary hydrogen bonding monomer units **2** and **3** that form strong linear supramolecular alternating block copolymers in dilute toluene (1 mM) solutions.

**FIGURE 12.5** Self-healing of a supramolecular polymer (a) cutting, (b) mending, (c) healing, and (d) stretching. (Reproduced from Tournilhac, F. et al., *Macromol. Symp.*, 291, 84–85. 2010. With permission.)

## Stimuli-Responsive Properties

The dynamic nature of the supramolecular polymerization process also endows supramolecular polymers with stimuli-responsive properties. Their properties can be actively modified and tuned by changes in temperature,[46,47] light,[48,49] solvent,[50–52] concentration,[53,54] and small guest molecules.[55–59] These stimuli trigger changes that shift the monomer–polymer equilibrium, leading to dramatically longer or shorter average chain lengths. An early example of this was the photoinduced depolymerization by Folmer et al.[60] The self-complementary hydrogen bonding bis(ureidopyrimidinone) units in monomer **4** forms lengthy supramolecular polymers even in the presence of the photoactive trigger molecule **5**. Irradiation of the mixture cleaves *o*-nitrobenzyl protecting group from **5**, unmasking a ureidopyrimidinone hydrogen bonding recognition unit. The competition from the end-capping

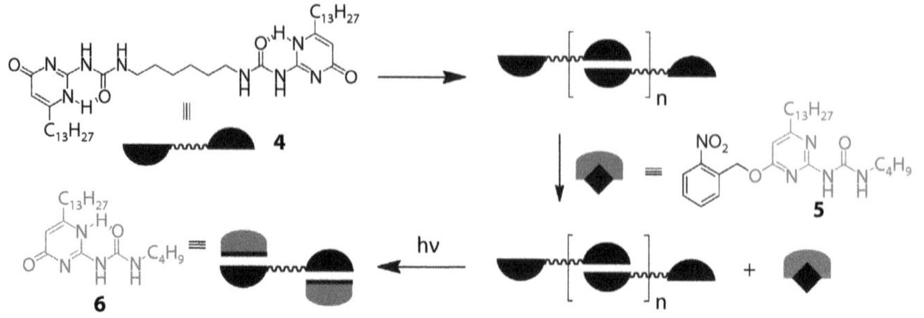

**SCHEME 12.3**  Meijer et al. photoinduced depolymerization process. Monomer **4** is self-complementary and forms lengthy polymer chains even in the presence of photoactive trigger molecule **5**. Once irradiated, end-capping monomer **6** is deprotected and dramatically shortens the average chain lengths of poly-**4**. (From Folmer, B. J. B. et al., *Chemical Communications*, 1847–1848, 1998.)

unit **6** leads to dramatic shortening of the average poly-**4** chain lengths as measured by a large drop in the viscosity of the solution (Scheme 12.3).

# KEY DESIGN PARAMETERS OF SUPRAMOLECULAR POLYMERS

The properties of supramolecular polymers can be controlled in much the same way as traditional polymers via the lengths of the polymer chains, branch/cross-link density, and compatibility with the solvent environment or copolymer matrix. However, the manner in which these parameters are tuned in supramolecular systems is often very different than in traditional polymers. Therefore, this section will provide an overview of the key design choices and parameters in designing a supramolecular polymer system for a particular application. This discussion is divided into (1) the selection of the recognition units and (2) the structure of the spacer unit between the recognition units.

## SUPRAMOLECULAR RECOGNITION UNIT

The three main factors that need to be taken into consideration when choosing a recognition unit are the (1) strength of the resulting monomer–monomer interactions, (2) the rate at which the monomer–monomer interactions are broken and reformed, and (3) the sensitivity of the interaction to different solvent and copolymer environments.

### Strength of Monomer–Monomer Interactions

The strength of the monomer–monomer interactions are directly correlated to the length of the supramolecular polymers. For linear supramolecular polymers, the number of monomer units connected together is proportional to the square root of the monomer–monomer association constant ($K_a$) and the monomer concentration ([M]) as shown in Equation 12.1.[2,34] Thus, the formation of supramolecular polymers containing 100 or more monomer units requires a minimum $K_a$ of $10^3$ M$^{-1}$ and a monomer concentration of 1.0 M. Supramolecular polymers that can form mechanically stable films requires much stronger interactions ($K_a > 10^6$ M$^{-1}$), such as the quadruple hydrogen bonding ureidopyrimidinone motif developed by Meijer et al.[34,42]

$$DP = (K_a \cdot [M])^{1/2} \tag{12.1}$$

### Rate of Formation and Breaking of Monomer–Monomer Interactions

The rate of the formation and breaking of the monomer–monomer interactions is another key factor in tuning the mechanical properties. The association constant is simply a ratio of the rates of the formation and breaking of the monomer–monomer interaction. Thus, a large association constant does not necessarily ensure that the resulting polymer will have dynamic self-repair and stimuli-responsive properties. This means that binding interaction with slow exchange rates behave differently from those assembled with fast exchange rates. Experimentally, it has been shown that kinetic rates play a paramount role in determining the polymer's properties. For example, Yount et al.[61,62] have shown that bulk mechanical properties such as viscoelastic response and sheer rates of supramolecular polymer networks can be controlled by attenuating the rate of formation and breaking of the recognition interactions.

### Compatibility of Monomers with Chosen Solvents and Polymer Environments

The compatibility of the monomers with the chosen solvents and polymer environments is also another key factor in the design of the monomer units. Specifically, the supramolecular interactions should be stable and the monomer units must also be soluble in the self-assembling environment. Hydrogen bonding interactions, despite their generally weak nature, possess much wider compatibilities with organic solvents and polymer matrices because of their lack of ionic charges, but are not viable in highly polar solvents such as water and alcohols. Reversible metal–ligand interactions, on the other hand, are typically much stronger, but the presence of charged metal ions, counterions, and ligands limit the solubility and compatibility of these interactions to polar solvent and polymer environments. For example, Beck et al. have developed oligoglycol units capped with 2,6-bis(1′-methylbenzidmidazolyl) pyridine (Mebip) ligands, which self-assemble in the presence of $Co^{II}$ and $Zn^{II}$ ions (Scheme 12.4).[83] The oligoglycol spacer units in the monomer provides not only flexibility but also serves to assist in solubilizing and maintaining phase stability with the highly charged metal ions and counterions. Thus, this ionic supramolecular

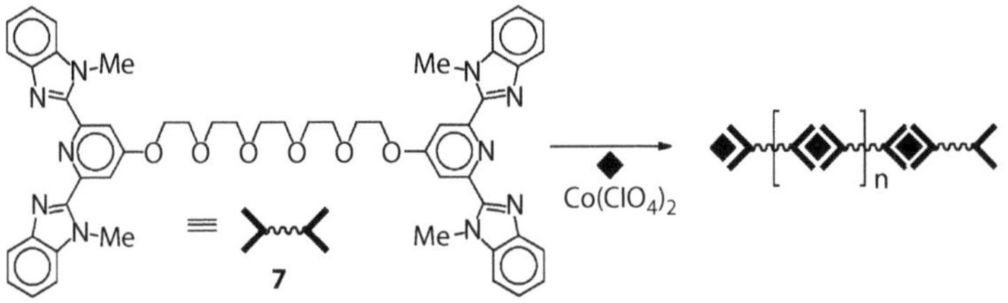

**SCHEME 12.4** Formation of a metallo-supramolecular gel using lanthanide and transition metal ions with bis-terpyridine monomer **7**. (From Beck, J. B. et al., *Macromolecules*, 38, 5060–5068, 2005.)

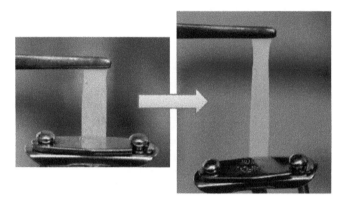

**FIGURE 12.6** Illustration of potential elastic properties of a metallo-supramolecular polymer. (Reprinted from Fox, J. D., Rowan, S. J., *Macromolecules*, 42, 6823–6835, 2009. With permission.)

polymer can form stable polymer films even without blending with a conventional covalent polymer (Figure 12.6).

A second aspect to consider is the compatibility of the supramolecular monomer and recognition units with conventional covalent polymers. The blending of supramolecular and conventional polymers represents the easiest and most practical way to implement unique dynamic properties into existing polymer systems. For example, supramolecular–covalent polymer blends have already been developed as rheology modifiers, adsorbents, surfactants, stabilizers, coating, adhesives and electro-optic technologies.[64–67] An alternative method to making hybrid supramolecular–covalent polymer systems is to cap the ends of short covalent polymers with recognition units. This is the strategy used by Folmer et al.[42] in which self-complementary hydrogen bonding recognition units were attached to the ends of short poly(ethylene butylene).

## APPLICATIONS OF SUPRAMOLECULAR POLYMERS

Although the field of supramolecular polymers is still in its infancy, a wide range of applications is rapidly emerging. Some recent examples are highlighted below, such as liquid-crystals, small molecule gelators, assembly of biological macromolecules, and tissue engineering. These examples were chosen to showcase the many different types of interactions and polymeric structures that can be formed via self-assembly. These examples also show that one of the main reasons for the rapid development of applications for supramolecular polymers has been the ability to directly integrate the supramolecular interactions into existing polymer platforms, which are already in common usage for these applications.

### LIQUID CRYSTALS

Liquid crystals can be classified as supramolecular polymers because they consist of small molecule or polymer units that self-assemble into higher-order structures.[68]

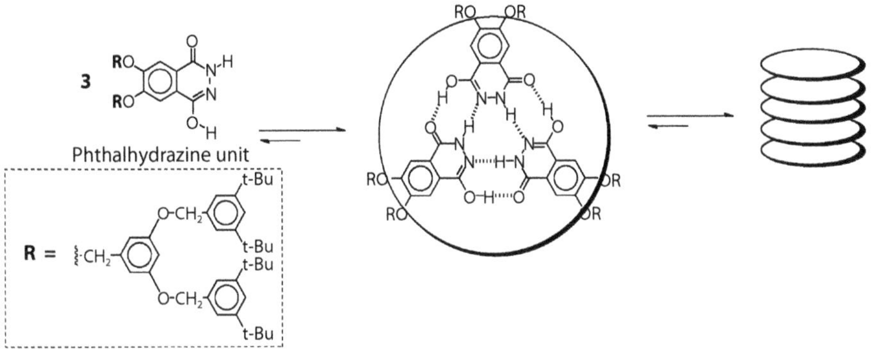

**SCHEME 12.5** Illustration of the self-assembly of three phthalhydrazine units via hydrogen bonding between self-complementary lactam–lactim groups to form a planar mesogen, which further self-assembles via π-stacking into columnar liquid crystals.

This ability to form and switch between different self-assembled liquid crystalline phases forms the basis for liquid crystalline display technology. In particular, liquid crystals that form columnar stacks can be thought of as main-chain supramolecular polymers. An example of a liquid crystalline monomer that was specially designed to use noncovalent hydrogen bonding and π-stacking interactions to form self-assembled one-dimensional columns was selected (Scheme 12.5). Most columnar liquid crystals are formed from disc-shaped molecules or polymers. However, Zimmerman et al. designed a small oligomeric unit that self-assembles into disc-shaped mesogen.[69] Cooperative hydrogen bonding interactions between three phthalhydrazide units forms a flat trimer, which self-assembles further via π-stacking interactions into columnar structures. In addition to the hydrogen-bonding head group, each phthalhydrazide unit is functionalized on the periphery (R groups) with two first-generation Frechet dendrons. The primary purpose of the dendrons is to form disordered "liquid" regions between the "crystalline" columnar cores. However, the steric bulk and incompatibility with the rigid polar head groups of the dendrons also help guide and stabilize the formation of the columnar structures. The primary advantage of using self-assembly to form the discotic monomer units is synthetic efficiency. Instead of having to synthesize the entire discotic monomer, units that are only a third of the size can be synthesized and then allowed to self-assemble into a disc-shaped trimer. This allows structural modification and optimization of the mesogen units to be carried out more quickly and efficiently.

## SMALL MOLECULE GELATORS

Another common self-assembly process that can be classified as a supramolecular polymerization process is the ability of certain small molecules to aggregate and form gel solutions.[70–72] One common gelation motif is the self-assembly of the small molecule gelator—a linear supramolecular polymer or ribbons which assemble further via main-chain to main-chain interactions into fiber bundles. The assembly of the low-molecular weight gelators (LMOG) is often temperature-dependent and reversible, which provides a method of controlling the viscosity and possibility of a solution. Recently, Bouteiller et al. used the dynamic nature of the LMOG as a hardening agent for asphalt.[73] Asphalt or bitumen has the highest boiling or nonvolatile fraction collected from petroleum distillation. This black viscous hydrophobic liquid is commonly used as a binding agent for paving roads and for tarring and sealing roofs. To improve the hardness and strength of asphalt, polymer additives are often added. However, these polymer additives also increase the melt temperature of the material, which has the undesirable consequence of requiring higher temperatures to process and apply asphalt films and concretes. Bouteiller et al. used a LMOG, which acts as a polymer at ambient temperatures, increasing viscosity and hardness but which is disassembled at higher temperatures and thus will not increase the melting temperature.

A wide range of readily accessible and inexpensive LMOG were initially screened using a crude fraction that is a liquid at room temperature. Among the most effective, dramatically increasing the viscosity via gelation, were simple alkane dicarboxylic acids ranging from glutaric acid (5 carbons) to tridecanedioic (13 carbons) acid. The diacid assembled into supramolecular polymers via hydrogen bonding between the carboxylic acid groups (Scheme 12.6), which self-assembled into bundled fibers leading to the formation of a gel. This ordered assembly process was confirmed by Scanning Transmission Electron Microscopy (STEM) analysis of the gel (Figure 12.7). Long crystalline fibers that are many microns long and 10 to 20 nm in width were observed. The widths of the fibers are considerably larger than the width of a single supramolecular strand, which is only 0.5 nm, and thus, the fibers must consist of a bundle of multiple supramolecular polymer strands.

Organogels were prepared by mixing bitumen and the diacids at 160°C and then cooling to room temperature. The gelators' ability to harden bitumen was assessed by measuring their softening temperatures. The addition of just 3% (w/w) of organogelator had a dramatic effect on the softening temperature (Figure 12.7b). For example, the softening temperature increased from 47°C (pure bitumen) to 107°C (bitumen with 3% (w/w) dodecanedioic acid). Also, the softening temperatures increased with increasing length of the diacids. This was attributed to the increasing interchain van der Waals' interactions between the alkyl spacers for the longer diacids and with the nonpolar, hydrophobic bitumen. The mechanical properties of the bitumen were also improved in the solid-state at room temperature by more than an order of magnitude. Despite the increase in the softening temperature, the diacid gelators do not also

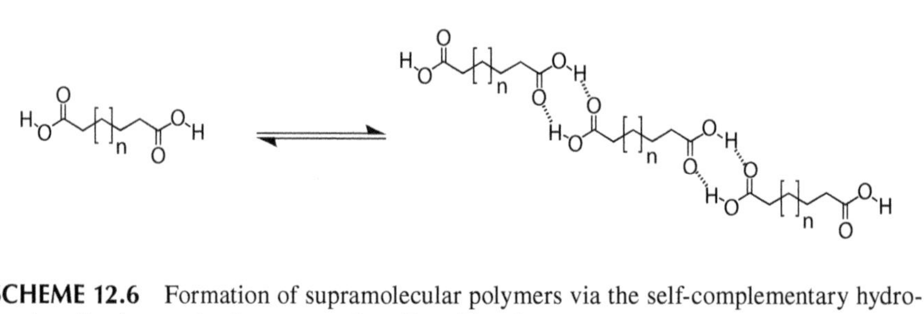

**SCHEME 12.6** Formation of supramolecular polymers via the self-complementary hydrogen bonding interaction between carboxylic acid end-groups.

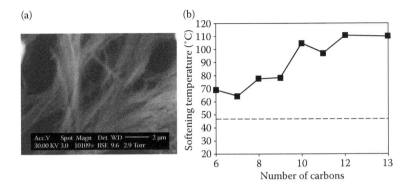

(a)

(b)

**FIGURE 12.7** (a) Fiber network of bitumen and 6% (w/w) dodecanedioic acid after dissolution of bitumen observed by STEM. (b) Softening temperature of pure bitumen (broken line) and bitumen containing 3% (w/w) of diacids of varying length (filled boxes). (Reproduced from Isare, B. et al., *Langmuir*, 25, 8400–8403, 2009. With permission.)

increase the viscosity of the melt as the supramolecular polymers and fibers disassemble at higher processing temperatures. Thus, the dynamic properties of the diacid gelators provided the desired balance of increased strength and hardness at room temperature without sacrificing processability at higher temperatures.

## USING SUPRAMOLECULAR POLYMERS TO ORGANIZE AND ARRANGE BIOLOGICAL MACROMOLECULES

Many biological structures are formed by utilizing the same self-assembly principles and noncovalent interactions used to form synthetic supramolecular polymers. For example, the DNA double helix is held together by hydrogen-bonding and π-stacking interactions between the nucleotide side chains of two complementary DNA strands. Thus, the use of naturally available biopolymers and oligomers as supramolecular monomers provides a potentially efficient route to larger nanoscale structures and materials, as well as providing a means to organize and pattern biomolecules and biomacromolecules. The challenge with this approach is integrating new supramolecular interactions that are compatible and orthogonal into the existing noncovalent interactions in the biomolecule units.

A recent example by Rybtchinski and Lewis, in which short double-stranded DNA units were assembled together into a supramolecular polymer, was selected to highlight the potential of this strategy.[74] Hydrophobic perylenediimide (PDI) units were installed at the ends of a short double-stranded DNA oligomer (Figure 12.8). The PDI units are the approximate length of a hydrogen-bonded nucleotide base pair and, thus, span the distance between opposing DNA strands without significant distortion of the DNA double helix.[75,76]

Under low salt conditions, the oligomers do not self-associate. However, with increasing salt concentrations, the end-to-end assembly of the PDI-capped oligomers was observed. The strong dependence of the assembly process with salt concentration was consistent with the formation of hydrophobic interactions of the large

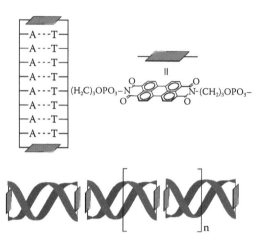

**FIGURE 12.8** PDI capped short, double-stranded DNA oligomers which assemble end-to-end under high salt concentrations because of the hydrophobic interactions between PDI units. (Reproduced from Neelakandan, P. P. et al., *J. Am. Chem. Soc.*, 132, 15808–15813, 2010. With permission.)

aromatic surfaces of the PDI end-groups. A sodium chloride concentration of 10 mM was required to assemble half of the oligomers into supramolecular polymers and almost complete assembly was observed at 100 mM of sodium chloride. This ability to trigger the assembly or disassembly of the supramolecular polymer simply by changing the solvent environments shows the advantages of the dynamic nature and reversibility of the supramolecular polymerization process.

The assembly process could easily be followed by the dramatic drop in fluorescence arising from quenching of closely associated PDI fluorophores. The thermodynamics of the assembly process were shown by an isodesmic model in which the first and each successive association constant were the same.[77] From the isodesmic model, an association constant of $3.2 \times 10^7$ M$^{-1}$ was estimated for the supramolecular polymer formed at 100 mM of sodium chloride. The DNA-based supramolecular polymers were also characterized by Transmission Electron Microscopy (TEM) and Atomic Force Microscopy (AFM). Isolated supramolecular polymer strands of up to 100 nm in length were observed by Cryo-TEM, which corresponds to a polymer containing 30 DNA-oligomer units. TEM and AFM obtained from the dried solutions show long polymer fiber bundles having the width of multiple strands. The cross-section of the fibers was 10 to 100 nm, which corresponds to a diameter of 4 to 40 individual polymer strands. The incorporation of a strong fluorophore and electron-accepting aromatic surface into the π-stacked core of DNA may yield interesting electronic properties and these are currently being investigated.

## BIOACTIVE SCAFFOLDS FOR TISSUE ENGINEERING

An exciting application of supramolecular polymers that is currently under investigation is tissue engineering.[78] Biological applications present a particular challenge for supramolecular polymers because of the reversibility of the self-assembly process. The monomer–monomer interactions must be compatible within a highly competitive aqueous environment containing high concentrations of ions, molecules, and macromolecules. Meijer et al., however, have shown that even supramolecular polymers formed using hydrogen bonding interactions can be used as matrices and surfaces for facilitating tissue growth. The self-complementary hydrogen bonding recognition groups were attached to the ends of short oligomers of biocompatible polymers (Figure 12.9a). Specifically, the 2-ureido-4[1H]-pyrimidinone (UPy) moieties were attached to both ends of oligo-caprolactone segments to form the PCLdiUPy monomer, which self-assemble into supramolecular polymers that form strong, stable films. By comparison, oligo-caprolactone segments without UPy end-groups form waxy brittle solids.

Although the primary cohesive interactions in the PCLdiUPy polymers are hydrogen bonds, the resulting films are surprisingly stable in buffered aqueous solutions and even when implanted subcutaneously. This stability enables their use in bioengineering applications. One of the advantages of using supramolecular polymers for tissue engineering applications is their ease of processability. Despite their strength

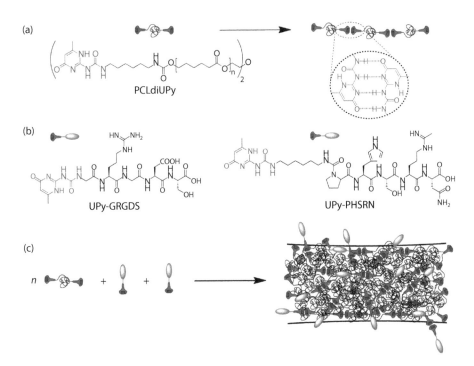

**FIGURE 12.9** Example of a supramolecular hydrogen bonding polymer for tissue engineering applications. (a) The monomer is an oligocaprolactone functionalized on either end with UPy hydrogen bonding UPy units (PCLdiUPy), which self-assemble to form a main chain supramolecular polymer. (b) Two supramolecular co-monomers (UPy-GRGDS and UPy-PHSRN) containing bioactive peptide units. (c) Formation of a functionalized polymer film via the self-assembly of the PCLdiUPy, UPy-GRGDS, and UPy-PHSRN monomers. (Reproduced from Dankers, P. Y. W. et al., *Nat. Mater.*, 4, 568–574, 2005. With permission.)

**FIGURE 12.10** The supramolecular UPy-functionalized biomolecular polymer is process-able into several scaffolds; films, fibers, meshes, and grids (SEM pictures; Philips XL30 FEG SEM). (Reproduced from Dankers, P. Y. W. et al., *Nat. Mater.*, 4, 568–574, 2005. With permission.)

and high apparent molecular weights, the PCLdiUPy supramolecular polymers melt and disassemble when heated to more than 80°C, becoming a low viscosity solution. The polymers can be solvent-cast into films, extruded or spun into fibers, or compression-molded into grids and meshes (Figure 12.10).

A second advantage is that functionalized polymers can be readily prepared and their materials and bioactivities optimized using the self-assembly process. For example, Meijer et al. were able to quickly prepare a functionalized polymer film containing bioactive peptide units that facilitate cell and tissue growth simply by mixing the monomers together (Figure 12.9b and c). Mouse 3T3 fibroblast cells were grown on grids formed from different supramolecular copolymer blends. Cells grown on blends containing both peptide-functionalized co-monomers UPy-GRGDS and UPy-PHSRN showed much higher degrees of adhesion and spreading than on blends containing only one or no peptide co-monomers. *In vivo* studies were also carried out. The supramolecular polymer blend containing both peptide co-monomers and a control blend containing only PCLdiUPy scaffold were implanted subcutaneously into rats. The results were analogous to those observed in the *in vitro* cell culture experiments. The blend containing UPy-GRGDS and UPy-PHSRN showed good cellular infiltration and biocompatibility, whereas the control polymer showed poor cellular compatibility and was encapsulated as a fibrous cyst.

In summary, this example demonstrates that supramolecular polymers can maintain their structural integrity even in aqueous biological environments, opening up many applications in bioengineering and biomaterials.

## CONCLUSION

Nature has long used self-assembly of smaller units to generate larger nano-scale structures such as virus protein shells, lipid membranes, and collagen fibrils.[79]

Chemists have only recently developed recognition systems with sufficient strength and fidelity to prepare stable supramolecular polymers with properties comparable to traditional covalent polymers. This field of supramolecular polymers opens up the possibility of "smart" or multifunctional materials, as supramolecular polymers possess dynamic self-healing and stimuli-responsive properties. Conventional polymers, in contrast, have static and fixed properties and thus have been primarily used in passive applications for mechanical structure and strength. Although this study perceptively examines the key design concepts and foremost accomplishments within the area of supramolecular polymers, we have just scratched the surface of their capabilities and applications. The potential of this field leans toward the development of "reworkable" adhesive, recyclable plastic, biomaterial, optoelectronic material, thermoelastomers, and many other possibilities.

## REFERENCES

1. Stevens, M. P. *Polymer Chemistry: An Introduction*, 3rd ed. New York: Oxford University Press, 1998.
2. Fox, J. D., S. J. Rowan. 2009. *Macromolecules*, 42, 6823–6835.
3. Brunsveld, L., B. J. Folmer, E. W. Meijer. 2001. *Chemical Reviews*, 101, 4071–4097.
4. Khlobystov, A. N., A. J. Blake, N. R. Champness, D. A. Lemenovskii, A. G. Majouga, N. V. Zyk, M. Schroder. 2001. *Coordination Chemistry Reviews*, 222, 155–192.
5. Binder, W., ed. 2007. *Hydrogen Bonded Polymers*; *Advances in Polymer Science 207*; Berlin Heidelberg: Springer-Verlag.
6. Yang, S. K., A. V. Ambade, M. Weck. 2011. *Chemical Society Reviews*, 40, 129–137.
7. Martina, M., D. W. Humacher. 2007. *Polymer International*, 56, 145–157.
8. Yoon, H.-J., W.-D. Jang. 2010. *Journal of Materials Chemistry*, 20, 211–222.
9. Goodbrand, H. B. 2003. Phase change ink composition. U.S. 02003/105185 A1.
10. Smith, T. W. 2003. Aqueous ink compositions. U.S. 2003/0079644 A1.
11. Vanmaele, L. 2003. Ink composition containing a particular type of dye, and corresponding ink jet printing process. EP1310533 A2.
12. Pappas, S. P. 2002. Imageable element and composition comprising thermally reversible polymers. WO02053626 A1.
13. Asawa, Y. 2002. Two-layer imageable element comprising thermally reversible polymers. WO02053627 A1.
14. Bosman, A. W., R. P. Sijbesma, E. W. Meijer. 2004. *Materials Today*, 7, 34–39.
15. Eling, B. 2002. Supramolecular polymer forming polymer. WO02046260 A1.
16. Mougin, N. 2002. Cosmetic composition forming after application a supramolecular polymer. WO02098377 A1.
17. Cooper, J. H. 2003. Cosmetic and personal care compositions. WO03032929 A2.
18. Goldoni, F. 2002. Laundry composition. WO02092744 A1.
19. Ciferri, A. *Supramolecular Polymers*, Boca Raton, FL: CRC Press, Taylor & Francis Group. 2005.
20. de Greef, T. F. A., M. M. J. Smulders, M. Wolffs, A. P. H. J. Schenning, R. P. Sijbesma, E. W. Meijer. 2009. *Chemical Reviews*, 109, 5687–5754.
21. Yang, H. H., ed. *Kevlar Aramid Fiber*. Chichester, England: Wiley & Sons, 1993.
22. Weaver, R. F. *Molecular Biology*, 4th ed.; New York: McGraw-Hill Science, 2007.
23. de Groot, F. M. H., G. Gottarelli, S. Masiero, G. Proni, G. P. Spada, N. Dolci. 1997. *Angewandte Chemie (International ed. in English)*, 36, 954–955.
24. Weck, M. 2007. *Polymer International*, 56, 453–460.
25. Hammond, M. R., R. Mezzenga. 2008. *Soft Matter*, 4, 952–961.

26. Rotello, V. M., S. Thayumanavan, eds. 2008. *Molecular Recognition and Polymers: Control of Polymer Structure and Self-Assembly*; Hoboken, NJ: John Wiley & Sons, Inc.
27. Shimizu, L. S. 2007. *Polymer International*, 56, 444–452.
28. Armstrong, G., M. Buggy. 2005. *Journal of Materials Science*, 40, 547–559.
29. ten Brinke, G., J. Ruokolainen, O. Ikkala. 2007. *Advances in Polymer Science*, 207, 113–177.
30. Harada, A., Y. Takashima, H. Yamaguchi. 2009. *Chemical Society Reviews*, 38, 875–882.
31. Neelakandan, P. P., Z. Pan, M. Hariharan, Y. Zheng, H. Weissman, B. Rybtchinski, F. D. Lewis. 2010. *Journal of the American Chemical Society*, 132, 15808–15813.
32. Akiba, I., H. Masunaga, K. Sasaki, Y. Jeong, K. Sakurai, S. Hara, K. Yamamoto. 2004. *Macromolecules*, 37, 1152–1155.
33. Burattini, S., H. M. Colquhoun, J. D. Fox, D. Friedmann, B. W. Greenland, P. J. F. Harris, W. Hayes, M. E. Mackay, S. J. Rowan. 2009. *Chemical Communications*, 6717–6719.
34. Sijbesma, R. P., F. H. Beijer, L. Brunsveld, B. J. B. Folmer, J. H. K. K. Hirschberg, R. F. M. Lange, J. K. L. Lowe, E. W. Meijer. 1997. *Science*, 278, 1601–1604.
35. Lehn, J.-M., M. Mascal, A. Decian, J. Fischer. 1992. *Journal of the Chemical Society, Perkin Transactions*, 2, 461–467.
36. Kotera, M., J. M. Lehn, J. P. Vigneron. 1995. *Tetrahedron*, 51, 1953–1972.
37. Lehn, J.-M. 1999. *Chemistry—A European Journal*, 5, 2455–2463.
38. Burattini, S., B. W. Greenland, W. Hayes, M. E. Mackay, S. J. Rowan, H. M. Colquhoun. 2011. *Chemistry of Materials*, 23, 6–8.
39. Burattini, S., H. M. Colquhoun, J. D. Fox, D. Friedmann, B. W. Greenland, P. J. F. Harris, W. Hayes, M. E. Mackay, S. J. Rowan. 2009. *Chemical Communications*, 6717–6719.
40. Harada, A., Y. Takashima, H. Yamaguchi. 2009. *Chemical Society Reviews*, 38, 875–882.
41. Cai, C., M. M. Bosch, B. Muller, Y. Tao, A. Kundig, C. Bosshard, Z. Gan, I. Biaggio, I. Liakatas, M. Jager, H. Schwer, P. Gunter. 1999. *Advanced Materials*, 11, 745–749.
42. Folmer, B. J. B., R. P. Sijbesma, R. M. Versteegen, J. A. J. van der Rijt, E. W. Meijer. 2000. *Advanced Materials*, 12, 874–878.
43. Berl, V., M. Schmutz, M. J. Krische, R. G. Khoury, J. M. Lehn. 2002. *Chemistry—A European Journal*, 8, 1227–1244.
44. Tournilhac, F., P. Cordier, D. Montarnal, C. Soulie-Ziakovic, L. Leibler. 2010. *Macromolecular Symposium*, 291, 84–85.
45. Tournilhac, F., P. Cordier, D. Montarnal, C. Soulie-Ziakovic, L. Leibler. 2008. *Nature*, 451, 977–980.
46. Folmer, B. J. B., R. P. Sijbesma, E. W. Meijer. 2001. *Journal of the American Chemical Society*, 123, 2093–2094.
47. Lillya, C. P., R. J. Baker, S. Hutte, H. H. Winter, Y.-G. Lin, J. Shi, L. C. Dickinson, J. C. W. Chien. 1992. *Macromolecules*, 25, 2076–2080.
48. Folmer, B. J. B., E. Cavini, R. P. Sijbesma, E. W. Meijer. 1998. *Chemical Communications*, 1847–1848.
49. Steinem, C., A. Janshoff, M. S. Vollmer, M. R. Ghadiri. 1999. *Langmuir*, 15, 3956–3964.
50. Rubio, M., D. Lopez. 2009. *European Polymer Journal*, 45, 3339–3346.
51. Sabadini, E., K. R. Francisco, L. Bouteiller. 2010. *Langmuir*, 26, 1482–1486.
52. Weng, W., Z. Li, A. M. Jamieson, S. J. Rowan. 2009. *Macromolecules*, 42, 236–246.
53. Loveless, D. M., S. L. Jeon, S. L. Craig. 2007. *Journal of Materials Chemistry*, 17, 56–61.
54. Sannigrahi, A., S. Ghosh, J. Lalnuntluanga, T. Jana. 2009. *Journal of Applied Polymer Science*, 111, 2194–2203.
55. Castellano, R. K., D. M. Rudkevich, J. Rebek, Jr. 1997. *Proceedings of the National Academy of Sciences of the United States of America*, 94, 7132–7137.
56. Castellano, R. K., C. Nuckolls, S. H. Eichhorn, M. R. Wood, A. J. Lovinger, J. Rebek, Jr. 1996. *Journal of the American Chemical Society*, 38, 2603–2606.

57. Castellano, R. K., J. Rebek, Jr. 1998. *Journal of the American Chemical Society*, 120, 3657–3663.
58. Castellano, R. K., R. Clark, S. L. Craig, C. Nuckolls, J. Rebek, Jr. 2000. *Proceedings of the National Academy of Sciences of the United States of America*, 97, 12418–12421.
59. Castellano, R. K., C. Nuckolls, S. H. Eichhorn, M. R. Wood, A. J. Lovinger, J. Rebek, Jr. 1999. *Angewandte Chemie (International ed. in English)*, 38, 2603–2606.
60. Folmer, B. J. B., E. Cavini, R. P. Sijbesma, E. W. Meijer. 1998. *Chemical Communications*, 1847–1848.
61. Yount, W. C., D. M. Loveless, S. L. Craig. 2005. *Angewandte Chemie (International ed. in English)*, 44, 2746–2748.
62. Yount, W. C., D. M. Loveless, S. L. Craig. 2005. *Journal of the American Chemical Society*, 127, 14488–14496.
63. Rowan, S. J., J. B. Beck. 2005. *Faraday Discuss*, 128, 43–53.
64. Stewart, S., C. T. Imrie. 1995. *Journal of Materials Chemistry*, 5, 223–228.
65. Muller, M., A. Dardin, U. Seidel, V. Balsamo, B. Ivan, H. W. Spiess, R. Stadler. 1996. *Macromolecules*, 29, 2577–2583.
66. Brovelli, S., F. Meinardi, G. Winroth, L. Zalewski, J. A. Levitt, F. Marinello, P. Schiavuta, K. Suhling, J. L. Anderson, F. Cacialli. 2010. *Advanced Functional Materials*, 20, 272–280.
67. Keizer, J. M., R. P. Sijbesma, J. F. G. A. Jansen, G. Pasternack, E. W. Meijer. 2003. *Macromolecules*, 36, 5602–5606.
68. Kato, T., N. Mizoshita, K. Kanie. 2001. *Macromolecular Rapid Communications*, 22, 797–814.
69. Suarez, M., J. M. Lehn, S. C. Zimmerman, A. Skoulios, B. Heinrich. 1998. *Journal of the American Chemical Society*, 120, 9526–9532.
70. Sangeetha, N. M., U. Maitra. 2005. *Chemical Society Reviews*, 34, 821–836.
71. George, M., R. G. Weiss. 2006. *Accounts of Chemical Research*, 39, 489–497.
72. Bouteiller, L. 2007. *Advances in Polymer Science*, 207, 79–112.
73. Isare, B., L. Petit, E. Bugnet, R. Vincent, L. Lapalu, P. Sautet, L. Bouteiller. 2009. *Langmuir*, 25, 8400–8403.
74. Neelakandan, P. P., Z. Pan, M. Hariharan, Y. Zheng, H. Weissman, B. Rybtchinski, F. D. Lewis. 2010. *Journal of the American Chemical Society*, 132, 15808–15813.
75. Bevers, S., T. P. O'Dea, L. W. McLaughlin. 1998. *Journal of the American Chemical Society*, 120, 11004–11005.
76. Abdalla, M. A., J. Bayer, J. O. Radler, K. Mullen. 2004. *Angewandte Chemie (International ed. in English)*, 43, 3967–3970.
77. Smulders, M. M. J., M. M. L. Nieuwenhuizen, T. F. A. de Greef, P. van der Schoot, A. P. H. J. Schenning, E. W. Meijer. 2010. *Chemistry—A European Journal*, 16, 362–367.
78. Dankers, P. Y. W., M. C. Harmsen, L. A. Brouwer, M. J. A. van Luyn, E. W. Meijer. 2005. *Nature Materials*, 4, 568–574.
79. Lindsey, J. S. 1991. *New Journal of Chemistry*, 15, 153–180.
80. Beijer, F. H., R. P. Sijbesma, H. Kooijman, A. L. Spek, E. W. Meijer. 1998. *Journal of the American Chemical Society*, 120, 6761–6769.
81. Folmer, B. J. B., E. Cavini, R. P. Sijbesma, E. W. Meijer. 1998. *Chemical Communications*, 1847–1848.
82. Lortie, F., S. Boileau, L. Bouteiller, C. Chassenieux, F. Laupretre. 2005. *Macromolecules*, 38, 5283–5287.
83. Beck, J. B., J. M. Ineman, S. J. Rowan. 2005. *Macromolecules*, 38, 5060–5068.
84. Schmatloch, S., A. M. J. van den Berg, A. S. Alexeev, H. Hofmeier, U. S. Schubert. 2003. *Macromolecules*, 36, 9943–9949.

# 13 Supramolecular Hydrogels for Soft Nanotechnology

*Hsin-Chieh Lin and Bing Xu*

## CONTENTS

## INTRODUCTION

In the last two and half decades, nanotechnology has grown explosively and evolved into a subfield of science. In the meantime, the nanoscale sizes and surface characters shared between physical science and biological science have led to the recognition that "future developments in nanoscience could provide the basis for artificial life"[1] because living cells consist of arrays of dynamically self-assembled nanostructures (e.g., enzymes, receptors, microtubules, ribosomes, or molecular motors) for carrying out metabolic processes, proliferation, differentiation, and other biological functions. These conceptual advancements have resulted in the development of soft nanotechnology, a branch of nanotechnology that focuses on the synthesis, properties, and functions of organic nanostructures for mimicking and interacting with naturally existing nanostructures. Because supramolecular interaction is one of the fundamental driving forces for the formation of cellular nanostructures[2] via self-assembly, it is reasonable to use it for generating organic nanostructures. Based on this simple logic, supramolecular hydrogels, consisting of water and molecular nanofibers formed by self-assembly, should be an ideal candidate for the development of soft nanotechnology. Therefore, in this chapter, we describe several representative cases of supramolecular hydrogels, especially their applications, to illustrate their promises for developing soft nanotechnology.

We arrange this chapter in the following way. First, we give the definition and characteristics of supramolecular hydrogels. Second, we present several representative examples of supramolecular hydrogels and their potential applications, hoping to establish a general sense regarding the impact of molecular structures in the design and functions of biomaterials based on supramolecular hydrogels. Third, we discuss the supramolecular hydrogels self-assembled from enzyme catalysis because enzymatic regulation promises a unique opportunity to integrate molecular self-assembly in water with natural biological processes. Fourth, we briefly describe supramolecular hydrogels for catalysis, which highlights their uniqueness for cascade catalytic reactions that may ultimately improve the efficiency and diversity for synthesis of new organic products. Finally, we offer our limited perspectives on the challenges and opportunities in supramolecular chemistry for the development of soft nanotechnology.

## DEFINITION AND CHARACTERISTICS OF SUPRAMOLECULAR HYDROGELS

Hydrogels, composed of three-dimensional, elastic networks whose interstitial spaces are filled with water, have received significant attention because of their broad applications in biomedicine. Conventional hydrogels use either biopolymers or synthetic polymers to cross-link into networks for immobilizing water (Scheme 13.1a). In the past two decades, partially because of the successful development of low–molecular mass organogelators,[3] small molecules have emerged as hydrogelators, which self-assemble in water to yield supramolecular chains that aggregate into nanofibers. These nanofibers further entangle into a three-dimensional network to imbibe water and to produce hydrogels (Scheme 13.1b). This gelation process is referred to as supramolecular hydrogelation,[4] the molecules as hydrogelators, and the resulting hydrogels as supramolecular hydrogels.

Compared with conventional polymers, hydrogelators possess several inherent advantages. First, hydrogelators offer a convenient system to achieve and study self-assembly of molecules in water. Second, unlike polymers, hydrogelators normally have well-defined configurations and persistent conformations. Third, being held together by noncovalent forces, the nanofibers of the hydrogelators are more biocompatible and biodegradable than those of polymers. Fourth, it is possible to convert clinically used therapeutics into hydrogelators to form hydrogels as "self-delivery" drugs. Therefore, the self-assembly of hydrogelators provides hierarchical

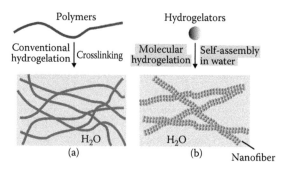

**SCHEME 13.1** Comparison of the formation of 3D networks of conventional polymeric hydrogelation (a) and molecular hydrogelation (b).

nanostructures as the matrices of a hydrogel, thus offering abundant opportunities for the development of soft nanotechnology. In the following section, we will start with selected examples of supramolecular hydrogels for biomedicine.

## SOME SUPRAMOLECULAR HYDROGELS AND APPLICATIONS

Because of their resemblance to extracellular matrices and the noncovalent supramolecular interactions among their components, supramolecular hydrogels have attracted considerable efforts that aim to design and synthesize novel hydrogelators as materials for biomedical applications. Among them, supramolecular hydrogels consisting of peptide nanostructures represent one of the most promising families of nanomaterials formed by molecular self-assembly. In general, hydrogel formation is a balance between crystallization and water solubilization. Noncovalent intermolecular interactions, including electrostatic interactions, hydrogen bonding, van der Waals forces, $\pi-\pi$ stacking, and other weak interactions, drive the formation of supramolecular hydrogels. Therefore, a common design of a supramolecular hydrogelator uses hydrophobic groups such as aliphatic chains or $\pi$-conjugated chromophores to facilitate aggregation in water and hydrophilic groups such as amide or acids to promote water solubility. In Figure 13.1, Stupp and coworkers have demonstrated peptide nanofibers that contain a high density of epitopes of laminin for regulating the differentiation of neural progenitor cells. The peptide derivative (**1**) contains an epitope sequence (IKVAV) and other residues ($A_4G_3$), followed by an alkyl tail of 16 carbons. The $A_4G_3$ sequence and alkyl components create a hydrophobic segment in **1** (Figure 13.1a). The addition of cell suspensions to the aqueous solution of **1** produces nanofibers (Figure 13.1b) with a high aspect ratio and large surface areas, and the entanglement of the nanofibers results in a supramolecular hydrogel.[5] These nanofibers, formed around cells, present the epitopes at an artificially high density relative to a natural extracellular matrix, thus leading to the growth of neurites instead of astrocytes. Although the up-regulation of laminin

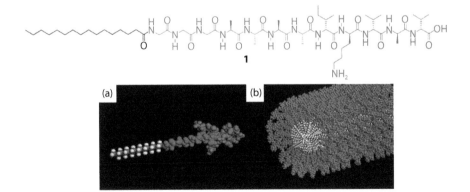

**FIGURE 13.1**  Molecular structure of the peptide amphiphile molecule **1**. (a) Graphic illustration of hydrogelator **1** and (b) self-assembly nanofiber. (From Silva, G. A. et al., *Science*, 303, 1352–1355, 2004.)

pathway remains to be confirmed, this work not only represents an important step in using supramolecular hydrogel as biomaterials, but also opens a new way to control cellular processes using supramolecular chemistry.

Besides the synthetic motifs for generating supramolecular nanofibers, the naturally existing protein motifs are also able to promote molecular self-assembly in water. Using the β-hairpin, the simplest motif that acts as a nucleation site for protein folding, Schneider and coworkers have devised an elegant way to use the β-hairpin motif as a trigger for supramolecular hydrogelation.[6] As shown in Figure 13.2, a 20-residue peptide (2) consisting of valine and lysine residues to flank a tetrapeptide segment (–V⁹PPT–) can adopt the β-hairpin secondary structure via a pH-promoted intramolecular folding. This design allows the self-assembly process to depend on its unimolecular folded state. Once the intramolecular folding occurs, the self-assembly of monomeric hairpins propagates laterally via hydrogen bonding between individual hairpins and associates facially via the hydrophobic, valine-rich region of folded peptides. Most interestingly, this unimolecular folding process is reversible after tuning the pH value. At a pH below the intrinsic $pK_a$ of the lysine side chains, the intrastrand charge repulsion unfolds individual hairpins, which ultimately provides a way for gating the hydrogel (Figure 13.2). Another important feature of this design is that the β-branched residue at the position of the turn (Val-9) enforces a *trans*-prolyl amide bond geometry at the next position, so that intramolecular folding of monomeric hairpins still holds in the self-assembly and hydrogel state. Thus, the design of intramolecularly folded peptides for self-assembly provides a useful strategy for developing new hydrogelators.

Besides amino acids, other biologically active molecules can also serve as building blocks for generating supramolecular hydrogels. Hamachi and coworkers reported a glycosylated amino acid derivative that acts as a hydrogelator (3) and took advantage of the resulting hydrogel as a protein microarray for the development of a rapid and

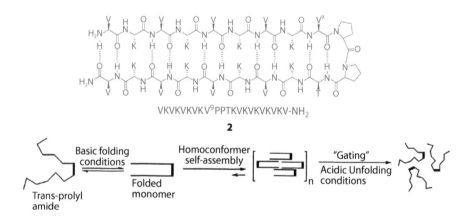

VKVKVKVKV⁹PPTKVKVKVKVKV-NH₂

**2**

**FIGURE 13.2**    Folding and self-assembly pathways of **2**. Valine (V9) enforces a *trans*-prolyl amide bond geometry favoring hairpin formation prior to the self-assembly. (Reprinted with permission from Schneider, J. P. et al., *Journal of the American Chemical Society*, 124, 15030–15037, 2002. Copyright 2002 American Chemical Society.)

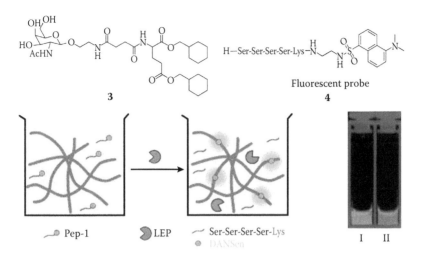

Fluorescent probe
4

3

Pep-1        LEP        Ser-Ser-Ser-Ser-Lys
                         DANSen                          I    II

**FIGURE 13.3** Chemical structures of **3** and **4** (upper). Illustration of an enzyme activity assay using the supramolecular hydrogel containing **4** and the direct observation of fluorescence change with the naked eye. (I) **4** after cleavage by LEP in the hydrogel; (II) **4** in the hydrogel (without LEP). (Reprinted by permission from Macmillan Publishers Ltd. Kiyonaka, S. et al., *Nature Materials*, 3, 58–64, Copyright 2004.)

high-throughput assay of enzymes (Figure 13.3).[7] Based on the observation that an environmentally sensitive fluorescent probe (DANSen, 5-dimethylaminonaphthalene-1-(N-2-aminoethyl)sulfonamide) mixed with the hydrogel produces stronger emissions compared with the same probe in an aqueous solution does, the researchers connected hydrophilic pentapeptide and lysine (Lys) and DANSen to produce **4** (Figure 13.3). When the enzyme lysyl-endopeptidase cleaves a peptide bond between Lys and DANSen, the resulting DANSen fragment moves to the hydrophobic domain of the nanofibers, thus inducing a significant increase of fluorescence intensity and an obvious color change (from a pinkish-yellow hydrogel into a light green; Figure 13.3). These results demonstrate that the hydrogel provides a unique medium for sensing enzymatic activity. Notably, the hydrogel is mechanically robust enough to be spotted on glass, which is the prerequisite for developing a protein microarray. Most importantly, the hydrogel prevents denature of proteins, thus allowing the realization of protein arrays in a relatively easy manner.

Dipeptide building blocks, the simplest peptide assemblies, also are able to self-assemble to form a supramolecular nanostructure. For example, Gazit et al. first reported the use of nanotubes of diphenylalanine, formed by diluting dipeptides from perfluoroalcohol into water, for casting metal nanowires. Xu and coworkers also found that Fmoc-protected dipeptides formed nanofibers as the matrices of supramolecular hydrogels. Shortly after that, Ulijn and coworkers demonstrated that the hydrogels of a series of Fmoc dipeptides (**5**), formed at physiological pH, were capable of supporting cell culture of chondrocytes in two and three dimensions (Figure 13.4).[8]

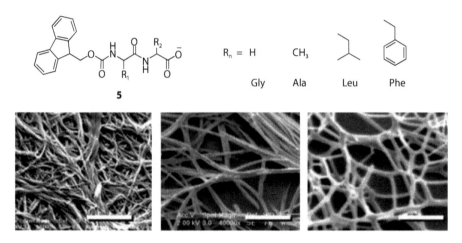

**FIGURE 13.4** The molecular structure of Fmoc dipeptides **5** and cryogenic scanning electron microscopy images of nanofibrous materials obtained by the self-assembly of Fmoc-Gly-Gly at pH 4, Fmoc-Ala-Ala at pH 4, and Fmoc-Phe-Phe at pH 7, respectively (left to right). Scale bar = 1 μm. (Jayawarna, V. et al.: *Advanced Materials*. 2006. 18. 611–614. Copyright Wiley-VCH Verlag GmbH & Co. KGaA. Reproduced with permission.)

Because amino acids or glycosamine have served as building blocks for hydrogelators, it is reasonable to integrate amino acids and glycosamine in a hydrogelator. Based on the unique wound-healing[9] activity of D-glucosamine and other low–molecular weight gelators made of carbohydrates,[7] Xu and coworkers linked Nap-L-phenylalanine or Nap- D-phenylalanine with D-glucosamine and transformed the D-glucosamine into molecular hydrogelators.[10] With better biocompatibility, the one linked with Nap-D-phenylalanine (**6**) (Figure 13.5a) promoted a much faster wound healing and resulted in smaller scars than that of the control (Figure 13.5c and d). This result further supports the notion that molecular hydrogels could confer beneficial effects for the biomedical applications that require high local concentrations of the therapeutic agents or bioactive epitopes.[5]

Hydrogelators may stem from some less expected building blocks. For example, vancomycin, as a broad-spectrum glycopeptidic antibiotic against Gram-positive

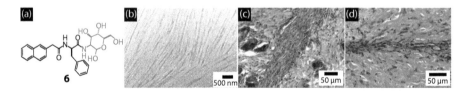

**FIGURE 13.5** (a) The molecular structure of Nap- D-phenylalanine-D-glucosamine **6**. (b) TEM image of the cryo-dried hydrogel of **6**. (c, d) Histological cross-section images of the dorsal skins of Balb/C mice on day 6 after wounding. (c) Negative control and (d) gel of **6** treated immediately after the incision was made. (Yang, Z. M. et al., *Chemical Communications*, 43, 843–845, 2007. Reproduced by permission of The Royal Society of Chemistry.)

bacteria, represents one of the most important antibiotics because of its ability to inhibit methicillin-resistant *Staphylococcus aureus*. Although it would be unlikely to attract interest as a building block for hydrogelators, Xu and coworkers modified vancomycin by introducing a pyrene group to the C-terminal of the backbone of vancomycin and serendipitously found vancomycin-pyrene (**7**) as a hydrogelator, which formed a hydrogel (Figure 13.6a) with minimum gelation concentration at 0.36 wt% (2.2 mM). According to spectroscopic analysis,[11] the π–π interaction between pyrenes and the hydrogen bonding between vancomycins drive the self-assembly of **7** in water (Figure 13.6b). Additional transmission electron micrographs (TEM) studies (Figure 13.6c) suggest that the molecules of **7** self-assemble to produce

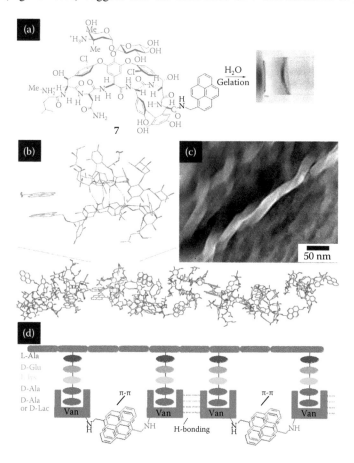

**FIGURE 13.6**   (a) Molecular structure of vancomycin-pyrene (**7**) with the optical image of the hydrogel at 0.36 wt%; (b) involved noncovalent interactions for the self-assembly of **7**; (c) TEM image of the helical nanofibers of **7**. (d) Illustration of plausible multivalent interactions between vancomycin and terminal peptides on the cell surface of VRE (e.g., VanA and VanB strains). (Reprinted with permission from Xing, B. G. et al., *Journal of the American Chemical Society*, 124, 14846–14847, 2002. Copyright 2002 American Chemical Society; Xing, B. G. et al., *Chemical Communications*, 39, 2224–2225, 2003. Reproduced by permission of The Royal Society of Chemistry.)

supramolecular polymers that aggregate further into nanofibers as the matrices of the hydrogel. Most structural modifications usually reduce the activities of a commercial drug; on the contrary, vancomycin-pyrene exhibited an unusual potency (nearly three orders of magnitude more potent than vancomycin) against vancomycin-resistant enteroccoci.[11] Further investigations indicate that the enhanced efficacy might have resulted from the self-assembly of vancomycin-pyrene on the surface of the bacteria cells.[12] If this proposed mechanism is validated, the correlation of supramolecular chemistry on a surface and in a solution might provide a useful guide and a starting point for designing drugs based on self-assembled multivalency.

## ENZYME-INSTRUCTED SELF-ASSEMBLY TO FORM SUPRAMOLECULAR HYDROGELS

Usually, the formation of a supramolecular hydrogel requires the transformation of a solution into a hydrogel on a physical or chemical perturbation (e.g., change of pH, temperature, or ionic strength).[13–18] Mimicking nature, in which enzymatic reactions govern self-assembly processes, enzymatic hydrogelation becomes a useful process for making supramolecular nanofibers/hydrogels.[19–25] This seemingly simple strategy allows the seamless integration of the event of self-assembly and its obvious assay (i.e., hydrogelation) with a wide range of biological processes that associate with enzymes, thus providing a powerful way to explore the promise of supramolecular hydrogels and a simple assay to investigate or model the self-assembly regulated by an enzyme. In this section, we discuss several examples to illustrate the concept and potential applications of enzyme-instructed molecular self-assembly which contains three basic steps (Figure 13.7).[21] The enzyme converts a precursor into a hydrogelator (normally via bond cleavage); the hydrogelator then self-assembles, usually forming nanofibers, and the nanofibers then entangle to serve as a matrix for the hydrogel.[19] Besides bond cleavage, enzyme-catalyzed bond formation[20] can also generate hydrogels.

Based on the catalytic dephosphorylation reaction of a simple derivative of tyrosine phosphate, Xu and coworkers used alkaline phosphatase for enzymatic supramolecular hydrogelation.[19] Their ease of accessibility and high activity render phosphatases an excellent choice for enzymes to control the self-assembly and hydrogelation of small molecules. As shown in Figure 13.8, alkali phosphatase

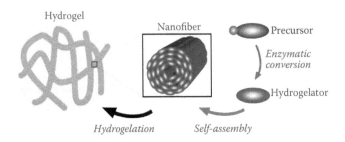

**FIGURE 13.7**  The essential steps in the enzymatic hydrogelation of small molecules. (Reprinted with permission from Yang, Z. et al., *Accounts of Chemical Research*, 41, 315–326, 2008. Copyright 2008 American Chemical Society.)

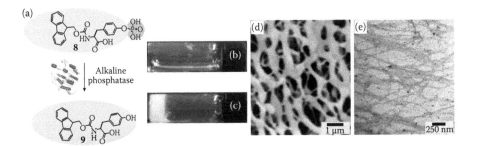

**FIGURE 13.8**    (a) Chemical structure of the precursor (**8**), its corresponding hydrogelator (**9**), and the schematic gelation process; (b) optical images of the solution of **8** in alkali phosphate buffer (pH = 9.8); (c) hydrogel of **9** formed by adding alkali phosphatase to the solution of **8**; (d) SEM image of the hydrogel of **9**; (e) TEM image of the hydrogel of **9**. (Yang, Z. M., et al.: *Advanced Materials*. 2004. 16. 1440–1444. Copyright Wiley-VCH Verlag GmbH & Co. KGaA. Reproduced with permission.)

turns a readily available precursor **8** (Figure 13.8a) into a more hydrophobic compound **9** after removing the hydrophilic phosphate group, and **9** forms a supramolecular hydrogel (Figure 13.8c), the matrices of which are made of the nanofibers of **9** (Figure 13.8d and e). Because enzymatic hydrogelation normally takes place at physiological conditions, it is particularly suitable for exploring applications in biology and biomedicine.

Unavoidably, more than one enzyme exists in the biological environment *in vivo*. Therefore, it would be helpful to understand the biological regulation involved in using a pair of enzymes to control the self-assembly of small molecules because it is quite common for pairs of enzymes to work counteractively to regulate biological functioning in nature.[2] The phosphatase/kinase pair, as a critical enzyme switch, regulates many signal transductions in cells. Kinases catalyze phosphorylation, and phosphatases catalyze dephosphorylation.[2] Thus, Xu et al. designed a small molecule that can undergo phosphorylation and dephosphorylation catalyzed by a kinase and a phosphatase.[22] As shown in Figure 13.9, the hydrogelator (**11**) self-assembles into nanofibers and forms a supramolecular hydrogel in an aqueous solution. As a substrate of tyrosine kinase, **11** undergoes phosphorylation catalyzed by tyrosine kinase in the presence of Adenosine-5′-triphosphate (ATP) to yield a more hydrophilic compound (**10**), which leads to the gel–sol transition. Upon the addition of the alkali phosphatase, the dephosphorylation of **10** restores the hydrogel (Figure 13.9c). Moreover, the subcutaneous injection of **10** in mice led to the formation of the supramolecular hydrogel *in vivo* (Figure 13.9d). This work validates the combination of the enzymatic switch with a small molecular hydrogelator as a new way to make and apply biomaterials for therapeutic interventions because many diseases are associated with the abnormal activity of kinases and phosphatases, or other types of enzymatic switches.

Xu et al. also explored the use of enzymatic hydrogelation as an ultra–low cost, easy assay to report the presence of β-lactamase.[23] As shown in Figure 13.10, consisting of a cephem nucleus as the linker connecting a hydrophilic group and a hydrogelator, the precursor (**12**) only forms a solution with water. Upon the action of a

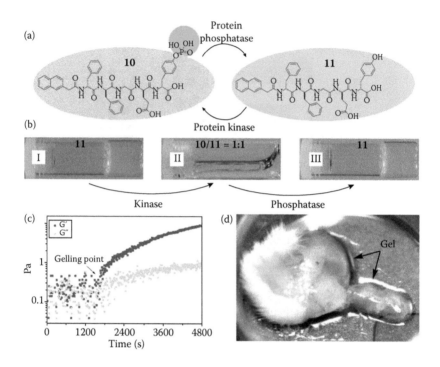

**FIGURE 13.9** (a) Chemical structures of the precursor (**10**) and its corresponding hydrogela-tor (**11**) and the diagram of a phosphatase/kinase enzyme switch. (b) Optical images of the hydrogel (panel I) formed by **11** in a HEPES buffer by adjusting the pH; the solution (panel II) obtained by treating the hydrogel with a kinase (at 50% conversion); and the hydrogel (panel III) restored by adding phosphatase into the solution. (c) Dynamic time sweep of a solution containing **10** and **11** after the addition of the phosphatase. The arrow indicates the apparent gelation point. (d) An optical image of a hydrogel formed subcutaneously 1 h after injecting **11** into the mice. (Reprinted with permission from Yang, Z. M. et al., *Journal of the American Chemical Society*, 128, 3038–3043, 2006. Copyright 2006 American Chemical Society.)

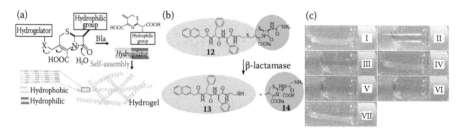

**FIGURE 13.10** (a) Illustration of the design of using β-lactamase to form supramolecular hydrogels. (b) Chemical structures of the compounds involved in the β-lactamase catalyzed hydrogelation. (c) Results of adding different types of cell lysates into the solution of **12** (the final concentration of **12** is 0.35 wt%, pH = 8.0): (I) *E Coli* with C600, (II) *E Coli* with CTX-M13, (III) *E Coli* with CTX-M14, (IV) *E Coli* with JP995, (V) *E Coli* with SHV-1, (VI) *E Coli* with TEM-1, and (VII) water. (Reprinted with permission from Yang, Z. M. et al., *Journal of the American Chemical Society*, 129, 266–267, 2007. Copyright 2007 American Chemical Society.)

β-lactamase, the β-lactam ring opens to release the hydrogelator (**13**), which self-assembles into nanofibers to produce a hydrogel. Because β-lactamase in a bacterial lysate converts the precursor (**12**) to its corresponding hydrogelator (**13**), resulting in the formation of a supramolecular hydrogel (II, III, V, and VI in Figure 13.10c), this facile process could detect β-lactamase in the lysates of bacteria.

Unlike phosphatases, which break a covalent bond to trigger hydrogelation, other enzymes catalyze the formation of a covalent bond between two compounds to form a supramolecular hydrogelator, which self-assembles into a three-dimensional fibril network to support the hydrogel. For example, Ulijn et al. have used thermolysin[20] to catalyze the reverse hydrolysis of hydrophobic amino acids and peptides, which greatly expand the scope of enzymatic hydrogelation.[21]

Enzyme catalysis also leads to the formation of supramolecular nanofibers within a cell, an unprecedented process. An endogenous enzyme should convert a soluble precursor, which does not necessarily self-assemble and gel extracellularly, into a hydrogelator that self-assembles into nanofibers or some other ordered nanostructure. To meet this requirement, Xu et al. designed and synthesized **15** as an esterase substrate (Figure 13.11).[24] Mammalian cells uptake **15** by diffusion, their endogenous esterases convert **15** to **16**, and the molecules of **16** self-assemble to form nanofibers, resulting in a supramolecular hydrogel inside the cells. The gelation changes the viscosity of the cytoplasm and causes cell death. Interestingly, at a certain concentration of the precursor, most HeLa cells died at day 3 after the addition of **15** to the culture medium, whereas most of the NIH3T3 cells remained alive and dividing at the same concentration of **15** (Figure 13.12). Although other factors (e.g., differences in the uptake of **15** by HeLa and NIH3T3 cells) might also contribute to the apparent low toxicity of **15** for NIH3T3 cells, this result indicates that the kinetic formation of the intercellular nanostructure can be specific to different types of cells. Whether intracellular hydrogelation leads to apoptosis or necrosis, or both, has yet be determined, but this approach clearly offers a new way to control the fate of cells.

To further confirm that supramolecular nanofibers inside cells can control their fate, Xu et al. designed the precursors for enzymatic hydrogelation inside bacteria. Two types of *Escherichia coli* strains were used in the experiments: the wild-type

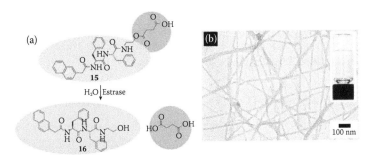

**FIGURE 13.11** (a) Illustration of using an esterase to convert the precursor molecule (**15**) and the hydrogelator (**16**), thus resulting in supramolecular hydrogelation. (b) TEM of the hydrogel (inset: optical image). (Yang, Z. M. et al.: *Advanced Materials*. 2007. 17. 3152–3156. Copyright Wiley-VCH Verlag GmbH & Co. KGaA. Reproduced with permission.)

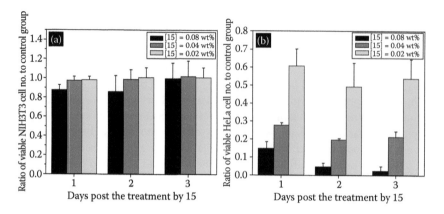

**FIGURE 13.12**   MTT assays of (a) NIH3T3 cells and (b) HeLa cells treated with **15** at the concentrations of 0.08 wt%, 0.04 wt%, and 0.02 wt%. (Yang, Z. M. et al.: *Advanced Materials*. 2007. 17. 3152–3156. Copyright Wiley-VCH Verlag GmbH & Co. KGaA. Reproduced with permission.)

BL21 (as the control) and the BL21 strain (BL21(P+)) bred to overexpress a human tyrosine phosphatase. The only difference between the two strains was their phosphatase expression. Thus, the same types of cell walls and cell membranes should eliminate any discrepancy in the uptake of the precursor. After the diffusion of the precursor (**17**) into *E. coli*, the phosphatase converted **17** into **18**, which resulted in the formation of three-dimensional networks of nanofibers and triggered supramolecular hydrogelation (Figure 13.13a). The BL21(P+) bacteria stopped growing after the addition of **17** ($IC_{50}$ = 20 µg/mL), but the wild-type BL21 bacteria grew normally ($IC_{50}$ > 2000 µg/mL).[25] High pressure liquid chromatography (HPLC) showed that **18** accumulated significantly in BL21(P+) (Figure 13.13b). After the bacteria had been isolated and broken apart by sonication, hydrogelation could be observed only in the sample prepared from BL21(P+) and treated with **17** (Figure 13.13c). The formation of nanofibers in the hydrogels was also evident in TEM images (Figure 13.13d). These results confirm that this enzyme-instructed formation of supramolecular nanofibers and hydrogels inside the bacteria inhibits bacterial growth. The principle and the strategy demonstrated in that work could lead to a new paradigm for developing therapeutic agents that take advantage of the kinetics of enzymatic reactions rather than a tight ligand-receptor binding.[26]

## SUPRAMOLECULAR HYDROGELS FOR ENCAPSULATION AND MIMICKING OF ENZYMES

Enzymes are a class of proteins that exhibit high efficiency and specificity for catalyzing a myriad of reactions in complex cellular fluids.[27,28] Several features of enzymes are particularly worthy to note: (i) enzymes perform catalysis by binding the substrates to the active sites mainly via noncovalent interactions, (ii) hydrophobic residues usually surround active sites despite enzymes catalyzing reactions in aqueous environments in most living organisms,[29] (iii) the path of the substrates entering usually differs from the path of the products leaving the active sites,[30] and (iv) enzymes

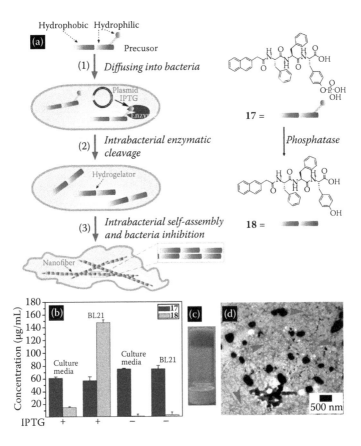

**FIGURE 13.13** (a) Schematic representation of intracellular nanofiber formation leading to hydrogelation and the inhibition of bacterial growth, the chemical structures, and graphic representations of the precursor (**17**) and the corresponding hydrogelator (**18**). (b) Concentrations of **17** and **18** in the culture medium and within the cells (BL21, IPTG+: overexpression of phosphatases, or IPTG−: normal expression of phosphatases). (c) Optical and (d) TEM images of the hydrogel formed inside the bacteria after culturing with **17** for 24 h (arrows indicate the nanofibers formed by **18**). The high electron density areas (dark black areas) likely are intracellular inclusions.[27] (Yang, Z. M. et al.: *Angewandte Chemie*. 2007. 46. 8216–8219. Copyright Wiley-VCH Verlag GmbH & Co. KGaA. Reproduced with permission.)

can alter their conformations to achieve maximal activities.[31] These features highlight the pivotal importance of supramolecular interactions in enzymes and supports the use of small peptide-based molecular hydrogels for mimicking and stabilizing enzymes. Because small peptide-based molecular hydrogels,[5,6,8,19,32–36] consisting of large amounts of water and nanofiber matrices formed via the self-assembly of the amphiphilic, small peptides, not only rely on supramolecular interaction to form ordered nanostructures (i.e., nanofibers),[4,21] but also provide the foundation to establish supramolecular interactions by chemical design. Therefore, small peptides in the molecular hydrogels can mimic nature's remarkable ability by providing the backbone of active sites and an easily tunable biological microenvironment

(a)                    (b)                        (c)

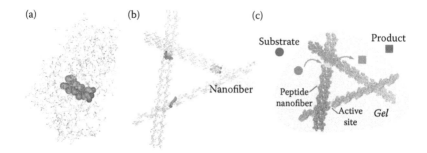

FIGURE 13.14   Similarity between (a) an enzyme (horseradish peroxidase) and (b) peptide nanofibers doped with active centers that are shown in CPK model. (c) Catalysis in gels. (Gao, Y. et al., *Chemical Society Reviews*, 2010, 39, 3425–3433. Reproduced by permission of The Royal Society of Chemistry.)

(Figure 13.14).[37–41] In the following section, we discuss the use of molecular hydrogels to improve the quantum yield of bioluminescence in aqueous phase and the use of nanofibers for enzyme mimicking.

The peptide nanofiber-based molecular hydrogels not only can serve as a medium to immobilize enzyme for improving catalysis in an organic solvent,[42] but also offer unique advantages for enhancing catalytic reactions in water, as recently demonstrated by the case of bioluminescence.[37] As shown in Figure 13.15a, simply mixing sodium carbonate, Fmoc-(Nε)-L-lysine (**19**), and Fmoc-L-phenylalanine (**20**) in water creates a suspension that turns into a clear solution on heating. The addition of luminol (**21**) and methemoglobin (Hb, **22a**) or horseradish peroxidase (HRP, **22b**) followed by cooling to room temperature yields the hydrogel (denoted as $Gel_{1+2}[\mathbf{21, 22}]$) containing the components for chemiluminescence, thus offering a system mimicking the bioluminescent environment. TEM of $Gel_{1+2}[\mathbf{21, 22a}]$ (Figure 13.15b) and $Gel_{1+2}$ reveal the width of the nanofibers to be approximately 16 nm (Figure 13.15c), indicating the encapsulation of **21** and **22a** in the nanofibers. The chemical luminescence (CL) spectra of $Gel_{1+2}[\mathbf{21, 22a}]$, $Gel_{1+2}[\mathbf{21, 22b}]$, the free **21** and **22a** solution (denoted as [**21, 22a**]) indicate that [**21, 22a**] had a CL quantum yield of 1.02%, agreeing well with the values reported in the literature. The CL quantum yields in $Gel_{1+2}[\mathbf{21, 22a}]$ and $Gel_{1+2}[\mathbf{21, 22b}]$ were 11.22% and 10.13%, respectively. These results imply that peptide nanofibers are able to mimic the luciferase to insulate the excited state products from the outside environment and to reduce the nonradiative decay, thus achieving a high quantum yield for artificial bioluminescence.

It is well known that the spatial arrangement of atoms or groups near the active center affect the activities of enzymes.[43] The two main advantages of native enzymes are the site-isolation of the prosthetic groups and the concentrating effect of reactants within enzyme nanodomains. Similar to the polypeptide as the backbone of active sites in natural enzymes, the self-assembled nanofibers of the derivatives of amino acids in molecular hydrogels should allow the incorporation of the model compounds of active centers to mimic an enzyme. Based on this simple notion, Xu and Chang made an artificial enzyme by mixing a hemin chloride (**25**) or a heme model (**25′** or **25″**) into the supramolecular hydrogel formed by the self-assembly of two amino

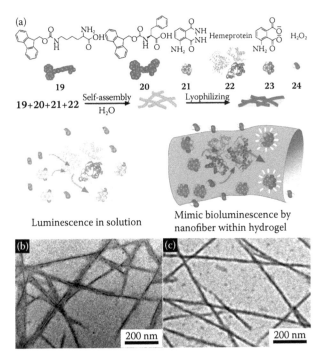

**FIGURE 13.15**  (a) Preparation of the molecular hydrogel and illustration of the mimicking bioluminescence by the molecular hydrogels. Typical TEM images of (b) Gel$_{1+2}$[**21, 22a**] and (c) the control Gel$_{1+2}$. (Wang, Q. G. et al.: *Chemistry—A European Journal*. 2009. 15. 3168–3172. Copyright Wiley-VCH Verlag GmbH & Co. KGaA. Reproduced with permission.)

acids.[38,39] The self-assembled network of the amino acid nanofibers acts as the protein-like structure and hemin as the active sites, thus the hydrogels serve at least two functions—as the skeletons of the artificial enzyme to aid the function of the active site in organic solvents and as immobilization carriers to facilitate their application.

Figure 13.16 illustrates the simple procedure that uses the peptide nanofibers of supramolecular hydrogels to encapsulate hemin. The TEM and atomic force microscopy (AFM) images confirmed a network structure (Figure 13.17), and the hemin molecules on the gray part around the nanofibers. Furthermore, UV/Vis spectroscopy indicated that the localized enzyme was monomeric hemin chloride. Being encapsulated in the supramolecular hydrogel to mimic peroxidase and to catalyze oxidation in organic media, hemin chloride (**25**) achieves catalytic activity at approximately 60% of the nascent activity of HRP, which is 10 times higher than that of free hemin or hemins entrapped by β-cyclodextrin in an aqueous buffer.[44] Instead of hemin chloride (**25**), the heme model compounds (**25′** or **25″**) encapsulated in the supramolecular hydrogel afford a new type of artificial peroxidase that exhibits higher catalytic activity in organic media, reaching approximately 90% of the nascent activity of HRP. These works[38,39] suggest that the use of supramolecular nanofibers as the structural component of artificial enzymes could provide a new and useful approach to the development of biomimetic catalysts, one of the Holy Grails in chemistry.[45–47]

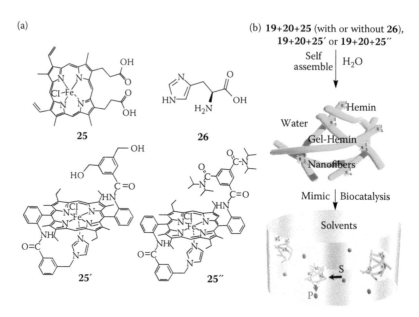

**FIGURE 13.16** (a) Structure of the molecular components. (b) Procedure for mimicking peroxidase by the molecular hydrogels containing hemin chlorides and heme models. (Wang, Q. G. et al.: *Chemistry—A European Journal.* 2008. 14. 5073–5078. Copyright Wiley-VCH Verlag GmbH & Co. KGaA. Reproduced with permission; Wang, Q. G. et al.: *Angewandte Chemie.* 2007. 46. 4285–4289. Copyright Wiley-VCH Verlag GmbH & Co. KGaA. Reproduced with permission.)

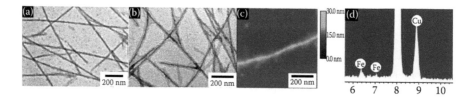

**FIGURE 13.17** TEM images of (a) the control, (b) the molecular hydrogel containing **25**, (c) the AFM image of the gel containing **25**, and (d) energy-dispersive X-ray spectroscopy (EDS) of the gray part in (b). (Wang, Q. G. et al.: *Angewandte Chemie.* 2007. 46. 4285–4289. Copyright Wiley-VCH Verlag GmbH & Co. KGaA. Reproduced with permission.)

## PERSPECTIVES AND CHALLENGES

After five decades of growth, supramolecular chemistry, which focuses on the weak and reversible noncovalent interactions between molecules, has become the foundation for many subfields of science, including soft nanotechnology. Therefore, supramolecular hydrogels not only provide a novel type of biomaterials, but also offer a simple model system to examine the use of noncovalent interactions for constructing nanoscale assemblies. Particularly, the enzyme-catalyzed formation of hydrogels of small molecules has become an effective approach for exploring a range of potential biomedical applications as diverse as screening enzyme inhibitors, assisting biomineralization,[48] typing bacteria, developing smart drug delivery systems, and controlling the fate of cells.[21] These progresses underscore the tremendous opportunities offered by using enzymes to regulate supramolecular interactions. Before realizing the full potential of supramolecular gels in soft nanotechnology, many challenges need to be addressed. For example, we can easily use molecular self-assembly in water to generate secondary structures (e.g., nanofibers, nanotubes, and nanospheres), but there is no satisfactory method to rapidly and accurately analyze the molecular arrangements in those nanostructures. Our understanding of cell compatibility in supramolecular nanofibers is still insufficient. Despite progress on the use of supramolecular nanofibers to mimic the extracellular matrix, the fate of supramolecular nanofibers inside cells remains unknown. Although we have recently been able to create supramolecular nanostructures inside cells, we are still far away from having reasonable control on the specificity and functions of those nanostructures. Clearly, future development of supramolecular chemistry in soft nanotechnology requires scientists from different disciplines to work together to address fundamental problems and develop practical applications. For example, tasks such as correlating the structures of the molecules and the molecular arrangement in the nanostructures and developing materials to solve real-world problems based on the molecular arrangement in the nanostructures.

## REFERENCES

1. Mann, S. 2008. *Angewandte Chemie (International ed. in English)*, 47, 5306–5320.
2. Lodish, H., A. Berk, P. Matsudaira et al. 2003. *Molecular Cell Biology*, 5th ed., New York: Freeman.
3. Terech, P., R. G. Weiss. 1997. *Chemical Reviews*, 97, 3133–3159.
4. Estroff, L. A., A. D. Hamilton. 2004. *Chemical Reviews*, 104, 1201–1217.
5. Silva, G. A., C. Czeisler, K. L. Niece, E. Beniash, D. A. Harrington, J. A. Kessler, S. I. Stupp. 2004. *Science*, 303, 1352–1355.
6. Schneider, J. P., D. J. Pochan, B. Ozbas, K. Rajagopal, L. Pakstis, J. Kretsinger. 2002. *Journal of the American Chemical Society*, 124, 15030–15037.
7. Kiyonaka, S., K. Sada, I. Yoshimura, S. Shinkai, N. Kato, I. Hamachi. 2004. *Nature Materials*, 3, 58–64.
8. Jayawarna, V., M. Ali, T. A. Jowitt, A. E. Miller, A. Saiani, J. E. Gough, R. V. Ulijn. 2006. *Advanced Materials*, 18, 611–614.
9. Shaunak, S., S. Thomas, E. Gianasi, A. Godwin, E. Jones, I. Teo, K. Mireskandari, P. Luthert, R. Duncan, S. Patterson, P. Khaw, S. Brocchini. 2004. *Nature Biotechnology*, 22, 977–984.
10. Yang, Z. M., G. L. Liang, M. L. Ma, A. S. Abbah, W. W. Lu, B. Xu. 2007. *Chemical Communications*, 843–845.
11. Xing, B. G., C. W. Yu, K. H. Chow, P. L. Ho, D. G. Fu, B. Xu. 2002. *Journal of the American Chemical Society*, 124, 14846–14847.
12. Xing, B. G., P. L. Ho, C. W. Yu, K. H. Chow, H. W. Gu, B. Xu. 2003. *Chemical Communications*, 2224–2225.
13. Timpl, R., J. C. Brown. 1996. *Bioessays*, 18, 123–132.
14. Weiss, R. G., P. Terech. 2006. *Molecular Gels: Materials with Self-Assembled Fibrillar Networks*. Dordrecht: Springer.

15. Hwang, I., W. S. Jeon, H.-J. Kim, D. Kim, H. Kim, N. Selvapalam, N. Fujita, S. Shinkai, K. Kim. 2007. *Angewandte Chemie (International ed. in English)*, 46, 210–213.
16. Menger, F. M., K. L. Caran. 2000. *Journal of the American Chemical Society*, 122, 11679–11691.
17. de Loos, M., A. Friggeri, J. van Esch, R. M. Kellogg, B. L. Feringa. 2005. *Organic & Biomolecular Chemistry*, 3, 1631–1639.
18. Zhang, Y., Z. M. Yang, F. Yuan, H. W. Gu, P. Gao, B. Xu. 2004. *Journal of the American Chemical Society*, 126, 15028–15029.
19. Yang, Z. M., H. W. Gu, D. G. Fu, P. Gao, K. J. K. Lam, B. Xu. 2004. *Advanced Materials*, 16, 1440–1444.
20. Toledano, S., R. J. Williams, V. Jayawarna, R. V. Ulijn. 2006. *Journal of the American Chemical Society*, 128, 1070–1071.
21. Yang, Z., G. Liang, B. Xu. 2008. *Accounts of Chemical Research*, 41, 315–326.
22. Yang, Z. M., G. L. Liang, L. Wang, B. Xu. 2006. *Journal of the American Chemical Society*, 128, 3038–3043.
23. Yang, Z. M., P. L. Ho, G. L. Liang, K. H. Chow, Q. G. Wang, Y. Cao, Z. H. Guo, B. Xu. 2007. *Journal of the American Chemical Society*, 129, 266–267.
24. Yang, Z. M., K. M. Xu, Z. F. Guo, Z. H. Guo, B. Xu. 2007. *Advanced Materials*, 17, 3152–3156.
25. Yang, Z. M., G. L. Liang, Z. F. Guo, Z. H. Guo, B. Xu. 2007. *Angewandte Chemie (International ed. in English)*, 46, 8216–8219.
26. Hardman, J. G., L. E. Limbird, P. B. Molinoff, R. W. Ruddon. 1995. *The Pharmacological Basis of Therapeutics*, 9th ed., New York: McGraw-Hill.
27. Bubel, A., C. Fitzsimons. 1989. *Microstructure and Function of Cells: Electron Micrographs of Cell Ultrastructure*, 1st ed., New York: Ellis Horwood Limited.
28. Walsh, C. 2001. *Nature*, 409, 226–231.
29. Zhou, G. W., J. C. Guo, W. Huang, R. J. Fletterick, T. S. Scanlan. 1994. *Science*, 265, 1059–1064.
30. Que, L., W. B. Tolman. 2008. *Nature*, 455, 333–340.
31. Henzler-Wildman, K., D. Kern. 2007. *Nature*, 450, 964–972.
32. Hartgerink, J. D., E. Beniash, S. I. Stupp. 2001. *Science*, 294, 1684–1688.
33. Zhang, S., T. Holmes, C. Lockshin, A. Rich. 1993. *Proceedings of the National Academy of Sciences of the United States of America*, 90, 3334–3338.
34. Holmes, T. C., S. de Lacalle, X. Su, G. S. Liu, A. Rich, S. G. Zhang. 2000. *Proceedings of the National Academy of Sciences of the United States of America*, 97, 6728–6733.
35. Haines-Butterick, L., K. Rajagopal, M. Branco, D. Salick, R. Rughani, M. Pilarz, M. S. Lamm, D. J. Pochan, J. P. Schneider. 2007. *Proceedings of the National Academy of Sciences of the United States of America*, 104, 7791–7796.
36. Gao, Y., Y. Kuang, Z. F. Guo, Z. H. Guo, I. J. Krauss, B. Xu. 2009. *Journal of the American Chemical Society*, 131, 13576–13577.
37. Wang, Q. G., L. H. Li, B. Xu. 2009. *Chemistry—A European Journal*, 15, 3168–3172.
38. Wang, Q. G., Z. M. Yang, M. L. Ma, C. K. Chang, B. Xu. 2008. *Chemistry—A European Journal*, 14, 5073–5078.
39. Wang, Q. G., Z. M. Yang, X. Q. Zhang, X. D. Xiao, C. K. Chang, B. Xu. 2007. *Angewandte Chemie (International ed. in English)*, 46, 4285–4289.
40. Wang, Q. G., Z. M. Yang, L. Wang, M. L. Ma, B. Xu. 2007. *Chemical Communications*, 1032–1034.
41. Wang, Q. G., Z. M. Yang, Y. Gao, W. W. Ge, L. Wang, B. Xu. 2008. *Soft Matter*, 4, 550–553.
42. Gao, Y., F. Zhao, Q. Wang, Y. Zhang, B. Xu. 2010. *Chemical Society Reviews*, 39, 3425–3433.
43. Chang, C. K., T. G. Traylor. 1973. *Proceedings of the National Academy of Sciences of the United States of America*, 70, 2647–2650.

44. Huang, Y., W. Ma, J. Li, M. Cheng, J. Zhao, L. Wan, J. C. Yu. 2003. *Journal of Physical Chemistry B*, 107, 9409–9414.
45. Breslow, R. 1995. *Accounts of Chemical Research*, 28, 146–153.
46. Breslow, R., S. D. Dong. 1998. *Chemical Reviews*, 98, 1997–2011.
47. Wolfe, J., A. Muehldorf, J. Rebek. 1991. *Journal of the American Chemical Society*, 113, 1453–1454.
48. Schnepp, Z. A. C., R. Gonzalez-McQuire, S. Mann. 2006. *Advanced Materials*, 18, 1869–1872.

# 14 Supramolecular Drug-Delivery Systems

*Andreas Mohr and Rainer Haag*

## CONTENTS

## SUPRAMOLECULAR DRUG-DELIVERY SYSTEMS BASED ON LINEAR AMPHIPHILES

Today, a vast number of low-molecular weight drug molecules (typically <500 g/mol) are commonly used as therapeutic agents. Most new drugs, however, are poorly soluble in water. As a consequence, relatively small amounts of the drug reach the target site. Thus far, numerous supramolecular drug-delivery systems are under investigation to circumvent these limitations and improve the potential of the respective drug.

Generally, these nanosized drug-delivery systems can be categorized into two distinct classes, that is, supramolecular drug-delivery systems and drug-polymer conjugates. This book chapter focuses on various types of supramolecular drug delivery and drug targeting systems such as nonionic polymeric materials, smart polymers, and dendritic polymers.

### BLOCK COPOLYMER MICELLES

In the past, nonionic polymeric materials have been extensively used as effective drug delivery and drug targeting systems. Until now, poor aqueous solubility has posed a serious problem in the formulation of hydrophobic drug molecules because the availability of biocompatible and biodegradable nanocarriers is especially important for clinical application of therapeutic compounds. Current approaches toward drug delivery and drug targeting systems involve the use of nonionic polymeric materials such as Cremophor, Tween, and Pluronics, which have been extensively used to formulate hydrophobic drug molecules (Figure 14.1).[1,2]

Pluronics are commercially available triblock (ABA-type) copolymers composed of alternating polyethylene oxide (PEO) and polypropylene oxide building blocks. Because of their amphiphilic structure, Pluronics have attracted particular interest in recent years, as they tend to form thermodynamically stable micelles in aqueous solution. The utility of polymeric micelles for delivery of therapeutic agents results from their particular core-shell architecture with a well-defined hydrophobic core and hydrophilic shell. The hydrophobic block generally makes up the inner core of the micelle and can accommodate poorly water-soluble drugs or dyes, whereas the hydrophilic block, usually polyethylene glycol (PEG), forms the corona. Polymeric micelles serve as appropriate reservoirs for hydrophobic dyes and drug molecules. Some typical examples of drugs loaded into polymeric Pluronic micelles are doxorubicin, cisplatin, carboplatin, epirubicin, and haloperidol.[3–5] Cremophor EL is obtained as a heterogeneous nonionic surfactant by the reaction of castor oil with ethylene oxide at a molar ratio of 1:35. Because of its surfactant properties, Cremophor EL has been widely used as a carrier of water-insoluble drug molecules including anaesthetics (Propofol), photosensitizers (C8KC), sedatives (Diazepam), antineoplastics (Aplidine), and immunosuppressive agents (Cyclosporin A) as well as antitumor agents (Taxol).[6–11] However, Cremophor EL is known to produce allergic reactions at the site of injection. Polysorbate-80, commercially known as Tween-80, is a viscous, water-soluble yellow liquid derived from polyethoxylated sorbitan and oleic acid. Tween-80 is structurally similar to the (polyethylene) glycols and used both in injections (0.8–8.0%) and in oral suspension (0.375% w/v). A number of anticancer drugs can be formulated by Tween-80. Typical examples include etoposide and minor groove-binding cyclopropylpyrroloindole analogues like carzelesin. However, substantial evidence has been generated in recent years that suggests

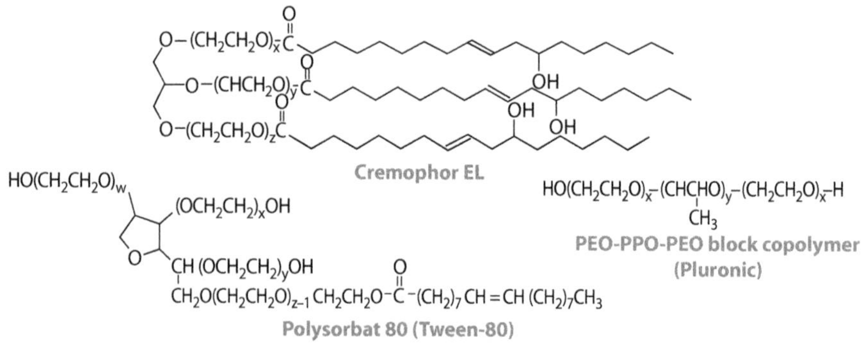

**FIGURE 14.1** Chemical structures of amphiphilic PEO-*b*-PPO-*b*-PEO block copolymers (Pluronics), polyoxyethylenesorbitan 20 monooleate (Tween-80) where $w + x + y + z = 20$, and polyoxyethylene-glycerol 35 triricinoleate (Cremophor EL) where $x + y + z = 35$.

Tween-80 may be a cause of hypersensitivity reactions using drugs with Tween-80 as a solvent.[12]

Polymeric micelles form spontaneously by self-assembly in water when the concentration of the copolymer is above the critical micellar concentration (CMC).[1,2] Compared with low–molecular weight surfactants, some of the major advantages of polymeric micelles are their high solubilization capacity without the need for additional solvents and low CMC values indicating their thermodynamic stability.[13] The CMC is the most important parameter and primarily depends on polymer architecture, the solvent, pH, presence of added salts, and temperature. Based on the functionality (polarity) of the individual building blocks, polymeric micelles can be divided into two main categories: hydrophobically assembled micelles[2,13] and polyion micelles.[2,14] In the former case, micellization is driven by hydrophobic interactions. The copolymer consists of a water-soluble hydrophilic part in combination with a hydrophobic segment group, which, for example, includes biodegradable polyesters such as poly(lactic acid), poly(e-caprolactone), and poly(glycolic acid). On the contrary, ionic interactions lead to the formation of polyion micelles. For instance, a positively charged poly(aspartate) block (PAsp), poly(ethyleneimine) (PEI), or poly(L-lysine) is complexed to a negatively charged biopolymer such as DNA.[14]

The size of block-copolymer micelles is determined by thermodynamic parameters, but partial control over the size is possible by varying the block length of the polymer. Typically, block-copolymer micelles are 10 to 50 nm in diameter with a relatively narrow distribution and are therefore similar in size to viruses, lipoproteins, and other naturally occurring transport systems.[2,15] A prolonged circulation time is usually achieved by using a relatively long PEG chain along with a sufficiently low critical micelle concentration. Compared with low-molecular-weight surfactants, block copolymers have a lower CMC. Therefore, polymeric micelles are generally more stable than micelles of small surfactant molecules and can retain the loaded drug for a longer time. On the other hand, polymeric micelles can be unstable under shear stress or at high dilutions because the formation of micelles is a thermodynamic phenomenon.

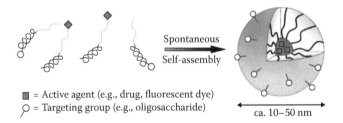

**FIGURE 14.2** Micellization of amphiphilic block-copolymers in water. The characteristic features are a pronounced core–shell architecture which can be controlled by the individual polymer blocks. Typical examples of block copolymers are PEO-*b*-PPO, PEO-*b*-PCl, and PEO-*b*-PAsp. (From Haag, R., *Angewandte Chemie (International ed. in English)*, 43, 278–282, 2004. With permission; Haag, R., and Kratz, F., *Angewandte Chemie (International ed. in English)*, 45, 1198–1215, 2006. With permission.)

As shown in Figure 14.2, a particular feature of a block copolymer micelle is the core-shell architecture. The inner core can be used to physically encapsulate or covalently attach active compounds, whereas terminal functionalities on the hydrophilic outer shell control biocompatibility and may incorporate potential targeting properties. The micellar corona often consists of a polar PEO layer and sufficiently protects the hydrophobic core from the aqueous bulk phase. More importantly, it has been demonstrated that PEO prevents the adsorption of proteins and hence forms a biocompatible polymeric shell.[16,17]

## DRUG-POLYMER CONJUGATES WITH CLEAVABLE LINKERS

The coupling of low–molecular weight anticancer drugs to polymers through a cleavable linker has been a convenient method for enhancing the therapeutic effect.[15] A representative example is a pH-sensitive, doxorubicin-conjugated micellar nano-carrier based on amphiphilic poly(ethylene glycol)–poly(aspartate-hydrazone-Adriamycin) block copolymer (Figure 14.3).[18] In contrast to drug–polymer conjugates, in which antitumor agents are covalently attached to a single macromolecule chain, doxorubicin was coupled with an acid-labile hydrazone linker to the end of the aspartate side chains of the polymer. The acid sensitivity of the micelles was evaluated by reversed-phase liquid chromatography. At pH < 6, the doxorubicin-conjugated micelles exhibited a significant pH-responsive drug release profile, thus showing the effective cleavage of the hydrazone bonds at acidic pH. On the contrary, the release of doxorubicin was negligible under physiological conditions (pH 7.4), clearly showing the potential of the polymer for controlled release. It has long been known that cancer or inflammation makes the extracellular pH at the disease site acidic because of the overproduction of acidic metabolites. Similarly, the endosomal and lysosomal compartments of cells are slightly more acidic than blood and normal tissues.[19,20]

In recent years, particular attention has been given to the formulation of poorly soluble natural products as novel anticancer agents. Paclitaxel (Taxol) is a diterpenoid obtained from the bark of the pacific yew tree, *Taxus brevifolia*,[21] and is one of the most promising chemotherapeutic agents acclaimed by the National Cancer

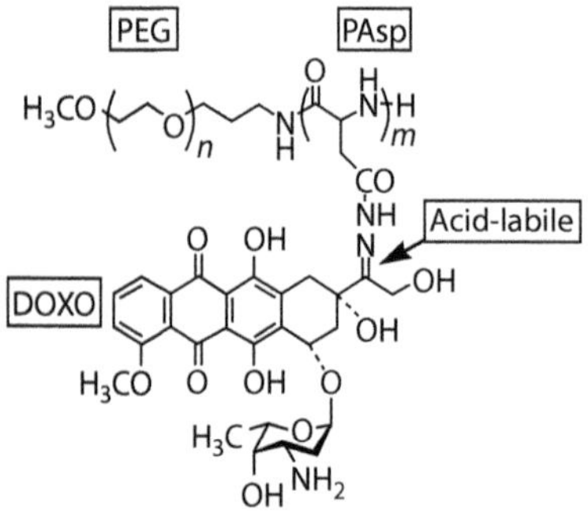

**FIGURE 14.3** A doxorubicin block-copolymer conjugate which self-assembles to pH-sensitive polymeric micelles in water. The acid-labile hydrazone bond is cleaved at pH < 6 and doxorubicin is released.

Institute.[22] Because of the low aqueous solubility of paclitaxel (0.3 µg/mL), numerous attempts are being made to increase the solubility of the drug by using various formulations and prodrug conjugates.[23–25] Among the latter, $N$-(2-hydroxypropyl)methacrylamide (HPMA) copolymer prodrugs have been the most promising because of an increased aqueous solubility, enhanced pharmacological activity, lowered systemic toxicity, and prolonged blood circulation and release of the active compound in target cells. HPMA copolymers are known to accumulate selectively in tumors.[26–28] Furthermore, a supramolecular aggregate of an amphiphilic polyglycerol has been reported to transport Taxol selectively into tumor tissue *in vivo* (see next section).[29]

## SMART POLYMERIC MATERIALS

Drug–polymer conjugates should be sufficiently stable in the bloodstream before the drug is released at the site of action. This is particularly important if the polymeric architectures have high surface charge, molecular weight, and a tendency to interact with biomacromolecules in blood.[30] However, polymers which respond to external stimuli such as temperature, pH, light, electric field, chemicals, and ionic strength are of great interest due to the wide range of applications.[31] Thus, current intensive research is focused on (i) responsive biointerfaces that are functionally similar to natural surfaces,[31,32] (ii) coatings that are capable of interacting with and responding to their (micro)environment,[31,33–35] (iii) composite materials that actuate and mimic the action of muscles,[31,36] (iv) thin films and particles that are capable of sensing very small concentrations of analytes,[31,37,38] and finally, (v) smart polymers for responsive drug-delivery systems.[31,39–41] In recent years, a significant amount of research work has been focused on poly($N$-isopropylacrylamide) (PNIPAM)[42,43] and its copolymers, because these materials display an entropy-driven conformational change from a water-soluble coil to a hydrophobic globule at 32°C. The coil-to-globule transition leads to an abrupt change in the physical properties of the polymer and takes place at the so-called lower critical solution temperature (LCST). PNIPAM is soluble in water below its LCST, as hydrogen-bonding interaction between amide groups of the polymer chain and water molecules is favorable. The polymer, however, becomes insoluble as the temperature rises above the LCST. As a result, the PNIPAM block becomes hydrophobic and phase separation takes place. The LCST of PNIPAM is very close to human body temperature, 37°C, and can be tuned by either changing the molecular weight or incorporating hydrophilic or hydrophobic comonomers, which, respectively, increase or decrease the LCST of the polymer. Because the temperature-dependent coil-to-globule transformation is a reversible process, PNIPAM-based materials show potential applications in drug delivery, bioengineering, and biotechnology.[42,43]

You and Oupický[44] and Heath et al.[45] prepared a Y-shaped block copolymer consisting of one block of PEG and one block of end-functionalized PNIPAM with lysine as the focal point. The terminus of the PNIPAM block is functionalized with biotin, and the specific role of this type of copolymer is to turn on and to turn off the biotin signal through temperature dependent coil-to-globule transition (Figure 14.4). Above the LCST, the polymer adopts a micelle-like structure, where the biotin ligands become hidden in the hydrophobic core. Disassembly, however, is observed

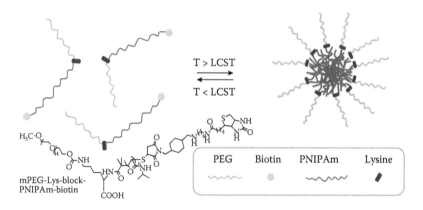

**FIGURE 14.4** Heterobifunctional mPEG-Lys-block-PNIPAm-biotin block copolymers associate upon heating above the LCST with PNIPAM chains collapsing into a globular dense core. The biotin ligands in the center of the micelles are protected from the aqueous bulk solution and binding of the biotin moiety to avidin is considerably inhibited. (From You, Y., and Oupický, D., *Biomacromolecules*, 8, 98–105, 2007. With permission; Heath, F. et al., *The AAPS Journal*, 9, 235–240, 2007. With permission.)

on cooling below the LCST. As a result, the biotin ligands are free and peptides can specifically bind to the biotin ligands. In this context, it is worthwhile to mention that the high affinity of avidin for biotin (vitamin H) is of considerable practical interest, especially for biomedical and diagnostic applications. Avidin is a positively charged glycoprotein (66 kDa) found in raw egg whites. It binds up to four molecules of biotin simultaneously with a high degree of affinity ($K_a = 10^{15}$ M$^{-1}$).[46]

Gene therapy aims to treat diseases by intracellular delivery of nucleic acids. For an efficient DNA/RNA delivery, suitable transfection agents are required. Gene transfection agents are usually classified into viral and nonviral carriers including liposome-forming cationic lipids,[47,48] polyethylenimines,[49] chitosan,[50] and synthetic cationic polymers.[51] The search for nonviral alternatives remains a challenge because of the problems associated with viral gene transfection such as undesired immune response and limited selectivity. Since the late 1990s, cationic functionalized PNIPAM polymers have been receiving increasing attention because of their hydrophilic-to-hydrophobic switching properties in aqueous media. Complexation of DNA molecules with, for example, cationic copolymers is based on the interaction between the globule (chain-collapsed conformation) that is formed above the LCST and negatively charged phosphate groups of the DNA backbone. Because of their compact, neutral structure, the resulting copolymer–DNA complexes can be easily taken up into the cell by endocytosis. The ability to efficiently condense DNA to small polyplexes was demonstrated by the initial work of Hinrichs et al.,[51] who synthesized a broad variety of cationic copolymers of 2-(dimethylamino) ethyl methacrylate and *N*-isopropylacryl amide. They found that above the LCST of the polymers, transfection was most effective for 200-nm-sized polyplexes.

In the last two decades, there has been much effort to develop stimuli-responsive polymers as potential drug-delivery systems. Thus far, polymers that respond to only a single stimulus have been studied extensively, whereas only a few investigations have focused on multisensitive compounds. Polymers that respond to multiple stimuli can indeed offer further control over phase behavior in aqueous solution. Arotçaréna et al.[52] synthesized a switchable diblock copolymer based on *N*-isopropylacrylamide (NIPA) in combination with the zwitterionic comonomer 3-[*N*-(3-methacrylamidopropyl)-*N*,*N*-dimethyl]-ammoniopropane sulfonate (SPP; Figure 14.5). A particular feature of this type of copolymer is the combination of two hydrophilic comonomers that exhibit double thermo-responsive behavior in water. Similar to neat PNIPAM, the poly-NIPA building block becomes insoluble in water as the temperature rises above the LCST. Interestingly, no phase separation was observed at high solution temperatures. Instead, the diblock copolymer forms micelle-like aggregates with nonionic poly-NIPAM chains collapsing into a globular dense core. Similar to classical block copolymer micelles, the hydrophilic corona consists of polar SPP chains and sufficiently protects the hydrophobic core from the aqueous bulk solution. At intermediate temperatures, however, both building blocks become water-soluble and no micelles exist. Because the zwitterionic poly-SPP block exhibits an upper critical solution temperature (UCST), further cooling resulted in polar aggregates, that is, reverse micelles that are kept in solution by the polymeric NIPA chains. Thus, the colloidal aggregates switch reversibly in water simply by altering the temperature below and above the UCST and LCST, respectively.[52] Depending on the solution temperature, the micellar assemblies provide nanosized cavities of different polarities, which can serve as potential reservoirs for polar and nonpolar drugs. Interestingly, the lengths of the individual building blocks were adjusted in such a way that the colloidal aggregates could stay in solution in the full temperature range between 0°C and 100°C.

A particular feature of the polysulfobetaine block is the UCST, as has been observed in water due to an abrupt change in the related physical properties on micelle formation. Below the UCST, in the absence of electrolyte, attractive (electrostatic) interactions between the cationic amino groups and the negatively charged sulfonate groups prevent the dissolution of polymer in water. Upon heating, however,

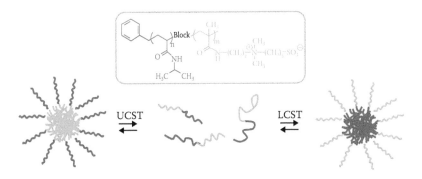

**FIGURE 14.5** Double thermoresponsivity of a water soluble diblock copolymer based on the two thermally-sensitive building blocks *N*-isopropylacrylamide (NIPA) and 3-[*N*-(3-methacrylamido-propyl)-*N*,*N*-dimethyl]ammoniopropane sulfonate (SPP). (From Arotçaréna, M. et al., *Journal of the American Chemical Society*, 124, 3787–3793, 2002. With permission.)

the thermal energy increases to break the relatively weak interchain interactions between the oppositely charged functional groups. As a result, the micelles disassemble into single polymer chains when the temperature rises above the UCST (Figure 14.5).

Apart from solution temperature, the aqueous phase behavior of such smart polymers can also be tuned by pH, the ionic strength, or the electric field. Bütün et al.[53] and Liu et al.[54] were the first to report on water-soluble AB-type diblock copolymers which self-assemble into schizophrenic micelles. The term *schizophrenic* acknowledges the ability of the polymers to reversibly form micellar assemblies with non-solvated cores comprising either the A block or the B block (Figure 14.6). In one of the reported works,[53] 2-(*N*-morpholino)ethyl methacrylate (MEMA) was copolymerized with 2-(diethylamino)ethyl methacrylate (DEAMA) using group transfer polymerization. Depending on pH solution and salt concentration, the MEMA-DEAMA diblock copolymer formed two types of switchable micelles. However, the block symmetry was the decisive factor for the formation of stable aggregates. Thus, the long-term stability of the MEMA core micelles was considerably decreased when the asymmetric 73:27 MEMA-DEAMA diblock copolymer was used. Here, the DEAMA block was relatively short. Accordingly, the micellar core was insufficiently protected by the thin poly-DEAMA corona. In another approach,[54] a symmetrical zwitterionic poly(4-vinylbenzoic acid-block-2-(diethylamino)ethyl methacrylate) (VBA$_{60}$-*b*-DEA$_{66}$) block copolymer was prepared by atom transfer radical polymerization. Contrary to the MEMA-DEAMA system described previously, the two micellar transitions of this type of copolymer were solely induced by changes of the pH (Figure 14.6). On the other hand, the same copolymer exhibited an isoelectric point at neutral pH. Thus, the VBA$_{60}$-*b*-DEA$_{66}$ diblock copolymer dissolved in either dilute acidic (pH 2) or dilute alkaline solution (pH 10), whereas precipitation was clearly observed in the pH range of 6.6–8.6. This behavior could be understood on the basis that the cationic charge on the DEA block just balanced the anionic charge on the VBA block. As a result, the VBA-*b*-DEA diblock copolymer became more hydrophobic and therefore phase separated from the aqueous bulk solution. It

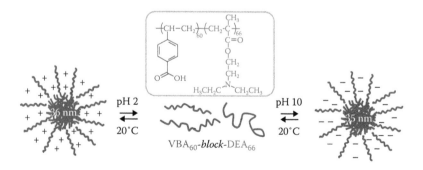

**FIGURE 14.6** Schematic representation of the precipitation, redissolution, and the pH-induced schizophrenic micellization behavior of the VBA$_{60}$-*b*-DEA$_{66}$ diblock copolymer in aqueous solution. (From Liu, S., and Armes, S. P., *Angewandte Chemie (International ed. in English)*, 41, 1413–1416, 2002. With permission.)

is well known that hydrophobically modified polymers can associate with each other to minimize the exposure of hydrophobic groups in water.[55]

## PROTEIN DRUG DELIVERY AND SUPRAMOLECULAR PROTEIN ASSEMBLIES

Protein drug delivery has also become an area of high interest in biomedical research.[56,57] Soon after the beginning of the biotechnology boom in the late 1990s, a number of therapeutic peptides and proteins were developed against a broad range of diseases including diabetes,[58–61] cancer,[62] rheumatoid arthritis,[63] multiple sclerosis, and neurodegenerative disorders such as Alzheimer disease and Parkinson disease.[64] Proteins are high–molecular weight macromolecules composed of unbranched polypeptide chains of different sizes. Peptide chains consist of a number of amino acid units covalently held together by peptide bonds. Each protein has a specific amino acid sequence also known as a primary structure. The function of a protein, however, is directly dependent on its three-dimensional structure. Proteins participate in a wide range of physiological processes[65] such as transport (hemoglobin) and storage (ferritin) of endogenous and exogenous molecules, maintenance of colloidal blood pressure and blood pH (serum albumin), transmission of nerve impulses, and control of cell growth as well as differentiation of cells. In this context, therapeutic peptides and proteins have become popular in recent years as a platform for drug targeting and delivery.[66]

Diabetes is a chronic metabolic disease caused by low levels of insulin in the blood or a decreased ability to use insulin.[59,60] Like many hormones, insulin is a protein secreted by the β-cells of the islets within the pancreas. Insulin controls the transport and storage of glucose in the body and allows the sugar to enter cells and be converted to energy. When insulin concentration is too low, however, glucose levels in the blood increase dangerously, thus leading to diabetes mellitus. Because of the very short half-life of insulin in the body, treatment of diabetes typically requires multiple daily injections of the hormone. For more than 80 years, this has been the most practical route and mimics the best secretion of insulin by a healthy pancreas.[59] To considerably decrease the frequency of injections, many research groups have been using biodegradable polymers as sustained injectable release systems for insulin.[67–70]

Supramolecular hydrogels based on block copolymers are the most promising candidates in the development of protein delivery systems. The polymeric network prevents irreversible protein denaturation during application, that is, the proteins retain their active forms inside the gel. Barichello et al.[67] used Pluronic F127 gels, polylactic-co-glycolic acid (PLGA) nanoparticles and their combination for parenteral delivery of insulin. The *in vitro* release profiles were evaluated under physiological conditions using a membraneless dissolution model. The results showed that the higher the concentration of the poloxamer in the gel, the slower the release of insulin, independent of the formulation used. Controlled release of insulin from the commercially known hydrogel, ReGel (MacroMed Inc., Sandy, Utah), was demonstrated by Kim et al.[68] and Choi and Kim.[69] ReGel is a thermally gelling, biodegradable triblock copolymer of poly(lactic-co-glycolic-acid)-*b*-poly(ethylene-glycol)-*b*-poly-lactic-co-glycolic acid (PLGA-*b*-PEG-*b*-PLGA) (Figure 14.7). At room temperature, ReGel behaves like a liquid and thus can be administered via a syringe. At body

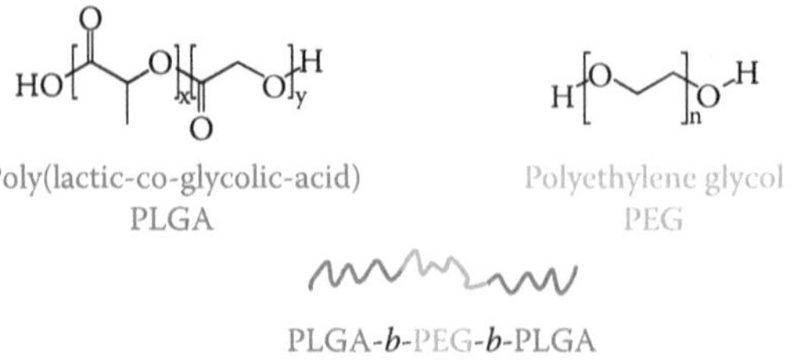

Poly(lactic-co-glycolic-acid)
PLGA

Polyethylene glycol
PEG

PLGA-*b*-PEG-*b*-PLGA

**FIGURE 14.7** Chemical structures of poly(lactic-co-glycolic-acid (PLGA), poly(ethylene-glycol) (PEG), and the biodegradable triblock copolymer of poly(lactic-co-glycolic-acid)-*b*-poly(ethylene-glycol)-*b*-poly(lactic-co-glycolic acid (PLGA-*b*-PEG-*b*-PLGA). (From Zentner, G. M. et al., *Journal of Controlled Release*, 72, 203–215, 2001. With permission.)

temperature, however, ReGel becomes a gel and can be used as an injectable depot for sustained release of insulin. Quite interestingly, the gelation temperature of the copolymer can be easily controlled by the PEG/PLGA and lactic/glycolic acid ratio as well as the molecular weight.[71]

Recently, Abu Hashim et al.[70] reported on the potential use of γ-cyclodextrin polypseudorotaxane (γ-CyD PPRX) hydrogels as an injectable, sustained release system for insulin. The supramolecular hydrogels were prepared in aqueous solution by simply mixing insulin, PEG (20.000), and γ-CyD. Keeping the mixture in a refrigerator at 4°C for 12 hours yielded the hydrogel as a viscous, polymeric network. Hydrogel formation, however, was based on physical cross-linking induced by the insertion of two PEG chains in the less polar γ-CyD cavity. In this way, insulin could be physically entrapped in the three-dimensional, hydrophilic network. After subcutaneous administration of insulin-loaded γ-CyD PPRX hydrogel to rats, the serum insulin level was significantly prolonged. As shown by high-performance

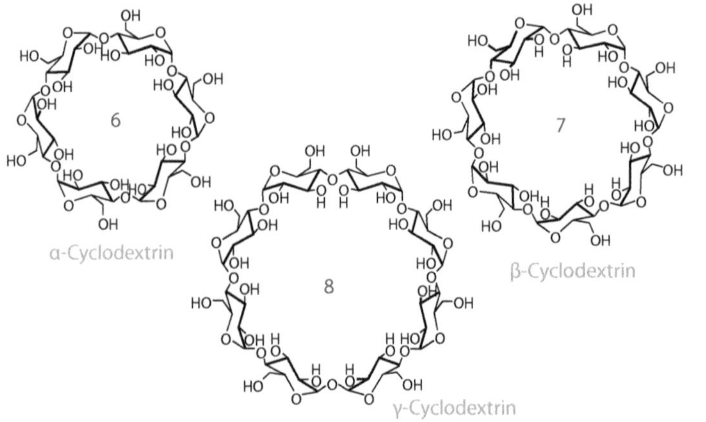

α-Cyclodextrin

β-Cyclodextrin

γ-Cyclodextrin

**FIGURE 14.8** Chemical structures of α-, β-, and γ-cyclodextrins (CDs). CDs are natural-based cyclic oligosaccaharides obtained by enzymatic conversion of starch. (From Szejtli, J., *Chemical Reviews*, 98, 1743–1754, 1998. With permission.)

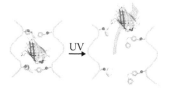

**FIGURE 14.9** Schematic drawing of photoresponsive protein release from the photo-responsive hydrogel composed of *trans* AB–Dex and CD–Dex. (From Peng, K. et al., *Chemical Communications*, 46, 4094–4096, 2010. With permission.)

liquid chromatography, the hydrogel significantly sustained the serum insulin level and gradually recovered to the basal level after 6 hours. When insulin was injected alone, the insulin level decreased to the basal level within 2 hours. Interestingly, the release rate of the hormone from the hydrogels could be controlled by adjusting the concentration of cyclodectrin in the polymeric network. Figure 14.8 shows the chemical structures of a-, b-, and g-cyclodextrins (CDs).

As mentioned previously, stimuli-responsive materials are very attractive for new applications in pharmaceutical formulations and are of great interest in biomedical research. Most commonly, temperature, pH, light, electric field, and chemicals are used as external stimuli. Light is of particular interest in this context because many proteins are usually not harmed when exposed to light. Furthermore, light can be controlled with temporal and spatial precision.[73] Peng et al.[74] obtained a dextran-based photoresponsive hydrogel system for the light-controlled release of green fluorescent proteins. As shown in Figure 14.9, the hydrogel is based on the inclusion complex of *trans* azobenzene and β-cyclodextrin and consists of azobenzene functionalized dextran (AB-Dex) and β-cyclodextrin modified dextran (CD-Dex). Upon exposure to UV light, the azobenzene moieties isomerize from *trans* to *cis* configurations. As a result, the polymeric gel network dissociates and allows the entrapped protein to migrate into the bulk solution.

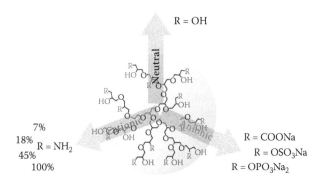

**FIGURE 14.10** Structures of (ionic) dendritic polyglycerol derivatives, wherein R = OH, $NH_2$, $OPO_3Na_2$, and $OSO_3Na$. The depicted polymer structure represents only one possible isomer and a small part of the PG scaffold. (From Khandare, J. et al., *Biomaterials*, 31, 4268–4277, 2010. With permission.)

## SUPRAMOLECULAR NANOCARRIERS BASED ON
## HYPERBRANCHED AND DENDRITIC POLYGLYCEROLS

Dendritic polymers are remarkable for their structural diversity and their wide range of applications in materials science, catalysis, and biomedical research.[75–79] Dendrimers have been known for almost 30 years and are extremely well-defined, globular, synthetic polymers with a number of characteristics that make them useful in biological systems. They are of particular interest because their regular, compartmentalized, and well-defined unimolecular architecture allows chemical modification at either the core or the shell. So far, only two types, polyamidoamine and poly(propylene imine), have been commercialized, because of their tedious, stepwise, and time-consuming synthesis.

Hyperbranched and dendritic polyglycerols (dPGs)[80–82] are currently attracting considerable interest as so-called dendritic nanocarriers for application in drug solubilization and delivery.[83] dPGs are structurally defined, consist of an aliphatic polyether backbone, and possess multiple functional end groups (Figure 14.10).[30,83,84] Several attempts have been made to mimic specific proteins, for example, histones or polysaccharides such as heparin.[15,84,85] Sulfated polyglycerols and sulfonates, for example, have been explored for inflammatory diseases as selectin inhibitors and indicators. The amine-terminated scaffolds could act instead like histones to electrostatically bind and compact DNA.[15,86,87] In this context, it is worthwhile to mention that dendritic polyamines and derivatives are clearly multivalent in gene complexation.

The size of dPGs can be defined precisely between 5 and 20 nm.[2,15] The encapsulation of guest molecules is driven by noncovalent interactions, including ionic, hydrogen bonding, and van der Waals interactions, and can be simultaneously tailored for various drugs. However, drug–polymer conjugates, in which the drugs are covalently attached to the polymer, have to be synthesized individually.

PG-based architectures demonstrate optimal biocompatibility on the cellular and systemic levels. They have similar toxicological properties as PEG and both are classified as highly nontoxic for *in vivo* applications.[83] Thus, the development of new generation PG architectures with an improved biocompatibility and designed drug release profile will further enhance our fundamental understanding of such systems and could potentially lead to new drug-delivery systems.

This chapter reviews, in particular, recent progress in research on chemical and physical properties of dendritic core-(multi)shell architectures accomplished by the Haag group. In the past few years, emphasis has been given to versatile architectures, which are all based on hyperbranched or dPGs and classified according to the chemical nature of the building units. Herein, we give a brief overview of dendritic core-(multi)shell architectures and nonionic dendritic amphiphiles.

### WATER-SOLUBLE POLYGLYCEROL DENDRITIC CORE-SHELL ARCHITECTURES

Polymeric micelles can be unstable under shear stress or at high dilutions because the formation of micelles is a thermodynamic phenomenon.[55] Indeed, a low critical micelle concentration prevents the drug-loaded micelle from dissociation on dilution in the bloodstream. Dendritic polymers can be a solution for these problems because star polymers are generally considered unimolecular polymeric micelles. Kurniasih

**TABLE 14.1**

**Solubilization of Pyrene, Nile Red, and Nimodipine in 0.1 wt.% Polymer Solution**

| 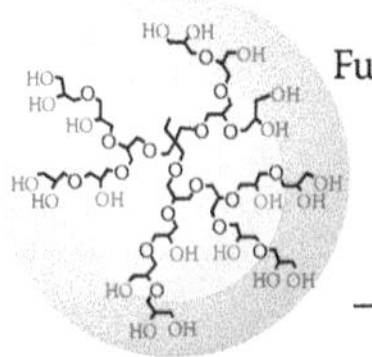 | Degree of Functionalization | Solubility of Pyrene  | | Solubility of Nile Red 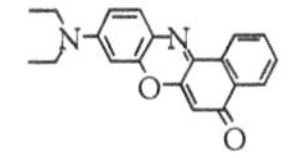 | | Solubility of Nimodipine 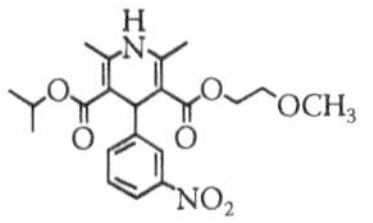 | |
|---|---|---|---|---|---|---|---|
| | % | mg·g$^{-1\,a}$ | mol·mol$^{-1\,b}$ | mg·g$^{-1\,a}$ | mol·mol$^{-1\,b}$ | mg·g$^{-1\,a}$ | mol·mol$^{-1\,b}$ |
| 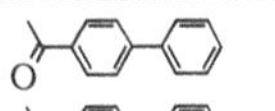 | 27 | 1.63 | 0.11 | 2.7 | 0.07 | 15.57 | 0.46 |
| 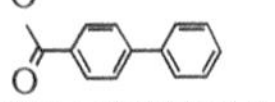 | 45 | 2.47 | 0.18 | 13.05 | 0.35 | 18.89 | 0.65 |
| $-(CH_2)_3S(CH_2)_2C_6F_{13}$ | 16 | 1.53 | 0.11 | 5.59 | 0.14 | 5.63 | 0.18 |
| $-(CH_2)_3S(CH_2)_2C_6F_{13}$ | 27 | 3.69 | 0.27 | 14.51 | 0.43 | 9.94 | 0.38 |
| $-(CH_2)_3S(CH_2)_2C_6F_{13}$ | 46 | 5.08 | 0.37 | 15.40 | 0.57 | 10.92 | 0.53 |
| 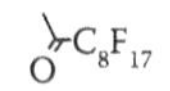 $C_8F_{17}$ | 39 | 1.59 | 0.12 | 20.25 | 0.80 | 5.2 | 0.23 |

*Source:* Kurniasih, I. N. et al., *Macromolecular Rapid Communications*, 31, 1516–1520, 2010.

$^a$ mg guest · g$^{-1}$ polymer.

$^b$ mol guest · mol$^{-1}$ polymer.

et al.[88] worked on the design of a new type of water-soluble core-shell architecture composed of dendritic hyperbranched polyglycerols containing hydrophobic biphenyl groups or perfluoronated chains in their core. Because of the increased hydrophobicity of the core, the species was used to selectively solubilize hydrophobic (aromatic) dyes and drug molecules such as pyrene, Nile red, and nimodipine. The encapsulation of the guest molecules was driven by noncovalent weak binding interactions such as hydrophobic and van der Waals interactions as well as π–π stacking (Table 14.1).

Encapsulation experiments in water revealed that cores functionalized with 45% of biphenyl groups showed better transport capacity than those with a lower degree of functionalization (27%). However, the polymer became insoluble in water if the degree of functionalization exceeded 50%. Hydrophobic interactions played an important role with Nile red as a guest molecule. Therefore, polymers with a perfluorinated core were more hydrophobic in nature thus showing significantly better transport capacity than other polymers.[88]

Encapsulation and transport of anticancer drugs by water-soluble dendritic nanocarriers have been studied by several research groups.[2] Relatively little is known, however, about the release of the encapsulated drugs by the cleavage of the shell in the physiological pH range. A general synthetic approach was presented in previous studies by Krämer et al.[89] and Xu et al.,[90,91] who synthesized a number of dendritic core-shell architectures with pH-labile linkers based on hyperbranched PEI cores and biocompatible PEG shells. In the latter approach,[90] the polymeric core-shell architectures were prepared by simply attaching monomethyl PEG (mPEG) shells to the PEI core using imine bond formation (Figure 14.11). To demonstrate the potential of these polymers for controlled release, the time-dependent releases of

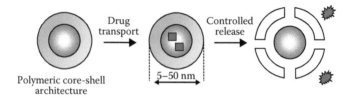

**FIGURE 14.11** Unimolecular dendritic nanocarrier for supramolecular encapsulation of biologically active compounds. Drugs can be released selectively in acidic media (such as tumor tissue) when the acid-labile linkers (connecting the shell to the core) are cleaved. (From Xu, S. et al., *Bioorganic & Medicinal Chemistry Letters*, 19, 1030–1034, 2009. With permission.)

three prototypal dyes (Congo red, rose bengal, and thymol blue) were evaluated at 37°C in buffered aqueous solutions of pH 5 and 7.4, respectively. The results showed that the acid-labile nanocarriers exhibited much higher transport capacities for dyes compared with unfunctionalized hyperbranched PEI. As determined by UV–visible spectroscopy, the measured half-life times of dye release were approximately two to five times faster at pH 5 than those determined at physiological pH.

## DENDRITIC CORE-MULTISHELL ARCHITECTURES AS UNIVERSAL NANOCARRIERS

A fundamental problem concerning conventional nanotransport systems, either of micellar origin or based on liposomes, is their limited matrix compatibility. They can either transport nonpolar guest molecules into an aqueous environment, or, in the case of inverted micelles, polar compounds are transported to a nonpolar environment (organic solvent). Fortunately, dendritic core-multishell architectures provide universal transport of hydrophilic and hydrophobic guest molecules in both polar and nonpolar solvents. Previously, Radowski et al.[92] reported on two dendritic multishell architectures based on PEI cores with different molecular weights ($M_n$ = 3600 g/mol, PD = 1.4; $M_n$ = 10,500 g/mol, PD = 2.0; Figure 14.9). These hyperbranched cores were functionalized with linear amphiphilic building blocks formed

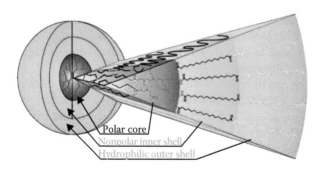

**FIGURE 14.12** Schematic representation of the multishell architecture based on a hyperbranched poly(ethylene imine) (PEI) core. This architecture mimics the structure of a liposome on a unimolecular basis. (From Radowski, M. et al., *Angewandte Chemie (International ed. in English)*, 46, 1265–1269, 2007. With permission.)

by alkyl diacids ($C_6$, $C_{12}$, or $C_{18}$) connected to mPEG (6, 10, and 14 glycol units on average) with different degrees of functionalization (70–100%). The terminal mPEG chains acting as an external polar layer provided good solubility in water as well as in organic solvents and were highly biocompatible.

These multishell architectures are fundamentally different in their structure and transport behavior compared to the simple detergent micelles and unimolecular carrier systems reported previously. Like hydrophobically modified dendrimers, the multishell nanocarriers self-assemble into supramolecular aggregates above a well-defined threshold concentration (critical aggregation concentration (CAC)). In addition, this new type of supramolecular aggregate only acts as a carrier for guest molecules after self-assembly and not as a unimolecular system. Surprisingly, multishell systems with PEI core could accommodate polar and nonpolar guest molecules and adapt to various environmental polarity conditions ranging from toluene to water. Because of these particular properties, they have to be considered chemical chameleons. For the evaluation of the solubilization behavior in aqueous solution, a representative selection of dyes (pyrene, Nile red, Congo red, rose bengal, and thymol blue) as well as commercial drug molecules, including nimodipine, were applied. Nimodipine is poorly water-soluble (0.4 mg/L),[93] and is used for the treatment of heart diseases and neurological deficits.

The multishell nanotransporters described here have a similar structure as liposomes and provided hydrophilic and hydrophobic microdomains on a unimolecular basis. The size of the hydrophobic inner shell is essential for the encapsulation of lipophilic guest molecules. Thus, nimodipine was not solubilized when the aliphatic inner shell was absent. On the other hand, encapsulation gradually increased up to a factor of 3 with an increase in the length of the aliphatic chain from $C_6$ to $C_{18}$. Surprisingly, the nonpolar microdomain also play an important role in the transport of polar molecules as deduced from the increasing transport capacity of Congo red with growing nonpolar chain length ($C_6$ to $C_{18}$) by a factor of 2 (Figure 14.12).

As has been shown by Calderon et al.[83] and Quadir et al.,[94] the multilamellar architectures were able to transport a broad variety of biologically active agents including

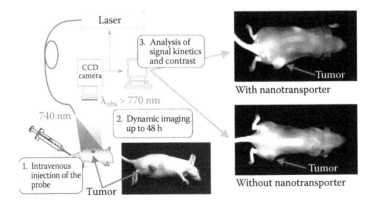

**FIGURE 14.13** Fluorescence image of a tumor-bearing mouse (F9 teratocarcinoma) 6 h after injection of NIR dye with and without the nanotransporter. (From Quadir, M. A. et al., *Journal of Controlled Release*, 132, 289–294, 2008. With permission.)

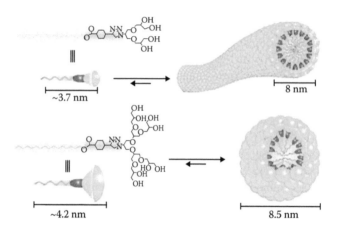

**FIGURE 14.14**   Schematic representation of micellization of various types of non-ionic dendritic glycerol-based amphiphiles in water. Due to the packing parameter [G1] amphiphiles preferably form cylindrical micelles. (From Trappmann, B. et al., *Journal of the American Chemical Society*, 132, 11119–11124, 2010. With permission.)

the antitumor drugs doxorubicin hydrochloride, methotrexate, and sodium ibandronate. Doxorubicin hydrochloride and methotrexate loaded carriers were soluble in both organic and aqueous media as determined by SEC and UV–visible spectroscopy. For the visible transparent sodium ibandronate, isothermal titration calorimetric experiments showed an exothermic interaction of the drug with the dendritic nanocarrier. The enthalpic stabilization ($\Delta H$) on encapsulation, however, was in the order of 7 kcal/mol, indicating attractive interactions between sodium ibandronate and the dendritic nanoparticle.

The ability of the nanocarriers to localize in tumors *in vivo* was demonstrated by fluorescence imaging of tumor-bearing mice (Figure 14.13) with the indotricarbocyanine–nanocarrier complex. Near-infrared fluorescence imaging allowed an easy visualization of the tumor-accumulation of the dye–PG complexes because near-infrared and far-red light (650–900 nm) can avoid strong absorption by RBC and water to allow light to pass through the body of the mice in the depth of several centimeters.

## Nonionic Dendritic Glycerol-Based Amphiphiles

Highly defined micellar constructs have been reported by Trappmann et al.[95] using nonionic dendritic glycerol-based amphiphiles (Figure 14.14). All amphiphiles feature a construction with dendronized hydrophilic headgroups and hydrophobic segment groups of aromatic and aliphatic units. The hydrophilic dendrons are in different generations of branching, unlike the hydrophobic tails that contain a phenyl group in combination with an alkyl side chain. The synthesis is based on a modular approach using click chemistry. A particular feature of the amphiphiles is their tendency to spontaneously self-assemble into micelles and thus their ability to accommodate hydrophobic active agents. Interestingly, the type of self-assembly, as well as the aggregation number is strongly influenced by the dendritic head group leading to structurally persistent and highly defined spherical micelles for [G2] and [G3] dendrons. Because of the packing parameter [G1] amphiphiles preferably form cylindrical micelles (Figure 14.14). Values for the critical micelle concentration, as determined by surface tension measurements, were low, typically lying in the micromolar concentration range. The micelles were tested as potential nanocarriers for hydrophobic compounds and were shown to entrap the solvatochromic dyes Nile red and pyrene. Additionally, it was shown that the aromatic spacer group in the hydrophobic part plays a crucial role in the binding and transport of aromatic guest molecules. Beside the hydrophobic effect, CH-π and π-π interactions of arenes with other aromatic units seem to be particularly important for solubilizing hydrophobic (aromatic) compounds. In fact, aromatic rings are common components of drug molecules, and the π-electrons on the face of the aromatic residues may enable π-π interactions with the aromatic spacer groups.

To evaluate the potential use of the micellar assemblies as drug-delivery vehicles, it is necessary to investigate the effect of different core structures with respect to solubilization, micelle (formulation) stability, and cytotoxicity. The micelle-enhanced solubilization of hydrophobic compounds is of major practical importance in drug-delivery

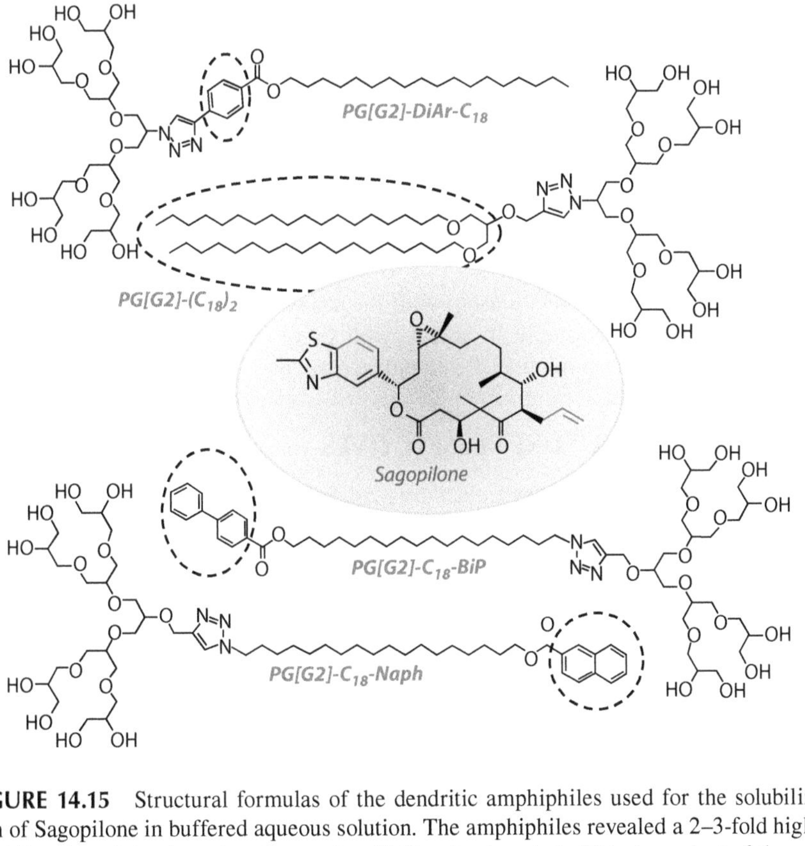

FIGURE 14.15 Structural formulas of the dendritic amphiphiles used for the solubilization of Sagopilone in buffered aqueous solution. The amphiphiles revealed a 2–3-fold higher solubilization of the drug than Cremophor ELP and polysorbate 80 independent of the core structure. (From Richter, A. et al., *European Journal of Pharmaceutical Sciences*, 2010, 40, 48–55. With permission.)

processes. Much attention has been given to both the extent of solubilization and the possible positions at which solubilization processes occur. Electrostatic and hydrophobic interactions are the main driving forces for micellar binding. Therefore, drug molecules can be solubilized at a number of different sites, due to the interplay of hydrophilicity and hydrophobicity. In aqueous solution, hydrophilic molecules are dissolved in water or slightly adsorbed at the micelle–water interface. As the solubility in water decreases, these molecules can be solubilized either between the hydrophilic head groups or in the palisade layer of the micelle between the hydrophilic head groups and the first few carbon atoms of the hydrophobic tail. Finally, substances that are poorly soluble in water are preferably located in the core of the micelle. In a recent study, Richter et al.[96] considered a number of nonionic dendritic amphiphiles composed of different hydrophobic modifications to evaluate the micelles as potential vehicles for the novel anticancer drug Sagopilone (Figure 14.15). Sagopilone is sparingly soluble in water with a solubility of 12 mg/L. With regard to the clinical applications of Sagopilone as an anticancer therapeutic, a final formulation concentration of at least 1 g/L is required.

However, in buffered aqueous solution, the amphiphiles formed spherical (7–10 nm), monodisperse (PDI 0.04–0.20) micelles with the exception of the double-chained amphiphile PG[G2]-($C_{18}$)$_2$. All micelles solubilized Sagopilone and, more importantly, showed superior solubilization behavior compared with standard excipients used in parenteral formulations. In general, the amphiphiles revealed a 2- to 3-fold higher solubilization of Sagopilone than Cremophor ELP and polysorbate 80 independent of the core structure. The amphiphile, comprising a diaromatic spacer (PG[G2]-DiAr-C18), provided the best solubilization effect for Sagopilone out of the different dendritic amphiphiles tested. This indicates preferential drug localization on the interface between the hydrophobic core and the hydrophilic corona of the micelle. Cytotoxicity studies with various functionalized amphiphiles showed a clear structure–response relationship with the structure comprising a naphthyl end group being the least cytotoxic. Its actual cytotoxicity values were comparable to the standard excipient Cremophor ELP and polysorbate 80.

## SUMMARY AND FUTURE PERSPECTIVES

In recent years, several nanoparticulate drug-delivery and drug-targeting systems have been developed to address some of the great challenges faced in antitumor therapy, gene therapy, and radiotherapy, in the delivery of dyes and drugs, proteins, antibiotics, virostatics, and vaccines. The aim of this book chapter has been to present recent approaches to supramolecular drug-delivery systems. Some of the carriers presented are already in clinical phase trials, whereas others are very attractive for applications in pharmaceutical formulations. However, there is still a need for biocompatible, nanosized materials that minimize toxicity and improve efficacy regarding drug delivery, targeting, and release. There are many factors that determine the delivering ability of a nanocarrier, for example, molecular size, molecular weight, surface charge, and chemical functionality. Thus, the safety of a nanoparticulate drug-delivery system is largely dependent on the physicochemical characteristics which are often not efficiently studied.

The coupling of low–molecular weight anticancer drugs to polymers through a cleavable linker has been an effective method for improving the therapeutic index of clinically established agents, and the first candidates of anticancer drug–polymer conjugates are being evaluated in clinical trials. A representative example is a pH-sensitive, doxorubicin-conjugated, micellar nanocarrier based on amphiphilic poly(ethylene glycol)–poly(aspartate-hydrazone-adriamycin) block copolymer.

A significant amount of research work has been focused on stimuli-responsive polymers, particularly the thermoresponsive behavior of PNIPAM and its derivatives in aqueous media. A challenging approach to smart polymeric materials is still the combination of thermally sensitive polymers with pH, redox, and light-responsive components.

dPGs consist of an aliphatic polyether backbone bearing a large number of primary and secondary hydroxy groups. The noncharged polymers can be chemically modified either at the core or at the shell, thus leading to a broad variety of functionalized dPGs. Because dPGs are synthesized in a controlled manner to obtain definite molecular weight and narrow molecular polydispersity, they have been evaluated for a broad variety of biomedical applications ranging from protein-resistant coatings (neutral species) to DNA transfection agents (polycationic systems), and anticoagulating and anti-inflammatory drugs (polyanionic systems). dPG-based architectures with an imine-linked PEG shell showed the necessary release profile for tumor tissue. The observed pH sensitivities are in the same range as for malignant tissues (e.g., tumor and infection) or endosomes, and hence, could be used as a trigger to selectively release encapsulated anticancer drugs from these nanocarriers. On the other hand, dendritic core (multi)shell architectures are versatile and, because of their chameleon-like behavior, of particular interest for a number of biomedical and technological applications. These polymers can adapt to various environmental polarity conditions ranging from toluene to water. Interestingly, above a well-defined threshold concentration (CAC), the multishell polymers spontaneously self-assemble into supramolecular aggregates. These supramolecular aggregates only act as a carrier for guest molecules after self-assembly in water and not as unimolecular scaffolds.

Nonionic dendritic glycerol-based amphiphiles undergo controlled aggregation at very low critical micelle concentrations to form small, well-defined aggregates in aqueous solution. In a recent study, the micellar assemblies were shown to encapsulate the hydrophobic dyes, Nile red and pyrene, thus indicating their potential use as drug-delivery vehicles. Interestingly, dendritic glycerol-based amphiphiles showed better solubilization behavior for the poorly water-soluble anticancer drug Sagopilone than the standard nonionic surfactants used in parenteral formulations.

Supramolecular drug-delivery systems will have a bright future in many biomedical applications, but more translational research is necessary to evaluate their safety in the clinics.

## REFERENCES

1. Schick, M. J. *Nonionic Surfactants: Physical Chemistry*, Boca Raton, FL: CRC Press, 1987.
2. Haag, R. 2004. *Angewandte Chemie (International ed. in English)*, 43: 278–282.
3. Adams, M. L., A. Lavasanifar, G. S. Kwon. 2003. *Journal of Pharmaceutical Sciences*, 92: 1343–1355.
4. Torchilin, V. P. 2006. *Pharmaceutical Research*, 24: 1–16.

5. Kabanov, A. V., V. Y. Alakhov. 2002. *Critical Reviews in Therapeutic Drug Carrier Systems*, 19: 1–72.
6. Authier, N., J. P. Gillet, J. Fialip, A. Eschalier, F. Coudore. 2000. *Brain Research*, 887: 239–249.
7. Windebank, A. J., M. D. Blexrud, P. C. de Groen. 1994. *Journal of Pharmacology and Experimental Therapeutics*, 268: 1051–1056.
8. Burstein, S. H., E. Friderichs, B. Kögel, J. Schneider, N. Selve. 1998. *Life Sciences*, 63: 161–168.
9. Tabarelli, Z., D. B. Berlese, P. D. Sauzem, C. F. Mello, M. A. Rubin. 2003. *Brazilian Journal of Medical and Biological Research*, 36: 119–123.
10. Gelderblom, H., J. Verweij, K. Nooter, A. Sparreboom. 2001. *European Journal of Cancer*, 37: 1590–1598.
11. Sparreboom, A., W. J. Loos, J. J. Verweij, A. I. de Vos, M. E. L. van der Burg, G. Stoter, K. Nooter. 1998. *Analytical Biochemistry*, 255: 171–175.
12. ten Tije, A. J., J. J. Verweij, W. J. Loos, A. Sparreboom. 2003. *Clinical Pharmacokinetics*, 42: 665–685.
13. Kataoka, K., A. Harada, Y. Nagasaki. 2001. *Advanced Drug Delivery Reviews*, 47: 113–131.
14. Kakizawa, Y., K. Kataoka. 2002. *Advanced Drug Delivery Reviews*, 54: 203–222.
15. Haag, R., F. Kratz. 2006. *Angewandte Chemie (International ed. in English)*, 45: 1198–1215.
16. Szleifer, I. 1997. *Current Opinion in Solid State Materials Science*, 2: 337–344.
17. Deng, L., M. Mrksich, G. M. Whitesides. 1996. *Journal of the American Chemical Society*, 118: 5136–5137.
18. Bae, Y., S. Fukushima, A. Harada, K. Kataoka. 2003. *Angewandte Chemie (International ed. in English)*, 42: 4640–4643.
19. Mellmann, I., R. Fuchs, A. Helenius. 1986. *Annual Review of Biochemistry*, 55: 663–700.
20. Engin, K., D. B. Leeper, J. R. Cater, A. J. Thistlethwaite, L. Tupchong, J. D. McFarlane. 1995. *International Journal of Hyperthermia*, 11: 211–216.
21. Wani, M. C., H. L. Taylor, M. E. Wall, P. Coggon, A. T. McPhail. 1971. *Journal of the American Chemical Society*, 93: 2325–2327.
22. Singla, A. K., A. Garg, D. D. Aggarwal. 2002. *International Journal of Pharmaceutics*, 235: 179–192.
23. Dorr, R. 1994. *The Annals of Pharmacotherapy*, 28: 11–14.
24. Feng, X., Y. Yuan, J. Wu. 2002. *Bioorganic & Medicinal Chemistry Letters*, 12: 3301–3303.
25. Ooya, T., J. Lee, K. Park. 2004. *Bioconjugate Chemistry*, 15: 1221–1229.
26. Erez, R., E. Segal, K. Miller, R. Satchi-Fainaro, D. Shabat. 2009. *Bioorganic & Medicinal Chemistry*, 17: 4327–4335.
27. Etrych, T., M. Sirova, L. Starovoytova, B. Rihova, K. Ulbrich. 2010. *Molecular Pharmaceutics*, 7: 1015–1026.
28. Sohn, J. S., J. I. Jin, M. Hess, B. W. Jo. 2010. *Polymer Chemistry*, 1: 778.
29. Mugabe, C., B. A. Hadaschik, R. K. Kainthan, D. E. Brooks, A. I. So, M. E. Gleave, H. M. Burt. 2009. *BJU International*, 103: 978–986.
30. Khandare, J., A. Mohr, M. Calderón, P. Welker, K. Licha, R. Haag. 2010. *Biomaterials*, 31: 4268–4277.
31. Stuart, M. A. C., W. T. S. Huck, J. Genzer, M. Müller, C. Ober, M. Stamm, G. B. Sukhorukov, I. Szleifer, V. V. Tsukruk, M. Urban, F. Winnik, S. Zauscher, I. Luzinov, S. Minko. 2010. *Nature Materials*, 9: 101–113.
32. Senaratne, W., L. Andruzzi, C. K. Ober. 2005. *Biomacromolecules*, 6: 2427–2448.
33. Motornov, M., S. Minko, K. Eichhorn, M. Nitschke, F. Simon, M. Stamm. 2003. *Langmuir*, 19: 8077–8085.
34. Luzinov, I., S. Minko, V. V. Tsukruk. 2008. *Soft Matter*, 4: 714.

35. Mendes, P. M. 2008. *Chemical Society Reviews*, 37: 2512.
36. Liu, Z., P. Calvert. 2000. *Advanced Materials*, 12: 288–291.
37. Anker, J. N., W. P. Hall, O. Lyandres, N. C. Shah, J. Zhao, R. P. Van Duyne. 2008. *Nature Materials*, 7: 442–453.
38. Tokarev, I., S. Minko. 2009. *Soft Matter*, 5: 511.
39. Jhaveri, S. J., M. R. Hynd, N. Dowell-Mesfin, J. N. Turner, W. Shain, C. K. Ober. 2009. *Biomacromolecules*, 10: 174–183.
40. Hoffman, A. S. 2008. *Journal of Controlled Release*, 132: 153–163.
41. Bayer, C. L. N., A. Peppas. 2008. *Journal of Controlled Release*, 132: 216–221.
42. Alarcon, C. D. L. H., S. Pennadam, C. Alexander. 2005. *Chemical Society Reviews*, 34: 276–285.
43. Schild, H. G. 1992. *Progress in Polymer Science*, 17: 163–249.
44. You, Y., D. Oupický. 2007. *Biomacromolecules*, 8: 98–105.
45. Heath, F., P. Haria, C. Alexander. 2007. *The AAPS Journal*, 9: 235–240.
46. Green, N. M. 1963. *The Biochemical Journal*, 89: 585.
47. Ropert, C. C. 1999. *Brazilian Journal of Medical and Biological Research*, 32: 163–169.
48. Balazs, D. A., W. T. Godbey. 2011. *Journal of Drug Delivery*, 1–12.
49. Ogris, M. In *Polymeric Gene Delivery: Principles and Applications*, Boca Raton, FL: CRC Press, 2004.
50. Kumar, M. N. V. R., R. A. A. Muzzarelli, C. Muzzarelli, H. Sashiwa, A. J. Domb. 2004. *Chemical Reviews*, 104: 6017–6084.
51. Hinrichs, W. L. J., N. M. E. Schuurmans-Nieuwenbroek, P. van de Wetering, W. E. Hennink. 1999. *Journal of Controlled Release*, 60: 249–259.
52. Arotçaréna, M., B. Heise, S. Ishaya, A. Laschewsky. 2002. *Journal of the American Chemical Society*, 124: 3787–3793.
53. Bütün, V., N. C. Billingham, S. P. Armes. 1998. *Journal of the American Chemical Society*, 120: 11818–11819.
54. Liu, S., S. P. Armes. 2002. *Angewandte Chemie (International ed. in English)*, 41: 1413–1416.
55. Holmberg, K., B. Jönsson, B. Kronberg, B. Lindman. *Surfactants and Polymers in Aqueous Solution*, 2nd ed., Chichester, UK: John Wiley & Sons, 2002.
56. Bartus, R. T., M. A. Tracy, D. F. Emerich, S. E. Zale. 1998. *Science*, 281: 1161–1162.
57. Brown, L. R. 2005. *Expert Opinion on Drug Delivery*, 2: 29–42.
58. Nayak, A. K. 2010. *International Journal of Pharmacy and Pharmaceutical Sciences*, 2: 1–5.
59. Joshi, S. R., R. M. Parikh, A. K. Das. 2007. *The Journal of the Association of Physicians of India*, 55 (Suppl): 19–24.
60. Das, A. K., S. Shah. 2011. *The Journal of the Association of Physicians of India*, 59 (Suppl): 6–7.
61. Shah, S. N., S. R. Joshi, D. V. Parmar. 1997. *The Journal of the Association of Physicians of India*, 45 (Suppl): 4–9.
62. Torchilin, V. P., A. N. Lukyanov. 2003. *Drug Discovery Today*, 8: 259–266.
63. Tarner, I. H., U. Müller-Ladner. 2008. *Expert Opinion on Drug Delivery*, 5: 1027–1037.
64. Giordano, C., D. Albani, A. Gloria, M. Tunesi, S. Batelli, T. Russo, G. Forloni, L. Ambrosio, A. Cigada. 2009. *The International Journal of Artificial Organs*, 32: 836–850.
65. Berg, J. M., J. L. Tymoczko, L. Stryer. *Biochemistry,* 5th ed., New York: W. H. Freeman, 2002.
66. MaHam, A., Z. Tang, H. Wu, J. Wang, Y. Lin. *Small* 2009, 5: 1706–1721.
67. Barichello, J. M., M. Morishita, K. Takayama, T. Nagai. 1999. *International Journal of Pharmaceutics*, 184: 189–198.
68. Kim, Y. J., S. Choi, J. J. Koh, M. Lee, K. S. Ko, S. W. Kim. 2001. *Pharmaceutical Research*, 18: 548–550.

69. Choi, S., S. W. Kim. 2003. *Pharmaceutical Research*, 20: 2008–2010.
70. Abu Hashim, I. I., T. Higashi, T. Anno, K. Motoyama, A. H. Abd-ElGawad, M. H. El-Shabouri, T. M. Borg, H. Arima. 2010. *International Journal of Pharmaceutics*, 392: 83–91.
71. Zentner, G. M., R. Rathi, C. Shih, J. C. McRea, M. Seo, H. Oh, B. Rhee, J. Mestecky, Z. Moldoveanu, M. Morgan, S. Weitman. 2001. *Journal of Controlled Release*, 72: 203–215.
72. Szejtli, J. 1998. *Chemical Reviews*, 98: 1743–1754.
73. Yagai, S., A. Kitamura. 2008. *Chemical Society Reviews*, 37: 1520–1529.
74. Peng, K., I. Tomatsu, A. Kros. 2010. *Chemical Communications*, 46: 4094–4096.
75. Frechet, J. M. J., D. A. Tomalia. *Dendrimers and Other Dendritic Polymers*, New York: John Wiley & Sons, Ltd., 2001.
76. Newkome, G. R., C. N. Moorefield, F. Vögtle. *Dendrimers and Dendrons: Concepts, Synthesis, Applications*, 1st ed., Weinheim: Wiley-VCH, 2001.
77. van Heerbeek, R., P. C. J. Kamer, P. W. N. M. van Leeuwen, J. N. H. Reek. 2002. *Chemical Reviews*, 102: 3717–3756.
78. Astruc, D., F. Chardac. 2001. *Chemical Reviews*, 101: 2991–3024.
79. Newkome, G. R., E. He, C. N. Moorefield. 1999. *Chemical Reviews*, 99: 1689–1746.
80. Sunder, A., R. Hanselmann, H. Frey, R. Mülhaupt. 1999. *Macromolecules*, 32: 4240–4246.
81. Haag, R., A. Sunder, J. Stumbé. 2000. *Journal of the American Chemical Society*, 122: 2954–2955.
82. Sunder, A., R. Mülhaupt, R. Haag, H. Frey. 2000. *Advanced Materials*, 12: 235–239.
83. Calderon, M., M. A. Quadir, S. K. Sharma, R. Haag. 2010. *Advanced Materials*, 22: 190–218.
84. Türk, H., R. Haag, S. Alban. 2004. *Bioconjugate Chemistry*, 15: 162–167.
85. Dernedde, J., A. Rausch, M. Weinhart, S. Enders, R. Tauber, K. Licha, M. Schirner, U. Zügel, A. von Bonin, R. Haag. 2010. *Proceedings of the National Academy of Sciences of the United States of America*, 107: 19679.
86. Fischer, W., M. Calderón, A. Schulz, I. Andreou, M. Weber, R. Haag. 2010. *Bioconjugate Chemistry*, 21: 1744–1752.
87. Fischer, W., B. Brissault, S. Prévost, M. Kopaczynska, I. Andreou, A. Janosch, M. Gradzielski, R. Haag. 2010. *Macromolecular Bioscience*, 10: 1073–1083.
88. Kurniasih, I. N., H. Liang, J. P. Rabe, R. Haag. 2010. *Macromolecular Rapid Communications*, 31: 1516–1520.
89. Krämer, M., J. F. Stumbé, H. Türk, S. Krause, A. Komp, L. Delineau, S. Prokhorova, H. Kautz, R. Haag. 2002. *Angewandte Chemie (International ed. in English)*, 41: 4252–4256.
90. Xu, S., Y. Luo, R. Haag. 2007. *Macromolecular Bioscience*, 7: 968–974.
91. Xu, S., Y. Luo, R. Graeser, A. Warnecke, F. Kratz, P. Hauff, K. Licha, R. Haag. 2009. *Bioorganic & Medicinal Chemistry Letters*, 19: 1030–1034.
92. Radowski, M., A. Shukly, H. von Berlepsch, C. Böttcher, G. Pickaert, H. Rehage, R. Haag. 2007. *Angewandte Chemie (International ed. in English)*, 46: 1265–1269.
93. Grunenberg, A., B. Keil, J. Henck. 1995. *International Journal of Pharmaceutics*, 118: 11–21.
94. Quadir, M. A., M. R. Radowski, F. Kratz, K. Licha, P. Hauff, R. Haag. 2008. *Journal of Controlled Release*, 132: 289–294.
95. Trappmann, B., K. Ludwig, M. R. Radowski, A. Shukla, A. Mohr, H. Rehage, C. Böttcher, R. Haag. 2010. *Journal of the American Chemical Society*, 132: 11119–11124.
96. Richter, A., A. Wiedekind, M. Krause, T. Kissel, R. Haag, C. Olbrich. 2010. *European Journal of Pharmaceutical Sciences*, 40: 48–55.

# 15 Protein and Nucleic Acid Targeting

*Laura Baldini, Francesco Sansone,*
*Alessandro Casnati, and Rocco Ungaro*

## CONTENTS

## INTRODUCTION

Supramolecular chemistry, the chemistry "beyond the molecules" or of the nonco-valent interactions between molecules, initially found its inspiration in the world of biology, from biomolecules such as proteins, nucleic acids, lipids, and oligosaccha-rides and their multimolecular aggregates. Subsequently, the study of supramolecu-lar interactions on synthetic models considerably simpler than natural ones greatly increased the knowledge of the noncovalent interactions that govern molecular rec-ognition and self-assembly processes. This has allowed, in the last 30 years, the design and development of novel molecular receptors often characterized by impres-sive recognition ability and self-assembly properties, which gave rise to sophisti-cated devices (sensors, ionophores, paramagnetic/fluorescence probes, etc.) and smart materials (switches, electronics, and polymers). Concurrently to these applica-tions in materials science, supramolecular chemistry also greatly developed in water and biological media, joining the field of chemical biology. In this context, supra-molecular chemistry certainly plays an important role because interactions between drugs and biomolecules are noncovalent, and the supramolecular concepts of self-assembly, dynamic combinatorial chemistry (DCC), and adaptive chemistry can be important tools for targeting off-regulated biological processes of several diseases and, at the same time, for understanding biological pathways and mechanisms.[1] In this chapter, we will mainly focus our attention on the interactions of small molecules with proteins and nucleic acids (the main biological targets thus far in this field), giving emphasis both to the supramolecular aspects of the assemblies and to the approaches used to discover the new ligands/drugs.

## PROTEIN TARGETING

The cell life strictly depends on a wide variety of enzymatic processes, protein–protein (PPI) and protein–carbohydrate interactions, which accelerate and regulate nearly every reaction and function taking place in living organisms. This section of the present chapter deals with the design and synthesis of artificial agents for targeting these processes.

### DESIGNED MOLECULES FOR THE INHIBITION OF ENZYMES AND OF PROTEIN–PROTEIN INTERACTIONS

The use of small molecules (MW < 500–700 Da) as PPI or enzyme inhibitors is especially attractive because they are usually cheap and can be orally administered.[2] In the past, enzyme targeting with low–molecular weight compounds led to inhibitors, similar to the substrates, that bind to the protein active site (competitive inhibition), or allosteric modulators that, occupying the regulatory site, alter the activity of the enzyme.[3] These strategies, mostly based on medicinal chemistry approaches, are nowadays rather well consolidated and will not be surveyed here. We will instead illustrate few examples of the most recent approaches for enzyme targeting, which deeply exploit supramolecular concepts and tools. The design of synthetic agents that disrupt PPIs by binding to the surface of one of the protein partners is not a simple task because the typical protein–protein interface regions are usually much larger (up to 1600–2000 Å$^2$) than the surface of small molecules (300–500 Å$^2$), are amphiphilic, flat or scarcely convex, and often lack well-defined grooves and pockets.[4] However, a number of synthetic small molecules targeting protein surfaces have been reported, thanks to the identification of "hot spots"[5] on the protein exteriors. These are areas or groups of residues that contribute a significant fraction of the favorable free energy for the formation of naturally occurring protein–protein complexes. Small molecules having chemical functionalities complementary to these hot spots could therefore bind with high affinity to the surface of one of the protein partners and prevent the formation of the protein–protein complex. Two main types of synthetic compounds designed for hot spot–targeting can be identified: protein surface receptors and mimetics of protein secondary structures. The first group (Figure 15.1, top) consists of synthetic molecules that are complementary to a relatively large and flat surface area, whereas the second (Figure 15.1, bottom) comprises nonpeptidic synthetic scaffolds that display key residues in a spatial arrangement similar to that of a α-helix, β-turn,

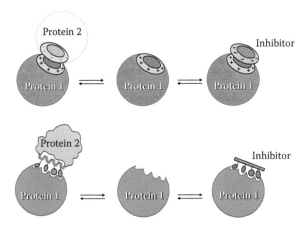

**FIGURE 15.1** Different strategies for the inhibition of PPIs. Protein surface receptors (top) and mimetics of protein secondary structures (bottom).

or β-strand which constitute a critical binding domain for one of the protein partners. Although both approaches have been successfully used to inhibit PPIs, the former may also lead to enzyme inhibition, if, by binding to the protein surface, the receptor sterically blocks the access of the substrate to the active site.

We examine here some of what we believe are the most significant results reported in the last few years. The examples reported will be grouped by the scaffolds involved to fully show the peculiarities of each synthetic platform.

## Protein Surface Receptors

The use of a "supramolecular" scaffold for protein surface recognition was pioneered by Hamuro et al.[6] who reported the synthesis of a family of receptors by linking to the upper rim of a calix[4]arene (calixarene ligands have been extensively proposed and studied in biomedical applications)[7–9] four constrained peptide loops, a design inspired by the six hypervariable peptide loops of the antigen recognition sites of antibodies. The recognition properties of these concave receptors, whose surfaces cover an area of 450 to 500 Å$^2$, can be varied by changing the amino acids of the peptide loops. The first member (**1**) of this family was reported in 1997 and contains the negatively charged GlyAspGlyAsp sequence. It binds strongly to a region of the surface of cytochrome $c$ (cyt $c$) constituted by the exposed surface of the heme edge surrounded by positively charged Lys and Arg residues. As a consequence of this binding, the approach of reducing agents to the Fe$^{III}$–cyt $c$ heme active site is inhibited, as shown by a significant decrease in the rate of reduction of cyt $c$ by ascorbate. Subsequently, **1** was shown to strongly bind also to the surface of α-chymotrypsin (ChT), a member of the serine protease family, which has several cationic residues near the active site. By binding to this region, **1** blocks the approach of substrates into the active site of the enzyme and thus inhibits (with submicromolar activity) the protein function.[10]

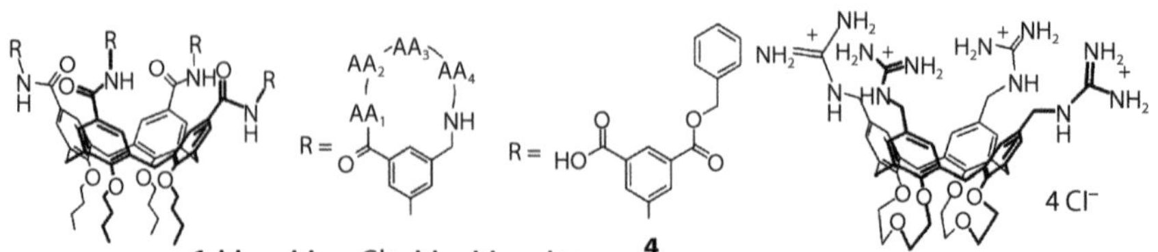

**1** $AA_1 = AA_3 = Gly, AA_2, AA_4 = Asp$
**2** $AA_1 = AA_3 = Gly, AA_2 = Asp, AA_4 = Tyr$
**3** $AA_1 = AA_3 = Gly, AA_2, AA_4 = Lys$

The high potentialities offered by the use of protein surface receptors to disrupt clinically relevant PPIs were definitely validated by Hamilton with the targeting of platelet-derived growth factor (PDGF) and of vascular endothelial growth factor (VEGF), two growth factors that are critically involved in angiogenesis (VEGF is crucial for the initiation and PDGF for the maintenance of new blood vessels) and oncogenesis. The signaling function of PDGF, which in malignant diseases eventually results in tumor growth, is triggered by the binding of PDGF to its cell surface receptor, PDGFR, a tyrosine kinase. Compound **2** was shown to bind with high affinity ($IC_{50} = 250$ nM) and selectivity to an area of the PDGF surface primarily composed of cationic and hydrophobic residues, the same region that is involved in the PDGF–PDGFR interaction. Indeed, on binding to PDGF, **2** prevents the binding of PDGF to PDGFR, blocks the receptor autophosphorylation and the subsequent cascade of biological events. *In vivo* studies also showed a significant inhibition of tumor growth and angiogenesis in nude mouse models.[11] Another member of the library, **3**, with analogous mechanism, blocks the binding of VEGF to its receptor and the receptor tyrosine phosphorylation, and consequently inhibits the VEGF-dependent signaling, angiogenesis, and tumorigenesis both *in vitro* and *in vivo*.[12] In a subsequent study, the same research group reported a second-generation targeted library of calix[4]arene derivatives in which the peptide loops are replaced by acyclic isophthalic acid groups variably functionalized. One member of the library, **4**, is a potent inhibitor of both VEGFR and PDGFR tyrosine phosphorylation, with $IC_{50}$ values of 190 nM (PDGF) and 480 nM (VEGF), resulting in the inhibition of angiogenesis and tumor growth in mice bearing human tumors.[13]

A further elegant example of rational design was reported by Gordo et al.,[14] who exploited rigid cone-shaped calix[4]arene (**5**) for the stabilization of the tetrameric assembly of protein p53. The wild-type protein p53 protects cells from tumors when DNA is damaged by triggering the expression of DNA repair machinery or by inducing cell arrest and apoptosis when the damage is irreversible. The activity of protein p53, however, is strongly related to its tetramerization domain (p53TD), in which four α-helices are held together by H-bonding, hydrophobic interactions, and salt bridges. Because of a mutation that induces the replacement of Arg-337 by Hys, the tetrameric assembly is destabilized because of weaker H-bonding and hydrophobic contacts and the protein (p53-R337H) misses its function of "genome guardian." Tetra(guanidinomethylene)calixarene **5** acts as a templating ligand that is able to bind and stabilize the tetramer assembly of mutant p53-R337H thanks to the simultaneous interaction of its four guanidinium groups to four Glu carboxylates of the protein, which leads to the recovery of the wild-type tetrameric stability. Important factors for the binding are the rigidity and the conical shape of the calix[4]arene scaffold that preorganize the guanidinium groups for charge–charge matching and H-bonding, together with the simultaneous operation of hydrophobic interactions between the macrocyclic scaffold and the lipophilic pocket of the protein.

Porphyrins are large, flat, hydrophobic scaffolds that can be easily functionalized at their periphery. Moreover, they are fluorescent, and variations in their emission spectrum can be easily used to detect binding phenomena. Aya and Hamilton[16] designed a series of porphyrin derivatives (6–9) based on the tetraphenylporphyrin (TPP) scaffold to target cyt *c*. Interestingly, the affinity of these receptors for cyt *c* increases by increasing the number of hydrophobic and negatively charged groups: $K_D$ (6) = 860 nM, $K_D$ (7) = 160 nM, $K_D$ (8) = 20 nM,[15] $K_D$ (9) = 0.67 nM.

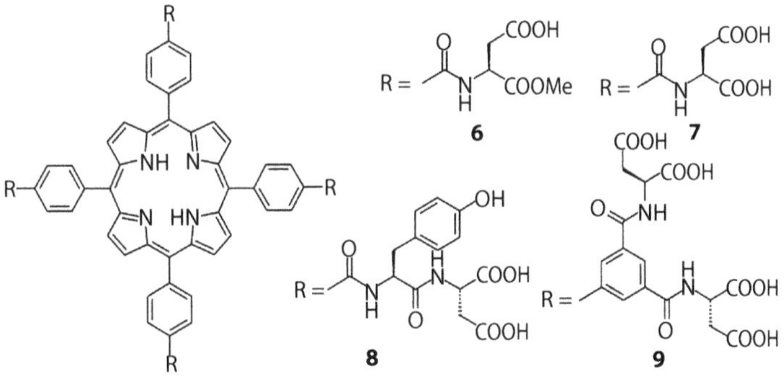

The TPP scaffold was later exploited to build a small library of protein surface receptors peripherally functionalized with substituents of different charge, size, hydrophobicity, and symmetry. The members of this library, having distinct binding characteristics, serve as components of a solution phase protein-detecting array. When incubated with different proteins, the array responds with a unique pattern of fluorescence changes that represent a characteristic fingerprint for each protein.[17,18]

An alternative approach to the functionalization of existing scaffolds, such as calixarenes or porphyrins, for the construction of large area protein surface receptors, is the noncovalent assembly of various functional subunits. Takashima et al.[19] obtained three dendritic polyanionic metal complexes (10–12) having a different number of negative charges by stepwise ligating functionalized bipyridine (bpy) units to a Ru(II) center. The affinity of 10–12 for various proteins was estimated by ultrafiltration binding assays, which showed a selectivity of binding to the positively charged surface of cyt *c* over proteins having different protein surface characteristics, and an increase in binding affinity to cyt *c* correlated to the number of carboxylate groups (10 < 11 < 12).

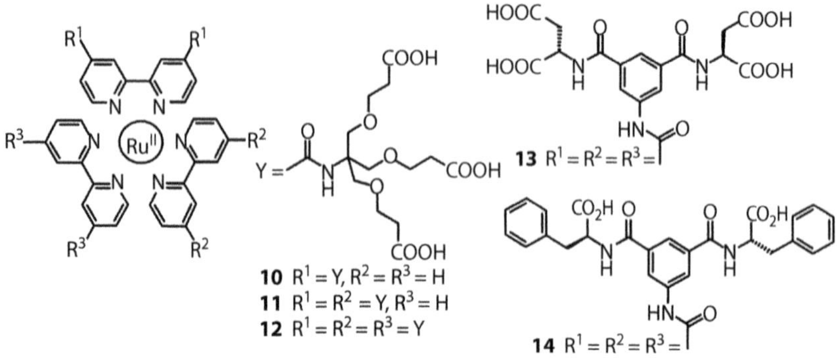

More recently, two new dendritic [Ru(bpy)$_3$] complexes with interesting properties, **13** and **14**, were reported by Muldoon et al.[20] and Ohkanda et al.,[21] respectively. Compound **13** is one of the highest affinity ($K_D$ = 2 nM) receptors reported to date for cyt $c$, whereas **14** binds with submicromolar affinity to the positively charged surface of ChT with a 1:2 (**14**: ChT) stoichiometry. Both compounds inhibit the activity of the proteins, but the mechanisms proposed by the authors are different. Compound **13** is supposed to bind to the surface area of cyt $c$ comprising the heme edge and to thus block the substrate access to the active site. On the contrary, the inhibiting effect of **14** on ChT activity is attributed to a conformational change of the protein induced by **14** on binding.

Other protein surface receptors with micromolar or submicromolar affinity for their targets based on the anthracene scaffold,[22] on multivalent Cu(II) and Zn(II) complexes,[23,24] on dendrimers,[25,26] or nanoparticles,[27] on DNA G-quadruplexes,[28,29] containing multiple units of resorcinarene[30] or cyclodextrins[31] have been reported.

## Mimetics of Protein Secondary Structures

The advantage of this approach relies on the chemical versatility offered by synthetic scaffolds, which may lead to inhibitors having good drug-like characteristics, such as low molecular weight, high stability in physiological conditions, and good transport properties across membranes. Probably one of the earliest successful efforts in the search of nonpeptidal peptidomimetics was reported by Hirschmann et al. in 1993. They introduced the use of a carbohydrate scaffold to design a β-turn mimetic of the peptide hormone somatostatin (SRIF), a cyclic tetradecapeptide that inhibits the release of several hormones. The D-glucose scaffold was chosen for its well-defined conformation and the ability to project substituents at C-2, C-1, and C-6 in a spatial orientation that mimics the disposition of the critical Phe-Trp-Lys side chains of SRIF. **15** and the 3-deoxy sugar **16** bind to the SRIF receptor in a dose-dependent manner, with affinity in the μM range.[32]

The α-helix is the secondary structure most often found at protein–protein interfaces. It is usually at least 10 residues long and displays interacting residues predominantly on one face, occupying the $i$, $i + 3$ or $i + 4$, and $i + 7$ positions.[33] Besides pioneering the supramolecular chemistry of synthetic protein surface receptors, the Hamilton group also reported the first true α-helix mimetic and, subsequently, a number of "next gen-eration" mimetics. The earliest scaffold they proposed was a terphenyl derivative that,

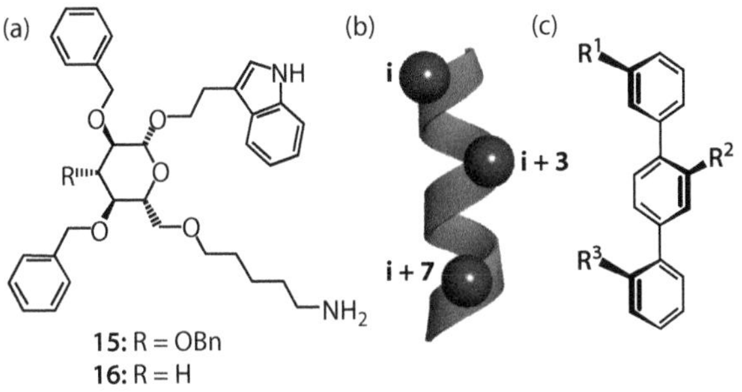

**FIGURE 15.2** Structure of the β-turn mimetics **15** and **16** (a), schematic representation of an α-helix (b) and of its terphenyl mimetic (c).

because of the staggered conformation adopted by the central aryl ring, displays the substituents on the ortho positions ($R^1$, $R^2$, and $R^3$) in the same spatial orientation as the $i$, $i + 3(4)$, and $i + 7$ residues on a α-helix (Figure 15.2) and thus inhibits a number of therapeutically relevant PPIs.[34,35] The drawbacks of the terphenyl scaffold, however, are the challenging synthesis, a scarce solubility, and excessive flexibility. To address these issues, Hamilton has proposed a number of alternative scaffolds to improve the solubility and drug-like character of the molecules, based on terephthalamide, oligoamide, oligopyridine, enaminone, and benzoylurea.[33]

## Fragment-Based Drug Discovery and Dynamic Combinatorial Chemistry

A quite interesting new approach to identify PPI or enzyme inhibitors is the so-called fragment-based drug discovery (FBDD).[36] After the identification of a series of low affinity (in the millimolar to micromolar range) "fragments" by using a combination of different techniques such as structure–activity relationship by nuclear magnetic resonance,[37] x-ray crystallography, mass spectrometry, and *in vivo* bioassays, these components are properly linked together. After optimization of the linker length (Figure 15.3), a high-affinity ligand may be obtained. After FBDD, ABT-737 (**17**) was identified as a potent inhibitor with subnanomolar activity of the Bcl-2 proteins. It is active in cell-based assays and on tumor models in animals, in which it also shows a synergy with other chemotherapeutics, and is currently being investigated in phase II clinical trials for cancer.[38]

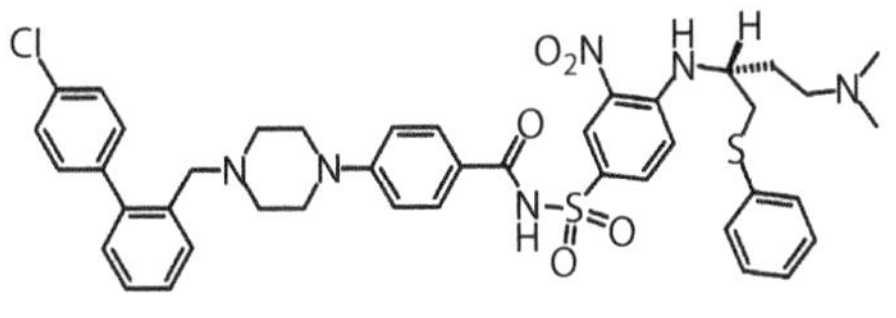

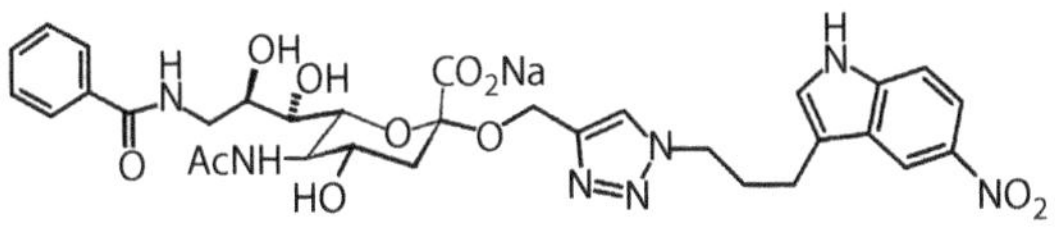

Compared with high-throughput screening, the fragment-based approach has the advantage of testing a much lower number of compounds. These fragments are

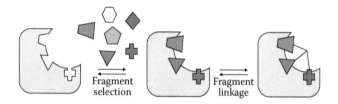

**FIGURE 15.3**   Schematic representation of the FBDD process.

smaller and less complex than those used in high-throughput screening and therefore they also have a higher probability to match the protein-binding site. Starting from small compounds, it is therefore also easier to end up with a lead compound possessing oral drug characteristics. Another quite important feature of small molecules in the inhibition of PPIs is related to the kinetic advantage they usually have, because they can penetrate the protein–protein interfaces faster and therefore accelerate the dissociation of the interacting proteins compared with a competing protein partner.

Quite interestingly, remarkable results can also be obtained in the absence of structural data of the protein and of its binding sites. In this case, it is fundamental to identify suitable fragments which bind to proximal binding sites and link them together without generating distortion in their binding to the protein. It was recently shown that the fragment-based in situ combinatorial approach could lead to high-affinity ligands via a three-step procedure. After the identification of a micromolar affinity first-site ligand by physiological assay or screening, an adjacent second-site ligand is sought within the member of a fragment library by enhanced paramagnetic relaxation caused by the spin-labeled first-site ligand (Figure 15.4). Finally, a series

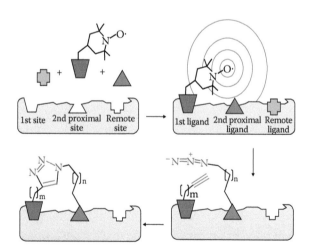

**FIGURE 15.4**   Identification of a ligand for an unknown binding site by the fragment-based in situ combinatorial approach. (Shelke, S. V. et al.: *Angew. Chem. Int. Ed. Engl.* 2010. 49. 5721–5725. Copyright Wiley-VCH Verlag GmbH & Co. KGaA. Reproduced with permission.)

of first-site and second-site ligands functionalized with methylene chains of different length, terminating with azido or alkyne moieties, is incubated with the protein.

Although 1,3-dipolar cycloadditions usually take place by thermal or copper catalysis, the protein mediates the formation of triazole (in situ Click Chemistry)[39] between those ligands that expose the azide and the acetylene groups at optimal distance and spatial orientation. Following this approach, Shelke et al.[40] identified an inhibitor (**18**) of myelin-associated glycoprotein (MAG, Sieglec-4), which exhibits a $K_D = 190$ nM, 1000-fold better than that of the natural MAG antagonist, the tetrasaccharide epitope (GQ1bα). This is of particular interest because of the ability of MAG to inhibit myelin component axonal regrowth after injury.

A rather interesting perspective in the discovery of new ligands for proteins and enzymes is offered by DCC. All the members of a dynamic combinatorial library (DCL) are in equilibrium and are interconverting one into the other through reversible chemical processes that imply labile covalent bonds or noncovalent interactions.[41–43] In the presence of the target protein, the member of the DCL possessing the highest affinity for the biomolecule is preferentially formed ("molecular amplification") and hence can be identified (Figure 15.5).

Of the different types of covalent and reversible bonds that might be used in the presence of biomolecules, mainly disulfides, imines, and acylhydrazone have been used.[44,45] One of the first proofs-of-principle aimed at demonstrating the feasibility of using DCC for the selection of inhibitors for enzymes was given by Huc and Lehn in an article on carbonic anhydrase (Figure 15.5).[46] In a 12-member DCL, made up of four amines and three aldehydes, they noticed the amplification of a sulfonamide member (**19**), which showed the strongest enzyme inhibition activity.

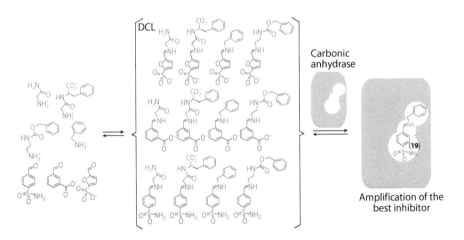

**FIGURE 15.5** The formation of a DCL and the molecular amplification process in the presence of carbonic anhydrase. (Reprinted from *Chem. Rev.*, 106, Corbett, P. T. et al., 3652–3711, Copyright 2006, with permisssion from Elsevier.)

## Multivalent Neoglycoconjugates for Lectin Binding

Lectins are proteins which bind carbohydrates but do not have enzymatic or immune activity.[47] They also play a crucial role in several physiological and pathological processes in humans, such as fertilization, cell growth and trafficking, virus, bacteria, and toxin adhesion to cell membrane, tumor proliferation, and metastasis diffusion.[48] To overcome the intrinsically low affinity of the protein–carbohydrate interactions ($K_D$ in the millimolar range), lectins are often present as multivalent aggregates that bind to polyglycosylated entities such as polysaccharides, glycoproteins, and glycolipids much more strongly ($K_D$ in the nanomolar range) and specifically. This particular aspect of multivalency[49] is known as the glycoside cluster effect.[50] In the last few years, a wide variety of scaffolds have been used for the construction of multivalent neoglycoconjugates[51-55] called glycoclusters, ranging from low valency compounds such as branched aliphatic scaffolds, benzene derivatives, monosaccharides, azamacrocycles, self-assembled transition metal complexes, fullerenes, and cavity containing macrocycles such as cyclodextrins, calixarenes, resorcinarenes, and cucurbiturils, to high-valency polyglycosylated systems such as dendrimers, polymers, peptoids, proteins, micelles, liposomes, and self-assembled monolayers on nanoparticles or planar surfaces.[53]

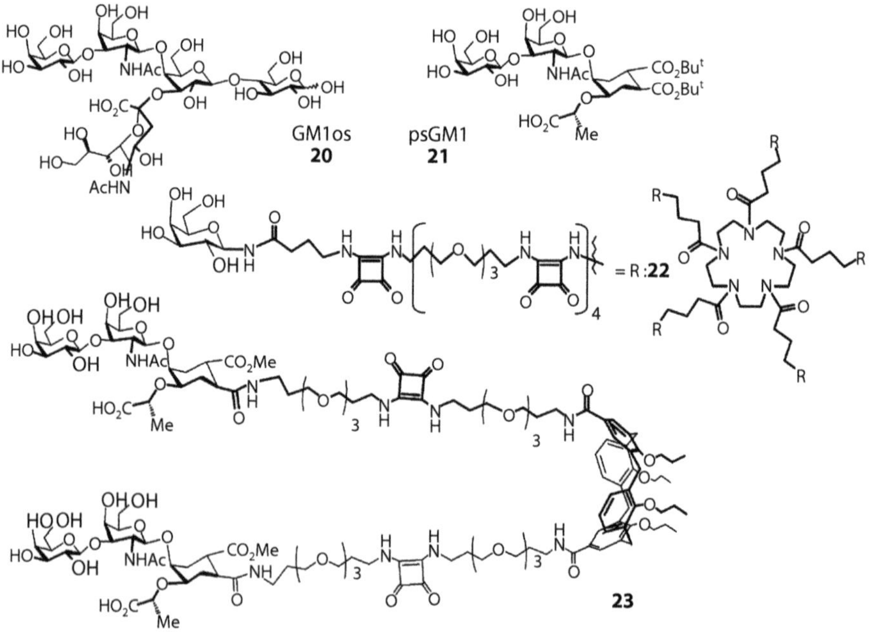

A textbook case of multivalency is given by the adhesion of cholera toxin (CT) from *Vibrio cholerae* to the gut epithelial cells. CT is a toroid-like pentavalent $AB_5$ lectin with its carbohydrate recognition domains pointing toward the same side of the torus. In this way, the toxin can firmly bind to GM1 gangliosides present on the cell surface and infect the cell with the toxic A subunit. Although the single interaction of a B subunit (CTB) with GM1os (**20**) is relatively weak ($K_D = 0.2$ μM), the pentavalent binding of CT to the cell reaches a subnanomolar affinity. This prompted several chemists to try to inhibit the adhesion of CTB to the cell by using multivalent neoglycoconjugates. Zhang et al. prepared a galactose-based pentavalent ligand (**22**) for CTB showing an affinity roughly one order of magnitude weaker than that of the natural monovalent GM1os, but displaying a remarkable affinity enhancement (20,000 per sugar unit) compared with that of a single galactose unit used in the glycocluster.[56] Arosio et al.[57] showed that a calixarene blocked in the cone conformation might also be an effective inhibitor of CT. The divalent glycoconjugate **23**, bearing the GM1 mimic psGM1 (**21**) having a fair affinity for CT ($K_D$ for psGM1 = 190 μM), exhibits an affinity ($K_D = 0.048$ μM) slightly better that that of the natural GM1os (**20**). Extremely efficient inhibitors were also obtained by De Smet et al.[58] using an octavalent dendritic scaffold (**24**) functionalized with GM1os and having an affinity enhancement of 47,500 per sugar unit.

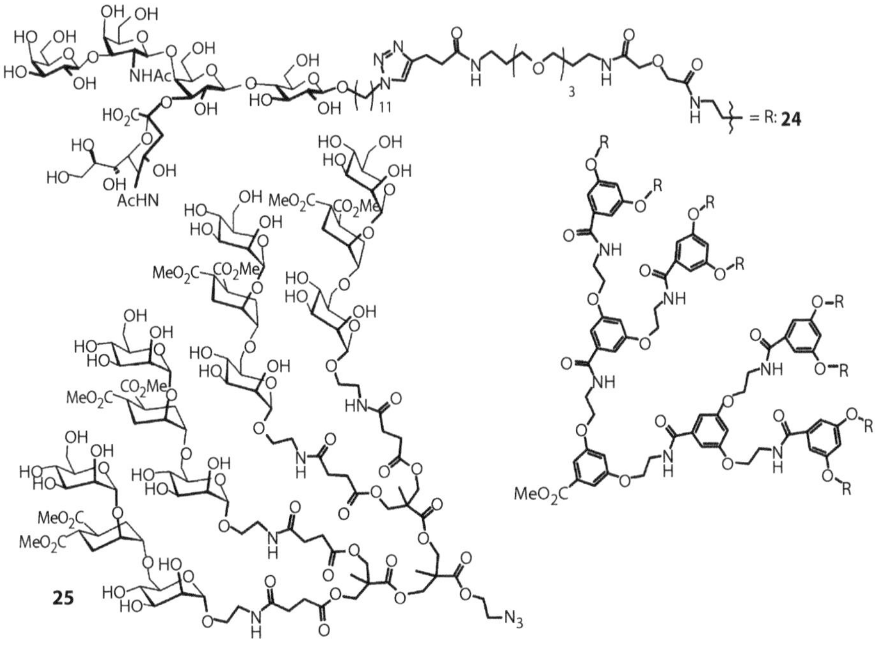

An important lectin to be targeted is dendritic cell–specific ICAM-3 grabbing nonintegrin (DC-SIGN), which is expressed at the surface of immature dendritic cells and is involved in the initial stages of HIV infection. In this context, Sattin et al.[59] showed that, thanks to the good affinity for DC-SIGN, the tetravalent dendron (**25**) containing four copies of a linear trimannoside mimic inhibits the *trans* HIV infection process of CD4+ T lymphocytes at the micromolar level. This compound also presents a high solubility, a negligible cytotoxicity, and a long-lasting effect and is a potential candidate to be used as a topical microbicide.

A critical class of medically relevant lectins is that of human galectins. These proteins are overexpressed by cancer cells and involved in tumor progression and migration. To inhibit the action of Gal-1, Gal-3, and Gal-4, representatives of the three different subgroups of the galectin family, André et al.[60] synthesized a small library of 14 upper rim glycosylthioureidocalix[n]arenes (e.g., **26–33**; Figure 15.6) characterized by different types of sugars (Gal or Lac), valency, size, conformational freedom, and epitope orientation.

Solid phase inhibition assays and *in vitro* experiments with well-established tumor lines indicated that especially lactoclusters (**30–33**) have remarkable inhibition efficiency. Quite interestingly, it was shown that in these processes, not only does a simple, although reinforced, interaction between the protein and the clustered saccharides take place, but also an active role is played by the macrocyclic platform that, depending on either its conformational adaptability or preorganized orientation of the epitopes, determines selectivity among the structurally different lectins. Interestingly, the presence in these potent galectin binders of a lipophilic cavity that is potentially able to include the organic residues of selected chemotherapeutics suggests their possible use as site-specific drug delivery systems.[55]

Aside from these few representative examples, a myriad of novel glycoconjugates have been reported having high affinity for human/plant lectins, toxins, and the protein receptors present on the surface of different types of pathogens such as viruses and bacteria, thus showing the future potential of the glycoside cluster effect for therapeutics and diagnostics in nanomedicine.[53,61,62]

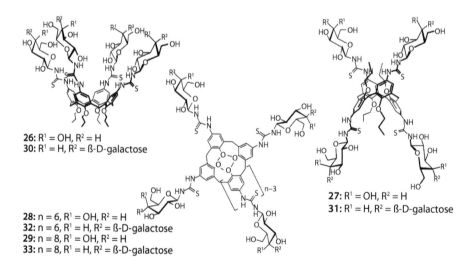

**26:** R¹ = OH, R² = H
**30:** R¹ = H, R² = ß-D-galactose

**28:** n = 6, R¹ = OH, R² = H
**32:** n = 6, R¹ = H, R² = ß-D-galactose
**29:** n = 8, R¹ = OH, R² = H
**33:** n = 8, R¹ = H, R² = ß-D-galactose

**27:** R¹ = OH, R² = H
**31:** R¹ = H, R² = ß-D-galactose

**FIGURE 15.6** Selected members of the library of *upper* rim galactosyl (**26–29**) and lactosyl (**30–33**) thioureidocalixarenes.

## NUCLEIC ACID TARGETING

### RNA Binding by Aminoglycosides

The primary function of RNA is protein synthesis. However, recent discoveries have expanded the cellular role of this natural macromolecule. It has been recognized that RNA is essential for transcriptional and translational regulation, protein function, and catalysis. Most of these biological functions have been targeted,[63–65] although validation has been performed only for a few of them. Compared with DNA, RNA targeting with small molecules is a more difficult task because RNA lacks repair mechanisms and is highly flexible, giving rise to a variety of secondary (Figure 15.7) and tertiary structures that complicate the identification of a general strategy for selective targeting.[66] Notwithstanding these difficulties, molecular recognition of RNA is nowadays considered a very attractive research topic, with very promising therapeutic perspectives. RNAs, such as ribosomal RNA (rRNA), viral RNAs, transfer RNAs, messenger RNAs, riboswitches, and even ribozymes are currently being targeted with small molecules for a better understanding of their functions and for therapeutic purposes.[63–68] We will describe here only a few examples referred to as rRNA targeting[63–65,69,70] in which the supramolecular interactions involved in the molecular recognition events have been clarified.

It has long since been recognized that ribosomes play a central role in the biosynthesis of proteins both in prokaryotic and eukaryotic cells and that, in the former case, this process can be inhibited by certain low–molecular weight natural compounds (aminoglycosides, macrolides, etc.), which therefore act as antibiotics. It is only in the last 10 years, thanks mainly to the studies of the 2009 Nobel laureates, V. Ramakrishnan, T. A. Steitz, and A. Yonath, that the structures of the ribosomal machinery for protein synthesis in bacteria have been elucidated at sufficient resolution to allow structure-based drug design.[71–73] It soon became clear that the fully functioning ribosome, called 70S, responsible for protein synthesis, is constituted by the association of the 50S and 30S subunits, which are composed of rRNA and many

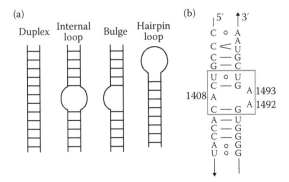

**FIGURE 15.7** Schematic representation of RNA secondary structures with regions potentially amenable to be targeted (a) and secondary structure of the bacterial decoding site (A site) in 16S rRNA (b). Aminoglycoside binding site is marked by a box.

individual ribosomal proteins. The main subunits of the 50S and 30S are 23S and 16S, respectively. With the availability of high-resolution crystal structures of the ribosome subunits, researchers quickly started solving the antibiotic complexes and designing new variants of the old compounds to face the problem of antibiotic resistance.[74–77] The aminoglycoside antibiotics are oligosaccharides containing variable numbers of sugar rings and ammonium groups. They all contain a central 2-deoxy-streptamine moiety (2-DOS or ring II) that has an ammonium group on either side of the deoxycarbon atom and is connected at position 4 to ring I, thus forming the neamine nucleus.

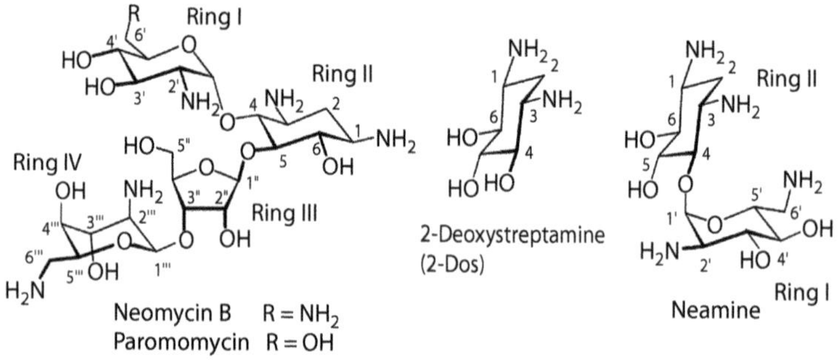

Neomycin B    R = NH$_2$
Paromomycin  R = OH

2-Deoxystreptamine
(2-Dos)

Neamine

    Active antibiotics bind to the 16S rRNA (A site) of the 30S ribosomal subunit (Figure 15.7b). This causes disruption of messenger RNA–decoding fidelity by displacing two flexible adenine residues at a bulge (A1492 and A1493, *Escherichia coli* numbering) in helix 44 of the 16S rRNA and inducing a conformational change which reduces discrimination against noncognate transfer RNAs and decreases translational accuracy.[69,70] Over time, the accumulation of erroneous proteins that are truncated or incorrectly folded leads to bacterial cell death. Because of the anionic nature of RNAs, the most important interactions between aminoglycoside antibiotics and rRNAs are electrostatic, and the binding affinities of these compounds often correlate with the number of their ammonium groups. The second most important interactions are H-bonds, which can be direct or water-mediated. The role of other noncovalent forces such as π–π stacking, C–H/π, or cation/π interactions is still under investigation.[64] Interesting details on the supramolecular aspects of the recognition of rRNA by aminoglycosides were obtained when it was shown that paromomycin, neomycin B, and other aminoglycoside antibiotics bind to small fragments of rRNA (27–49 nucleotides in size) in the same manner as they would do to the complete ribosome.[64,78] The bound paromomycin adopts an L shape, which represents its bioactive conformation (Figure 15.8). Ring I intercalates into the A site and forms two direct H-bonds to the Watson–Crick sites of A 1408 and definite C–H/π interactions to G 1491. The neamine core makes six direct or water-bridged H-bonds to phosphate oxygen atoms of adenines 1492 and 1493, whereas the invariant nitrogen atoms of ring II form constant interactions with A 1493, G 1494, and U 1495. Overall, paromomycin forms 15 direct and 7 water-mediated H-bonds with the A site. The binding of rings I and II force A 1492

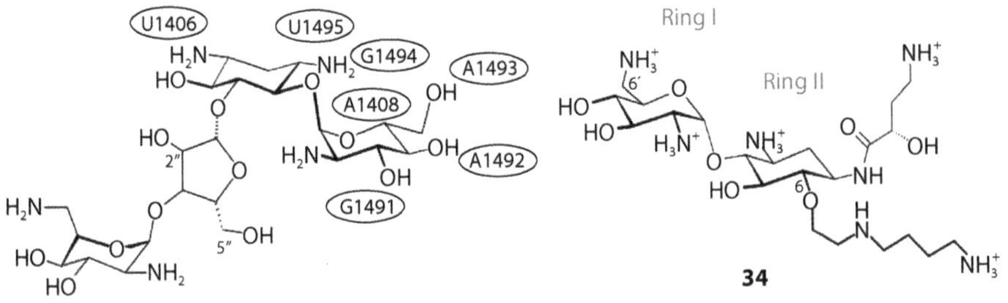

**FIGURE 15.8** The bioactive conformation of paromomycin and a simplified analogue (**34**) of paromomycin.

and A 1493 to bulge out of the deep/major groove. Rings III and IV interact with the lower stem of the A site mainly through nonvariable charge and H-bonding interactions with neighboring nucleotides. Similar structures were also found for the complexes of other active aminoglycosides, which form a comparable number of H-bonds, thus explaining the very similar dissociation constant ($K_D \approx 1.5$ μM) found for the complexes of these antibiotics with rRNA.[64]

These hallmark structural results give a solid ground to the structure-based design of more potent rRNA binders with enhanced antibiotic properties and are possibly able to overcome aminoglycoside toxicity and resistance. For example, the substitution of an amino group at position 6′ in paromomycin, leading to the chemical structure of neomycin B, confers to the latter an increased binding affinity of approximately 10-fold at pH 7. Among the most promising compounds are analogues (e.g., **34**) of paromomycin that contain unique basic extended arms in lieu of some of the rings.[79,80] These side chains are involved in additional H-bonds and electrostatic interactions with the RNA, which account for the superior antibiotic activity (e.g., >90% of translational inhibition at 200 mM of **34**) and the absence of susceptibility to bacterial resistance enzymes. Compound **34** is highly active against various resistant and pathogenic bacteria, including *Pseudomonas aeruginosa* and *Staphylococcus aureus*. The crystal structure of compound **34** complexed to the rRNA A site construct exhibited the same extrahelical conformation of the critical A 1492 and A 1493 residues as in complexes with natural aminoglycosides. This observation confirmed that the molecular basis of action of compound **34** is the same as the one of natural aminoglycosides. Several authors have tried to lock the "bioactive conformation" of aminoglycosides with the aim of obtaining antibiotics that are more potent and specific and less susceptible to enzymatic inactivation compared with natural compounds.[74-77] For example, Bastida et al.[81] synthesized two conformationally restricted neomycin B derivatives having a covalent bond between positions O5$_{Rib}$ and N2$_{Glc}$ that, in natural antibiotics, are H-bonded in the ribosome-bound state. This simple modification, leading to **35a** and **35b**, provided an effective protection against aminoglycoside inactivation by *S. aureus* ANT4 and *Mycobacterium tuberculosis* AAC(2′) both *in vitro* and *in vivo*, while maintaining a reasonable affinity to the target RNA and a significant antibiotic activity, comparable to that of neomycin B.

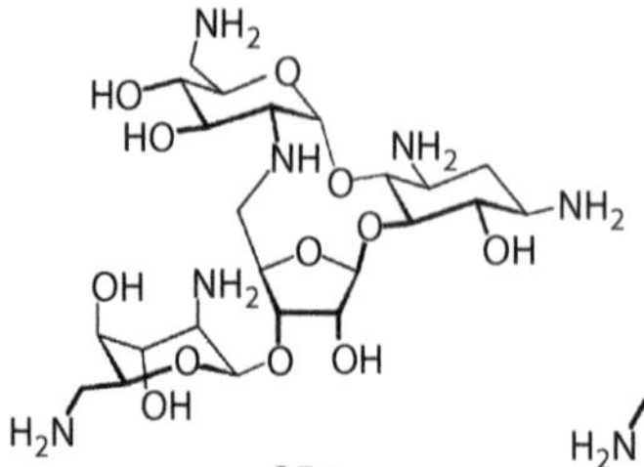

**35a** 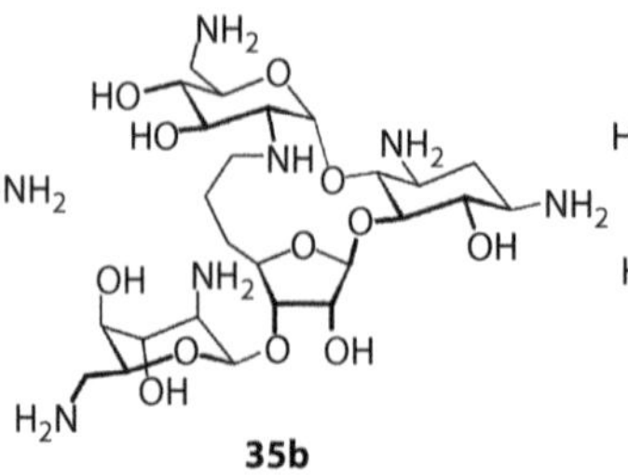          **35b** 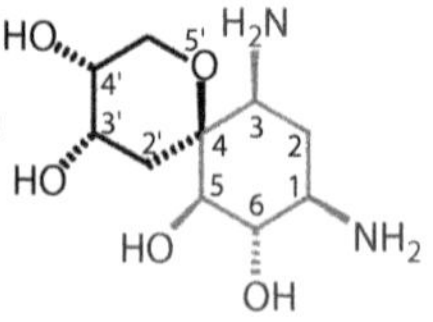          **36**

Katsoulis et al.[82] have recently been inspired by both the concepts of "electro-static complementarity" and "bioactive conformation" to design and synthesize low–molecular weight compounds based on unnatural rigid scaffolds for targeting the bacterial ribosome. Molecular modeling indicated that the spiro compound **36** docked at a bacterial A site model forms H-bonding networks very close to those found in the x-ray crystal structure of the corresponding neamine supramolecular complex. Interestingly, the RNA binding affinity (RNA ligand $EC_{50}$) of the rigid synthetic analogue **36** ($EC_{50} = 6.5$ μM) is lower than that observed for neamine ($EC_{50} = 9.5$ μM), even if **36** possesses half of the electrostatic load present in the neamine, with only two amino groups embedded in its structure. Biological activity, which refers to the inhibition of protein production, is also encouraging.

## DNA TARGETING WITH SMALL MOLECULES

The well known relevance of DNA in determining the growth and life of organisms, and the tight relationship between many diseases and the alterations in genetic data make DNA an obvious and attractive target for therapeutic and diagnostic strategies. A significant number of these is based on the supramolecular noncovalent inter-actions that molecules properly equipped with H-bonding and positively charged groups, aromatic, and lipophilic moieties can establish with nucleic acids, thus alter-ing or stopping their functions.[83] Considering the Watson–Crick double helix, which corresponds to the prevalent B structure, the molecular recognition of DNA can take place mainly in four different regions: (i) in the major groove, (ii) in the minor groove, (iii) through intercalation between the adenine–thymine (A–T) and gua-nine–cytosine (G–C) base pairs, and (iv) through metal coordination to the bases.

Binding to the major groove is performed by molecules that form triplex systems through H-bonding with the Hoogsteen face of the nucleobases. Proteins frequently form complexes with this region of DNA, especially exploiting their α-helices' sec-ondary structures, zinc fingers, and leucine zippers. The complexity and the exten-sion of the interaction, in these cases, are so large that it is nearly impossible to exhaustively reproduce them with fully synthetic molecules. Triplex structures are also formed with natural and synthetic oligonucleotides,[84] among which PNAs,[85] which give rise to Hoogsteen and reverse-Hoogsteen patterns by interaction with the major groove edges of nucleobases. This biologically inspired approach will not be treated in the present chapter.

Differently from the major groove binding, minor groove targeting and intercala-tion can be successfully achieved by using small molecules and this has boosted the research in this field. Following this strategy, a number of biologically active compounds were found and used in clinical applications.[86] Few selected examples will be illustrated herein. In general, the pharmacological activity of these DNA ligands is obtained by inducing changes in the DNA structure, which impairs the correct functions of DNA-processing enzymes, or by forming ternary DNA–ligand–enzyme complexes in which normally transient and labile enzyme–DNA interactions are stabilized. The recognition of the DNA minor groove is performed in general by cationic molecules, often having a bent shape and dimensions that allow them to fit the width of the groove and occupy it.[87] Distamycin is a natural polyamide that binds to DNA in this structural motif.

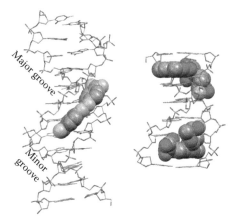

Together with some of its synthetic analogues, such as berenil, pentamidine, and Hoechst 33258, it is used for the treatment of cancers, protozoal diseases, and viral and bacterial infections.

These substantially flat molecules insert parallel to the groove walls (Figure 15.9, left), with a strong preference for A–T-rich regions, where they can penetrate more deeply with respect to sequences where guanine is present. Amidinium or protonated amines in the structure of these molecules drive the binding to DNA by

**FIGURE 15.9** Structure of berenil (space filled) bound in the DNA minor groove (left, PDB ref. 1D63) and structure of two molecules of doxorubicin intercalated into an oligonucleotide (right, PDB ref. 1D12). (From Brown, D. G. et al., *Journal of Molecular Biology*, 226, 481–490, 1992. With permission; Frederick, C. A. et al., *Biochemistry*, 29, 2538–2549, 1990. With permission.)

generating charge–charge interactions with the anionic phosphates. Additional $\pi$–$\pi$ stacking, van der Waals contacts, and H-bonding further stabilize the complexes. The important work by Lown et al.[90] and Dervan and Burli,[91] which addressed the total or partial replacement of the pyrrole rings of distamycin with imidazoles and hydroxypyrroles, produced a series of analogues (e.g., **37**) with a selectivity shifted to G–C reach sequences, resulting from a different pattern of H-bonding acceptors and donors. Dervan defined a set of "heterocyclic pairing rules" for the interaction of these polyamides with DNA base pairs.[92] Imidazole instead of pyrrole also gives these derivatives the ability to recognize thymine–guanine mismatches[93] (Figure 15.10) that are at the origin of numerous tumors in humans. The presence of the additional nitrogen on the imidazole ring, in fact, allows H-bonding with the guanine N2 amino group residing in the minor groove and not involved in binding with thymine which, on the contrary, happens with cytosine. Distamycin and these synthetic polyamides were also shown to bind to DNA as $\pi$-stacked dimers, widening, if necessary, the groove width.[94] The noncovalent dimer is constituted of two polyamides lying head-to-tail to push away the two cationic groups. In this way, they simultaneously recognize the bases of both DNA strands, improving selectivity and efficiency. The covalent linkage of these ligand dimers[95] resulted in hairpins with enhanced affinity for DNA with respect to the two independent monomers, decreasing the entropic costs of the complex formation.

Planar, small organic molecules based on at least two fused aromatic rings tend to bind to DNA as intercalators (Figure 15.9, right). They occupy the space between two stacked base pairs, filling this hydrophobic pocket and creating a "sandwich" held together by $\pi$–$\pi$ interactions with the bases above and below. When the space between two base pairs is occupied by a ligand molecule, the adjacent pocket cannot be filled because of the distortion of the structure. This effect is identified as the neighbor group exclusion principle. The first example of DNA binding by intercalation was described by Lerman,[96] using the acridine derivative proflavine as ligand which was since then adopted as the basic structure for the development of a wide variety of synthetic intercalators. Among the most used, studied, and characterized DNA intercalators are the natural anthracyclines, for example doxorubicin (Adriamycin), and indolocarbazoles, for example staurosporine and rebeccamycin.

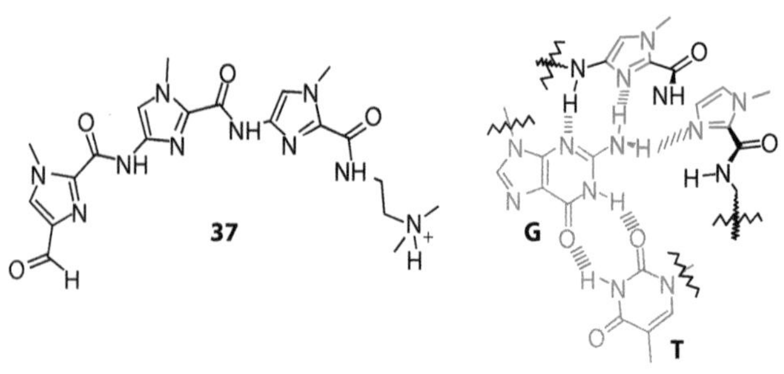

**FIGURE 15.10** Example of polyimidazole polyamide (**37**) and schematic representation of T–G mismatch recognition by its imidazole rings. (From Yang, X. L. et al., *Nucleic Acids Research*, 27, 4183–4190, 1999. With permission.)

Doxorubicin

Proflavin

Voreloxin

**38**

Together with their synthetic analogues, these natural compounds are used in cancer therapy (anticancer antibiotics) as drugs with a broad spectrum of activity. The intercalation process primarily determines a distortion of the double helix that impairs the normal cleavage and relegation functions of topoisomerases. Intercalation is also the mode of binding of voreloxin,[97] a novel naphthyridine analogue, currently under investigation for the treatment of different types of malignancies and for acute myeloid leukemia. Extended fused aromatic molecules can also bind to triplex and quadruplex DNA structures. The binding to triple strands is favored with respect to duplex DNAs when the size of the ligand would leave part of its surface exposed to the aqueous environment in the complex with the double helix.[98] With guanine quartets, the interaction can also involve more than one molecule of smaller duplex intercalators. In most cases, however, the interaction with quadruplex structures is not a real intercalation but rather a stacking between the aromatic surface of the ligands and the four guanines at the end of the quartet motif, in the proximity of the loops that connect the four strands involved in the quadruplex structure.[99] As in the case of the minor groove binders, polar or positively charged groups are frequently present in the structure of both natural and synthetic intercalators, ensuring an important contribution to the whole binding energy by interacting with complementary groups in either the major or the minor groove. In some cases, the positive charge consists of a transition metal cation incorporated through coordination to a polyaromatic ligand, such as terpyridine (e.g., **38**).[100] The transition metal, mainly Ru and Rh, furnishes interesting additional properties such as luminescence for detection and cleavage activity through photoinduced oxidation of the bound DNA molecule. Metallointercalators with large aromatic surfaces due to more than one polyaromatic ligands can penetrate the space between stacked bases where a base is missing, that is when a defect in the DNA double helix is present. The strand position lacking a base is an important therapeutic target because, when occurring in a gene of a pathogen or of a cancer cell, it is possible to prevent its repair causing relevant damage to them. Synthetic heterodimers have been specifically designed for this purpose, bearing an acridine and a nucleic base connected through a linker. The nucleic base inserts into the space left by the defect in the double helix and interacts through H-bonds with the complementary base of the other strand. The acridine unit acts as an intercalator and the linker can establish additional interactions in the groove or confer nuclease-like activity to the ligand.[101] The coordination of a metal to nitrogen atoms of nucleic bases to affect DNA transcription and replication is at the basis of the successful class of anticancer agents constituted by platinum compounds. Cis-platinum is the lead compound of this family and the most famous and potent drug among all the chemotherapeutics against ovarian and testicular tumors or other carcinomas. The bond formed with purines, preferentially guanines, is at the borderline between a noncovalent and a covalent interaction, and "platinated" DNAs can be substantially considered, as alkylated nucleic acids, inert species. Moreover, the long history and the extraordinary clinical efficacy of this simple, inorganic compound and its derivatives have produced a wide and extensive description of its mode of action.[102]

## DNA CONDENSATION AND CELL TRANSFECTION

Another important aspect related to the supramolecular recognition of DNA is that of gene delivery, an essential process for the development of gene therapy. Gene therapy represents a method with great potentiality for the treatment of inherited diseases, neuropathies, tumors, cardiovascular affections, cystic fibrosis, and many other severe health deficiencies and disorders. It is based on the delivery of specific DNAs inside the cells for the correction of altered genetic sequences, the inhibition of defective genes, the transport of suicide genes, the expression of therapeutic or prophylactic proteins and antigens. In general, gene delivery is achieved by a carrier that interacts with nucleic acids, masks their negative charges, and condenses their elongated forms in compacted species which can be uptaken by the cells. However, the success of DNA transport into the cells depends on so many aspects and parameters that these properties, although necessary, cannot be considered sufficient. Viruses are the most effective vectors for the introduction of genetic material into cells,[103] but their use in therapy is limited by serious drawbacks, in particular, high immunogenicity, violent inflammatory response, and potential for mutagenesis.[104] As an alternative, many research groups have been developing nonviral vectors for several years now.[105–107] Although the compounds designed thus far with this objective have not reached the efficiency and, more importantly, the selectivity of viruses, some of them have been successfully used in transfection protocols. In general, being constituted by positively charged heads linked, through a spacer, to lipophilic segments, the synthetic vectors have amphiphilic character and tend to self-aggregate in water forming liposomes, micelles, and vesicles. The cationic groups bind to the phosphate units of the nucleic acid molecules and to the negatively charged cell surfaces. The lipophilic tails participate in the condensation process, in the membrane fusion and crossing, and in the nucleic acid escape from the endosomes by bringing into play hydrophobic interactions (Figure 15.11).

Because of the complexation process, the original elongated and relaxed DNA folding dramatically changes, resulting in particles of nanometric spherical, toroidal, or rod shapes.[108] The zeta potential of the complex, due to the balance between negative and positive charges of DNA and vector molecules, is also an important factor influencing the delivery process. The uptake into cells of the vector–DNA complex most frequently occurs by endocytosis,[109] but the pathway to the cytosol is not always clear and completely identified. Nonviral vectors can be divided into different classes depending on their main structural features. The most important, successful, and widely used family is that of the cationic lipids, whose complexes with the nucleic acid filaments are called lipoplexes and the related DNA transfection

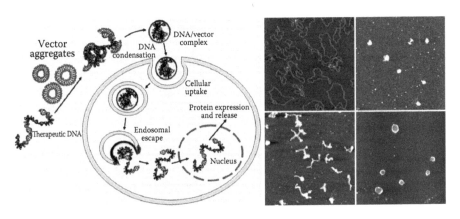

**FIGURE 15.11** (Left) Schematic representation of gene delivery process performed by amphiphilic nonviral vectors. (Right) AFM images (2 × 2 μm) in tapping mode on air showing a circular plasmid DNA in its so-called relaxed form (top left) and the same plasmid modified in its folding by the presence of different nonviral vectors.

procedure is called lipofection. The development of such compounds for gene delivery was pioneered by Felgner et al.[110] who, in 1987, reported on the preparation and gene delivery properties of DOTMA. Since then, many other derivatives have been produced, by varying the nature and number of the polar heads, the structure of the lipophilic portions of the vector, and the features of the spacer between them. Beyond DOTMA, some relevant examples having high efficiency and low toxicity are DOTAP, DOSPA, DORIE, DMRIE, DOGS, and DC-Chol.[105–107]

Most often, these nonviral vectors are used in formulation with a so-called helper lipid, such as 1,2-dioleoyl phosphatidylethanolamine (DOPE) or cholesterol, involved through supramolecular interactions in the formation of mixed liposomes, DNA compaction and membrane fusion. Some of these cationic lipids reached advanced phases in clinical trials for the treatment of numerous malignant cancers and are currently the most used commercially available products for transfection protocols.[111] Some cationic lipids present multiple cationic heads, as in the case of DOSPA and DOGS, which contribute to enhance the transfection efficiency. The presence in the lipid structure of several protonable amines can help the efficiency of transfection thanks to the "proton sponge effect," which favors DNA release from endosomes and protects the biopolymer from lysosomal attack and degradation. The decoration of the cationic lipid structure with moieties acting as specific ligands for cell surface receptors can produce vectors for targeted gene delivery. The presence, for example, of galactose units in the proximity of the ammonium groups allowed some cationic lipids the ability to transfect hepatic cells[112] that expose to the extracellular matrix receptors selective for asialoglycoproteins having galactosides as epitopes. A cationic amphiphile modified with the RGDK motif in its polar head group was successfully used in formulation with proper genes for targeted transfection of cells exhibiting on their surface integrin receptors for this peptide motif.[113] The presence in the cationic lipid structure of two mono- or polycationic head groups connected by a tether represents the structural feature characterizing the Gemini surfactants.[114] A high number of cationic groups can increase the affinity of the vector for the DNA molecule and, in principle, the transfection efficiency, but it is often accompanied by a higher toxicity. Nevertheless, polymers bearing positively charged groups are also successfully used as vectors.[115] With respect to cationic lipids, they lack a hydrophobic chain. Polyethylene imine (PEI) represents one of the most prominent examples of this class, since the first successful experiment of transfection mediated by this polymer was reported in 1995 by Boussif et al.[116] The complex between a cationic polymer and DNA molecules is called a polyplex. Sometimes, cationic lipids and cationic polymers are used in mixture and the ternary supramolecular complex is called a lipopolyplex. Considering that the limits in efficiency, toxicity, and selectivity related to the use of these nonviral vectors are not yet definitely overcome, the development of effective modifications in their structure and formulation, and the search for new molecular entities for improving gene delivery efficiency and reducing cytotoxicity is still active and lively. Other molecular architectures have been and are currently explored as further possible alternatives to viruses. It is the case of the dendrimer-based vectors,[117] with the most relevant built on polyamidoamine (PAMAM) and polypropylenimine structures, of the polypeptide vectors,[118] in particular, polylysines and polyarginines, of nanoparticles, carbon nanotubes, and macrocycles. Concerning the latter ones, although cyclodextrins have been used since the 1990s as "components" of gene delivery systems,[119] with the aim of improving the vector biocompatibility and overcoming cytotoxicity, only recently they have been exploited as downright vectors adorning their two rims with positively charged groups, such as ammonium and guanidinium, and lipophilic tails.[120–122] Also, calixarenes were identified as interesting scaffolds for the preparation of a new class of vectors using guanidinium as a charged head.[52,54] Derivatives with different structural and conformational features, with a different number of cationic groups and lipophilic/hydrophilic ratios were tested as nonviral vectors and showed transfection ability and toxicity levels strictly depending on finely tuned structural modifications.[123] One of them (**39**), in particular, mixed with DOPE, resulted more efficient than Lipofectamine LTX™ in the treatment of human Rhabdomyosarcoma and Vero cells, and also showed a remarkably low toxicity.[124]

Analogously to DNA delivery, the transport inside cells of RNA is an emerging field with therapeutic purposes through the targeted posttranscriptional silencing of undesired gene expression.[125] This process is known as RNA interference.[126] Also in this case, suitable vectors are needed that perform the delivery of the nucleic acid molecules, specifically *ds*RNAs, which inside the cells are properly modified in the so-called short interfering RNAs.[106] The first application of this approach that reached clinical trials was for the treatment of macular degeneration. The suppression of genes essential for the life of viral and bacterial pathogens could be another promising application for this therapeutic strategy.

## CONCLUSIONS

Since its birth in the late 1960s, supramolecular chemistry has been continuously inspired by the biological world, and by the many natural processes regulated by noncovalent interactions. Nowadays, the two fields of biosupramolecular chemistry and chemical biology, which developed in parallel in the last 30 years, are merging to give important contributions in the understanding of biological processes and in the development of new drugs. Proteins and nucleic acids represent the two most important classes of biological macromolecules which encode information essential for the life of organisms, and their targeting by small molecules has already produced valuable results. Protein–protein and protein–carbohydrate interactions, which regulate many physiological or pathological processes in living organisms, can be efficiently inhibited by a variety of synthetic compounds developed through the application of new supramolecular concepts such as protein surface recognition, mimetics of protein secondary structures, multivalent presentation of binding sites, and DCC. The molecular mechanism of aminoglycoside antibiotics has been clarified by solving the structure of rRNA and new antibiotics with improved properties have been obtained by the structure-based design that complements the combinatorial approach. DNA targeting by small molecules has allowed a deeper understanding of DNA intercalation and recognition, guiding the synthesis of a variety of efficient drugs used in the treatment of cancer and viral/bacterial infections. On the other hand, the study of cell internalization of DNA condensates allowed the development of efficient synthetic vectors that replace viruses in gene therapy. Aside from the therapeutic aspects, supramolecular systems are also addressed to diagnosis and follow-up, thus promising to be important actors in nanomedicine[127] and theranostics.[128,129]

## ACKNOWLEDGMENTS

The authors are grateful to the Ministero dell'Istruzione, dell'Università e della Ricerca (MIUR, PRIN 2008 project nos. 2008HZJW2L and 200858SA98).

## REFERENCES

1. Uhlenheuer, D. A., K. Petkau, L. Brunsveld. 2010. *Chemical Society Reviews*, 39, 2817–2826.
2. Wells, J. A., C. L. McClendon. 2007. *Nature*, 450, 1001–1009.
3. Babine, R. E., S. L. Bender. 1997. *Chemical Reviews*, 97, 1359–1472.
4. Yin, H., A. D. Hamilton. 2005. *Angewandte Chemie (International ed. in English)*, 44, 4130–4163.
5. DeLano, W. L. 2002. *Current Opinion in Structural Biology*, 12, 14–20.
6. Hamuro, Y., M. Crego-Calama, H. S. Park, A. D. Hamilton. 1997. *Angewandte Chemie (International ed. in English)*, 36, 2680–2683.
7. Casnati, A., F. Sansone, R. Ungaro. 2003. *Accounts of Chemical Research*, 36, 246–254.
8. Coleman, A. W., F. Perret, A. Moussa, M. Dupin, Y. Gu, H. Perron. 2007. *Creative Chemical Sensor Systems*, 277, 31–88.
9. Rodik, R. V., V. I. Boyko, V. I. Kalchenko. 2009. *Current Medicinal Chemistry*, 16, 1630–1655.
10. Park, H. S., Q. Lin, A. D. Hamilton. 1999. *Journal of the American Chemical Society*, 121, 8–13.
11. Blaskovich, M. A., Q. Lin, F. L. Delarue, J. Sun, H. S. Park, D. Coppola, A. D. Hamilton, S. M. Sebti. 2000. *Nature Biotechnology*, 18, 1065–1070.
12. Jain, R., J. T. Ernst, O. Kutzki, H. S. Park, A. D. Hamilton. 2004. *Molecular Diversity*, 8, 89–100.
13. Sun, J. Z., D. A. Wang, R. K. Jain, A. Carie, S. Paquette, E. Ennis, M. A. Blaskovich, L. Baldini, D. Coppola, A. D. Hamilton, S. M. Sebti. 2005. *Oncogene*, 24, 4701–4709.
14. Gordo, S., V. Martos, E. Santos, M. Menendez, C. Bo, E. Giralt, J. de Mendoza. 2008. *Proceedings of the National Academy of Sciences of the United States of America*, 105, 16426–16431.
15. Jain, R. K., A. D. Hamilton. 2000. *Organic Letters*, 2, 1721–1723.
16. Aya, T., A. D. Hamilton. 2003. *Bioorganic & Medicinal Chemistry Letters*, 13, 2651–2654.
17. Zhou, H. C., L. Baldini, J. Hong, A. J. Wilson, A. D. Hamilton. 2006. *Journal of the American Chemical Society*, 128, 2421–2425.
18. Baldini, L., A. J. Wilson, J. Hong, A. D. Hamilton. 2004. *Journal of the American Chemical Society*, 126, 5656–5657.
19. Takashima, H., S. Shinkai, I. Hamachi. 1999. *Chemical Communications*, 2345–2346.
20. Muldoon, J., A. E. Ashcroft, A. J. Wilson. 2010. *Chemistry—A European Journal*, 16, 100–103.
21. Ohkanda, J., R. Satoh, N. Kato. 2009. *Chemical Communications*, 6949–6951.
22. Wilson, A. J., J. Hong, S. Fletcher, A. D. Hamilton. 2007. *Organic & Biomolecular Chemistry*, 5, 276–285.
23. Ojida, A., M. Inoue, Y. Mito-oka, I. Hamachi. 2003. *Journal of the American Chemical Society*, 125, 10184–10185.
24. Ojida, A., M. Inoue, Y. Mito-oka, H. Tsutsumi, K. Sada, I. Hamachi. 2006. *Journal of the American Chemical Society*, 128, 2052–2058.
25. Paul, D., H. Miyake, S. Shinoda, H. Tsukube. 2006. *Chemistry—A European Journal*, 12, 1328–1338.
26. Azuma, H., Y. Yoshida, D. Paul, S. Shinoda, H. Tsukube, T. Nagasaki. 2009. *Organic & Biomolecular Chemistry*, 7, 1700–1704.
27. Fischer, N. O., C. M. McIntosh, J. M. Simard, V. M. Rotello. 2002. *Proceedings of the National Academy of Sciences of the United States of America*, 99, 5018–5023.
28. Tagore, D. M., K. I. Sprinz, S. Fletcher, J. Jayawickramarajah, A. D. Hamilton. 2007. *Angewandte Chemie (International ed. in English)*, 46, 223–225.

29. Cai, J. F., B. A. Rosenzweig, A. D. Hamilton. 2009. *Chemistry—A European Journal*, 15, 328–332.
30. Hayashida, O., N. Ogawa, M. Uchiyama. 2007. *Journal of the American Chemical Society*, 129, 13698–13705.
31. Leung, D. K., Z. W. Yang, R. Breslow. 2000. *Proceedings of the National Academy of Sciences of the United States of America*, 97, 5050–5053.
32. Hirschmann, R., K. C. Nicolaou, S. Pietranico, E. M. Leahy, J. Salvino, B. Arison, M. A. Cichy, P. G. Spoors, W. C. Shakespeare, P. A. Sprengeler, P. Hamley, A. B. Smith, T. Reisine, K. Raynor, L. Maechler, C. Donaldson, W. Vale, R. M. Freidinger, M. R. Cascieri, C. D. Strader. 1993. *Journal of the American Chemical Society*, 115, 12550–12568.
33. Cummings, C. G., A. D. Hamilton. 2010. *Current Opinion in Chemical Biology*, 14, 341–346.
34. Fletcher, S., A. D. Hamilton. 2006. *Journal of the Royal Society Interface*, 3, 215–233.
35. Saraogi, I., A. D. Hamilton. 2008. *Biochemical Society Transactions*, 36, 1414–1417.
36. Rees, D. C., M. Congreve, C. W. Murray, R. Carr. 2004. *Nature Reviews Drug Discovery*, 3, 660–672.
37. Shuker, S. B., P. J. Hajduk, R. P. Meadows, S. W. Fesik. 1996. *Science*, 274, 1531–1534.
38. Oltersdorf, T., S. W. Elmore, A. R. Shoemaker, R. C. Armstrong, D. J. Augeri, B. A. Belli, M. Bruncko, T. L. Deckwerth, J. Dinges, P. J. Hajduk, M. K. Joseph, S. Kitada, S. J. Korsmeyer, A. R. Kunzer, A. Letai, C. Li, M. J. Mitten, D. G. Nettesheim, S. Ng, P. M. Nimmer, J. M. O'Connor, A. Oleksijew, A. M. Petros, J. C. Reed, W. Shen, S. K. Tahir, C. B. Thompson, K. J. Tomaselli, B. L. Wang, M. D. Wendt, H. C. Zhang, S. W. Fesik, S. H. Rosenberg. 2005. *Nature*, 435, 677–681.
39. Kolb, H. C., K. B. Sharpless. 2003. *Drug Discovery Today*, 8, 1128–1137.
40. Shelke, S. V., B. Cutting, X. H. Jiang, H. Koliwer-Brandl, D. S. Strasser, O. Schwardt, S. Kelm, B. Ernst. 2010. *Angewandte Chemie (International ed. in English)*, 49, 5721–5725.
41. Corbett, P. T., J. Leclaire, L. Vial, K. R. West, J. L. Wietor, J. K. M. Sanders, S. Otto. 2006. *Chemical Reviews*, 106, 3652–3711.
42. Lehn, J. M., A. V. Eliseev. 2001. *Science*, 291, 2331–2332.
43. Ramstrom, O., T. Bunyapaiboonsri, S. Lohmann, J. M. Lehn. 2002. *Biochim. Biophys. Acta, Gen. Subj*, 1572, 178–186.
44. Erlanson, D. A., S. K. Hansen. 2004. *Current Opinion in Chemical Biology*, 8, 399–406.
45. Bhat, V. T., A. M. Caniard, T. Luksch, R. Brenk, D. J. Campopiano, M. F. Greaney. 2010. *Nature Chemistry*, 2, 490–497.
46. Huc, I., J. M. Lehn. 1997. *Proceedings of the National Academy of Sciences of the United States of America*, 94, 2106–2110.
47. Lis, H., N. Sharon. 1998. *Chemical Reviews*, 98, 637–674.
48. Gabius, H.-J., ed. 2009. *The Sugar Code*, Weinheim: Wiley-Blackwell.
49. Mammen, M., S.-K. Choi, G. M. Whitesides. 1998. *Angewandte Chemie (International ed. in English)*, 37, 2755–2794.
50. Lee, Y. C., R. T. Lee. 1995. *Accounts of Chemical Research*, 28, 321–327.
51. Choi, S.-K. 2004. *Synthetic Multivalent Molecules*, Choi, S.-K. ed., Hoboken, NJ: John Wiley & Sons, Inc.
52. Baldini, L., A. Casnati, F. Sansone, R. Ungaro. 2007. *Chemical Society Reviews*, 36, 254–266.
53. Chabre, Y. M., R. Roy. 2010. In *Advances in Carbohydrate Chemistry and Biochemistry*, Derek, H., ed., Vol. 63, 165–393, Academic Press.
54. Sansone, F., L. Baldini, A. Casnati, R. Ungaro. 2010. *New Journal of Chemistry*, 34, 2715–2728.
55. Sansone, F., G. Rispoli, A. Casnati, R. Ungaro. 2011. In *Synthesis and Biological Applications of Glycoconjugates*. Renaudet, O., Spinelli, N., eds., 36–63, Dordrecht: Bentham Science Publishers, doi:10.2174/978160805277611101010036.

56. Zhang, Z. S., E. A. Merritt, M. Ahn, C. Roach, Z. Hou, C. L. M. J. Verlinde, W. G. J. Hol, E. Fan. 2002. *Journal of the American Chemical Society*, 124, 12991–12998.
57. Arosio, D., M. Fontanella, L. Baldini, L. Mauri, A. Bernardi, A. Casnati, F. Sansone, R. Ungaro. 2005. *Journal of the American Chemical Society*, 127, 3660–3661.
58. De Smet, L. C. P. M., A. V. Pukin, G. A. Stork, C. H. Ric de Vos, G. M. Visser, H. Zuilhof, E. J. R. Sudhoelter. 2004. *Carbohydrate Research*, 339, 2599–2605.
59. Sattin, S., A. Daghetti, M. Thepaut, A. Berzi, M. Sanchez-Navarro, G. Tabarani, J. Rojo, F. Fieschi, M. Clerici, A. Bernardi. 2010. *ACS Chemical Biology*, 5, 301–312.
60. André, S., F. Sansone, H. Kaltner, A. Casnati, J. Kopitz, H. J. Gabius, R. Ungaro. 2008. *ChemBioChem*, 9, 1649–1661.
61. de Paz, J. L., P. H. Seeberger. 2006. *QSAR & Combinatorial Science*, 25, 1027–1032.
62. Doores, K. J., D. P. Gamblin, B. G. Davis. 2006. *Chemistry—A European Journal*, 12, 656–665.
63. Thomas, J. R., P. J. Hergenrother. 2008. *Chemical Reviews*, 108, 1171–1224.
64. Vicens, Q. 2009. *Journal of Inclusion Phenomena and Macrocyclic Chemistry*, 65, 171–188.
65. Aboul-Ela, F. 2010. *Future Medicinal Chemistry*, 2, 93–119.
66. Chow, C. S., F. M. Bogdan. 1997. *Chemical Reviews*, 97, 1489–1513.
67. Westhof, E. 2010. *Genome Biology*, 11.
68. Fulle, S., H. Gohlke. 2010. *Journal of Molecular Recognition*, 23, 220–231.
69. Sherer, E. C. 2010. *Annual Reports in Computational Chemistry*, 6, 139–166.
70. Matt, T., R. Akbergenov, D. Shcherbakov, E. C. Bottger. 2010. *Israel Journal of Chemistry*, 50, 60–70.
71. Yonath, A. 2010. *Angewandte Chemie (International ed. in English)*, 49, 4340–4354.
72. Ramakrishnan, V. 2010. *Angewandte Chemie (International ed. in English)*, 49, 4355–4380.
73. Steitz, T. A. 2010. *Angewandte Chemie (International ed. in English)*, 49, 4381–4398.
74. Houghton, J. L., K. D. Green, W. J. Chen, S. Garneau-Tsodikova. 2010. *Chembiochem*, 11, 880–902.
75. Hainrichson, M., I. Nudelman, T. Baasov. 2008. *Organic & Biomolecular Chemistry*, 6, 227–239.
76. Hermann, T. 2007. *Cellular and Molecular Life Sciences*, 64, 1841–1852.
77. Zhou, J., G. Wang, L. H. Zhang, X. S. Ye. 2007. *Medicinal Research Reviews*, 27, 279–316.
78. Vicens, Q., E. Westhof. 2003. *Biopolymers*, 70, 42–57.
79. Haddad, J., L. P. Kotra, B. Llano-Sotelo, C. Kim, E. F. Azucena, M. Z. Liu, S. B. Vakulenko, C. S. Chow, S. Mobashery. 2002. *Journal of the American Chemical Society*, 124, 3229–3237.
80. Russell, R. J. M., J. B. Murray, G. Lentzen, J. Haddad, S. Mobashery. 2003. *Journal of the American Chemical Society*, 125, 3410–3411.
81. Bastida, A., A. Hidalgo, J. L. Chiara, M. Torrado, F. Corzana, J. M. Perez-Canadillas, P. Groves, E. Garcia-Junceda, C. Gonzalez, J. Jimenez-Barbero, J. L. Asensio. 2006. *Journal of the American Chemical Society*, 128, 100–116.
82. Katsoulis, I. A., C. Pyrkotis, A. Papakyriakou, G. Kythreoti, A. L. Zografos, I. Mavridis, V. R. Nahmias, P. Anastasopoulou, D. Vourloumis. 2009. *ChemBioChem*, 10, 1969–1972.
83. Hannon, M. J. 2007. *Chemical Society Reviews*, 36, 280–295.
84. Da Ros, T., G. Spalluto, M. Prato, T. Saison-Behmoaras, A. Boutorine, B. Cacciari. 2005. *Current Medicinal Chemistry*, 12, 71–88.
85. Nielsen, P. E. 2010. *Current Opinion in Molecular Therapeutics*, 12, 184–191.
86. Demeunynck, M., C. Bailly, W. D. Wilson, eds. 2003. *DNA and RNA Binders*, Weinheim: Wiley-VCH Verlag GmbH & Co. KGaA.

87. Baraldi, P. G., A. Bovero, F. Fruttarolo, D. Preti, M. A. Tabrizi, M. G. Pavani, R. Romagnoli. 2004. *Medicinal Research Reviews*, 24, 475–528.
88. Brown, D. G., M. R. Sanderson, E. Garman, S. Neidle. 1992. *Journal of Molecular Biology*, 226, 481–490.
89. Frederick, C. A., L. D. Williams, G. Ughetto, G. A. Vandermarel, J. H. Vanboom, A. Rich, A. H. J. Wang. 1990. *Biochemistry*, 29, 2538–2549.
90. Lown, J. W., K. Krowicki, U. G. Bhat, A. Skorobogaty, B. Ward, J. C. Dabrowiak. 1986. *Biochemistry*, 25, 7408–7416.
91. Dervan, P. B., R. W. Burli. 1999. *Current Opinion in Chemical Biology*, 3, 688–693.
92. White, S., E. E. Baird, P. B. Dervan. 1997. *Chemistry & Biology*, 4, 569–578.
93. Yang, X. L., R. B. Hubbard, M. Lee, Z. F. Tao, H. Sugiyama, A. H. J. Wang. 1999. *Nucleic Acids Research*, 27, 4183–4190.
94. Pelton, J. G., D. E. Wemmer. 1989. *Proceedings of the National Academy of Sciences of the United States of America*, 86, 5723–5727.
95. Mrksich, M., P. B. Dervan. 1994. *Journal of the American Chemical Society*, 116, 3663–3664.
96. Lerman, L. S. 1961. *Journal of Molecular Biology*, 3, 18–30.
97. Advani, R. H., H. I. Hurwitz, M. S. Gordon, S. W. Ebbinghaus, D. S. Mendelson, H. A. Wakelee, U. Hoch, J. A. Silverman, N. A. Havrilla, C. J. Berman, J. A. Fox, R. S. Allen, D. C. Adelman. 2010. *Clinical Cancer Research*, 16, 2167–2175.
98. Escude, C., C. H. Nguyen, S. Kukreti, Y. Janin, J. S. Sun, E. Bisagni, T. Garestier, C. Helene. 1998. *Proceedings of the National Academy of Sciences of the United States of America*, 95, 3591–3596.
99. Haider, S. M., G. N. Parkinson, S. Neidle. 2003. *Journal of Molecular Biology*, 326, 117–125.
100. Howe-Grant, M., S. J. Lippard. 1979. *Biochemistry*, 18, 5762–5769.
101. Belmont, P., M. Jourdan, M. Demeunynck, J. F. Constant, J. Garcia, J. Lhomme, D. Carez, A. Croisy. 1999. *Journal of Medicinal Chemistry*, 42, 5153–5159.
102. Jamieson, E. R., S. J. Lippard. 1999. *Chemical Reviews*, 99, 2467–2498.
103. Hacein-Bey-Abina, S., F. Le Deist, F. Carlier, C. Bouneaud, C. Hue, J. De Villartay, A. J. Thrasher, N. Wulffraat, R. Sorensen, S. Dupuis-Girod, A. Fischer, M. Cavazzana-Calvo, E. G. Davies, W. Kuis, W. H. K. Lundlaan, L. Leiva. 2002. *New England Journal of Medicine*, 346, 1185–1193.
104. Check, E. 2003. *Nature*, 423, 573–574.
105. Mintzer, M. A., E. E. Simanek. 2009. *Chemical Reviews*, 109, 259–302.
106. Srinivas, R., S. Samanta, A. Chaudhuri. 2009. *Chemical Society Reviews*, 38, 3326–3338.
107. Bhattacharya, S., A. Bajaj. 2009. *Chemical Communications*, 4632–4656.
108. Ullner, M. 2008. In *DNA Interactions with Polymers and Surfactants*. Dias, R. S., Lindman, B., eds, 1–39, Hoboken, NJ: John Wiley & Sons, Inc.
109. Zabner, J., A. J. Fasbender, T. Moninger, K. A. Poellinger, M. J. Welsh. 1995. *Journal of Biological Chemistry*, 270, 18997–19007.
110. Felgner, P. L., T. R. Gadek, M. Holm, R. Roman, H. W. Chan, M. Wenz, J. P. Northrop, G. M. Ringold, M. Danielsen. 1987. *Proceedings of the National Academy of Sciences of the United States of America*, 84, 7413–7417.
111. Dass, C. R. 2004. *Journal of Molecular Medicine*, 82, 579–591.
112. Mukthavaram, R., S. Marepally, M. Y. Venkata, G. N. Vegi, R. Sistla, A. Chaudhuri. 2009. *Biomaterials*, 30, 2369–2384.
113. Pramanik, D., B. K. Majeti, G. Mondal, P. P. Karmali, R. Sistla, O. G. Ramprasad, G. Srinivas, G. Pande, A. Chaudhuri. 2008. *Journal of Medicinal Chemistry*, 51, 7298–7302.
114. Bell, P. C., M. Bergsma, I. P. Dolbnya, W. Bras, M. C. A. Stuart, A. E. Rowan, M. C. Feiters, J. B. F. N. Engberts. 2003. *Journal of the American Chemical Society*, 125, 1551–1558.

115. Midoux, P., G. Breuzard, J. P. Gomez, C. Pichon. 2008. *Current Gene Therapy*, 8, 335–352.
116. Boussif, O., F. Lezoualch, M. A. Zanta, M. D. Mergny, D. Scherman, B. Demeneix, J. P. Behr. 1995. *Proceedings of the National Academy of Sciences of the United States of America*, 92, 7297–7301.
117. Guillot-Nieckowski, M., D. Joester, M. Stohr, M. Losson, M. Adrian, B. Wagner, M. Kansy, H. Heinzelmann, R. Pugin, F. Diederich, J. L. Gallani. 2007. *Langmuir*, 23, 737–746.
118. Wu, G. Y., C. H. Wu. 1988. *Journal of Biological Chemistry*, 263, 14621–14624.
119. Li, J., X. J. Loh. 2008. *Advanced Drug Delivery Reviews*, 60, 1000–1017.
120. Srinivasachari, S., K. M. Fichter, T. M. Reineke. 2008. *Journal of the American Chemical Society*, 130, 4618–4627.
121. Mellet, C. O., J. M. Benito, J. M. G. Fernandez. 2010. *Chemistry—A European Journal*, 16, 6728–6742.
122. Mendez-Ardoy, A., M. Gomez-Garcia, C. O. Mellet, N. Sevillano, M. D. Giron, R. Salto, F. Santoyo-Gonzalez, J. M. G. Fernandez. 2009. *Organic & Biomolecular Chemistry*, 7, 2681–2684.
123. Sansone, F., M. Dudic, G. Donofrio, C. Rivetti, L. Baldini, A. Casnati, S. Cellai, R. Ungaro. 2006. *Journal of the American Chemical Society*, 128, 14528–14536.
124. Bagnacani, V., F. Sansone, G. Donofrio, L. Baldini, A. Casnati, R. Ungaro. 2008. *Organic Letters*, 10, 3953–3956.
125. Desigaux, L., M. Sainlos, O. Lambert, R. Chevre, E. Letrou-Bonneval, J. P. Vigneron, P. Lehn, J. M. Lehn, B. Pitard. 2007. *Proceedings of the National Academy of Sciences of the United States of America*, 104, 16534–16539.
126. Fire, A., S. Q. Xu, M. K. Montgomery, S. A. Kostas, S. E. Driver, C. C. Mello. 1998. *Nature*, 391, 806–811.
127. Riehemann, K., S. W. Schneider, T. A. Luger, B. Godin, M. Ferrari, H. Fuchs. 2009. *Angewandte Chemie (International ed. in English)*, 48, 872–897.
128. Warner, S. 2004. *Scientist*, 18, 38–39.
129. Lammers, T., F. Kiessling, W. E. Hennink, G. Storm. 2010. *Molecular Pharmaceutics*, 7, 1899–1912.

# 16 Gadolinium(III) Complexes–Based Supramolecular Aggregates as MRI Contrast Agents

*Antonella Accardo, Diego Tesauro, and Giancarlo Morelli*

## CONTENTS

## INTRODUCTION

A magnetic resonance imaging (MRI) scan is a noninvasive medical diagnostic imaging procedure currently used in clinical practice. This technique provides images that detect tiny changes in the structures within the body. The objective is achieved by measuring parameters related to the relaxation of hydrogen nuclei of water excited by a magnetic field.[1,2] This process is affected by three parameters: proton density, longitudinal relaxation time ($T_1$), and transverse relaxation time ($T_2$). The variation in the proton density between tissues is not very relevant; therefore, the pulse sequence that measures the $T_1$ and $T_2$ provides the necessary contrast for diagnosis. The image resolution could be enhanced using a contrast agent (CA) that improves both the sensitivity and specificity of MRI and provides anatomical and physiological information beyond the impressive resolution commonly obtained in uncontrasted images.[3,4] The CAs can be classified into two categories, $T_1$ agents and the $T_2$ agents. Both CA categories, based on paramagnetic or superparamagnetic agents, shorten both $T_1$ and $T_2$, and the designation of a CA to one category or the other depends on the ratio between the longitudinal relaxivity ($r_1$) and the transverse relaxivity ($r_2$). $T_1$ agents have $r_2/r_1$ ratios slightly higher than 1, whereas $T_2$ agents have much higher (>10) $r_2/r_1$ ratios. $T_1$ agents in the usual image acquisition mode give a positive contrast.[5] This class of agent is usually based on paramagnetic ions; primarily, the gadolinium(III) complexes. On the contrary, superparamagnetic compounds based on iron oxide are $T_2$ agents; they give a negative contrast in the usual image acquisition mode.[6]

Relaxivity describes the efficacy of the paramagnetic CA, at 1 mM concentration, in changing the rate of water proton relaxation.[7] The relaxation is due to dipole–dipole interactions between the proton nuclear spins and the fluctuating local magnetic field that results from the paramagnetic metal center. The most common paramagnetic compounds are based on gadolinium complexes. Because of the high magnetic moment of the paramagnetic $Gd^{3+}$ ion (with its seven unpaired electrons), the relaxation time of water molecules in the proximity of $Gd^{3+}$ ions is greatly reduced and signal intensity is thereby enhanced. The Gd CAs are manufactured by a chelating process reducing the toxicity that could result from exposure to gadolinium. The chelating agent, a Lewis base, forms a stable complex around the gadolinium saturating seven or eight coordination positions, leaving one or two free positions for water coordination in the nine-position coordination sphere of gadolinium ion. They can be sorted into two classes: branched structures such as diethylenetriaminepentaacetic acid (DTPA) or cyclic structures such as 1,4,7,10-tetraazacyclododecane-N,N,N,N-tetraacetic acid (DOTA), which will be described in the next section. The water molecules that are coordinated with the metal center give a direct contribution to the relaxivity, whereas the bulk solvent molecules experience a paramagnetic effect when they diffuse around a metal center. These two interactions give the most important contributions to the observed relaxivity and are known as the inner sphere and outer sphere relaxation rates, respectively. In addition, water molecules may be retained in the periphery of the metal center by hydrogen bonds for a relatively long time without binding to the metal; this is known as the second sphere relaxation effect.[8]

The overall measured relaxivity ($R_1^{obs}$) is, thus, a result of different contributions as indicated by Equation 16.1:

$$R_1^{obs} = R_{1\rho}^{IS} + R_{1\rho}^{SS} + R_{1\rho}^{OS} + R_1^{W}, \qquad (16.1)$$

where $R_{1\rho}^{IS}$, $R_{1\rho}^{SS}$, and $R_{1\rho}^{OS}$ are the inner sphere, the second sphere, and outer sphere relaxation enhancements in the presence of the paramagnetic complex at 1 mM concentration, respectively, and $R_1^{W}$ is the relaxation rate of the water solvent in the absence of a paramagnetic complex.[9–11]

According to Equation 16.2, the inner sphere contribution depends on four factors: (i) the molar concentration of the paramagnetic complex ($C$), (ii) the number of water molecules coordinated to the paramagnetic center, $q$, (iii) the mean residence lifetime, $\tau_M$, of the coordinated water protons, and (iv) their relaxation time, $T_{1M}$.[7,9,10]

$$R_{1\rho}^{IS} = \frac{q[C]}{55.5(T_{1M} + \tau_M)} \tag{16.2}$$

From Equation 16.2, it is clear that increasing the hydration number, $q$, will increase the inner sphere relaxivity. However, an increase in $q$ is often accompanied by a decrease in thermodynamic stability and kinetic inertness. $T_{1M}$ is directly proportional to the sixth power of the distance, $r_{GdH}$, between the metal center and the coordinated water protons. It was estimated that a decrease of 0.2 Å in $r_{GdH}$ distance leads to a 50% increase in inner sphere relaxivity.[11,12] Moreover, $T_{1M}$ depends on the molecular reorientation time, $\tau_R$, of the chelate, on the applied magnetic field strength, and on the electronic relaxation times, $T_{iE}$ ($i = 1, 2$), of the unpaired electrons of the metal (which also depend on the applied magnetic field strength) according to the modified Solomon–Bloembergen equations (Equations 16.3 and 16.4):

$$\frac{1}{T_{1M}} = \frac{2}{15}\left(\frac{\mu_0}{4\pi}\right)\left[\frac{\gamma_I^2 g^2 \mu_B^2 S(S+1)}{r_{GdH}^6}\right] \times \left(\frac{7\tau_{c2}}{1+\omega_s^2\tau_{c2}^2} + \frac{3\tau_{c1}}{1+\omega_I^2\tau_{c1}^2}\right) \tag{16.3}$$

$$\frac{1}{\tau_{ci}} = \frac{1}{\tau_R} + \frac{1}{\tau_M} + \frac{1}{T_{iE}} \quad \text{(where } i = 1, 2) \tag{16.4}$$

in which $\gamma_I$ is the nuclear giromagnetic ratio, $S$ is the electron spin quantum number (7/2 for Gd(III) ions), $g$ is the electron g-factor, $\mu_B$ is the Bohr magneton, $r_{GdH}$ is the electron spin–proton distance, $\omega_I$ and $\omega_s$ are the nuclear and electron Larmor frequencies, respectively, $\tau_R$ is the rotational correlation time, related to the reorientation of the metal ion-solvent nucleus vector, and $\tau_{c1}$ and $\tau_{c2}$ are longitudinal and transverse electron spin relaxation times.

Relaxivity of the CAs can be increased by introducing structural modifications on ligands with the aim to influence the three parameters $\tau_M$, $\tau_R$, and $T_{iE}$.[11,13–15] The exchange lifetime $\tau_M$ can represent a limiting factor to the attainable relaxation enhancement promoted by the gadolinium complexes. Experimentally, an accurate determination of the $\tau_M$ value of the water molecules coordinated at the gadolinium center can be pursued through the measurement of the transverse [17]O nuclear magnetic resonance (NMR) relaxation time at variable temperature. The $\tau_M$ is inversely proportional to the water exchange rate ($k_{ex} = 1/\tau_M$). Thus, an increase in the relaxivity can be achieved by increasing the water exchange rate up to an optimal value. In complexes having an octadentate chelate with one inner sphere water molecule ($q = 1$), such as DTPA or DOTA chelating agents, the water exchange mechanism is dissociative and its values are particularly small ($<10^6$ s$^{-1}$).[13] Complexes with two water molecules in the first coordination shell ($q = 2$) can have faster water exchange than those with $q = 1$. In other trivalent gadolinium complexes, the exchange mechanism should be distinguished into two different situations: the first one is a slow water exchange ($T_{1M} \ll \tau_M$) in which the residence lifetime $\tau_M$ represents the dominant factor, whereas the second one is a fast water exchange ($T_{1M} \gg \tau_M$). In this case, the relaxivity is directly dependent on the proton exchange and the rotation and electronic relaxation. The tuning of the steric environment in the proximity of the Gd(III) center can increase the dissociative water exchange rate as well as increase the relaxivity.[16] At the frequencies most commonly used in commercial tomographs (20–60 MHz), the molecular reorientational time $\tau_R$ of the Gd(III) chelates is the most important determinant of the observed relaxivity. Therefore, the achievement of higher water proton relaxation rates may be pursued through an increase of this parameter because an increase in the number of the metal-bound water molecules (q), which would lead to the same result, is likely accompanied by a decrease in the stability of the complex.

In the case of $q = 0$, for ligands that occupy all nine coordination positions around the gadolinium atom, there is no contribution from the inner sphere relaxation enhancement, $R_{1\rho}^{IS}$, and the overall measured relaxivity ($R_1^{obs}$) depend on (i) second sphere relaxation ($R_{1\rho}^{SS}$) in which water molecules hydrogen-bonded to the carboxylate oxygen atoms are relaxed via dipolar mechanisms and (ii) outer sphere relaxation ($R_{1\rho}^{OS}$) which arises due to diffusion of water molecules in the bulk near the Gd(III) complex.[8,13,17] In the latter case, the parameters focused on are the electronic relaxation time of the metal, the distance of the closest approach of solvent and solute, and the sum of their diffusion coefficients.[14] The outer sphere relaxivity is usually estimated by equations proposed by Freed.[18] This model is only an approximation for the polyaminocarboxylate ligands used in CAs because it does not take into account interactions of water with the complex which, for these ligands, are important.[19]

Several strategies have been developed to lengthen $\tau_R$ from a typical range of 50 to 90 picoseconds to hundreds of picoseconds. It has been shown that the effectiveness of Gd(III) complexes as CA may be significantly improved by using protein-chelate conjugates in which the metal complex is covalently attached to amino acid residues of the protein,[20] or by noncovalent binding of the complex to macromolecules.[18] This kind of approach permits the coupling of the strong chelation of the metal ion with the slow molecular tumbling of the macromolecules. According to the Solomon–Bloembergen–Morgan theory, an optimization of the all parameters previously reported, including the water exchange rate, rotation, and electron paramagnetic relaxation, reaches a relaxivity of 100 mM$^{-1}$ s$^{-1}$ for a complex with $q = 1$ at 20 MHz and 25°C.[14]

The outer sphere contribution ($R_{1\rho}^{OS}$) depends on $T_{iE}$, on the distance of maximum approach between the solvent and the paramagnetic solute, on the relative diffusion coefficients and again on the magnetic field strength.[14] The contribution of the magnetic field strength to the $R_{1\rho}^{IS}$, $R_{1\rho}^{SS}$, and $R_{1\rho}^{OS}$ values is very important; in fact, the analysis of the magnetic field dependence allows the determination of the principal parameters characterizing the relaxivity of a Gd(III) complex. This information can be obtained through a field-cycling relaxometer, an NMR instrument in which the magnetic field is changed to obtain the measure of $r_1$ on a wide range of frequencies, typically 0.01 to 50 MHz.[10]

The relaxivity value is the principal parameter in the development of CAs. Nevertheless, *in vivo* safety remains the most important issue toward having the gadolinium complex on the market for clinical use. The high spin state (seven unpaired electrons) and the long electronic relaxation times make the paramagnetic gadolinium atom the best candidate among the $T_1$ agents. However, the free Gd(III) ions accumulates in the liver, spleen, and bones, and is highly toxic, with a $LD_{50}$ of less than 0.2 mmol $kg^{-1}$ in mice.[21,22] Therefore, chelates are designed to form complexes with Gd(III), minimizing the toxicity to a biologically tolerable level of $LD_{50}$ to less than 10 mmol $kg^{-1}$. Complexes should be both thermodynamically and kinetically stable.[22] The transmetallation of gadolinium with endogenous metal ions, such as zinc, is less common in cyclic chelates than acyclic chelates.[23] For the clinical agents, the $LD_{50}$ falls into a range of 8 to 25 mmol $kg^{-1}$,[24–26] and the reported adverse effects for these gadolinium-based CAs are rare.[27–29]

## CLASSICAL CONTRAST AGENTS BASED ON GADOLINIUM ION

The first gadolinium-based MRI CA, the Gd(III) complex Gd-DTPA (Magnevist), was prepared by Schering AG in 1981 and approved for clinical use in 1987. It has been administered to more than 20 million patients in more than 15 years of clinical experimentation.

Other Gd(III) complexes which soon became available are Gd-DOTA (Dotarem, Guerbert SA, France), Gd-DTPA-BMA (Omniscan, GE Health, USA), Gd-HPDO3A (Prohance, Bracco Imaging, Italy), Gd-MS-325 (Ablavar, BayerSchering Pharma, Berlin, Germany), and Gd-DTPA-BMEA (OptiMARK, Mallinckrodt, USA; Figure 16.1).[5]

These compounds, as all clinical CAs, are regarded as nonspecific agents or extracellular fluid space agents. They present very similar pharmacokinetic properties because they are distributed in the extracellular fluid and are eliminated via glomerular filtration. They are particularly useful to delineate lesions as a result of the disruption of the blood–brain barrier. Successively, two other derivatives of Gd-DTPA have been introduced in the market, Gd-EOBDTPA[30] (Eovist, Schering AG, Germany) and gadolinium-4-carboxy-5,8,11-tris(carboxymethyl)-1-phenyl-2-oxa-5, 8,11-triazatridecan-13-oate (Gd-BOPTA; MultiHance, Bracco Imaging, Italy).[31] They are characterized by an increased lipophilicity because of the introduction of an aromatic substituent on the carbon backbone of the DTPA ligand. This modification significantly alters the pharmacokinetics and the biodistribution of these CA as compared with the parent Gd-DTPA. These hepatobiliary agents have affinity toward HAS and are specifically uptaken by the hepatocytes. They are partially excreted through the biliary system and the kidneys. By comparing the structure of hepatic agents with that of extracellular fluid space agents, the liver specificity can be ascribed to the pendent hydrophobic phenyl rings. Both MultiHance and Eovist have a faster water exchange rate ($k_{ex} = 7.0 \times 10^6$ $s^{-1}$) and a shorter $r_{GdH}$ than those of Magnevist because of a more steric environment for water exchange.[17]

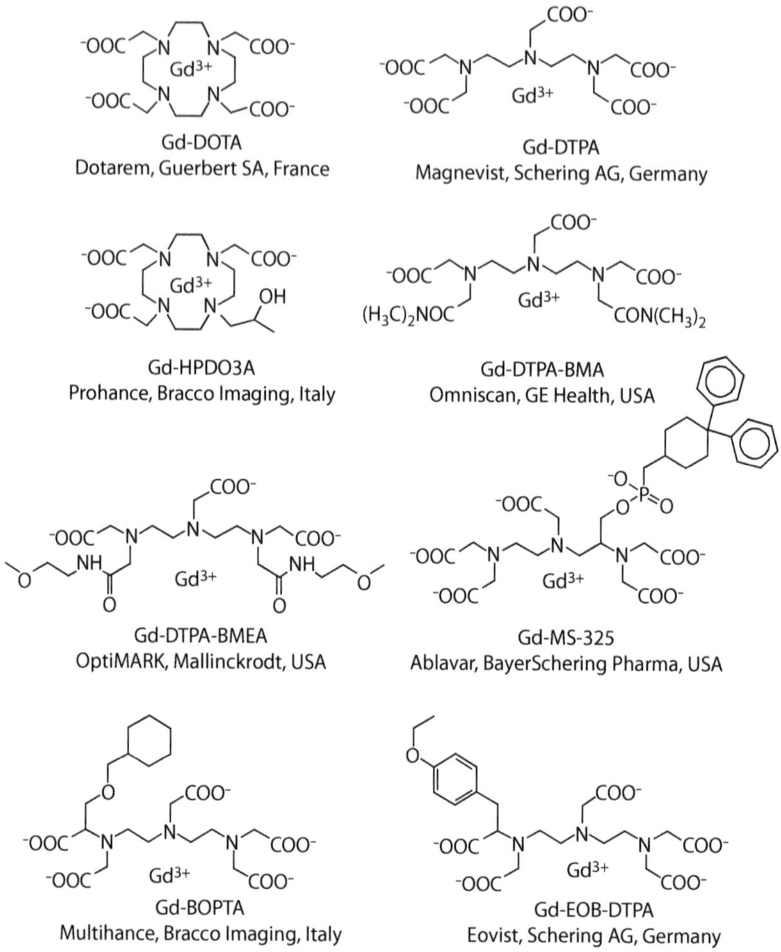

**FIGURE 16.1** Structures of the six Gd(III)-based MRI CAs (Dotarem, Magnevist, Prohance, Omniscan, MS-325 Ablavar, and OptiMARK) currently used in clinical practice, and of the two hepatobiliary agents (MultiHance and Eovist) which present pendent hydrophobic phenyl rings as aromatic substituents on the carbon backbone of the DTPA ligand.

## IMPROVED CONTRAST AGENTS

The reported gadolinium complexes clinically have promising relaxivity behavior and diffusion and penetration that can be modified according to different physiological states. However, they act as aspecific CAs; therefore, the aim of this research was to increase their relaxivity properties and their targeting ability.[32]

New CAs with improved properties were developed to reach this goal:

- *"Multimeric or macromolecular contrast agents"* in which a large number of gadolinium complexes are combined together and the total relaxivity results from the single contribution of each gadolinium ion.[33,34]

- *"Smart contrast agents"* in which a change in relaxivity is observed on activation in the *in vivo* environment in which they act.[35] Smart CAs, also referred to as responsive or activated CAs, are agents that react to variables in their environment, such as temperature,[36] pH,[37–39] partial pressure of oxygen,[40,41] metal ion concentration,[42] or enzyme activity,[43,44] giving a strong increase or decrease of the observed relaxivity. They are also used as CAs in tumor diagnosis for their responsive ability to tumor cell environment characterized by lower pH or by the presence of different amounts of enzymes and proteins with respect to nonpathological environments.
- *"Target-selective contrast agents"* in which the gadolinium complex is delivered in a selective way to cells or tissues of interest by bioactive molecules such as peptides and antibodies; in this approach, the presence of specific receptors or membrane proteins overexpressed by cancer cells is responsible for the target ability of the gadolinium complexes derivatized with peptides or antibodies.[45,46]

## MULTI-GADOLINIUM COMPLEXES AS CONTRAST AGENTS

The classical gadolinium complexes are characterized by low molecular weight and rapid *in vivo* elimination and extravasation out of the vasculature that reduces contrast from their neighbor's tissue.[47] Macromolecular gadolinium complexes have been designed with a molecular weight greater than 30 kDa, limiting extravasation but favoring enhanced permeability and retention. A relevant aspect of these compounds is the increase of the relaxivity per unit dose of the paramagnetic ion. Because of their size, the $\tau_R$ is longer than the low–molecular weight agents such as Magnevist or Dotarem. An additional advantage is the opportunity to append to a macromolecular platform multiple chelates reducing the dose of agent needed for obtaining well-resolved images.

A plethora of MR macromolecular CA have been reported over the last 30 years such as dendrimers,[48] linear polymers,[49] gadofullurenes,[50] gadonanotubes,[51] or large protein derivatives obtained by using the strong interaction present in the avidin-biotin or in the β-cyclodextrin–dextran complexes (Figure 16.2).[52]

The new properties of these compounds managed to increase the number of MRI applications. The monodisperse character of dendrimers and a prolonged vascular retention time due to the large size allow their use into clinics to visualize intratumoral vasculature (e.g., Gadomer, Schering AG, Germany).[53]

The gadolinium-based polylysine was studied and tested *in vivo* as a possible blood pool agent to monitor blood flow in the extremities.[54] The confinement of a Gd(III) ion within the fullerene cage prevents dissociation of the metal ion *in vivo*.[55–58] Specifically, derivatized Gd@C60 nanoscale materials (1.0 nm diameter) offer new nanoscale paradigms for the design of high-performance MRI CA probes that are up to 20 times more efficacious than current clinical CAs.[57] In addition, *in vivo* biodistribution studies of the water-soluble Gd@C60 derivative have shown decreased uptake by the reticuloendothelial system (RES) and facile excretion via the urinary tract.[58]

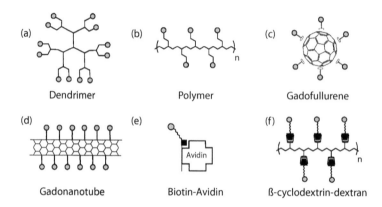

**FIGURE 16.2** Schematic representation of macromolecular adducts: (a) dendrimers, (b) linear polymers, (c) and (d) carbon nanotubes such as gadofullerenes and nanotubes externally derivatized by stable gadolinium complexes, respectively, and (e) macromolecular adducts obtained by noncovalent interactions between high–molecular weight molecules such as avidin or β-cyclodextrin polymers with monomeric gadolinium complexes of biotin or dextran.

Another recent example of macromolecular systems is nanotubes. With their nanoscalar, superparamagnetic Gd(III) ion clusters (1 × 5 nm) confined within ultrashort (20–80 nm) single-walled carbon nanotube capsules, gadonanotubes, are high-performance $T_1$-weighted CAs.[51] A successful strategy concerns the insertion, in the gadolinium complex, of a substituent able to bind serum proteins. For example, the attachment of the protein-binding group diphenylcyclohexyl to a gadolinium(III) chelate via a phosphodiester linkage, such as in the MS-325 CA, results in reversible binding of MS-325 to human serum albumin in plasma (e.g., Angiomark, Epix Medical, Cambridge, MA, USA or MP-2269, Mallinckrodt, USA). The binding to human serum albumin reduces the extravasation of the CA out of the vascular system and also leads to a high increase in relaxivity.[59–61]

Other macromolecular systems have been obtained by derivatizing a gadolinium complex with biotin, the biotinylated compound is then associated with the protein avidin, giving a macromolecular stable adduct containing four gadolinium complexes. This approach has also been used to develop target-selective MRI CAs in which the supramolecular adduct is delivered to tumor cells endowed with avidin receptors.[62]

## SUPRAMOLECULAR AGGREGATES AND RELAXIVITY BEHAVIOR

A special feature in the MRI CA delivery is recovered by supramolecular aggregates. Aggregates, such as micelles and liposomes, can be obtained by spontaneous assembly in aqueous solution of amphiphilic molecules consisting of a hydrophobic and a hydrophilic moiety. The driving forces of the aggregation into well-defined structures derive from the hydrophobic van der Waals associative interactions[63,64] and the repulsive interactions between the charged hydrophilic moieties.[65] In more detail, micellar aggregates are characterized by a core-shell formed by the hydrophobic fragment of the amphiphilic molecules and the outer shell or corona formed by the hydrophilic part. Two parameters characterize micelles: the critical micellar concentration (CMC) and the aggregation number. The CMC represents the lower concentration of a monomeric amphiphile at which micelles appear. Optimal CMC values should be in a micromolar or lower millimolar region. Micelle size normally varies between 5 and 50 nm and fills the gap between such drug carriers as individual macromolecules (antibodies, albumin, find a dextran) with a size smaller than 5 nm and nanoparticulates (liposomes, microcapsules) with a size approximately 50 nm and bigger.

Liposomes are self-enclosed bilayer-like structures that vary in size from 50 to 500 nm. They can be formed by one (unilamellar) or more concentric lipid bilayers (multilamellar) with an aqueous phase inside and between the lipid bilayers.[66]

The ease of preparation, the ability to tune the size and the shape of supramolecular assemblies, and their permeabilities all contributed to producing excellent candidates as carriers of Gd(III)-chelates to enhance the contrast efficacy and to change the pharmacokinetic properties of MRI CA. In the early 1980s, the first studies on the use of aggregates as carriers of MRI CA appeared in the literature (Figure 16.3).[67–69] The CA carriers could be classified as micelles and liposomes. There have been two main approaches in the development of liposomal CAs: in the first one (Figure 16.3a), CAs are entrapped within the internal aqueous space of liposomes (ensomes).[67] In the second approach (Figure 16.3b), lipophilic CAs are incorporated into the lipid bilayer of the liposome conjugating the metal-bonding site to the hydrophilic heads of the membrane molecule (memsomes).[68,69] In ensomes, the relaxivity of the small molecule Gd(III) complexes are influenced by membrane composition, aggregate size (smaller ensome, higher relaxivity), and water permeability. The $T_1$ relaxation times of bulk water solvent are lower for the reduced exchange of water across the lipid membrane. The limitation of water exchange could be circumvented in memsome. In this case, approximately half of the gadolinium complexes are well exposed to water exchange on the surface whereas the remaining one points to the aqueous inner core.

Classical gadolinium complexes, such as Gd-DTPA, Gd-DTPA-BMA, and Gd-HPDO3A, have been loaded within the internal aqueous space of liposomes: the main target for such supramolecular aggregates is the RES, given the avid accumulation of these aggregates by Kupffer cells, and the relatively slow clearance of the

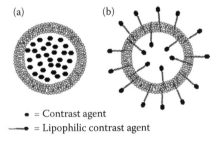

(a)　　　　(b)

• = Contrast agent
⌁● = Lipophilic contrast agent

**FIGURE 16.3** Schematic representation of liposomes as carriers of MRI CA: (a) the CAs are entrapped within the internal aqueous space of liposomes (ensomes), (b) lipophilic CAs are incorporated in the lipid bilayer of the liposome (memsomes).

gadolinium complexes once internalized.[70] It was observed that their relaxivity is two to five times lower compared with the same concentrations of free gadolinium complexes in solution because of the low permeability of the liposomal membrane to water,[71] which limits the exchange of bulk water with the CAs under these conditions.[66,72] Thus the second approach, where a hydrophilic chelating agent is covalently linked to a hydrophobic chain, may be more effective. In this scheme, the lipid part of the molecule is anchored to the liposome bilayer whereas the more hydrophilic gadolinium complex is localized on the liposome surface.[73,74] Several chelating probes of this type have been developed for liposome membrane incorporation studies: DTPA-PE,[75] DTPA-SA,[76] amphiphilic-acylated paramagnetic complexes of Mn and Gd.[70] This approach results in an improved ionic relaxivity of the metal compared with the approach of encapsulating the paramagnetic molecules in the aqueous interior liposomal space, and compared with low–molecular weight complexes.[77]

The nuclear medicine resonance dispersion profile of supramolecular CAs shows a typical peak at higher frequencies, in agreement with the increase in the rotational correlation times as compared with low–molecular weight Gd(III) chelates. This means that, at clinically relevant field strengths, these CAs have the highest gadolinium relaxivity. Furthermore, the amount of Gd(III) complexes per particle is high (varying from 50 atoms for small micelles to several hundred or several thousand atoms for liposomes). As for classical gadolinium complexes, the most important parameters for understanding the relaxivity of supramolecular CAs are the rotational correlation time ($\tau_R$), the coordination number ($q$), the exchange rate ($\tau_M$), and the electronic relaxation time ($T_{iE}$).

The rotational correlation time, $\tau_R$, is strictly related to the size and rigidity of the investigated system. For the analysis of the longitudinal $^{17}O$ and $^1H$ relaxation rates of the aggregates, the Solomon–Bloembergen–Morgan model, modified according to the Lipari–Szabo approach, should be used.[78,79] According to the Lipari–Szabo approach, the modulation of the interaction that causes relaxation is the result of two statistically independent motions: a rapid local motion of the Gd(III) complex, with a local rotational correlation time, $\tau_l$, and a slower global motion of the entire micellar aggregate, with a global rotational correlation time, $\tau_g$. The degree of spatial restriction of the local motion with regard to the global rotation is given by an additional model-free parameter, $S^2$. The coordination number, $q$, and the exchange rate, $\tau_M$, determine the amount of water molecules that can effectively coordinate the Gd(III) ion and thereby increase the relaxation rate ($T_{1M}$). In micellar structures, the gadolinium complexes are entirely exposed on the external surface of the aggregate, whereas in liposomes, gadolinium complexes are distributed between the inner and the outer compartment of the liposomes. The number of paramagnetic complex molecules embedded in the inner and outer layers of the liposomes could be different, and two different contributions arising respectively from the complexes in the inner and in the outer layers have to be considered. If the water exchange rate through the membrane is very slow, the main relaxation effect is expected to be due to the complex in the outer layer.[66,67] On the contrary, if this water exchange is extremely fast, complexes in the inner and in the outer layer will both contribute to the observed paramagnetic relaxation rate. The water exchange rate is highly dependent on the membrane permeability.[80]

## SUPRAMOLECULAR AGGREGATES AS MRI CONTRAST AGENTS

CAs based on supramolecular aggregates reported in the literature can be divided into at least three classes (Figure 16.4). Aggregates obtained directly by the self-assembly of an amphiphilic chelating agent belong to the first class; aggregates obtained by mixing a synthetic amphiphilic chelating agent with one or more commercial phospholipids belong to the second class; and finally, formulations based on the self-assembly of polymeric amphiphiles functionalized with the CA belong to the third class.

Amphiphilic gadolinium complexes (first class) are represented by a chelating agent covalently bound to a hydrophobic moiety such as a long alkyl chain or an organic molecule that promotes the aggregation process in a water-based solution. Generally, the kind of aggregate obtained by starting from amphiphilic gadolinium complexes is micelles. In micelles, the Gd(III) complexes are exposed to the hydrophilic exterior space; therefore, there is easy access of the bulk water to the paramagnetic center. Most of the amphiphilic Gd(III) complexes are composed of a gadolinium complex covalently bound to one or more alkylic chains. The relaxivity behavior of micelles can dramatically change as a function of the hydrophobic moiety. Several studies were carried out on amphiphilic chelating agents to determine the most accurate mathematical approach to study the rotational dynamics of the aggregates, and to justify the relaxivity values.

Between 1999 and 2002, Merbach and coworkers studied several DOTA-based Gd(III) complexes with alkyn chains of different lengths.[81,82] They demonstrated that relaxivity values increase with increasing chain length. This effect can be explained by the different rotational motions (global and local) and by the model-free parameter, $S^2$. As expected, the global rotational correlation time, $\tau_g$, increases with increasing length of the side chain. On the other hand, the local motions do not follow the expected behavior due to reduced internal flexibility of the micelles (short $\tau_l$ value and low $S^2$). Moreover, relaxivity values are influenced by the chelating agent on the head group. Clearly, the micellization of amphiphilic gadolinium complexes gives rise to a system with slower molecular tumbling with respect to the monomeric complex, but the resulting relaxivity still seems lower than the expected values as internal motions are faster than the overall tumbling of the micellar system.

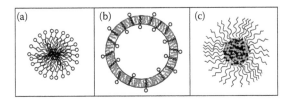

**FIGURE 16.4**  Schematic representation of the three classes of CAs based on supramolecular aggregates: (a) aggregates obtained directly by the self-assembly of an amphiphilic chelating agent, (b) aggregates obtained by mixing a synthetic amphiphilic chelating agent with one or more commercial phospholipids, and (c) formulations of polymeric amphiphiles functionalized with the CA.

Merbach and co-workers demonstrated in 2003 that interactions between nearby paramagnetic centers in micellar systems increase the transverse electronic relaxation of the electronic spin of Gd(III) and, therefore, reduce the attainable water proton relaxivity.[83]

Very recently, Geraldes et al. reported an *in vitro* characterization and *in vivo* animal imaging studies on self-assembling micelles of gadolinium-hydroxymethyl-(hexadecanoyl ester)ethylene-propylenetriaminepentaacetic acid (Gd-EPTPA-C16) monomer (see Table 16.1). The *in vivo* results were compared with the commercially available low–molecular mass Magnevist.[84] The *in vivo* studies showed a persistent hepatic positive-contrast effect in $T_1$-weighted images, which is qualitatively similar to that of the clinically established Gd(III)-based hepatobiliary agents (Gd-EOB-DTPA and Gd-BOPTA).[85,86] The possibility of using this type of micellar compound for imaging disease depends on the degree of stability of the imaging agent in the body relative to its CMC. The amphiphilic gadolinium complexes described previously have only one hydrophobic chain. CMC of mono-tailed surfactants is usually in the range of 0.1 to 0.01 mM. Upon dilution in the blood after injection, these aggregates may not be sufficiently stable and disassemble immediately after administration. Hence, there is a need to find a new class of surfactant molecules that are able to form more stable micelles with lower CMC values. One of the possible candidates for this role could be represented by aggregates obtained by self-assembling of monomers with two or more alkyl chains.

Recently, Paduano et al. reported supramolecular aggregates, constituted basically by the DTPAGlu moiety bound to a hydrophobic double-tail (Table 16.1). The amphiphilic molecule behaves as an anionic surfactant, and is capable of forming aggregates of different sizes and shapes in aqueous solution by varying the method of preparation and environmental conditions such as pH and ionic strength.[87] A micelle-to-vesicle transition was observed by decreasing the pH value from 7.4 to 3.0, and by increasing the ionic strength. An alternative approach to preparing physiologically stable aggregates includes the use of Gd-labeled polymerized liposomes.[88] These systems show increased membrane rigidity and avoid uptake by RES.

The literature reports only two examples of supramolecular aggregates that were also obtained by starting from amphiphilic gadolinium complexes in which the hydrophobic moiety is represented by an organic molecule such as cholesterol (Chol): Gd-DTPA·Chol[89] and Gd-DOTA Chol (see Table 16.1).[90] Gd-DTPA·Chol shows a relaxivity value in water (27.2 mM$^{-1}$ s$^{-1}$), that is extremely high compared with the relaxivity of the Gd-DTPA (3.7 mM$^{-1}$ s$^{-1}$), which is an indirect evidence of its micellar self-organization. On the other hand, Gd-DOTA·Chol shows a relaxivity of 4.42 mM$^{-1}$ s$^{-1}$, which is in the same order of values of the clinically used Dotarem (5.25 mM$^{-1}$ s$^{-1}$), thus indicating that the Gd-DOTA·Chol is unable to self-aggregate. The incapability of Gd-DOTA·Chol, with respect to Gd-DTPA·Chol, to self-aggregate can be ascribed to the lack of anionic charges on the complex that reduce the amphiphilic character of the molecule.

In the second class are reunited supramolecular aggregates obtained by coassembling of amphiphilic gadolinium molecules with one or more commercial surfactants such as phospholipids, nonionic surfactants, or cholesterol. Classical phospholipids can be divided into saturated and unsaturated molecules. The most

**TABLE 16.1**

**Amphiphilic Gd(III) Complexes and Relaxivity Values ($r_1$ at 20 MHz and 25°C) of Supramolecular Aggregates**

| Compounds | Schematic representation of amphiphilic Gd(III) complexes | $r_1$ (mM$^{-1}$s$^{-1}$) |
|---|---|---|
| Gd-EPTPA-C16 | | 22.6 |
| Gd-(C18)$_2$DTPAGlu | | 21.5 |
| Gd-DTPA·Chol | | 27.2 |
| Gd-DOTA·Chol | | 4.42 [a] |
| Gd-DOTA·DSA | | 4.10 [b] |
| Gd-DTPA·PLL-NGPE | | c |
| PEG$_x$-P(Asp$_y$(DTPA-Gd)$_z$) | | 10 ÷ 11 |
| PSI-mPEG-C16-DTPA(Gd) | | d |

[a] Value measured at 25°C.

[b] The relaxivity values are given for compounds in a monomeric state: concentration below their CMC.

[c] Relaxivity values reported as a function of liposomal lipid concentration.

[d] Values not reported.

common components of phospholipids are phosphatidyl choline, phosphatidyl glycerol, and phosphatidyl ethanolamine. Generally, the presence of phospholipids within the supramolecular aggregate favors the formation of bilayer structures such as vesicles or liposomes. Recently, Gløgård et al. investigated the effect on the relaxivity of several parameters including the incorporation degree of the Gd(III) amphiphilic chelates on membrane packing.[91,92] The incorporation efficacy seems to be directly correlated to the lipophilic moiety of the chelates. Moreover, larger liposomes showed only a minor positive effect on the incorporation efficacy, whereas the cholesterol content had no effect on the Gd-chelate loading.[93] In disagreement with the expected results, the relaxivity value decreased with the increase of incorporation degree of Gd(III) complex. An explanation for this result could be the negative influence of the Gd(III)-chelate on the membrane packing in the sense of creating disorder. A likely impact of this is an increase in the lateral surface motion on the liposome surface, leading to a shortening of $\tau_R$ for the Gd(III)-chelates and thereby a decrease in the relaxivity.

Strictly related to the incorporation degree is the degree of mobility of the amphiphilic gadolinium complex, which depends on the length and location of the alkylic chains on the Gd(III) complex. The other parameters influencing relaxivity are represented by the membrane composition of aggregates, such as the saturation level of membrane, the transition phase temperature of phospholipid, and the cholesterol content. Gadolinium complexes incorporated in liposome bilayers can distribute them on both sides of the phospholipidic membrane and the two contributions arising respectively from the complexes in the inner and outer layers have to be considered. The two contributions depend on liposome membrane permeability, which is strictly related to the bilayer composition: saturated phospholipids present a lower membrane permeability with respect to the liposome obtained by unsaturated phospholipids.[80,94] The reason for such a behavior is due to the different packing of the hydrophobic chains in the bilayer. The presence of an unsaturation in the hydrocarbon chain facilitates the water flux across the bilayer and improving the relaxivity.[95] On the contrary, if the membrane permeability is low, water exchange rate is very slow and the main relaxation effect will be due to the complex in the outer layer.[96] The water permeability ($P_w$) of liposomes can also be strongly influenced by the incorporation of amphiphilic gadolinium complexes in the lipid bilayer.[97]

Another extremely important parameter influencing relaxivity is represented by the transition phase temperature. The liposome bilayer, which presents a liquid-crystalline state, has a high water exchange rate between the interior and exterior liposome compartments, allowing bulk water to experience dipolar interaction with the Gd-chelates located on the inner surface. On the contrary, liposome bilayer in the solid–gel state have a low water exchange rate, making the inner surface chelates less accessible for the bulk water and thereby decreasing their contribution to the overall relaxivity.[77] On the other hand, the lateral motion on the liposome surface is about twice as high in a liquid-crystalline membrane compared to a gel state membrane.[98] This should give rise to a longer $t_R$ for the Gd-chelates incorporated into a gel state-membrane with respect to complexes incorporated in liquid-crystalline state, thereby increasing their relaxivity.

The introduction of cholesterol in the formulation always has a positive effect on relaxivity. The influence of cholesterol on the liposome membrane is associated with the phase transition temperature of the latter. With the incorporation of cholesterol into a solid–gel liposome fluidizes the membrane, leading to increased transmembrane water permeability and hence an increased relaxivity of the Gd-chelates present inside the liposomes. In the liquid-crystalline liposomes, the positive effect observed on cholesterol incorporation is most likely related to an increase in the membrane rigidity. The rotational correlation time ($\tau_R$) of the Gd-chelate is prolonged, without adversely affecting the conditions of fast water exchange.[99]

Nowadays, the research is devoted to finding new multifunctional liposomes that are able to address at the same time two or more objectives, such as pDNA transfection or diagnosis and treatment of diseases. For example, last year, Miller et al. reported the utility of liposomes as bimodal paramagnetic and fluorescent imaging systems both for *in vitro* cell labeling and *in vivo* tumor imaging.[100] They synthesized the amphiphilic gadolinium complex Gd-DOTA·DSA (see Table 16.1), incorporating it into 1,2-dioleoyl-*sn*-glycero-3-phosphocholine (DOPC)/Chol/ 1,2-distearoyl-*sn*-glycero-3-phosphoethanolamine-N-[methoxy(polyethylene glycol)- 2000] (DSPE-PEG$_{2000}$) liposomes.[100] A small amount of the fluorescent lipid phosphatidylethanolamine lissamine rhodamine (0.5 ÷ 1.0 mol%) was also incorporated in the liposome formulation to obtain fluorescent aggregates. This method permits us to quantify uptake and internalization processes of liposomes into the cell cytosol. The liposomal surface was modified with PEG to prolong the presence of contrast liposomes in the body.[101] In fact, surface modification with polymers has been shown to produce changes in biodistribution and body retention of liposomes.[102] In greater detail, modification with PEG is known to prolong circulation times after intravenous administration of the coated liposomes[103] due to the prevention of the liposomes' opsonization with macrophage-recognizable.[104] The liposome formulation was intravenously injected into nude mice in which IGROV-1 (human ovarian cancer cells) xenografts were previously implanted. After intravenous injection of Gd-liposomes, xenograft tumors were monitored using MRI over a 24-hour period. During this period, a substantial 60% reduction in tumor $T_1$ values compared with the situation with control liposomes (without Gd) and visibly enhanced tumor image brightness was observed. This outcome suggested that liposomal half-life time ($t_{1/2} >$ 4 hours) is long enough to allow passive targeting to the xenograft tumor by an enhanced permeation and retention mechanism due to the porous nature of the endothelial cell layer.[105] The fluorescence imaging of tumor tissue slices (postmortem) supported liposome accumulation in intravascular spaces and in surrounding viable tumor tissue.

The third class of supramolecular aggregates as MRI CAs is represented by polymeric micelles. They are obtained by self-assembling block copolymers consisting of hydrophilic and hydrophobic monomer units with the length of the hydrophilic block exceeding, to some extent, that of the hydrophobic one. If the hydrophilic block length is high, copolymers exist in water as individual molecules, whereas copolymers with very long hydrophobic blocks form lamellar structures.[106] In different amphiphilic polymers, monomer units with different hydrophobicity can be arranged into two conjugated blocks, each consisting of monomers of the same type

(A-B type copolymers), or they can form alternating blocks with different hydrophobicities (A-B-A type copolymers). The hydrophilic copolymers can also represent a backbone chain to which hydrophobic blocks are attached as side chains (graft copolymers). Polymeric micelles have recently attracted much attention as CAs in MRI for their higher stability with respect to micelles prepared from conventional surfactants. Usually, amphiphilic micelle-forming monomers include PEG blocks (1–15 kDa) as hydrophilic corona-forming blocks[107] for its low toxicity and its capability of shielding micelles from biologically active macromolecules.[108] At the same time, a variety of monomers may be used to build hydrophobic core-forming blocks: propylene oxide,[109] L-lysine,[110] aspartic acid,[111] β-benzoyl-L-aspartate,[112] γ-benzyl-L-glutamate,[113] spermine,[114] and some others. Some of these hydrophobic-forming blocks, such as poly(lysine) and poly(aspartic acid), are largely used in the synthesis of copolymers as MRI CAs for the presence of reactive functions on their side chains for the coupling of chelating agents. Nevertheless, only a few examples of supramolecular aggregates (Table 16.1), obtained by using gadolinium-containing copolymers, are reported as MRI CAs. The low number of proposed compounds makes it difficult to take an appropriate survey of the relationship between their structures and relaxivities.

Two examples of A-B type copolymers are represented by the poly(lysine)-based polychelating polymer (Gd-DTPA-PLL-NGPE)[115] incorporated into liposome membranes and by cationic polymers (polyallylamine or protamine) and several anionic blocks (PEG$_x$-P(Asp)$_y$) that bound Gd(III).[116] Polymeric micelles are able to increase the signal intensity at the targeted site and to lower, at the same time, the signal intensity in the vasculature space. In fact, it is well known that high concentrations of the CA in the vascular space create a disadvantageous background because blood is supplied both to the targeted tissues (or organs) and to the nontargeted ones. Another example of polymeric micelle-type MRI CA are obtained by self-assembling of poly-succinimide (PSI)-mPEG-C16-(DTPA-Gd), an A-B-A type polymeric compound prepared by conjugating a hydrophilic methoxy-poly(ethylene glycol) (mPEG), a hydrophobic group (C16) and a gadolinium complex (DTPA-Gd) to a biodegradable polymer (PSI).[117] As an alternative to the polymeric aggregates previously reported, and obtained by self-assembling of polychelating polymers, Turner et al. proposed a procedure in which gadolinium complexes are covalently bound on the external surface of the polymeric aggregate only after the assembling process. The authors developed shell cross-linked knedel-like nanoparticles (SCKs) as the scaffolding from which to produce contrast-enhancing agents. SCKs are self-assembled core-shell materials, originating from amphiphilic block copolymer micelles that are covalently stabilized by a cross-linking reaction.[118]

## Supramolecular Aggregates Derivatized with Peptides or Antibodies

The micelles and liposomes reported here have been developed with the aim of enhancing contrast efficacy by increasing the relaxivity of each gadolinium atom and by increasing the number of gadolinium complexes for each aggregate. Nevertheless, a challenge in diagnosis is the detection of the disease at an early stage. This goal could be reached using targeted CAs, which are able to recognize a molecular entity of interest, for example, endothelial cell surface receptors that are overexpressed as a consequence of disease processes. The labeling of gadolinium-containing micelles and liposomes with bioactive molecules such as antibodies and peptides combines the high relaxivity of the supramolecular aggregate with target selectivity due to the presence of the bioactive moiety. In fact, even if the number of surface receptors on target cells is expected to be lower than requested to have good contrast, the use of target-selective supramolecular aggregates carrying a high number of high-relaxivity gadolinium complexes could overcome this problem.

The bioactive molecule on the external surface of the supramolecular aggregate can be coupled before or after assembling procedure of the supramolecular aggregate.[119] The obvious goal of each approach is to achieve high coupling efficiency and for the ligand to retain full binding affinity for its target receptor. In the first case, the coupling of a ligand to the aggregate requires the introduction of suitable activated functional groups onto the terminus of one of the aggregate components. Activated functional groups must be compatible with the aggregation process and should remain available on the aggregate surface for efficient chemical ligation. This strategy has proven to be particularly successful for the coupling of large ligands such as monoclonal antibodies. As schematized in Figure 16.5a, an example of this

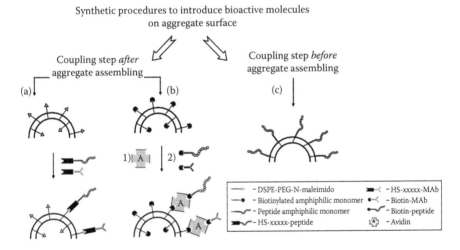

**FIGURE 16.5** Examples of labeling procedures to introduce the bioactive molecule on the external surface of supramolecular aggregates. In the first procedure, labeling is achieved by a coupling step after aggregate assembling. According to the first procedure (a), a DSPE-PEG monomer derivatized with an N-maleimido function is incorporated into the mixed aggregate; after aggregate formation, the externally exposed N-maleimido function reacts with the bioactive molecule to produce the labeled supramolecular aggregate. In another approach (b), biotinylated aggregates are obtained by using a biotin bearing lipophilic tail in the surfactant mixture; successively, in a two-step process, the biotinylated supramolecular aggregates react with the protein avidin, then biotinylated peptides or antibodies interact with free avidin sites giving a noncovalently labeled compound. In the second procedure (c), labeled aggregates are obtained, assembling a mixture of amphiphilic peptides and amphiphilic gadolinium complexes.

approach consists of the preparation of liposomes or micelles containing a DSPE-PEG monomer derivatized with an N-maleimide that is able to react with the bioactive molecule, according to the sulfhydryl-maleimide coupling method. An example of this coupling procedure was used by Mulder et al.[45] to prepare immunoliposome (Gd-DTPA-bis(stearylamide)/Chol/DOPC/N-maleimide-DSPE-PEG$_{2000}$) decorated with antibodies able to detect E-selectin expression on human umbilical vein endothelial cells. This is an attractive model for receptor expression on endothelial cells because E-selectin levels can be regulated by tumor necrosis factor-$\alpha$ and represents a model system for a variety of endothelial cell surface receptors that are of physiological and therapeutic interest, including $\alpha_v$ $\beta_3$ integrins and vascular endothelial growth factor receptors.

The same strategy was used by Brandwijk et al.[120] to obtain paramagnetic liposomes decorated by the Anginex, a synthetic 33-mer angiostatic peptide capable of homing to activated endothelium, as a targeting ligand. The coaggregation process of Gd-DTPA-BC18, maleimide-PEG2000-DSPE, distearoylphosphatidylcholine, and Chol is followed to the linkage with the Anginex by sulfhydryl-maleimide coupling to maleimide-PEG-DSPE.

Another opportunity to link the bioactive portion to amphiphilic monomer in aggregates is based on the interaction of biotin and avidin. A biotinylated liphophilic monomer in the surfactant mixture assembles in aggregates; in a two-step process, the supramolecular aggregates react with avidin and then with the biotinylated peptide or antibody giving a noncovalently labeled compound (Figure 16.5b). Mulder et al.[121–124] obtained mixed micelles combining the amphiphilic gadolinium complex, Gd-DOTA-C16, the 1-hexadecanoyl-2-(9Z-octadecenoyl)-*sn*-glycero-3-phosphocholine phospholipids, and the Tween-80 surfactant; moreover, a little amount of DPPE-biotin was added to produce biotinylated micelles. These aggregates were used for *ex vivo* images[121] and *in vivo* experiments[122] to detect atherosclerosis, targeting macrophage scavenger receptors. The same approach, based on biotin-avidin affinity interaction, has been used by Li and co-workers in their pioneering work on immunoliposomes. They prepared polymerized vesicles of 300 to 350 nm in diameter by UV irradiation of a mixture of gadolinium-DTPA polymerizable lipid, a polymerizable biotinylated lipid and diacetylene phosphatidylcholine filler lipid,[123,124] then derivatized the liposomes with a specific antibody. Very interesting results have been obtained for *in vivo* detection of tumor angiogenesis targeting the integrin $\alpha_v$ $\beta_3$.[124] In the second approach, the coupling of a bioactive ligand to an aggregate component before aggregation is, in principle, chemically less complicated, but has the disadvantage that, at least in the case of liposomes, after final assembly of the aggregate a fraction of the conjugated bioactive ligand remains entrapped in the inner region and is not more available for receptor binding. This labeling procedure, based on the use of amphiphilic peptides that assemble together with amphiphilic gadolinium complexes in the peptide-labeled supramolecular aggregates, is schematized in Figure 16.5c.

Bull et al.[46,125] by using a solid-phase technique, synthesized peptide amphiphilic molecules containing the RGD peptide sequences, important for cell adhesion, or a cysteine-rich peptide sequence for cross-linking of the self-assembled systems, and a DOTA-Gd complex covalently bound to the N-terminus of the peptide.

These amphiphilic molecules self-assemble into spherical and fiber-like nanostructures, which show an enhanced $T_1$ relaxation time with a relaxivity increased up to 22.8 mM$^{-1}$ s$^{-1}$ after assembling in supramolecular aggregates.[46] Moreover, liposomes conjugated with the $\alpha_v \beta_3$-specific RGD peptide attached to PEG moieties have been developed by Mulder et al.[105,126] These liposomes were used as MRI CAs in *in vitro* and *in vivo* studies for the detection of $\alpha_v \beta_3$-integrin and visualization of proliferating human activated endothelium tumor.

Several supramolecular aggregates derivatized with regulatory peptide and metal complexes on their surface have been developed by Morelli and Paduano as target-selective CA systems.[127–133] The selected peptides are analogues of cholecystokinin and somatostatin and are able to target tumoral cells overexpressing their respective receptors. Four generations of supramolecular aggregates have been designed and studied. In the first and second generations, the monomers were obtained by covalently binding the peptide moiety or the metal complex to one (I generation) or two (II generation) hydrophobic tails at 18 carbon atoms. Peptides were spaced from the alkyl chains by using PEG units of different lengths.[127–131] The presence in the aggregates of two separate monomers allows the tuning of the ratio between the active components to find the right compromise between the number of peptide-targeting molecules on the aggregate surface, and of the metal complexes. The size and the shape of aggregates, investigated by several physical chemical methodologies, were found to be dependent on a large number of parameters (monomer ratio, pH, ionic strength, temperature, and ratio between the hydrophobic and hydrophilic moieties in the molecule).

The design of the molecules has been developed with the main aim of being able to realize supramolecular aggregates (micelles or liposomes) suitable for *in vivo* use as target-selective CAs: low CMC values of the aggregates in physiological environment, no toxicity of the entire aggregate and of the single monomers, and availability of the bioactive peptide toward the receptors adequately exposed above the surface.

All systems showed low CMC values compatible for their use *in vivo* and high relaxivity values ranging between 17 and 24 mM$^{-1}$ s$^{-1}$. After the first step in which aggregates were fully characterized, many systems have also been studied to define their *in vitro* and *in vivo* binding activity. For this purpose, the gadolinium atom was replaced with an $^{111}$In gamma-emitting radioisotope and the aggregate targeting and biodistribution were studied by using a nuclear medicine technique.[132] In the third generation, a unique monomer with an upsilon shape (MonY) has been developed.[133] MonY contains, in the same molecule, all three fundamental tasks that are required: (a) the hydrophobic moiety, (b) the bioactive peptide for target recognition, and (c) the gadolinium complexes. The relaxivity value for each gadolinium complex of self-assembling micelles of MonY(Gd) is 15.03 mM$^{-1}$ s$^{-1}$. Finally, in the fourth generation, two monomers based on gemini surfactants were developed.[134] The two gemini surfactants contain, respectively, CCK8 peptide and Gd complex. After the aggregation process, mixed micelles with cylindrical shape were obtained at 37°C. The high relaxation value (21.5 mM$^{-1}$ s$^{-1}$) and the excellent exposition of the bioactive peptide on hydrophilic surface are signs of the high potential performance of these aggregates as MRI CAs.

## CONCLUSIONS

The *in vivo* applicability of supramolecular aggregates, such as micelles and liposomes, containing Gd(III) complexes as CAs in MRI has been well documented. There are several reports that have approached and optimized issues of stability, relaxivity, and overall safety for *in vivo* applications. The possibility of using micellar or liposomal compounds for imaging purposes depends on the degree of stability of the imaging agent in the body. Liposomes show high stability in biological fluid, whereas micellar stability depends on its CMC. Physiologically stable micelles (CMC value in a micromolar or low millimolar region) can be performed by using either monomers with two or more alkyl chains or by polymerizable amphiphilic Gd(III) complexes. High contrast efficacy can be achieved by designing supramolecular paramagnetic MR-CAs characterized by high rigidity and good access of bulk water to the paramagnetic center. High rigidity can be obtained by combining a restricted motion of the CA of the amphiphilic gadolinium complexes in the aggregate and a high molecular weight of the particle. These two effects will normally increase the contrast efficacy by prolonging the rotational correlation time of the Gd(III) complex. Fast exchange between the water molecule coordinated to the gadolinium ion and the bulk water can be easily achieved if the Gd(III) complexes are exposed on the external surface of the aggregate, that is, micelles, or in the case of liposomes, by using a fluid membrane that allows high water transport across the membrane. The water flux across the liposome bilayer depends on the length of the phospholipid alkyl chain and on the presence of unsaturations in the hydrocarbon chain and by the addition of cholesterol. Future work will also have to address how to optimize issues pertaining to the relationship of size and relaxivity. As it stands, the required size of the aggregates currently used that can produce relaxivity changes in a range suitable for current MRI scanners is fairly large. Issues concerning safety for *in vivo* use of such an approach need further investigation. Applications in which large amounts of high–molecular weight contrast media can freely diffuse to selective targets, such as contrast-enhanced imaging of the vascular system or molecular targeting of integrins present on endothelial cells, have been very successful in animal models and may very well be successful in clinical use. On the other hand, permeability issues are likely to pose a major impairment for applications of molecular imaging in tissues where the target receptor or protein is not freely accessible. Reaching targets on tumor cells, for example, or in other tissues with permeability barriers, will be very difficult and adequate concentrations of such aggregates for application in MRI are not likely to be achieved with currently available technology. However, given the rapid rate of progress in this field, it is likely that optimization of aggregate size and relaxivity will allow the widening of the range of molecular targeting applications with such an approach.

In conclusion, the development of novel MRI CAs based on supramolecular aggregates derivatized on the external surface by peptides or antibody should permit us to couple high relaxivity and targeting selectivity on the same particle. Their envisioned use for the *in vivo*, noninvasive visualization of molecular markers should allow us to visualize diseases, for example cancers, in the early stages of the transformation toward the malignant phenotype and to improve the quality of life. Moreover, administration of the targeted CA should substantially reduce the deleterious side effects related to the imaging procedure.

## REFERENCES

1. Young, I. R. *Methods in Biomedical Magnetic Resonance Imaging and Spectroscopy*, Chichester: John Wiley & Sons Ltd., 2000.
2. Rinck, P. A. *Magnetic Resonance in Medicine*, Berlin: ABW Wissenschaftsverlag GmbH, 2003.
3. Tweedle, M. F. In *Lantanide Probes in Life, Chemical and Earth Sciences: Theory and Practice*, J.-C. G. Bünzli, G. R. Choppin (eds.), 127, Amsterdam: Elsevier, 1989.
4. Brücher, E., A. D. Sherry. In *The Chemistry of Contrast Agents in Medical Magnetic Resonance Imaging*, A. E. Merbach, E. Toth (eds.), Chichester: Wiley, 243, 2001.
5. Weinmann, H. J., A. Mühler, B. Radüchel. In *Biomedical Magnetic Resonance Imaging and Spectroscopy*, I. R. Young (ed.), Chichester: John Wiley & Sons Ltd., 705, 2000.
6. Weissleder, R., M. Papisov. *Rev. Magnetic Resonance in Medicine* 1992, 4, 1–20.
7. Banci, L., I. Bertini, C. Luchinat. *Nuclear and Electronic Relaxation*, Weinheim: VCH, 1991.
8. Botta, M. 2000. *European Journal of Inorganic Chemistry*, 399–407.
9. Peters, A., J. J. Huskens, D. J. Raber. 1996. *Progress in Nuclear Magnetic Resonance Spectroscopy*, 28, 283–350.
10. Koenig, S. H., R. D. Brown III. 1991. *Progress in Nuclear Magnetic Resonance Spectroscopy*, 22, 487–567.
11. Lauffer, R. B. 1987. *Chemical Reviews*, 87, 901–927.
12. Caravan, P., A. V. Astashkin, A. M. Raitsimring. 2003. *Inorganic Chemistry*, 42, 3972–3974.
13. Caravan, P., J. J. Ellison, T. J. McMurry, R. B. Lauffer. 1999. *Chemical Reviews*, 99, 2293–2552.
14. Aime, S., M. Botta, M. Fasano, E. Terreno. 1998. *Chemical Society Reviews*, 27, 19–29.
15. Aime, S., M. Botta, M. Fasano, S. Geninatti Crich, E. Terreno. 1999. *Coordination Chemistry Reviews*, 321, 185–186.
16. Helm, L., A. E. Merbach. 2001. *Chemical Reviews*, 105, 1923–1960.
17. Aime, S., A. S. Batsanov, M. Botta, J. A. K. Howard, D. Parker, K. Senanayake, G. Williams. 1994. *Inorganic Chemistry*, 33, 4696–4706.
18. Freed, J. H. 1978. *Journal of Chemical Physics*, 68, 4034–4037.
19. Aime, S., M. Botta, E. Terreno. 2005. *Advances in Inorganic Chemistry*, 57, 173–237.
20. Deal, K. A., R. Motekaitis, A. E. Martell, M. J. Welch, 1996. *Journal of Medicinal Chemistry*, 39, 3096–3106.
21. Tweedle, M. F., V. M. Runge. 1992. *Drugs Future*, 17, 187.
22. Cacheris, W. P., S. C. Quary, S. M. Rocklage. 1990. *Magnetic Resonance Imaging*, 8, 467–481.
23. Puttagunta, N. R., W. A. Gibby, G. T. 1996. *Investigative Radiology*, 31, 739–742.
24. Vogler, H., J. Platzek, G. Schuhmann-Giampieri, T. Frenzel, H. J. Weinmann, B. Radüchel, W. R. Press. 1995. *European Journal of Radiology*, 21, 1–10.
25. Oksendal, A., P. Hals. 1993. *Journal of Magnetic Resonance Imaging*, 3, 157–165.
26. Kanal, E., F. G. Shellock. *Safety Manual on Magnetic Resonance Imaging Contrast Agents*, Cedar Knolls, NJ: Lippincott-Raven Healthcare, 1996.
27. Runge, V. M. 2001. *Topics in Magnetic Resonance Imaging*, 12, 309–314.
28. Bellin, M. F., J. A. W. Webb, A. J. Van der Molen, H. S. Thomsen, S. K. Morcos. 2005. *European Radiology*, 15, 1607–1614.

29. Kirchin, M. A., V. M. Runge. 2003. *Topics in Magnetic Resonance Imaging*, 14, 426–435.
30. Schmitt-Willich, H., M. Brehm, C. L. J. Evers, G. Michl, A. Müller-Farnow, O. Petrov, J. Platzek, B. Radüchel, D. Sülzle. 1999. *Inorganic Chemistry*, 38, 1134–1144.
31. Uggeri, F., S. Aime, P. L. Anelli, M. Botta, M. Brocchetta, C. de Haën, G. Ermondi, M. Grandi, P. Paoli. 1995. *Inorganic Chemistry*, 34, 633–642.
32. Bottrill, M., L. Kwok, N. J. Long. 2006. *Chemical Society Reviews*, 35, 557–571.
33. Battistini, E., E. Gianolio, R. Gref, P. Couvreur, S. Fuzerova, M. Othman, S. Aime, B. Badet, P. Durand. 2008. *Chemistry—A European Journal*, 14, 4551–4561.
34. Aime, S., M. Botta, E. Garino, S. Geninatti Crich, G. Giovenzana, R. Pagliarin, G. Palmisano, M. Sisti. 2000. *Chemistry—A European Journal*, 6, 2609–2617.
35. Lowe, M. P. 2002. *Australian Journal of Chemistry*, 55, 551–556.
36. Frich, L., A. Bjornerud, S. Fossheim, T. Tillung, I. Gladhaug. 2004. *Magnetic Resonance in Medicine*, 52, 1302–1309.
37. Lowe, M. P., D. Parker, O. Reany, S. Aime, M. Botta, G. Castellano, E. Gianolio, R. Pagliarin. 2001. *Journal of the American Chemical Society*, 123, 7601–7609.
38. Zhang, S. R., K. C. Wu, A. D. Sherry. 1999. *Angewandte Chemie (International ed. in English)*, 38, 3192–3194.
39. Aime, S., M. Botta, S. Geninatti Crich, G. Giovenzana, G. Palmisano, M. Sisti. 1999. *Chemical Communications*, 1577–1578.
40. Aime, S., M. Botta, E. Gianolio, E. Terreno. 2000. *Angewandte Chemie (International ed. in English)*, 39, 747–750.
41. Burai, L., E. Toth, S. Seibig, R. Scopelliti, A. E. Merbach. 2000. *Chemistry—A European Journal*, 6, 3761–3770.
42. Que, E. L., C. J. Chang. 2006. *Journal of the American Chemical Society.* 128, 15942–15943.
43. Anelli, P. L., I. Bertini, M. Fragai, L. Lattuada, C. Luchinat, G. Parigi. 2000. *European Journal of Inorganic Chemistry*, 625–630.
44. Nivorozhkin, A. L., A. F. Kolodziej, P. Caravan, M. T. Greenfield, R. B. Lauffer, T. J. McMurry. 2001. *Angewandte Chemie (International ed. in English)*, 40, 2903–2906.
45. Mulder, W. J. M., G. J. Strijkers, A. W. Griffioen, L. van Bloois, G. Molema, G. Storm, G. Koning, K. Nicolay. 2004, *Bioconjugate Chemistry*, 15, 799–806.
46. Bull, S. R., M. O. Guler, R. E. Bras, T. J. Maede, S. I. Stupp. 2005. *Nano Letters*, 5, 1–4.
47. Mühler, A. 1995. *MAGMA*, 3, 21–33.
48. Bosman, A. W., H. M. Janssen, E. W. Meijer. 1999. *Chemical Reviews*, 99, 1665–1688.
49. Ladd, D. L., R. Hollister, X. Peng, D. Wei, G. Wu, D. Delecki, R. A. Snow, J. L. Toner, K. Kellar, J. Eck, V. C. Desai, G. Raymond, L. B. Kinter, T. S. Desser, D. L. Rubin. 1999. *Bioconjugate Chemistry*, 10, 361–370.
50. Laus, S., B. Sitharaman, E. Toth, R. D. Bolskar, L. Helm, L. J. Wilson, A. E. Merbach. 2007. *Journal of Physical Chemistry C*, 111, 5633–5639.
51. Hartman, K. B., S. Laus, R. D. Bolskar, R. Muthupillai, L. M. Helm, E. Toth, A. E. Merbach, L. J. Wilson. 2008. *Nano Letters*, 8, 415–419.
52. Geninatti Crich, S., A. Barge, E. Battistini, C. Cabella, S. Coluccia, D. Longo, V. Mainero, G. Tarone, S. Aime. 2005. *Journal of Biological Inorganic Chemistry*, 10, 78–86.
53. Dong, Q., D. R. Hurst, H. J. Weinmann, T. L. Chenevert, F. J. Londy, M. R. Prince. 1998. *Investigative Radiology*, 33, 699–708.
54. Marchal, G., H. Bosmans, P. Van Hecke, U. Speck, P. Aerts, P. Vanhoenacker, A. Baert. 1990. *American Journal of Roentgenology*, 155, 407–411.
55. Kato, H., Y. Kanazawa, M. Okumura, A. Taninaka, T. Yokawa, H. Shinohara. 2003. *Journal of the American Chemical Society*, 125, 4391–4397.
56. Fatouros, P. P., F. D. Frank, Z. Chen, W. C. Brosddus, J. L. Tatum, B. Kettenmann, Z. Ge, H. W. Gibson, J. L. Russ, A. P. Leonard, J. C. Duchamp, H. C. Dorn. 2006. *Radiology*, 204, 756–764.

57. Toth, E., R. D. Bolskar, A. Borel, G. Gonzales, L. M. Helm, A. E. Merbach, B. Sitharaman, L. J. Wilson. 2005. *Journal of the American Chemical Society*, 127, 799–805.

58. Bolskar, R. D., A. F. Benedetto, L. O. Husebo, R. E. Price, E. F. Jackson, S. Wallace, L. J. Wilson, J. N. Alford. 2003. *Journal of the American Chemical Society*, 125, 5471–5478.

59. Caravan, P., N. J. Cloutier, M. T. Greenfield, S. A. McDermid, S. U. Dunham, J. W. M. Bulte, J. C. Amedio, Jr., R. J. Looby, R. M. Supkowski, W. D. Horrocks, T. J. McMurry, R. B. Lauffer. 2002. *Journal of the American Chemical Society*, 124, 3152–3162.

60. Lauffer, R. B., D. J. Parmelee, H. S. Ouellet, R. P. Dolan, H. Sajiki, D. M. Scott, P. J. Bernard, E. M. Buchanan, K. Y. Ong, Z. Tyeklar, K. S. Midelfort, T. J. McMurry, R. C. Walovitch. 1996. *Academic Radiology*, 3, S356–358.

61. Grist, T. M., F. R. Korosec, D. C. Peters, S. Witte, R. C. Walovitch, R. P. Dolan, W. E. Bridson, E. K. Yucel, C. A. Mistretta. 1998. *Radiology*, 207, 539–544.

62. Dirksen, A., S. Langereis, B. F. M. de Waal, M. H. P. van Genderen, T. M. Hackeng, E. W. Meijer. 2005. *Chemical Communications*, 2811–2813.

63. Jones, M., J. Leroux. 1999. *European Journal of Pharmaceutics and Biopharmaceutics*, 48, 101–111.

64. Martin A. (ed.), *Physical Pharmacy*, Philadelphia: Lippincott, Williams & Wilkins, 1993.

65. De Giorgio, V., M. Corti (eds), *Physics of Amphiphiles: Micelles, Vesicles and Microemulsions*, Amsterdam: Elsevier, 1985.

66. Torchilin, V. P. 2005. *Nature Reviews Drug Discovery*, 4, 145–160.

67. Tilcock, C., E. Unger, P. Cullis, P. MacDougall. 1989. *Radiology*, 171, 77–81.

68. Kabalka, G., E. Buonocore, K. Hubner, T. Moss, N. Norley, L. Huang. 1987. *Radiology*, 163, 255–258.

69. Grant, C. W. M., S. Karlik, E. A. Florio. 1989. *Magnetic Resonance in Medicine*, 11, 236–243.

70. Unger E., T. Fritz, G. Wu, D. Shen, B. Kulik, T. New, M. Crowell, N. Wilke. 1994. *Journal of Liposome Research*, 4, 811–834.

71. Koenig, S. H., Q. F. Ahkong, R. D. Brown III, M. Lafleur, M. Spiller, E. Unger, C. Tilcock. 1992. *Magnetic Resonance in Medicine*, 23, 275–286.

72. Unger, E., D. K. Shen, G. L. Wu, T. Fritz. 1991. *Magnetic Resonance in Medicine*, 22, 304–308.

73. Kabalka, G. W., M. A. Davis, E. Buonocore, K. Hubner, E. Holmberg, L. Huang. 1990. *Investigative Radiology*, 25, S63–64.

74. Kabalka, G. W., E. Buonocore, K. Hubner, M. Davis, L. Huang. 1988. *Magnetic Resonance in Medicine*, 8, 89.

75. Grant, C. W. M., S. Karlik, E. Florio. 1989. *Magnetic Resonance in Medicine*, 11, 236–243.

76. Kabalka, G., E. Buonocore, K. Hubner, M. Davis, L. Huang. 1989. *Magnetic Resonance in Medicine*, 8, 89–95.

77. Kim, S. K., G. M. Pohost, G. A. Elgavish. 1991. *Magnetic Resonance in Medicine*, 22, 57–67.

78. Lipari, G., A. Szabo. 1982. *Journal of the American Chemical Society*, 104, 4546–4559.

79. Lipari, G., A. Szabo. 1982. *Journal of the American Chemical Society*, 104, 4559–4570.

80. Koenig, S. H., Q. F. Ahkong, R. D. Brown III, M. Lafleur, M. Spiller, E. Unger, C. Tilcock. 1992. *Magnetic Resonance in Medicine*, 23, 275–286.

81. Andre, J. P., E. Toth, H. Fischer, A. Seelig, H. R. Macke, A. E. Merbach. 1999. *Chemistry—A European Journal*, 5, 2977–2983.

82. Nicolle, G. M., E. Toth, K. P. Eisenwiener, H. R. Macke, A. E. Merbach. 2002. *Journal of Biological Inorganic Chemistry*, 7, 757–769.

83. Nicolle, G. M., L. Helm, A. E. Merbach. 2003. *Magnetic Resonance in Chemistry*, 41, 794–799.

84. Torres, M. I., M. Prata, A. C. Santos, J. P. André, J. A. Martins, L. Helm, É. Tóh, M. L. García-Martín, T. B. Rodrigues, P. López-Larrubia, S. Cerdán, C. F. G. C. Geraldes. 2008. *NMR in Biomedicine*, 21, 322–336.
85. Schuhmann-Giampieri, G., H. Schmitt-Willich, W. R. Press, C. Weinmann, H. J. Negishi, U. Speck. 1992. *Radiology*, 183, 59–64.
86. Vittadini, G., E. Felder, C. Musu, C. Tirone. 1990. *Investigative Radiology*, 5, S59–S62.
87. Vaccaro, M., A. Accardo, D. Tesauro, G. Mangiapia, D. Löf, K. Schillén, O. Söderman, G. Morelli, L. Paduano. 2006. *Langmuir*, 22, 6635–6643.
88. Storrs, R. W., F. D. Tropper, H. Y. Li, C. K. Song, J. K. Kuniyoshi, D. A. Sipkins, K. C. P. Li, M. D. Bednarski. 1995. *Journal of the American Chemical Society*, 117, 7301–7306.
89. Lattuada, L., G. Lux. 2003. *Tetrahedron Letters*, 44, 3893–3895.
90. Morag, O., A. Ahmad, N. Kamaly, E. Perouzel, A. Caussin, M. Keller, A. Herlihy, J. Bell, A. D. Miller, M. R. Jorgensen. 2006. *Organic & Biomolecular Chemistry*, 4, 3489–3497.
91. Gløgård, C., S. L. Fossheim, R. Hovland, A. J. Aasen, J. Klaveness. 2000. *Journal of the Chemical Society Perkin Transactions*, 2, 1047–1052.
92. Gløgård, C., G. Stensrud, R. Hovland, S. L. Fossheim, J. Klaveness. 2002. *International Journal of Pharmaceutics*, 233, 131–140.
93. Gløgård, C., G. Stensrud, J. Klaveness. 2003. *International Journal of Pharmaceutics*, 253, 39–48.
94. Huster, D., A. J. Jin, K. Arnold, K. Gawrisch. 1997. *Biophysical Journal*, 73, 855–864.
95. Alhaique, F., I. Bertini, M. Fragai, M. Carafa, C. Luchinat, G. Parigi. 2002. *Inorganica Chimica Acta*, 331, 151–157.
96. Laurent, S., L. V. Elst, C. Thirifays, R. N. Muller. 2008. *Langmuir*, 24, 4347–4351.
97. Terreno, E., A. Sanino, C. Carrera, D. Delli Castelli, G. B. Giovenzana, A. Lombardi, R. Mazzon, L. Milone, M. Visigalli, S. Aime. 2008. *Journal of Inorganic Biochemistry*, 102, 1112–1119.
98. Tilcock, C. 1999. *Advanced Drug Delivery Reviews*, 37, 33–51.
99. Strijkers, G. J., W. J. M. Mulder, R. B. van Heeswijk, P. M. Frederik, P. Bomans, P. C. M. M. Magusin, K. Nicolay. 2005. *Magnetic Resonance Materials in Physics*, 18, 186–192.
100. Kamaly, N., T. Kalber, A. Ahmad, M. H. Oliver, P. W. So, A. H. Herlihy, J. D. Bell, M. R. Jorgensen, A. D. Miller. 2008. *Bioconjugate Chemistry*, 19, 118–129.
101. Weissig, V., J. Babich, V. P. Torchilin. 2000. *Colloids and Surfaces B: Biointerfaces*, 18, 293–299.
102. Trubetskoy, V. S., J. A. Cannillo, A. Milshtein, G. L. Wolf, V. P. Torchilin. 1995. *Magnetic Resonance Imaging*, 13, 31–37.
103. Klibanov, A. L., K. Maruyama, V. P. Torchilin, L. Huang. 1990. *FEBS Letters*, 268, 235–237.
104. Patel, H. M., S. M. Moghimi. In *Targeting of Drugs—Optimization Strategies*, G. Gregoriadis, A. C. Allison, G. Poste (eds.), 87, New York: Plenum Press, 1990.
105. Mulder, W. J. M., G. J. Strijkers, J. W. Habets, E. J. W. Bleeker, D. W. J. van der Schaft, G. Storm, G. A. Koning, A. W. Griffioen, K. Nicolay. 2005. *FASEB Journal*, 19, 2008–2010.
106. Torchilin, V. P. 2001. *Journal of Controlled Release*, 73, 137–172.
107. Kwon, G. S. 2003. *Critical Reviews in Therapeutic Drug Carrier Systems*, 20, 357–403.
108. Harris, J. M., N. E. Martin, M. Modi. 2001. *Clinical Pharmacokinetics*, 40, 539–551.
109. Miller, D. W., E. V. Batrakova, T. O. Waltner, V. Yu. Alakhov, A. V. Kabanov. 1997. *Bioconjugate Chemistry*, 8, 649–657.
110. Katayose, S., K. Kataoka. 1998. *Journal of Pharmaceutical Sciences*, 87, 160–163.
111. Harada, A., K. Kataoka. 1998. *Macromolecules*, 31, 288–294.

112. Kwon, G. S., M. Naito, M. Yokoyama, T. Okano, Y. Sakurai, K. Kataoka. 1997. *Journal of Controlled Release*, 48, 195–201.
113. Jeong, Y. I., J. B. Cheon, S. H. Kim, J. W. Nah, Y. M. Lee, Y. K. Sung. 1998. *Journal of Controlled Release*, 51, 169–178.
114. Kabanov, A. V., V. A. Kabanov. 1998. *Advanced Drug Delivery Reviews*, 30, 49–60.
115. Torchilin, V. P. 2000. *Current Pharmaceutical Biotechnology*, 1, 183–215.
116. Nakamura, E., K. Makino, T. Okano, T. Yamamoto, M. Yokoyama. 2006. *Journal of Controlled Release*, 114, 325–333.
117. Lee, H. Y., H. W. Jee, S. M. Seo, B. K. Kwak, G. Khang, S. H. Cho. 2006. *Bioconjugate Chemistry*, 17, 700–706.
118. Turner, J. L., D. Pan, R. Plummer, Z. Chen, A. K. Whittaker, K. L. Wooley. 2005. *Advanced Functional Materials*, 15, 1248–1254.
119. Drummond, D. C., O. Meyer, K. Hong, D. B. Kirpotin, D. Papahadjopoulos. 1999. *Pharmacological Reviews*, 51, 691–743.
120. Brandwijk, R. J. M. G. E., W. J. M. Mulder, K. Nicolay, K. H. Mayo, V. L. J. L. Thijssen, A. W. Griffioen. 2007. *Bioconjugate Chemistry*, 18, 785–790.
121. Lipinski, M. J., V. Amirbekian, J. C. Frias, J. G. S. Aguinado, V. Mani, K. C. Briley-Saebo, V. Fuster, J. T. Fallon, E. A. Fisher, Z. A. Fayad. 2006. *Magnetic Resonance in Medicine*, 56, 601–610.
122. Amirbekian, V., M. J. Lipinski, K. C. Briley-Saebo, S. Amirbekian, J. G. S. Aguinado, D. B. Weinreb, E. Vucic, J. C. Frias, F. Hyafil, V. Mani, E. A. Fisher, Z. A. Fayad. 2007. *Proceedings of the National Academy of Sciences*, 104, 961–966.
123. Li, K. C., M. D. Bednarski. 2002. *Journal of Magnetic Resonance Imaging*, 16, 388–393.
124. Sipkins, D. A., D. A. Cheresh, M. R. Kazemi, L. M. Nevin, M. D. Bednarski, K. C. P. Li. 1998. *Nature Medicine*, 4, 623–626.
125. Bull, S. R., M. O. Guler, R. E. Bras, P. N. Venkatasubramanian, S. I. Stupp, T. J. Maede. 2005. *Bioconjugate Chemistry*, 16, 1343–1348.
126. Mulder, W. J. M., D. W. J. van der Schaft, P. A. I. Hautvast, G. J. Strijkers, G. A. Koning, G. Storm, K. H. Mayo, A. W. Griffioen, K. Nicolay. 2007. *FASEB Journal*, 21, 378–383.
127. Accardo, A., D. Tesauro, P. Roscigno, E. Gianolio, L. Paduano, G. D'Errico, C. Pedone, G. Morelli. 2004. *Journal of the American Chemical Society*, 126, 3097–3107.
128. Mangiapia, G., A. Accardo, F. Lo Celso, D. Tesauro, G. Morelli, A. Radulescu, L. Paduano. 2004. *Journal of Physical Chemistry B*, 108, 17611–17617.
129. Accardo, A., D. Tesauro, G. Morelli, E. Gianolio, S. Aime, M. Vaccaro, G. Mangiapia, L. Paduano, K. Schillen. 2007. *Journal of Biological Inorganic Chemistry*, 12, 267–276.
130. Tesauro, D., A. Accardo, E. Gianolio, L. Paduano, J. Teixeira, K. Schillen, S. Aime, G. Morelli. 2007. *ChemBioChem*, 8, 950–955.
131. Morisco, A., A. Accardo, E. Gianolio, D. Tesauro, E. Benedetti, G. Morelli. 2009. *Journal of Peptide Science*, 15, 242–250.
132. Accardo, A., D. Tesauro, L. Aloj, L. Tarallo, C. Arra, G. Mangiapia, M. Vaccaro, C. Pedone, L. Paduano, G. Morelli. 2008. *ChemMed Chem*, 3, 594–602.
133. Vaccaro, M., G. Mangiapia, L. Paduano, E. Gianolio, A. Accardo, D. Tesauro, G. Morelli. 2007. *Chem. Phys. Chem*, 8, 2526–2538.
134. Accardo, A., D. Tesauro, A. Morisco, G. Mangiapia, M. Vaccaro, E. Gianolio, K. Heenan, L. Paduano, G. Morelli. 2009. *Journal of Biological Inorganic Chemistry*, 14, 577–589.

# 17 Applications in the Food and Textile Industries

*Hans-Jürgen Buschmann*

## CONTENTS

Supramolecular ligands are already used in food and textile industries. In the case of usage in foods, governmental regulations have already been expressed. Up to now, no comparable standards have been assigned in the textile industry. However, the regulations for other applications can be used for the evaluation of these molecules. For example, cosmetic products and clothing textiles are in contact with human skin. Thus, comparable standards may be used for the appraisal of risk exposure. These aspects are more or less not relevant for technical textiles because there is no direct contact with human skin.

Supramolecular compounds may be used during the processing of the product to improve their properties, or they may be a part of the final product, giving this product a new functionality. These aspects will be discussed for the food and the textile industries.

## APPLICATIONS IN FOOD INDUSTRY

The only supramolecular compounds tested and approved, at the moment, for use in foodstuff are cyclodextrins. Cyclodextrins are natural cyclic oligosaccharides formed by the enzymatic degradation of starch. They consist of six ($\alpha$-cyclodextrin, $\alpha$-CD), seven ($\beta$-cyclodextrin, $\beta$-CD), or eight ($\gamma$-cyclodextrin, $\gamma$-CD) glucose monomers. The cyclodextrin molecules have a torus-shaped structure with a rigid cavity (see Figure 17.1).

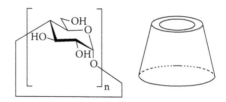

**FIGURE 17.1**   Chemical and schematic structure of cyclodextrins.

In these cavities, guest molecules can be enclosed because of hydrophobic inter-actions. A match between the size of the cavity and of the guest molecule is not essential. Because of the complex formation, the physical and chemical properties of the enclosed guest molecules change, for example, the vapor pressure is reduced. The most important advantages for the use of cyclodextrins in food are as follows: increased solubility; increased rate of solubilization; increased stability of emulsions; changes in the rheological properties; protection against oxidation, light, and heat-induced decomposition; stabilization against evaporation; and masking or reduction of undesired taste.

The use of cyclodextrins as food additives has been reviewed in the literature extensively. Detailed information about the specific usage of cyclodextrins can be found in the literature given in the cited books and review articles.[1–9] Just recently, the use of α- and β-cyclodextrin for the binding of flavors and the factors affecting complex stabilities have been discussed in detail.[10]

However, the industrial use of cyclodextrins is subject to governmental regulatory requirements and differs between countries. Since November 2000, β-cyclodextrin has been authorized in Germany as a food additive (E 459). In other countries, for example, the United States and Japan, β-cyclodextrins have been used as a food addi-tive for several years. The actual regulations of the WHO Food Standards are found in the Codex Alimentarius.[11] In the United States, the Food and Drug Administration decides on the safety of chemicals added to food (generally recognized as safe). The details of all actual regulations can be easily obtained from the Internet.[11]

During the last few years, cyclodextrins have also been used in food packaging. The cyclodextrins in these films are able to complex substances migrating through these barriers.[12] Using the complexes of cyclodextrins with natural antioxidants leads to controlled-release active packaging that is able to extend the oxidative stability of foods.[13–16] Complex β-cyclodextrin with 1-methylcyclopropen is used to increase the storage or transport times of fruits.[17] Low concentrations of 1-methylcyclopropen block the action of ethylene, which is essential for the maturation of fruits. Thus, the use of cyclodextrin complexes might lead to new active packaging with previously unknown properties.

Other supramolecular ligands, for example, crown ethers, calixarenes, or cucurbi-turils, are not used in the food industry because of their unknown toxicological data. Just recently, few toxicological data about cucurbiturils have been reported.[18] These data are required for even thinking about the use of these molecules in the food indus-try or other industries. As long as they are missing, no applications will be realized.

## APPLICATIONS IN TEXTILE INDUSTRY

### Use in Textile Processes

When considering applications of supramolecular ligands in textile industry processes, information about the toxicity of these substances is very important. During the production process, these substances may end up in the wastewater and cause environmental problems. Thus, the use of a supramolecular ligand as an auxiliary is restricted to a few ligands, for example, cyclodextrins and cucurbiturils.

Different applications of cyclodextrins in textile processes have already been discussed in the literature.[19-22] These examples will not be discussed in detail again.

One important usage of cyclodextrins in textile finishing is the dyeing process. The formation of complexes between β- and γ-cyclodextrins and different dyes in an aqueous solution is well known. Therefore, the different cyclodextrins can act as retarding, migrating, and leveling agents.[23-29] The use of cyclodextrins is not restricted to a special class of dye stuff because all dyes form complexes with cyclodextrins. However, in the case of reactive dyes, no clear results about the influence of the cyclodextrins on the dyeing process are observed.[29] In contrast, the intensity and the uniformity of the coloration are increased using preformed cyclodextrin complexes with disperse dyes.[27,30] During a conventional dyeing process, dye dispersers, surfactants, and other chemicals are used for the stabilization of the liquor. These auxiliaries are responsible for the large amounts of disperse dyes that remain in the liquor after finishing the dyeing process. Their removal from the wastewater is difficult because of the presence of the auxiliaries. Using cyclodextrin–disperse dye complexes, no additional chemicals are needed. The cyclodextrin complex is dissolved or dispersed in the aqueous solution. The dye concentration in the aqueous phase is limited by the solubility product of the dye molecule in water. The uncomplexed dye adsorbs at the fiber surface and migrates into the fiber. This process lowers the concentration of the uncomplexed dye in the solution and as a result the cyclodextrin–dye complex dissociates. On the other hand, no nucleation of the dye takes place because of the low concentration of the uncomplexed dye molecules. These equilibria are shown in Figure 17.2.

At the end of the dyeing process, the dye concentration in the liquor is given only by the solubility product of the dye. This concentration is generally very low. On the other hand, if the cyclodextrin–dye complex is present in the solution in a large amount, the dye uptake of the fiber takes place as long as the fiber is able to

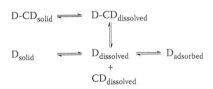

**FIGURE 17.2** Schematic presentation of all equilibria during a dyeing process using cyclodextrin complexes with dye molecules.

accommodate the dye molecules. The uptake of the dye stops if the fiber is saturated with the dye. Thus, more intense colors of the dyed material are possible compared with the conventional dyeing process.[31] Plotting the absorbance of the liquor after the dying process as a function of the $K/S$ values (Kubelka–Monk equation defines a relationship between the spectral reflectance $R$ (%) of a probe and its absorption $K$ and scattering $S$: $K/S = (1 - R)^2/2R$), one can compare the depth of the color of the colored textile with the amount of dye left in the solution after the dyeing process. This is shown in Figure 17.3.

These results clearly show the advantages of the use of cyclodextrin complexes with disperse dyes in dyeing processes compared with the conventional dyeing procedure.

Ion-selective dyes have been described in the literature for a long time. Mainly crown ethers and cryptands have been used as the ion-selective part of dye molecules. Only very few examples have been published in which dyes containing cryptand or calixarenes have been used for the dyeing of fibers.[33-35] Ion-selective dyes with a cryptand part of the molecule can be used for the dyeing of polyamide. These dyes change their colors in the presence of different salts.[33] The calixarene-based dyes are only known at the moment that they show good fastness properties.[34,35]

The dyes remaining in the liquor after finishing the dyeing process end up in the wastewater. This is a great ecological problem. Because of the demands of the customers, the stability of the dye molecules against UV irradiation or against oxidation has to be very high. However, this leads to difficulties for the decoloration of the wastewater. Several methods are used in practice to reduce the color of the wastewater.[36,37] However, some of them are relatively expensive or not very effective. Thus, new methods for the reduction of the chromaticity are of industrial interest. In 1905, Behrend et al.[38] described the synthesis of a chemical compound from glycolurile and formaldehyde able to form insoluble complexes with some indicator dye molecules. The structure of the chemical compound synthesized by Behrend was resolved many years later, and the compound was named cucurbituril because of the molecular appearance.[39] This molecule is built from six monomers; thus, the more correct name is cucurbit[6]uril (see Figure 17.4).

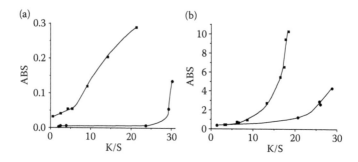

**FIGURE 17.3** Absorbance ABS of the liquor after finishing the dyeing process of PET yarn as plotted against the color intensity of the dyed yarn: (a) with Cellitonred GG and (b) with Dorosperseblue BLF (■) conventional dyeing; (●) dyeing with ß-cyclodextrin-dye complex. (From Buschmann, H.-J. et al., *Textilveredlung*, 31, 115–117, 1996. With permission.)

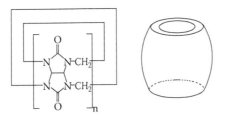

**FIGURE 17.4**  Chemical and schematic structure of cucurbit[$n$]urils.

Because of the formation of insoluble complexes with dyes and many other organic substances, cucurbit[6]uril can be used to remove them from the aqueous solution.[40–48] The potential of cucurbituril for the removal of different dyes has been tested in detail.[49–52] Even solid cucurbituril can be used as filling materials in columns to remove different dyes from the aqueous solution, and the solid cucurbituril can be regenerated using ozone.[43,46–48] Ozone destroys the complexed dye molecules, and the resulting reaction products are water soluble. In the solid state, cucurbit[6] uril is stable against ozone treatment.

However, other organic host molecules have also been used to remove organic dyes from the aqueous solution. It is well known that cyclodextrins form complexes with organic dyes as described previously. However, these complexes are water soluble and, therefore, only polymeric-bounded cyclodextrins can be used as adsorbent.[53–55] A recycling of these adsorbents is only possible by extraction with organic solvents. Water-insoluble calix[$n$]arenes have been used for the extraction of carcinogenic direct dyes from aqueous solutions.[56]

The selective removal of dyes from the wastewater after the dyeing process is still an important industrial problem. While doing research about this topic, one has to keep in mind that not only dyes but also surfactants, dispersers, and salts are present in the wastewater after dyeing. Thus, the selective removal of dyes in the presence of other auxiliaries has economical advantages over the conventional methods used for wastewater treatment.

After all single textile finishing processes, a washing step normally follows to remove chemicals from the fabric because they may cause problems during the next process. Normally, the fabrics pass through several baths containing pure water. However, washing adds costs and the number of baths is therefore limited. In most finishing processes, surfactants are used to improve the wettability of the fabric. Because of the chemical structure, surfactants have a high affinity for synthetic polymers. They are adsorbed on the fiber surface and their removal by common washing processes is not possible. The presence of surfactants has a negative influence on the adhesion of coatings, for example, the air bag used in automotives is made from coated polyamide 6.6. For this function, it is essential that the adhesive strength between fabric and coating is extremely high. The ability of cyclodextrins to form complexes with surfactants is well known.[57] With the knowledge of these results, cyclodextrins have been successfully used to remove adsorbed surfactants from the surface of polyester fabrics.[58] After the dyeing processes, adsorbed dye molecules are also present on the surface of the fibers. They have also been removed with the use of cyclodextrins.[24]

## USE FOR SURFACE MODIFICATION OF TEXTILE MATERIALS

### Cyclodextrins

The permanent fixation of supramolecular host molecules on the surface of different polymeric materials enables the creation of textile materials with new properties if the complexation behavior is maintained even after fixation. Depending on the polymeric raw materials, different methods similar to the dyeing processes of these materials can be used for the fixation of supramolecular host compounds. These methods are presented in Table 17.1, and for a better understanding, cyclodextrins will be discussed as an example.

Without any chemical modification, cyclodextrins are only bounded to cellulosic materials using epichlorhydrin[59] or other bifunctional reactants.[60] Because of the presence of hydroxyl groups in cellulosic materials and cyclodextrins bifunctional or polyfunctional reactants,[61] cross-linking between these groups takes place. A cyclodextrin derivative with a reactive group (e.g., the monochlorotriazinyl group) is able to react with the hydroxyl groups of cellulosic fibers like a reactive dye.[62–64] In the case of all other synthetic polymers, either cyclodextrin derivatives have to be used or the fiber surface has to be modified chemically to enable the fixation of cyclodextrins. Permanent fixation on fibers made from polyester is only possible with cyclodextrin derivatives with long alkyl chains or other hydrophobic groups. Comparable with disperse dyeing, the hydrophobic part of the substituted cyclodextrins migrates into the fiber above the glass transition temperature of the polymers. The polar cyclodextrin molecules remain on top of the fiber surface.[65]

Cyclodextrin derivatives with anionic groups have been used for fixation on polyamide fibers.[66] In cases in which no permanent fixation is necessary, cyclodextrin derivatives with alkyl or, for example, polyvinylamine chains, can be attached on the surfaces.[67] The existence of the cyclodextrins on the surface of the polymers and

---

## TABLE 17.1
## Strategies for the Permanent Fixation of Supramolecular Hosts on Different Materials

| Fixation Method | Material | | | | |
|---|---|---|---|---|---|
| | Cellulose | Wool | Polyamide | Polyester | Polyacrylnitrile |
| Cross-linking | + | − | − | − | − |
| Ionic interactions | − | + | + | − | + |
| van der Waals interactions | − | − | + | + | + |
| Covalent bonds | + | + | + | − | − |
| Penetration | − | − | − | + | − |

their ability to form complexes can be shown easily from the reaction with some indicator dyes. In the case of phenolphthalein, a red solution turns colorless on contact with the cyclodextrins fixed on a polymer surface. The methods for the fixation of cyclodextrins on the surface of different natural or synthetic polymers are shown in Figure 17.5. They are also suitable for other supramolecular hosts.

Textile materials with cyclodextrins fixed onto the fiber surfaces get new properties. As for instance, the comfort of clothing increases. The organic components of sweat are complexed, and therefore, the possibility of body odor development is reduced. The chemical analysis of sweat components can also be used for medical applications. The identification of organic compounds from patients enables new ways of medical diagnosis. Textile materials with cyclodextrins have also been used as filter material to remove organic substances from the air. In combination with activated charcoal, it can be used for the construction of protective cloths.[68]

At the moment, cyclodextrins are the most used supramolecular compounds in the textile industry. The cyclodextrin cavities can act as a depot for cosmetic and pharmaceutical substances and perfumes. As long as these substances are included in the cyclodextrins, they are stable and do not evaporate. In contact with the skin, the substances are released because of the water present at the skin surface, and they evaporate or migrate into the skin. A different approach to deposit these substances on textiles is the use of microcapsules, which are chemically fixed on the textiles.[69,70] The encapsulated substances are released after the capsules are mechanically ruptured. The most important difference between the usage of cyclodextrins and microcapsules is that in contrast to microcapsules, cyclodextrins can be reloaded after the release of the complexed substances.

Textiles with cosmetic properties can be provided using microcapsules as well as cyclodextrins.[71] The active compounds used include anticellulite, antimicrobial, or whitening agents.[72] Different textile products with microencapsulated cosmetic substances are already offered worldwide.[73] Textiles with cyclodextrins and complexed cosmetic agents are also well known.[74,75] In Japan, underwear containing a cyclodextrin complex with squalane to prevent the skin from drying and with linolenic acid for the suppression of atopic dermatitis are commercially available.[76]

Scented fabrics are of interest for clothes or household linen. These textiles are able to release fragrances or flavors. Again, both microcapsules[69,70,77] or cyclodextrins[78–81] can be used to store these substances. The cyclodextrin molecules are

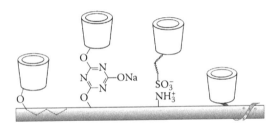

**FIGURE 17.5** Schematic fixation of cyclodextrin derivatives on different fiber.

covalently bound to cotton fibers, and therefore, they resist a certain number of washing cycles. During the washing process, all substances are removed from the cyclodextrin cavities. The empty cavities can be refilled by spraying with a fragrance or perfume solution. The formed complexes with cyclodextrins are very stable, and the storage time is not limited. However, one has to keep in mind that some fragrances are allergens. Contact of the scented clothes with human skin may cause, in a few cases, allergic contact eczema.[82]

Textiles with cyclodextrins are also used as medical textiles. In this case, the cyclodextrins act as a reservoir for pharmaceutical substances.[83,84] Wound dressings with cyclodextrins for the release of drugs have been tested.[85,86] However, the mere presence of cyclodextrins in wound dressings already has a positive influence on the healing process.[87,88] Chronic wounds can possess an offensive odor because of infections. Odor-absorbing dressings with cyclodextrins are already used in practice.[89–91] More aspects and details of textile materials permanently finished with cyclodextrins have been reviewed recently.[20,21]

## Calixarenes

Following the fixation strategies given in Table 17.1, it is possible to fix calixarene derivatives using modifications at the upper (Figure 17.5: R) or lower rim (Figure 17.5: X) of the molecules. Depending on the later use of the surface-modified fabrics, one has to decide which modification (at the upper rim or lower rim) is the best. Calixarenes bonded to stationary phases via the hydroxyl groups (lower rim) or substituents at the upper rim are already used in liquid chromatography.[92–94] For the complexation of cations, the hydroxyl groups (Figure 17.5: X) are replaced, for example, by ester, amides, or acid substituents.[92,95] For this use, the molecular flexibility of the calixarene molecules has to be maintained because the most stable complexes are formed with the cone conformation of the calixarenes. Therefore, the fixation of the calixarene on the surface of the fabric should be achieved with only one connection between fiber and calixarene. This is shown in Figure 17.6.

If the groups R are tert-butyl groups and $R_A$ is an alkyl chain, then the alkyl chain penetrates much easier into a polyester matrix than the bulky tert-butyl groups. Using a mixture of tert-butylphenol and nonylphenol, a suitable calixarene derivative

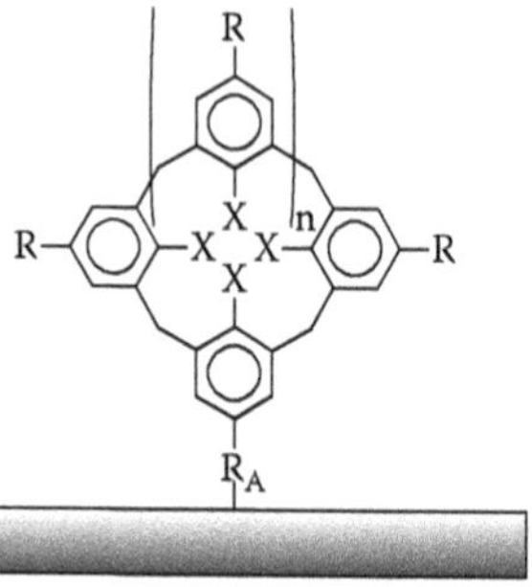

**FIGURE 17.6**  Fixation of calixarenes with the help of one chemical bond between fiber and the group $R_A$ at the upper rim of a calixarene.

can be synthesized.[96] The calix[6]arene derivatives are fixed on polyester, and the complexation of $Cu^{2+}$, $Co^{2+}$, and $Ni^{2+}$ from the aqueous solution is obvious from the color changes of the fabrics used and the decreasing concentration of these cations in the solution.[97,98] Such modified textile materials can be used as textile filters for the removal of specific cations from an aqueous solution in the presence of other cations. Choosing suitable substituents at the lower rim of the calixarenes, their selectivity against one cation can be increased dramatically. Thus, the ability of a calix[6]arene substituted at the lower rim with six carboxylic groups to complex uranium ions with an extremely high selectivity has been described in the literature.[99,100] The selectivity of the calixarenes used for $UO_2^{2+}$ over other cations is $K_{uranyl}/M_{cation} = 10^{12} - 10^{17}$ (cation: $Mg^{2+}$, $Ni^{2+}$, $Zn^{2+}$, and $Cu^{2+}$). Therefore, immobilized calixarenes on a polymer resin have been used to complex uranyl ions from seawater.[101] Also, the clean-up of very small amounts of uranyl ions present in the environment and groundwater near uranium mining sites is an important problem. Fleece with a permanent fixed uranophile calix[6]arene has been successfully used to remove even very small amounts of uranyl ions from aqueous solution[102,103] (see Figure 17.7).

Even at very low concentrations, more than 95% of the uranium ions present in the solution are complexed. Washing the filter with hydrochloric acid removes the complexed uranium ions from the filter, and the filter can be used again. For the selective complexation of additional cations, other derivatives of calixarenes can be fixed on the fleece to obtain suitable filter materials. The industrial use of these filter materials is of great importance, for example, for the accumulation of metals, for the production of ultrapure metals, and for the treatment of industrial wastewater. A more detailed discussion about the treatment of wastewater or the separation of cations is given in the chapter by B. Moyer in this book.

Because the calixarene molecules obtained from the condensation of tert-butyl-phenole and formaldehyde have a hydrophobic upper rim and a hydrophilic lower rim, they can also be used to influence the surface properties of synthetic polymers. The adhesion between a fabric and a polymer coating depends on the interactions between them. In the case of a coating of a nonpolar polyester fabric with a polar polyurethane coating, the adhesion strength between them is low. After the fixation of tert-butylcalixarenes on polyester, the fiber surface becomes more polar

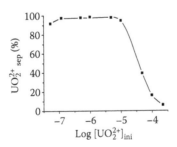

**FIGURE 17.7**  Uranium separation (in %) as function of the initial concentration using a fleece with a permanent fixed uranophile calix[6]arene at pH = 5 and 25°C.

compared with the untreated fiber. As a result, the adhesion between the coating and the polyester fabric increases.[104]

The antimicrobial activity of calixarenes have recently been reported.[105] Some calixarene derivatives examined even show fungicidal behavior. These effects are useful for geotextiles, which are in contact with soil. However, no textile product has been tested or even used thus far.

## Cucurbit[*n*]urils

Cucurbit[*n*]urils are rigid molecules possessing a hydrophobic cavity accessible by two identical portals formed by the carbonyl groups of the urea subunit.[39,106] Because of their structure, cucurbit[*n*]urils are able to include hydrophobic molecules, or part of those molecules, within their cavity. They also form complexes with cations due to electrostatic interactions with the carbonyl groups. However, these cucurbit[*n*]urils do not show any affinity to different fiber materials. As a result, no interactions between cucurbit[*n*]urils and fiber surfaces take place. Therefore, only temporary effects can be achieved using cucurbit[*n*]urils in textile finishing. In the literature, cucurbit[6]uril as a depot for perfumes or as a deodorant in textile finishing has been described.[107]

Because of their chemical inertness, the fixation of unsubstituted cucurbit[*n*]urils on polymer surfaces is only possible via the formation of so-called surface rotaxanes. After a surface pretreatment followed by the fixation of spermine[108,109] or allylamine,[110,111] cucurbit[6]uril forms an inclusion complex with the amino compound. The reaction of the second amino group of the amino compound with a bulky molecule finally leads to the formation of a rotaxane (see Figure 17.8).

The rigid cucurbit[6]uril acts as a distance piece between the polymer surface and the dye molecule acting as stopper group. As a result, the interactions between the polymer surface and the dye molecule are minimized. If a fluorophore is used as the stopper group, a polymer surface is obtained showing fluorescence. Until now, nothing is known about the changes in the surface properties due to the formation of surface rotaxanes. Depending on the stopper groups used, the hydrophobic or hydrophilic behavior can be influenced together with the roughness of the modified polymer surface.

No fixation on organic polymers of any cucurbit[*n*]uril derivative by the formation of covalent bonds or by other interactions (see Table 17.1) has been described so

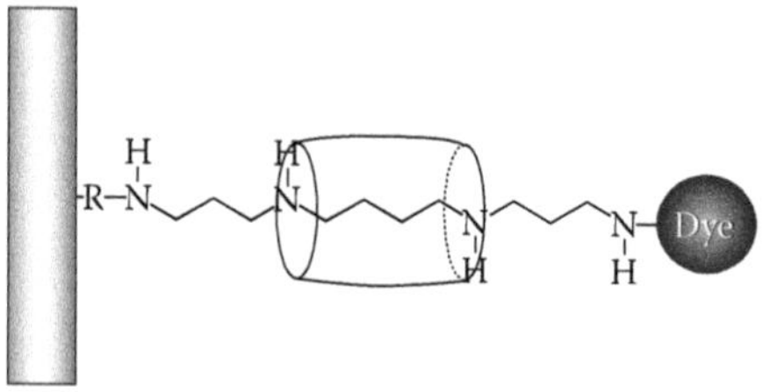

**FIGURE 17.8** Schematic structure of a surface rotaxane with cucurbituril and spermine stopped by a dye molecule.

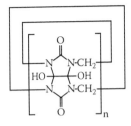

**FIGURE 17.9**  Schematic structure of perhydroxyl cucurbit[*n*]uril derivatives.

far. Cucurbituril derivatives suitable for this purpose have been reported.[112,113] One of these derivatives has hydroxyl groups at the equatorial plane of the cucurbituril molecule (see Figure 17.9).

The reactivity of the hydroxyl groups is comparable with those of cyclodextrins. Up to now, only the fixation of these cucurbit[*n*]uril derivatives on silica gel for use as a stationary phase in chromatography has been reported.[114,115] Therefore, possibilities for the fixation of perhydroxyl cucurbit[*n*]urils on fiber surfaces are related to those used for cyclodextrins. One expects that textile materials with permanently linked cucurbiturils can be used for the separation of gases[116] or for the selective complexation of organic molecules.

## Dendrimers

Dendrimers are repeatedly branched and are large spherical three-dimensional molecules. They are highly branched and have cavities of different sizes in the interior. Dendrimers are constructed by chemically identical subsequent growth steps. Each step represents a new "generation." No direct fixation of dendrimers via covalent bonds or other interactions with the fiber surface have been reported.

Different dendrimers as part of a sol–gel coating for textiles have been reported.[117] The following effects resulting from the presence of dendrimers have been claimed: "breathable, waterproof, oil repellency, soil repellency, antistatic, anti-UV irradiation, antibacterial, dying acceleration, and slow release of drug and perfume." However, from the literature, it is well known that most of these effects can be achieved by simple sol–gel coatings.[118] Poly(amidoamine) dendrimers loaded with silver have been deposited on nylon/cotton fabrics in a process similar to coating.[119] The treated fabrics show antimicrobial properties. Polymers with incorporated dendrimers for textile coating are commercially offered.

Spherical dendrimers are not suitable for permanent fixation on fibers. Parts of the dendrimer molecules, the so-called dendrons, are more suitable. The molecules look like trees having a crown at the top, a tree trunk, and at least a root. If this root is a reactive monomer with an attached dendron, linear polymers have been synthesized with a fully dendronized polymer surface.[120] For the modification of fiber surfaces, a different strategy has to be used. Dendrons with suitable chemical groups

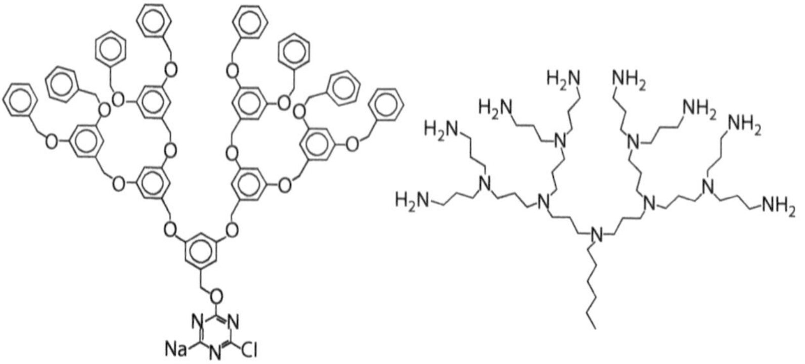

**FIGURE 17.10**  Chemical structures of a third-generation non-polar dendron (left) and a polar dendron (right).

(according to Table 17.1) enable the permanent surface modification of different fiber materials (see Figure 17.10).

The fixation of a nonpolar dendron on cotton results in a hydrophobization and the fixation of a polar dendron on polyester in a hydrophilization of the textile material.[121,122] Both dendrons are able to store and release molecules within their cavities. The storage of water molecules within the polar nitrogen-containing dendrons may be important for the wearing comfort of textiles for humans with sensitive skin.[123] The nitrogen-containing dendrons also show biostatic effects. The combination of storage and antimicrobial effects may be important for the development of textiles against atopic dermatitis.

### Further Developments

Although the self-organization of polyelectrolyte layers on different substrates has been well known for many years,[124] this technique is already used in different applications.[125] Up to now, mainly hard materials have been coated with polyelectrolyte layers. Very few results are known for the coating of flexible materials like fabrics, and even this technique can easily be used for these materials too.

The fabrics are dipped into solutions containing a polycation or polyanion followed by a washing process (see Figure 17.11).

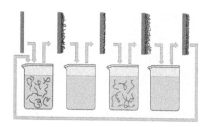

**FIGURE 17.11**  Layer-by-layer deposition process.

In this way, the thickness of the layers can be controlled very accurately. Each layer has a dimension within the nanometer scale. Because of the simplicity of this technique, it can be easily applied in textile finishing. Very recently, textile finishing using polyelectrolytes has been reported.[126–130] The incorporation of colloidal platinum particles results in a significant increase in thermal resistance in the case of poly(propylene) nonwovens.[126] Furthermore, the water sorption, thermal resistance, and surface resistance of different textile materials are changed if polyelectrolyte layers are deposited.[127,129,130]

In the future, important developments are expected from this technology because a large number of different materials can be incorporated within these layers. However, the relevance of this technology for textile finishing can only been seen in the future.

## ACKNOWLEDGMENTS

Financial support by the Ministerium für Innovation, Wissenschaft und Forschung des Landes Nordrhein-Westfalen (Department of Innovation, Science and Research of the State of Nordrhein-Westfalen) is thankfully acknowledged.

## REFERENCES

1. Szejtli, J. 1982. *Starch*, 34, 379–385.
2. Szejtli, J. *Cyclodextrin Technology*, Dordrecht: Kluwer, 1988.
3. Hashimoto, H. In *Cyclodextrins*, Szejtli, J., T. Osa, (eds.), *Comprehensive Supramolecular Chemistry*. Vol. 3, 483–502. Oxford: Elsevier, 1996.
4. Hedges, A. R. 1998. *Chemical Reviews*, 98, 2035–2044.
5. Hedges, A. R., C. McBride. 1999. *Cereal Foods World*, 44, 700–704.
6. Buschmann, H.-J., D. Knittel, C. Jonas, E. Schollmeyer. 2001. *Lebensmittelchemie*, 55, 54–56.
7. Szente, L., J. Szejtli. 2004. *Trends in Food Science and Technology*, 15, 137–142.
8. Cravotto, G., A. Binello, E. Baranelli, P. Carraro, F. Trotta. 2006. *Current Nutrition & Food Science*, 2, 343–350.
9. Astray, A., C. Gonzalez-Barreiro, J. C. Mejuto, R. Rial-Otero, J. Simal-Gandara. 2009. *Food Hydrocolloids*, 23, 1631–1640.
10. Astray, A., J. C. Mejuto, R. Rial-Otero, J. Simal-Gandara. 2010. *Food Research International*, 43, 1212–1218.
11. www.codexalimentarius.net.
12. Fenyvesi, E., K. Balogh, I. Siro, J. Orgovanyi, J. M. Senyi, K. Otta, L. Szente. 2007. *Journal of Inclusion Phenomena and Macrocyclic Chemistry*, 57, 1–4.
13. Siró. I., E. Fenyvesi, L. Szente, B. De Meulenaer, F. Devlieghere, J. Orgoványi, J. Sényi, J. Barta. 2006. *Food Additives & Contaminants: Part A*, 23, 845–853.
14. Koontz, J. L., J. E. Marcy, S. F. O'Keefe, S. E. Duncan. 2009. *Journal of Agricultural and Food Chemistry*, 57, 1162–1171.
15. Koontz, J. L., J. E. Marcy, S. F. O'Keefe, S. E. Duncan, T. E. Long, R. D. Moffitt. 2010. *Journal of Applied Polymer Science*, 117, 2299–2309.
16. Koontz, J. L., R. D. Moffitt, J. E. Marcy, S. F. O'Keefe, S. E. Duncan, T. E. Long. 2010. *Food Additives & Contaminants: Part A*, 27, 1598–1607.
17. NN *Food Technologie Magazin* 2009, June 8–9.
18. Uzunova, V. D., C. Cullinane, C. Brix, W. M. Nau, A. I. Day. 2010. *Organic & Biomolecular Chemistry*, 8, 2037–2042.

19. Buschmann, H.-J., U. Denter, D. Knittel. E. Schollmeyer. 1998. *Journal of the Textile Institute*, 89 Part 1, 554–561.
20. Buschmann, H.-J., D. Knittel. E. Schollmeyer. 2001. *Journal of Inclusion Phenomena and Macrocyclic Chemistry*, 40, 169–172.
21. Szejtli, J. 2003. *Starch*, 55, 191–196.
22. Andreaus, J., M. C. Dalmolin, I. B. de Oliveira Junior, I. O. Barcellos. 2010. *Quimica Nova*, 33, 929–937.
23. Savarino, P., G. Viscardi, P. Quagliotto, E. Montoneri, E. Barni. 1999. *Dyes and Pigments*, 42, 143–147.
24. Cireli, A., B. Yurdakul. 2006. *Journal of Applied Polymer Science*, 100, 208–218.
25. Perrin, E., A. Kumbasar, R. Atav, A. Yurdakul. 2006. *Journal of Applied Polymer Science*, 103, 2660–2668.
26. Vončina, B., V. Vivod, D. Jaušovec. 2007. *Dyes and Pigments*, 74, 642–646.
27. Parlati, S., R. Gobetto, C. Barolo, A. Arrais, R. Buscaino, C. Medana, P. Savarino. 2007. *Journal of Inclusion Phenomena and Macrocyclic Chemistry*, 57, 463–470.
28. Chalaya, N. E., V. V. Safonov. 2007. *Fibre Chemistry*, 39, 218–220.
29. Weber, T. *Über die Eignung von Cyclodextrinen als Hilfsmittel für das Färben mit Reaktivfarbstoffen*, Dissertation, Universität Stuttgart, 1995.
30. Buschmann, H.-J. In *Proceedings of the Eighth International Symposium on Cyclodextrins*, J. Szejtli, L. Szente (eds.), 547–552, Dordrecht: Kluwer, 1996.
31. Buschmann, H.-J., D. Knittel, E. Schollmeyer. 1996. *Textilveredlung*, 31, 115–117.
32. Buschmann, H.-J. Unpublished results.
33. Buschmann, H.-J., E. Schollmeyer. 1990. *Chemiefasern/Textilindustrie*, 40/92, 449–452.
34. Jain, V. K., S. G. Pillai, P. H. Kanaiya. 2008. *E-Journal of Chem*, 5, 1037–1047.
35. Menon, S. K., R. V. Patel, J. G. Panchal. 2010. *Journal of Inclusion Phenomena and Macrocyclic Chemistry*, 67, 73–79.
36. Robinson, T., G. McMullan, R. Marchant, P. Nigam. 2001. *Bioresource Technology*, 77, 247–255.
37. Sanghi, R., B. Bhattacharya. 2002. *Color Technology*, 118, 256–269.
38. Behrend, R., E. Meyer, F. Rusche. 1905. *Justus Liebigs Annalen der Chemie*, 339, 1–37.
39. Freeman, W. A., W. L. Mock, N.-Y. Shih. 1981. *Journal of the American Chemical Society*, 103, 7367–7368.
40. Buschmann, H.-J., H. Fink. 1990. Patent DE 4,001,139.
41. Buschmann, H.-J., A. Gardberg, E. Schollmeyer. 1991. *Textilveredlung*, 26, 153–157.
42. Buschmann, H.-J., D. Rader, E. Schollmeyer. 1991. *Textilveredlung*, 26, 157–160.
43. Buschmann, H.-J., A. Gardberg, D. Rader, E. Schollmeyer. 1991. *Textilveredlung*, 26, 160–162.
44. Buschmann, H.-J., C. Carvalho, U. Drießen, E. Schollmeyer. 1993. *Textilveredlung*, 28, 176–179.
45. Buschmann, H.-J., A. Gardberg, D. Rader, E. Schollmeyer. 1993. *Textilveredlung*, 28, 179–182.
46. Buschmann, H.-J., E. Schollmeyer. 1994. *Textilveredlung*, 29, 58–60.
47. Buschmann, H.-J., C. Jonas, W. Saus. 1994. DE Patent 4,412,320.
48. Buschmann, H.-J., C. Jonas, E. Schollmeyer. 1996. *European Water Pollution Control*, 6, 21–24.
49. Taketsuji, K., H. Tomioka. 1998. *Nippon Kagaku Kaishi*, 670–678.
50. Karcher, S., A. Kornnmüller, M. Jekel. 1999. *Water Science and Technology*, 40, 425–433.
51. Karcher, S., A. Kornmüller, M. Jekel. 1999. *Acta Hydrochimica et Hydrobiologica*, 27, 38–42.
52. Karcher, S., A. Kornmüller, M. Jekel. 2001. *Water Research*, 35, 3309–3316.
53. Crini, G., H. N. Peindy. 2006. *Dyes and Pigments*, 70, 204–211.

54. Crini, G., H. N. Peindy, F. Gimbert, C. Robert. 2007. *Separation and Purification Technology*, 53, 97–110.
55. Crini, G. 2008. *Dyes and Pigments*, 77, 415–426.
56. Ozmen, E. Y., S. Erdemir, M. Yilmaz, M. Bahadir. 2007. *Clean*, 35, 612–616.
57. Rekharsky, M. V., Y. Inoue. 1998. *Chemical Reviews*, 98, 1875–1917.
58. Buschmann, H.-J., R. Benken, D. Knittel, E. Schollmeyer. 1995. *Melliand Textilber*, 76, 732–734.
59. Szejtli, J. J., B. Zsadon, O. K. Horvath, A. Ujhazy, E. Fenyvesi, 1991. Patent HU 54,506.
60. Buschmann, H.-J., D. Knittel, E. Schollmeyer. 1991. *Melliand Textilber*, 72, 198–199.
61. Voncina, B., A. M. Le Marechal. 2005. *Journal of Applied Polymer Science*, 96, 1323–1328.
62. Reuscher, H., R. Hirsenkorn. 1996. *Journal of Inclusion Phenomena and Macrocyclic Chemistry*, 25, 191–196.
63. Denter, U., E. Schollmeyer. 1996. *Journal of Inclusion Phenomena and Macrocyclic Chemistry*, 25, 197–202.
64. Schmidt, A., D. Knittel, H.-J. Buschmann, E. Schollmeyer. 2001. Patent DE 10,155,781.
65. Ruppert, S., D. Knittel, H.-J. Buschmann, G. Wenz, E. Schollmeyer. 1997. *Starch/ Stärke*, 49, 160–164.
66. Knittel, D., H.-J. Buschmann, E. Schollmeyer. 1991. *Textilveredlung*, 26, 92–95.
67. Knittel, D., H.-J. Buschmann. 2006. Patent DE 10,200,601,0561.
68. N. N. 2006. Patent DE 102,006,001,528.
69. Nelson, G. 2002. *International Journal of Pharmaceutics*, 242, 55–62.
70. Rodrigues, S. N., I. M. Martins, I. P., Fernandes, P. B. Gomes, V. G. Mata, M. F. Barreiro, A. E. Rodrigues. 2009. *Chemical Engineering Journal*, 149, 463–472.
71. Shi, H., J. H. Xin. 2007. *9th Asian Textile Conference*, Taiwan, 2007, paper no: D01-14.
72. Schrader, K., A. Domsch. *Cosmetology—Theory and Practice. Research, Test Methods, Analysis, Formulas*, Augsburg: Verlag für chemische Industrie, 2005.
73. Mathis, R., A. Mehling. 2010. *COSSMA*, 31, 22–25.
74. Buschmann, H.-J., D. Knittel, E. Schollmeyer. 2004. *Cosmetics & Toiletries*, 119, 105–112.
75. Buschmann, H.-J., E. Schollmeyer, J. M. Quadflieg, S. Richert, A. Schrader. 2009. *Textilveredlung*, 44, 4–9.
76. Hashimoto, H. 2002. *Journal of Inclusion Phenomena and Macrocyclic Chemistry*, 44, 57–62.
77. Wang, C. X., Sh. L. Chen. 2005. *Fibres & Textiles in Eastern Europe*, 13, 41–44.
78. Martel, B., M. Morcellet, D. Ruffin, F. Vinet, M. Weltrowski. 2002. *Journal of Inclusion Phenomena and Macrocyclic Chemistry*, 44, 439–442.
79. Buschmann, H.-J., D. Knittel, E. Schollmeyer. 2002. *Perfumes Flavor*, 27, 36–38
80. Wang, C.-X., S.-L. Chen. 2004. *AATCC Review*, 5, 25–28.
81. Wang, C.-X., S.-L. Chen. 2005. *Journal of Industrial Textiles*, 34, 157–166.
82. Heydorn, S., T. Menné, J. D. Johansen. 2003. *Contact Dermatitis*, 48, 59–66.
83. Tonelli, A. E. 2003. *Journal of Textile and Apparel Technology and Management*, 3, 1–12.
84. Knittel, D., H.-J. Buschmann, C. Hipler, P. Elsner, E. Schollmeyer. 2004. *Aktuelle Dermatologie*, 30, 11–17.
85. Nichifor, M., M. Constantin, G. Mocanu, G. Funducanu, D. Branisteanu, M. Costuleanu, C. D. Radu. 2009. *Journal of Materials Science: Materials in Medicine*, 20, 975–982.
86. Montazer, M., E. B. Mehr. 2010. *Journal of the Textile Institute*, 101, 373–379.
87. Hipler, U. C., U. Schönfelder, C. Hipler, P. Elsner. 2007. *Journal of Biomedical Materials Research*, 83A, 70–79.
88. Buschmann, H.-J., U. C. Hipler. 2010. *Biomaterialien als Wundauflagen für chronisch stagnierende Wunden*, 1. AiF-Anwenderforum Medizintechnik. http://www.fms-dresden .de/index.php?id=57801&site=fms&lang=de&path=1%2C57800 (accessed November 25, 2011).

89. Lipman, R. D. A. 2002. Patent WO 02/09782.
90. Fleck C. A. 2006. *Advances in Skin & Wound Care*, 19, 242–245.
91. Lipman, R. D. A. 2007. *Wounds*, 19, 138–146.
92. Glennon, J. D., E. Horne, K. Hall, D. Cocker, A. Kuhn, S. J. Harris, M. A. McKervey. 1996. *Journal of Chromatography A*, 731, 47–55.
93. Ding, C., K. Qu, Y. Li, K. Hu, H. Liu, B. Ye, Y. Wu, S. Zhang. 2007. *Journal of Chromatography A*, 1170, 73–81.
94. Ludwig, R. 2000. *Fresenius' Journal of Analytical Chemistry*, 367, 103–128.
95. Danil de Namor, A. F., R. M. Cleverley, M. L. Zapata-Ormacher. 1998. *Chemical Reviews*, 98, 2495–2526.
96. Jansen, K., H.-J. Buschmann, E. Schollmeyer, A. M. Richter, D. Keil. 2002. Patent DE10,210,115.
97. Jansen, K., A. Wego, H.-J. Buschmann, E. Schollmeyer. 2004. *Melliand Textilber*, 85, 66–68.
98. Jansen, K., A. Wego, H.-J. Buschmann, E. Schollmeyer. 2002. *Vom Wasser*, 99, 119–130.
99. Shinkai, S., H. Koreishi, K. Ueda, T. Arimura, O. Manabe. 1987. *Journal of the American Chemical Society*, 109, 6371–6376.
100. Nagasaki, T., S. Shinkai. 1991. *Journal of the Chemical Society, Perkin Transactions 2*, 1063–1066.
101. Shinkai, S., H. Kawaguchi, O. Manabe. 1988. *Journal of Polymer Science—Part C: Polymer Letters*, 26, 391–369.
102. Schmeide, K., K. H. Heise, G. Bernhard, D. Keil, K. Jansen, D. Praschak. 2004. *Journal of Radioanalytical and Nuclear Chemistry*, 261, 61–67.
103. Schollmeyer, E., K. Jansen, H.-J. Buschmann, K. Schmeide. In *Polymer Surface Modification: Relevance to Adhesion*, K. L. Mittal, (ed.), Vol. 3, 353–366. Utrecht: VSP, 2004.
104. Buschmann, H.-J., E. Schollmeyer. 2010. *Journal of Adhesion Science and Technology*, 24, 113–121.
105. Lamartine, R., M. Tsukada, D. Wilson, A. Shirata. 2002. *Comptes Rendus Chimie*, 5, 163–169.
106. Lagona, J., P. Mukhopadhyay, S. Chakrabarti, L. Isaacs. 2005. *Angewandte Chemie (International ed. in English)*, 44, 4844–4870.
107. Döring, S., S. Kainz, R. Roesmann. 2003. Patent WO 2004,055,258.
108. Schollmeyer, E., H.-J. Buschmann, K. Jansen, E. Wego. 2002. *Progress in Colloid and Polymer Science*, 121, 39–42.
109. Schollmeyer, E., H.-J. Buschmann, K. Jansen, A. Wego. In *Polymer Surface Modification: Relevance to Adhesion*, K. L., Mittal (ed.), Vol. 3, 341–351. Utrecht: VSP, 2004.
110. Mix, R., K. Hoffmann, H.-J. Buschmann, J. F. Friedrich, U. Resch-Genger. 2007. *Vakuum*, 19, 31–37.
111. Hoffmann, K., R. Mix, J. F. Friedrich, H.-J. Buschmann, U. Resch-Genger. 2009. *Journal of Fluorescence*, 19, 229–237.
112. Kim, K., J. Kim, I.-S. Jung, S.-Y. Kim, E. Lee, J.-K. Kang. 2000. Patent EP1,094,065.
113. Jon, S. Y., N. Selvapalam, D. H. Oh, J.-K. Kang, S.-Y. Kim, Y. J. Jeon, J. W. Lee, K. Kim. 2003. *Journal of the American Chemical Society*, 125, 10186–10187.
114. Kim, K., D.-H. Oh, E. R. Nagarajan, Y.-H. Ko, S. Samal. 2004. Patent EP1,651,685.
115. Cheng, W. J., J. H. Go, Y. S. Baik, S. S. Kim, E. R. Nagarajan, N. Selvapalm, Y. H. Ko, K. Kim. 2008. *Bulletin of the Korean Chemical Society*, 29, 1941–1945.
116. Day, A. I., A. P. Arnold, R. J. Blanch. 2003. Patent US 20,030,140,787.
117. Hu, J., K. Nie, Q. Meng, G. Zheng. 2007. Patent US 20,080,282,480.
118. Mahltig, B., T. Textor. *Nanosols and Textiles*, Singapore: World Scientific Publishing Co., 2009.
119. Ghosh, S., S. Yadav, N. Vasanthan, G. Sekosan. 2010. *Journal of Applied Polymer Science*, 115, 716–722.

120. Schlüter, A. D. 2005. *Topics in Current Chemistry*, 245, 151–191.

121. Buschmann, H.-J., E. Schollmeyer. 2003. Patent DE10,340,573.

122. Buschmann, H.-J., E. Schollmeyer. In *Polymer Surface Modification: Relevance to Adhesion, Dendrons for Surface Modification of Polymeric Materials*, K. L. Mittal, (ed.), Vol. 4, 209–218, Utrecht: VSP, 2007.

123. Buschmann, H.-J., E. Schollmeyer. 2008. *Orthopädie-Technik*, 8, 628–632.

124. Decher, G. 1997. *Science*, 277, 1232–1237.

125. Decher, G., J. B. Schlenoff. *Multilayer Thin Films: Sequential Assembly of Nano-composite Materials*, Weinheim: Wiley-VCH, 2010.

126. Polowinski, S., D. Stawski. 2007. *Fibres & Textiles in Eastern Europe*, 15, 82–85.

127. Stawski, D., C. Bellmann. 2009. *Colloids and Surfaces A*, 345, 191–194.

128. Wang, Q., P. J. Hauser. 2009. *Cellulose*, 16, 1123–1131.

129. Benken, R., H.-J. Buschmann, E. Schollmeyer. 2010. *Melliand Textilber*, 91, 43–45.

130. Jantas, R., S. Polowinski, D. Stawski, J. Szumilewicz. 2010. *Fibres & Textiles in Eastern Europe*, 18, 87–91.

# Index